Behaviormetrics: Quantitative Approaches to Human Behavior

Volume 17

Series Editor

Akinori Okada, Professor Emeritus, Rikkyo University,
Tokyo, Japan

This series covers in their entirety the elements of behaviormetrics, a term that encompasses all quantitative approaches of research to disclose and understand human behavior in the broadest sense. The term includes the concept, theory, model, algorithm, method, and application of quantitative approaches from theoretical or conceptual studies to empirical or practical application studies to comprehend human behavior. The Behaviormetrics series deals with a wide range of topics of data analysis and of developing new models, algorithms, and methods to analyze these data.

The characteristics featured in the series have four aspects. The first is the variety of the methods utilized in data analysis and a newly developed method that includes not only standard or general statistical methods or psychometric methods traditionally used in data analysis, but also includes cluster analysis, multidimensional scaling, machine learning, corresponding analysis, biplot, network analysis and graph theory, conjoint measurement, biclustering, visualization, and data and web mining. The second aspect is the variety of types of data including ranking, categorical, preference, functional, angle, contextual, nominal, multi-mode multi-way, contextual, continuous, discrete, high-dimensional, and sparse data. The third comprises the varied procedures by which the data are collected: by survey, experiment, sensor devices, and purchase records, and other means. The fourth aspect of the Behaviormetrics series is the diversity of fields from which the data are derived, including marketing and consumer behavior, sociology, psychology, education, archaeology, medicine, economics, political and policy science, cognitive science, public administration, pharmacy, engineering, urban planning, agriculture and forestry science, and brain science.

In essence, the purpose of this series is to describe the new horizons opening up in behaviormetrics — approaches to understanding and disclosing human behaviors both in the analyses of diverse data by a wide range of methods and in the development of new methods to analyze these data.

Eric J. Beh · Rosaria Lombardo · Jose G. Clavel
Editors

Analysis of Categorical Data from Historical Perspectives

Essays in Honour of Shizuhiko Nishisato

Springer

Editors
Eric J. Beh
National Institute for Applied Statistics
Research Australia (NIASRA)
University of Wollongong
Wollongong, NSW, Australia

Centre for Multi-Dimensional Data
Visualisation (MuViSU)
Stellenbosch University
Matieland, South Africa

Jose G. Clavel
Department of Quantitative Methods
University of Murcia
Murcia, Spain

Rosaria Lombardo
Department of Economics
University of Campania "Luigi Vanvitelli"
Capua, Caserta, Italy

ISSN 2524-4027 ISSN 2524-4035 (electronic)
Behaviormetrics: Quantitative Approaches to Human Behavior
ISBN 978-981-99-5328-8 ISBN 978-981-99-5329-5 (eBook)
https://doi.org/10.1007/978-981-99-5329-5

This Springer imprint is published by the registered company Springer Nature Singapore Pte Ltd.
The registered company address is: 152 Beach Road, #21-01/04 Gateway East, Singapore 189721, Singapore

Paper in this product is recyclable.

Shizuhiko Nishisato in Toronto, Canada, aged 32 years (1967). *Source* Courtesy of the Toronto Public Library, Toronto, Canada

Foreword

"What is so special about someone's 88th birthday?" might a non-Japanese person ask. Sure, it is a venerable age, and remaining active in academic life is worth a tremendous compliment, but to publish a special book to honour a person on his 88th birthday seems a bit excessive. Doesn't it?

However, this is not true in Japan. In Japanese culture, the 88th birthday, or Beiju (米寿), is the celebration of a long life and represents purity and wholesomeness. The first kanji character of Beiju can be deconstructed as 8, 10, and 8 on top of each other:

Shizuhiko Nishisato is the ideal person to celebrate Beiju for. During his academic lifetime that spans nearly 60 years, he has been an influential scientist inspiring an innumerable number of colleagues in categorical data analysis, the academic love of his life. He also produced many content-related results by sharing his insights into categorial data through teaching and executing analyses in numerous disciplines. That he was well loved and appreciated as a person in the scientific community shines through in the pictorial tribute in this book.

His life has played itself out primarily in North America. Clearly, he foresaw a glittering career on that continent, but this book shows that he never lost his love for his native Japan. A Canadian document on the Internet referred to him as

> Dr. Shizuhiko Nishisato. The expert from Japan is involved in research in psychological scaling theory. (Ph.D., North Carolina). Professor in psychometrics and analysis of categorical data [dual scaling] [Toronto Star 28/9/1967]

In the book *Modern Quantification Theory* (Springer, 2021) written by Nishisato together with the editors of the present *Festschrift*, a full account is provided by him of his academic career, which started of course in Japan with Masanao Toda and Chikio Hayashi, continued with a thesis at the University of North Carolina under the direction of R. Darrell Bock. After a brief interlude back in Japan, he made the definitive step to North America, in particular Canada. After only about a year at McGill University, he made his final move to the University of Toronto, where he and his career came to a full bloom.

Nishi published a wealth of books both on his own and with his colleagues, as is elaborated in his chapter in this *Festschrift*. Next to his books, Nishi (as known to his friends) produced over 100 other academic publications. His early book *Analysis of Categorical Data: Dual Scaling and its Applications* (University of Toronto Press, 1980) has around 1000 citations (search date: April 2023) and shows that Nishisato has made a lasting mark in the world of categorical data analysis.

Detailed accounts of the various aspects of his work on quantification theory and applications thereof are dealt with in the above-mentioned books and of course in the ensuing papers. The appreciation for his work goes on relentlessly and the authors of the papers in this *Festschrift* acknowledge this abundantly and wholeheartedly.

Leiden, The Netherlands
April 2023

Pieter Kroonenberg

Preface

This book marks a celebration of the career and influence of our dear colleague and friend, Prof Shizuhiko Nishisato, or "Nishi" as we call him, in honour of his 88th birthday. Such a milestone deserves a moment to sit back and reflect upon a life filled with happiness, hope, at times sadness, but with love and passion for all that drives us forward. So it is with this *Festschrift* that we all celebrate Nishi's career and the influence (both personal and professional) he has had on us all. It is also our opportunity to thank him for all he has done for us as editors, and for everyone who was able to contribute to his *Festschrift* and those who were unable to do so.

Our connection to Nishi dates back about 20 years and so it is relatively young, certainly in comparison with many of those who have contributed to this book. A key moment was the first face-to-face meeting of the Nishi/Clavel and Beh/Lombardo teams at the IFCS (International Federation of Classification Societies) Conference in Tokyo in 2017. From this meeting came the 2021 Springer book that we had the pleasure to co-author with Nishi titled *Modern Quantification Theory: Joint Graphical Display, Biplots and Alternatives*. So, it came as a pleasant surprise in late February 2022 that we were invited to edit his *Festschrift*. Of course, we said "yes". We would like to acknowledge the early involvement of Prof. Yasumasa Baba who is a long-term dear friend of Nishi and who was originally committed to this project but, unfortunately, was unable to continue in the role.

Nishi's career spans a great many achievements that are laid out in many of the papers of this book, and so we leave it to you to peruse the pages and appreciate the depth of work he has committed a lifetime of passion to. It is safe to say his impact on quantification theory, and the vast array of research avenues this covers, is profound. Therefore, this *Festschrift* is a celebration of many of these avenues and is divided into four broad topics. The first, "Data Theory" provides a mix of written and pictorial accounts of Nishi's life and his work. It also gives some perspective of Nishi's influence in the context of the career of Prof. Chikio Hayashi, an early and highly influential pioneer of quantification theory. The next major part of this book is titled "On Associations and Scaling Issues" and includes papers that celebrate Nishi's impact on the numerical issues concerned with dual scaling and its related methods, as well as providing new insights into this area of research. A more visual appreciation of the scaling issues is explored in "On Correspondence Analysis and Related Methods". Here, the papers discuss a range of issues including the naming conventions used in the past, exploring the anatomy of correspondence analysis and detailing extensions of correspondence analysis for analysing various data structures. The final part of Nishi's *Festschrift* is titled "General Topics" and includes papers that are not necessarily related to issues concerned with quantification theory but are here because of the close professional and personal connections that Nishi has shared with the authors over the years.

This celebration of Nishi's career is a collection of 29 papers where all the corresponding authors and some of their co-writers were all personally invited to contribute. We must also acknowledge those who were invited to contribute to this collection and prepared an early version of their work for inclusion but, ultimately, were unable to do so. Every one of those who have been invited to contribute to this collection all share a personal and professional bond with Nishi. These papers are written by 45 authors that come from all corners of the globe; in alphabetical order, Australia, Barbados, Canada, England, France, Germany, India, Italy, Japan, the Netherlands, Scotland, Spain, Switzerland, and the USA.

The preparation of Nishi's *Festschrift* would not have been possible without the help and support of Springer. So, we thank them, and especially Sridevi Purushothaman, for the many email queries that were patiently responded to. We also extend our heartfelt appreciation to Pieter Kroonenberg, our close personal friend and of Nishi's, for writing the Foreword to this book. Our biggest thanks goes to each of the authors who have contributed to this collection of celebratory papers. It has been an immense pleasure communicating with each and every one of you and the email conversations that have followed. We thank you for your commitment to this book and for helping to celebrate Nishi as the kind and endearing person that he is and for the role he has played over the decades as a researcher who has committed himself to the development of quantification theory and its related methods.

No one succeeds in life without the love of those around them. While we have all provided various degrees of professional and/or personal help and support over many years (and decades), his successes rest primarily with his wife Lorraine. So, on a personal note, we thank Lorraine for her support and love as Nishi has carved a wide long path through his academic career, a path that many of us have travelled alongside Nishi or behind him. Whether you subscribe to the phrase of Scott Fitzgerald *behind every great man there is a great woman* or Tariq Ramadan's *behind every great man is not a woman, she is beside him, she is with him, not behind him*, or even Jim Carrey's *behind every great man is a woman rolling her eye*, Lorraine's influence has been a blessing to us all. So, thank you Lorraine. Finally, we thank Nishi for inviting us to edit his *Festschrift* and wish him continued health, happiness and love as he and those around him continue to mould and direct the next generation of researchers.

Newcastle, Australia — Eric J. Beh
Naples, Italy — Rosaria Lombardo
Murcia, Spain — Jose G. Clavel
June 2023

Contents

Contributors

Shizuhiko Nishisato Ontario Institute for Studies in Education, University of Toronto, Toronto, ON, Canada

Hervé Abdi School of Behavioral and Brain Sciences, University of Texas at Dallas, Richardson, TX, USA

Kohei Adachi Graduate School of Human Sciences, Osaka University, Osaka, Japan

Daniel Baier Marketing & Innovation, University of Bayreuth, Bayreuth, Germany

Roberto de la Banda Universidad Nacional de Educación a Distancia (UNED), Madrid, Spain

Eric J. Beh National Institute for Applied Statistics Research, Australia (NIASRA), University of Wollongong, Wollongong, NSW, Australia;
Centre for Multi-Dimensional Data Visualisation (MuViSu), Stellenbosch University, Stellenbosch, South Africa

Ulf Böckenholt Kellogg School of Management, Northwestern University, Evanston, IL, USA

Hamparsum Bozdogan Department of Business Analytics and Statistics, The University of Tennessee, Knoxville, TN, USA

Naohito Chino Aichi-Gakuin University, Nissin, Aichi, Japan

Vartan Choulakian Département de Mathématiques et de Statistique, Université de Moncton, Moncton, NB, Canada

Jose G. Clavel Department of Quantitative Methods, University of Murcia, Murcia, Spain

Jan de Leeuw Department of Statistics, University of California Los Angeles (UCLA), Los Angeles, CA, USA

Agostino Di Ciaccio Dipartimento Scienze Statistiche, Sapienza Università di Roma, Rome, Italy

Edwin Diday (Deceased) CEREMADE Laboratory, University of Paris Dauphine, Paris, France

Wolfgang Gaul Institute of Decision Theory and Management Science, Karlsruhe Institute of Technology, Karlsruhe, Germany

Michael Greenacre Department of Economics and Business, Universitat Pompeu Fabra, Barcelona, Spain;
Barcelona School of Management, Barcelona, Spain

Patrick J. F. Groenen Econometric Institute, Erasmus University Rotterdam, Rotterdam, The Netherlands

Willem J. Heiser Institute of Psychology, Leiden University, Leiden, The Netherlands

Lawrence Hubert Department of Psychology, University of Illinois, Champaign, IL, USA

Tadashi Imaizumi School of Management and Information Sciences, Tama University, Tokyo, Japan

Shuichi Iwatsubo Research Division, The National Center for University Entrance Examinations, Tokyo, Japan

Suman Katragadda HEAPS.ai, Bengaluru, Karnataka, India

Henk A. L. Kiers Department of Psychology, University of Groningen, Groningen, The Netherlands

Se-Kang Kim Psychology Division, Department of Pediatrics, Baylor College of Medicine, Texas Children's Hospital, Houston, TX, USA

Pieter M. Kroonenberg Faculty of Social and Behavioural Sciences, Leiden University, Leiden, The Netherlands

Ludovic Lebart Centre National de la Recherche Scientifique (CNRS), Paris, France

Sébastien Loisel Department of Mathematics, Heriot-Watt University, Edinburgh, Scotland

Rosaria Lombardo Department of Economics, University of Campania "L. Vanvitelli", Capua, Italy

Smail Mahdi Department of Computer Science, Mathematics and Physics, Faculty of Science and Technology, University of the West Indies, CaveHill Campus, Barbados

Jacqueline J. Meulman LUXs data science BV, Leiden, The Netherlands; Department of Statistics, Stanford University, Stanford, CA, USA

Boris Mirkin Department of Data Analysis and Artificial Intelligence, NRU HSE Moscow, Moscow, Russian Federation;
School of Computing and Mathematical Sciences, Birkbeck University of London, London, UK

Masahiro Mizuta The Institute of Statistical Mathematics, Tachikawa, Tokyo, Japan

Takashi Murakami Institute of Cultural Science, Chukyo University, Nagoya, Japan

James Ramsay Department of Psychology, McGill University, Montreal, QC, Canada

Gilbert Saporta Center for Studies and Research in Computer Science and Communication (Cédric), Conservatoire National des Arts et Métiers, Paris, France

Yoshio Takane Department of Psychology, University of Victoria, Victoria, BC, Canada

Jos M. F. ten Berge Department of Psychology, University of Groningen, Groningen, The Netherlands

Jun Tsuchida Department of Data Science, Kyoto Women's University, Kyoto, Japan

Jan L. A. van Rijckevorsel Amsterdam School of Economics, University of Amsterdam, Amsterdam, The Netherlands

Michel van de Velden Econometric Institute, Erasmus University Rotterdam, Rotterdam, The Netherlands

Helmut Vorkauf Bern, Switzerland

Ting-Wu Wang School of Information and Physical Sciences, University of Newcastle, Newcastle, NSW, Australia

Hiroshi Yadohisa Department of Culture and Information Science, Doshisha University, Kyoto, Japan

Ryozo Yoshino The Institute of Statistical Mathematics, Tokyo, Japan

Susu Zhang Departments of Psychology and Statistics, University of Illinois, Champaign, IL, USA

Data Theory

Gratitude: A Life Relived

Shizuhiko Nishisato

1 To Begin with

First of all, my heartfelt appreciation goes to Prof. Akinori Okada for his considerate thoughtfulness in conceiving my *Festschrift*. I was overjoyed with his proposal, but frankly speaking, this excitement was followed with mixed feelings of great honour and a definitive sense of the finale of my research career. It was indeed a long and enjoyable career spanning over 60 years. What a wonderful group of researchers I have had the privilege to meet and know!

I am also very grateful to the co-editors Eric J. Beh, Rosaria Lombardo, and José G. Clavel who kindly accepted the very time-consuming task of editing all those contributions into a fine book. My work with the three co-editors resulted in our joint book, published in 2021 (age 86), which was one of the highlights of my career. Of course, I am exceptionally grateful to all the contributors, too. One of them said: "What a wonderful feeling it is to be brought back together with those old timers!" Yes—many of them represent my good old days.

I am not only fortunate to have been surrounded by many wonderful researchers, but also blessed with the luxury of my family's tireless support, those in Canada and those in Japan.

In an attempt to acknowledge all those people, I would like to reflect on my personal life with photos (only those which I managed to find—many pre-digital photos, old negatives and slides are long gone).

2 Bygone Days and Memories

I was born on June 9, 1935, in Sapporo, Hokkaido, Japan (Fig. 1). The Second World War started when I was 6 years old. My family moved from Sapporo to Obihiro and then to a small mountain village of Urahoro in Hokkaido, where my father was born

S. Nishisato (✉)
Ontario Institute for Studies in Education, University of Toronto, Toronto, ON, Canada
e-mail: shizuhiko.nishisato@utoronto.ca

E. J. Beh et al. (eds.), *Analysis of Categorical Data from Historical Perspectives*, Behaviormetrics: Quantitative Approaches to Human Behavior 17,
https://doi.org/10.1007/978-981-99-5329-5_1

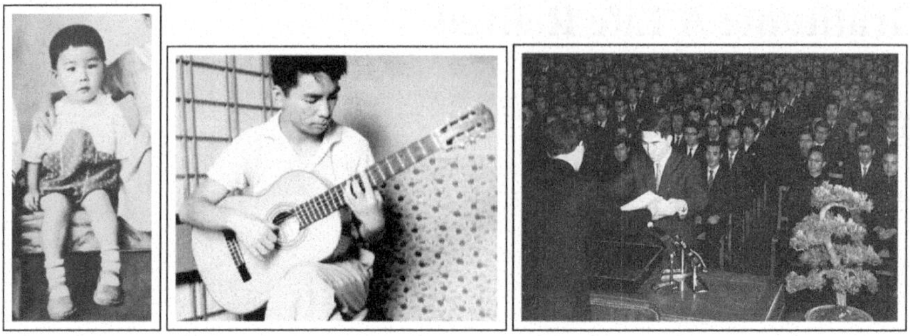

Fig. 1 Childhood; hobby; graduation from Hokkaido University

and passed away when I was only in grade 1 (age 6). Those ten years in Urahoro exposed me to an extreme hardship of life as well as firmly established lifelong friends who still meet annually at a nearby hot spring hotel (except during the COVID-19 pandemic). The museum in Urahoro has a special corner of permanent exhibits of my books, papers, and photos.

On April 3, 1952, a sunny spring day, an earthquake of magnitude 8.3 struck Urahoro and its neighborhood, destroying almost everything in sight. It changed my life forever. Moving back to my birth place of Sapporo served as a springboard for the next stage of my life.

After graduating from Sapporo Minami High School, I entered Hokkaido University in 1955 (age 20). Foreign languages were my major interest, and I took courses in English, French, German, Latin, Greek, and Esperanto. In my first year at the university, I founded the Esperanto Association and became its first President. I was the Esperanto interpreter when a Yugoslavian anthropologist gave a lecture on his lifework on the dawn of the human race at Sapporo City Hall.

Another equally strong hobby of mine was classical guitar, and I played the instrument in *Circolo Mandolinistico Aurora* of the university: I used to enjoy playing such pieces as *Recuerdos de la Alhambra* (Tarrega), *DanzaEspañola* N° 5 (Granados) and *Asturias* (Albeniz) (Fig. 1). Much later, we old-timers met in Sapporo (Fig. 2). I continued this hobby until some 30 years ago. My old colleague R. P. McDonald once called it *Nishi's latent ability*.

In choosing my major field of study, an English professor discouraged me from pursuing linguistics with the view that I would never be able to compete with those Europeans who were raised bilingual or multilingual. In 1959, with my BA thesis on *Factor Analysis of Anxiety* (Supervisors M. Toda, Y. Takada, and Y. Sugiyama), I finished my undergraduate program in experimental psychology and represented my graduating class at the graduation ceremony (Fig. 1). Two years later, I completed my MA thesis there titled *Human Reaction Time as a Function of Anxiety and Stress* (Supervisor Y. Sugiyama). A short paper based on this thesis (Nishisato, 1966) is one of my most frequently cited papers. I was blessed with excellent mentors (M. Toda, Y. Takada, T. Oyama, and Y. Sugiyama) and friends (Fig. 3).

Fig. 2 Ex-musicians of Hokkaido University

Fig. 3 Friends in experimental psychology, Hokkaido University

Fig. 4 1963 International Congress of Psychology; Spring in Chapel Hill

On September 6, 1961 (age 26), thanks to a fulbright scholarship, I arrived at the Raleigh-Durham airport in North Carolina, where my host family Mr and Mrs A. Ringwalt met me and drove me to Chapel Hill. I can never thank them and their family enough for their kind care and support during my four-year stay in Chapel Hill. The father of Mrs. Ringwalt, Dr. Rudolph Teusler, was the founder of St. Luke's Hospital in Tokyo, where coincidentally I had my physical examination for my entrance into the USA. The University of North Carolina in Chapel Hill (UNC) (six smaller photos in Figs. 4 and 5) was an academic paradise for me with super mentors (R. D. Bock, L. V. Jones, D. Adkins-Woods, T.G. Thurstone, E. Shuford, and H. F. Kaiser)

Fig. 5 1965 Williamsburg International Assembly; UNC; Graduation; Back Home

and wonderful fellow students (A. Rapoport, E. Abbe (née Niehl), S. Zyzanski, D. Messick, L. Gordon, N. Cole (née Stooksberry), B. Mukherjee, S. Das Gupta, T. Smith, J. and M Nakahara, S. Suzuki, M. Novick and H. Kusama).

During my Chapel Hill days, memorable events took place: In the fall of 1961, the newly elected President of the USA, John F. Kennedy, gave his famous *neither red nor dead* speech at the University Day in Chapel Hill; the Cuban Missile Crisis in 1962 (age 27); Dr. Martin Luther King's March to Washington with a quarter of a million demonstrators gathered in front of the Lincoln Memorial in 1963 (on that day, I was attending the International Congress of Psychology in Washington, D.C. (Fig. 4) and we were cautioned not to go outside the Mayflower Hotel); John F. Kennedy's assassination in 1963; the Tokyo Olympics in 1964; the Annual Williamsburg International Assembly of Foreign Students in 1965 (I was one of the two Japanese representatives; the student delegate from Norway was Gro Harlem

Fig. 6 Old Friends at OISE, University of Toronto, Canada

Brundtland who later served three terms as the prime minister of Norway and the director general of the World Health Organization) (Fig. 5). After completing my PhD thesis titled *Minimum Entropy Clustering of Test Items* (Supervisor R. D. Bock) and the final oral examination, I returned to Japan in September of 1965 (age 30) (Fig. 5).

An unexpected failure in finding a job anywhere in Japan was extremely devastating, but it was an unbelievable fortune in disguise. I was immediately offered a position in the Department of Psychology, McGill University, Montreal, Canada, by G. A. Ferguson, D. Bindra, and W. Lambert. There I met two persons who steered my life and career toward fulfillment: my future wife Lorraine A. M. Ford from South Africa, and Ross E. Traub.

Thanks to R. E. Traub, I was recruited in 1967 (age 32) to a new research center, the Ontario Institute for Studies in Education (OISE) at the University of Toronto, my home base until my retirement on June 9, 2000 (Fig. 6, some old timers). In 1967, I married Lorraine Ford, who continued to help me and my students with editorial work, the task she used to do for Bindra at McGill University.

At OISE, R. E. Traub (test theory; Princeton University), R. P. McDonald (factor analysis and structural equation modeling; University of Queensland, Australia), R. P. Bhargava (multivariate analysis of discrete and continuous variables; Stanford University) and myself (measurement theory and scaling; University of North Carolina) together established one of the world centers of psychometrics. In this process, I also served as the chairman of the Department of Measurement and Evaluation, OISE from 1971 to 1976 (age 36–41) (Fig. 6).

In those days, the East and the West were politically divided without much scientific communication between them. On the Western side from 1970 (age 35) onwards, I participated in many international meetings in France, Germany, the Netherlands, Italy, Spain, and Japan (Figs. 7, 8, and 9). I greatly benefited from those INRIA (Institut National de Recherche en Informatique et en Automatique) meetings in France, annual meetings of the German Classification Society, French-Japanese meetings, German-Japanese meetings, and M. J. Greenacre's CARME (Correspondence Analysis and Related Methods) conferences. I extend my sincere appreciation to the organisers of those conferences.

Fig. 7 International Conferences in European Countries

Fig. 8 German-Japanese Conference in Kyoto

Fig. 9 International Conference of Psychometric Society in Tokyo

On the Eastern side, I was lucky to be invited to Moscow, USSR, in December of 1990 (age 55). B. Mirkin and S. Adamov were hosts to a group of international researchers that included W. Gaul, H. H. Bock, H. Bozdogan, W. Day, and myself. Those were the last days of the Soviet Union since only a few weeks later the Soviet Union collapsed in 1991.

Another experience in the Eastern world preceeded the Moscow visit. In the summer of 1986 (age 51), V. Zudravkov of Sofia, Bulgaria, invited me, together with J. C. Gower, P. van der Heijden, E. van der Burg, and T. Saito, to give lectures on quantification theory to Bulgarian researchers. I was happily surprised when several Bulgarian researchers asked me to autograph copies of my 1980 book, entitled *Analysis of Categorical Data: Dual Scaling and Its Applications* (University of Toronto Press). Bulgaria was still a communist country.

International trips were not easy then with visa restrictions and limited funds, but surprisingly researchers knew many others abroad through exchanging postcards to request reprints of published papers, a custom we no longer have.

As for international conferences, I organised three major ones: the annual meeting of the Psychometric Society in Toronto with R. E. Traub, the annual meeting of the Psychometric Society in Banff, Alberta, Canada, and the International Conference on Measurement and Multivariate Analysis in Banff with Y. Baba (Fig. 10).

My professional services for academic organizations and awards are:

- Psychometric Society: President, Editor of Psychometrika and trustee.
- Classification Society of North America (CSNA): Trustee.
- German Classification Society (GfKl): Editorial Board for Studies in Classification, Data Analysis, and Knowledge Organization, Springer-Verlag.
- International Federation of Classification Societies (IFCS): Chair of Award Committee.
- American Statistical Association: Fellow.
- Japanese Classification Society: Fellow, Lifelong Achievement Award.
- Behaviormetric Society: Honorary Member, Lifelong Achievement Award, Publication Award.

Fig. 10 International Banff Conference, Alberta, Canada

- The University of North Carolina: Distinguished Alumnus Award of the UNC Psychology Alumni Association.

Outside the academic world, I engaged with several organizations, including:

- Metropolitan Toronto Japanese Family Services (JFS): First President (The JFS was established by my close friend S. Thurlow, a recipient of the Nobel Peace Prize).
- Volunteer of the Year Award from the Government of Ontario.
- Toronto Hokkaido Association: First President (T. Fuse and I founded this organization in 1972 (age 37).

So, together with academic and non-academic work, I have lived a busy life.

In retrospect, there are a few matters that come to my mind:

[1] **Greatest regret**: The translation of my 1980 book (Fig. 14) into Russian never materialised, solely due to the collapse of the Soviet Union (USSR). There was a signed agreement between the University of Toronto Press and Finansi Statistika Publisher in Moscow on its publication, and the translation had been completed by B. Mirkin and S. Adamov by 1990 when the aforementioned Moscow meeting took place. This translation included an addendum to the original 1980 book, namely some key developments since 1980.

Fig. 11 Mentor Bock, his wife and student; distinguished alumnus with friends

[2] **Proud moments in research**: (a) In 1997 (age 62), The American Psychological Association awarded the Distinguished Contribution Award to my mentor R. D. Bock and I chaired his memorial lecture; on the same day, Bock chaired my invited lecture. I celebrated the occasion with my mentor and Mrs. Bock (Fig. 11); (b) in 2000 (age 65), I was honoured as Distinguished Alumnus by the UNC Psychology Alumni Association (Fig. 11).

[3] **Busy life after retirement**: In 2000, I retired as Professor Emeritus from the University of Toronto and then worked part time for one to six months a year as Visiting Professor at Kwansei Gakuin University and Doshisha University in Japan and University of Murcia in Spain until 2007 (age 72) (Fig. 12). My wife and I enjoyed many international travels.

[4] **Blessed with co-authorships.** There are seven books written with co-authors (Fig. 13) and ten books by myself (Fig. 14). As for the co-authored books, I joined Iwamoto and Nakahara for the translation of the book by Penfield and Rasmussen *Cerebral Cortex of Man* into Japanese (the main work was done when we were students at Hokkaido University). Three books were written with my son Ira Nishisato who wrote the entire package of "DUAL3" dual scaling software with me. The Banff conference resulted in the proceedings with Baba, Bozdogan, and Kanefuji as the co-editors. And although this is not co-authorship, I translated my grandson Lincoln Dugas-Nishisato's first book, written when he was 8 years old, into Japanese. To solve the perennial controversy over the joint graphical display of quantification theory, I was joined for the book by Beh, Lombardo, and Clavel (Fig. 13).

Fig. 12 Activities of life after retirement

Fig. 13 Co-authors and books; the three co-authors of the last book are the editors of the current *Festschrift*

Fig. 14 Single-authored books; the last book is Nishisato, S. *Measurement, Mathematics and New Quantification Theory*: Springer (2023) (age 88)

[5] **Edwin Diday**: Regarding a great contribution of the late Edwin Diday, I once made the following remark; see page 560 of Nishisato, S.: Gleaning in the field of dual scaling. Psychometrika **61**, *61*, 559–599 (1996):

> Considering the abundance of publications and the outstanding contributions to the field by French researchers, it was hardly surprising when Edwin Diday, a leading French statistician, casually remarked that correspondence analysis had been exhaustively investigated by 1975, and that he and his colleagues were moving to the next stage of innovation, symbolic data analysis (Diday, personal communication, September 23, 1991). To his mind and many others, correspondence analysis must have seemed mathematically transparent; hence there was nothing more to discover about it.
>
> Whether Diday was right is a matter of opinion. There is still plenty of grain left by the reapers in the field of quantification, and continued gleaning can shed further light on a number of missing links between the mathematics of quantification theory and the validity of its applications. The main object of this paper is therefore to further the current understanding of quantification theory by bringing to the surface a number of its buried or implicit characteristics.

This paper was based on my Presidential address of the Psychometric Society, delivered at its annual meeting held in Banff, Canada, in 1996.

[6] **Lifelong Support 1**: My wife Lorraine greatly helped me with editorial work in my early days; my son Ira wrote a major portion of my computer programs; my grandson Lincoln Dugas-Nishisato kept his grandfather working (i.e., translation) until recently. I am very proud of both my family and extended families. Lincoln has been extensively involved in volunteer work with his mother Samantha and his other grandparents André and Gillian Dugas to help a countless number of those less fortunate in our community. Far from Canada, my siblings (Fig. 15) lived in Japan, to whom I also owe very much.

[7] **Lifelong Support 2**: I would like to mention another lifelong support of two groups of friends from my elementary school (Fig. 15) and Hokkaido University (Fig. 16).

[8] **Lifelong Support 3**: Expatriates all share hard experiences through a foreign language. I would like to mention four close friends who left Japan many years ago and overcame language barriers to achieve super careers: Setsuko Thurlow (the first photo, a recipient of the Nobel Peace Prize, Executive Director of the Japanese Family Services where I served as its first President), Yoshio Takane (the second photo of Fig. 17, Professor, University of Victoria, Former President of the Psychometric Society, one of the most influential psychometricians of the past 100 years, also a graduate of my alma mater University of North Carolina in Chapel Hill, whose first job was at McGill University as mine was), Akira Kobasigawa (the third photo, an internationally famous academic in child development; Professor Emeritus, University of Windsor, a gate-ball (croquet) buddy), Takashi Asano (the fourth photo, Recipient of the Stockholm Water Prize, Professor Emeritus, University of California at Davis, my high school and university friend who played the guitar together with me at Hokkaido University; see Fig. 2).

Fig. 15 With my siblings; elementary school friends in Urahoro

Fig. 16 Friends of Hokkaido University from the first year

Fig. 17 Super expatriates from Japan, my dear friends

[9] **Lifelong Support 4**: There are many researchers to whom I owe much gratitude, in particular those involved in quantification theory. I am pleased to share a few photos of them which I managed to find (Fig. 18). I must admit with regret, however, that I could not find photos of many other equally important friends.

Before I end my thanks and gratitude to my family and friends, I acknowledge that I have been exceptionally lucky to have lived my tumultuous early life in Japan, my hardworking student days at Hokkaido University, Sapporo, Japan, and at the University of North Carolina, Chapel Hill, North Carolina, USA, and my highly fulfilling research and teaching life first in Montreal and then Toronto, Canada. In May 2000, one month before my retirement from the University of Toronto, John C. Gower told me "Life exists after retirement" (see *14.1.1 John C. Gower* in Nishisato, 2022). His words encouraged me to work all the time until today.

In retrospect, when I first arrived in the USA as a student, the most remarkable novelty was the interdisciplinary pursuit of research, as exemplified by the joint seminars at the University of North Carolina's Departments of Statistics and Biostatistics and the Psychometric Laboratory. How lucky I was to be able to listen to talks by such eminent scholars as Hotelling, Roy, Bose, Madow, Chakrabarti, Grizzle, Sen, Koch, Hoeffding, Bock, Jones, Kaiser, Thurstone, Gabriel, Quade, Glaser, Donnely, and Shuford from my own university campus! A countless number of top invited speakers from all over the world also enriched my student life. The new trend of interdisciplinary research spread to other countries as well. The birth of the Japanese Behaviormetric Society fifty years ago is another timely example in response to interdisciplinary research. My academic life is deeply rooted in this revolutionary change of academia when researchers with different academic backgrounds worked together.

Thank you all for a wonderful life for this very lucky person!

Fig. 18 With mentors and colleagues in my research domain, past and present

3 Lifelong Publications: 1960 (age 25)–2023 (age 88)

Books

1. Nishisato, S.: Applications of Psychological Scaling: Analysis of Qualitative Data and Interpretations. Seishin Shobo Press, Tokyo (1975) (in Japanese).
2. Nishisato, S.: Analysis of Categorical Data: Dual Scaling and Its Applications. The University of Toronto Press Mathematical Expositions No. 24, Toronto (1980). ISBN 0-8020-5489-7.
3. Nishisato, S., Nishisato, I.: An Introduction to Dual Scaling. MicroStats, Toronto (1984). ISBN 0-9691785-0-6.
4. Nishisato, S.: Quantification of Qualitative Data: Dual Scaling and Its Applications. Asakura Shoten, Tokyo (1984) (in Japanese).
5. Iwamoto, T., Nakahara, J., Nishisato, S.: (Japanese translation of Penfield, W., Rasmussen, T.: Cerebral Cortex of Man. McMillan, New York (1950)). Fukumura Shuppan, Tokyo (in Japanese).
6. Nishisato, S., Nishisato, I.: DUAL3 Users' Guide. MicroStats, Toronto (1986). ISBN 0-9691785-2-6.
7. Nishisato, S.: Quantification of Categorical Data: A Bibliography 1975–1986. MicroStats, Toronto (1986). ISBN 0-9691785-2-6.
8. Nishisato, S., Nishisato, I.: Dual Scaling in a Nutshell. MicroStats, Toronto (1994). ISBN 0-9691785-3-6.
9. Nishisato, S.: Elements of Dual Scaling: An Introduction to Practical Data Analysis. Lawrence Erlbaum Associates, Hillsdale, NJ (1994). ISBN 0-8058-1209-1. (Retirement, June, 2000)
10. Nishisato, S., Baba, Y., Bozdogan, H., Kanefuji, K. (eds.): Measurement and Multivariate Analysis. Springer, Tokyo (2002). ISBN 4-431-70338-1.
11. Nishisato, S.: Insight into Data Analysis: The Necessity of Quantification. Kwansei Gakuin University Press (2007) (in Japanese). ISBN 978-4-86283-014-2.
12. Nishisato, S.: Multidimensional Nonlinear Descriptive Analysis of Categorical Data. Chapman & Hall, London (2007). ISBN 1-58488-612-9.
13. Nishisato, S.: Data Analysis for Behavioral Sciences: Applications of Methods Appropriate for Information Retrieval. Baifukan, Tokyo (2010) (in Japanese). ISBN 978-4-563-05218-8.
14. Nishisato, S.: Japanese translation of (Dugas-Nishisato, L.: Finding Greatness. Kids for Kids Books, Toronto (2018)). ISBN 978-1-926863-93-1). Hokkaido Shuppan Kikaku Center Press, Sapporo (2019). ISBN 978-4-8328-1911-5.
15. Nishisato, S., Beh, E.J., Lombardo, R., Clavel, J.G.: Modern Quantification Theory: Joint Graphical Display, Biplots and Alternatives. Springer, Singapore (2021). ISBN 978-981-16-2469-8.
16. Nishisato, S.: Optimal Quantification and Symmetry. Springer, Singapore (2022). ISBN 978-981-16-9160-0.
17. Nishisato, S.: Measurement, Mathematics and New Quantification Theory. Springer, Singapore (2023).

Selected Research Papers

- Nishisato, S.: Factor analytic study of anxiety. Jpn. J. Psychol. **31**, 228–236 (1960) (in Japanese).
- Oyama, T., Sugiyama, Y., Nishisato, S.: Discrimination between schizophrenic patients and neurotic patients: proposal of a simplified method and proposal of a new RRS score. Rorschachiana Japonica **4**, 65–79 (1961) (in Japanese).
- Nishisato, S.: A simple method of time series analysis. Festschrift for Professor Kin-ichi Yuki, pp. 102–111. Yamafuji Press, Sapporo (1965) (in Japanese).
- Nishisato, S.: Reaction time as a function of arousal and anxiety. Psychonomic Sci. **6**, 157–158 (1966).
- Nishisato, S., Wise, J.S.: Relative probability, inter-stimulus interval and speed of same-different judgment. Psychonomic Sci. **7**, 59–60 (1967).
- Bindra, D., Donderi, D.C., Nishisato, S.: Decision latencies of *same* and *different* judgments. Percept. Psychophys. **3**, 121–130 (1968).
- Nishisato, S.: Probability estimation of dichotomous response patterns by logistic fractional-factorial representation. Jpn. Psychol. Res. **12**, 87–95 (1970).
- Nishisato, S.: Structure and probability distribution of dichotomous response pattern. Jpn. Psychol. Res. **12**, 62–74 (1970).
- Nishisato, S.: Transform factor analysis: a sketchy presentation of a general approach. Jpn. Psychol. Res. **13**, 155–166 (1971).
- Nishisato, S., Torii, Y.: Effects of categorizing continuous normal variables on product-moment correlation. Jpn. Psychol. Res. **13**, 45–49 (1971).
- Nishisato, S., Torii, Y.: Assessment of information loss in scoring monotone items. Multivariate Behav. Res. **6**, 91–103 (1971).
- Nishisato, S.: Information analysis of binary response patterns. In: Takagi, S. (ed.) Modern Psychology and Quantification, Chap. 2. Theory of Measurement and Applications, pp. 73–92. University of Tokyo Press, Tokyo (1971) (in Japanese).
- Nishisato, S.: Analysis of variance through optimal scaling. In: Proceedings of the First Canadian Conference in Applied Statistics, pp. 306–316. Sir George Williams University Press, Montreal (1971).
- Nishisato, S.: Analysis of variance of categorical data through selective scaling. In: Proceeding of the 20th International Congress of Psychology, p. 279. Science Council of Japan, Tokyo (1972).
- Nishisato, S.: Optimal Scaling and Its Generalizations, I: Methods. Measurement and Evaluation of Categorical Data Technical Report (MECDTR) No. 1. Department of Measurement & Evaluation (ME), Ontario Institute for Studies in Education (OISE), Toronto (1972).
- Nishisato, S., Inukai, Y.: Partially optimal scaling of items with ordered categories. Jpn. Psychol. Res. **14**, 109–119 (1972).
- Nishisato, S.: Optimal Scaling and Its Generalizations, II: Applications. MECDTR No. 2, ME, OISE, Toronto (1973).
- Nishisato, S.: Elements of Applied Scaling. Department of Measurement & Evaluation, OISE, Toronto (1973).

- Nishisato, S., Yamauchi, H.: Principal components of deviation scores and standardized scores. Jpn. Psychol. Res. **16**, 162–170 (1974).
- Nishisato, S., Arri, P.S.: Nonlinear programming approach to optimal scaling of partially ordered categories. Psychometrika **40**, 525–548 (1975).
- Nishisato, S., Leong, K.S.: OPSCAL: A FORTRAN IV Program for Analysis of Qualitative Data by Optimal Scaling. MECDTR No. 3, ME, OISE, Toronto (1975).
- Nishisato, S.: Optimal Scaling as Applied to Different Forms of Data. MECDTR No. 4, ME, OISE, Toronto (1976).
- Nishisato, S.: Recent developments in scaling and related areas: a bibliographic overview. Jpn. J. Behav. **4**, 74–95 (1977).
- Nishisato, S.: Recent developments in scaling and related areas: multidimensional scaling. Jpn. J. Behav. **5**, 37–55 (1978).
- Nishisato, S.: Psychometrics: international trends. Math. Sci. **183**, 69–73 (1978).
- Nishisato, S.: Optimal scaling of paired comparison and rank order data: an alternative to Guttman's formulation. Psychometrika **43**, 263–271 (1978).
- Nishisato, S.: An Introduction to Dual Scaling. MECDTR No. 5, ME, OISE, Toronto. (1979).
- McDonald, R.P., Torii, Y., Nishisato, S.: Some results on proper eigenvalues and eigenvectors with applications to scaling. Psychometrika **44**, 211–227 (1979).
- Nishisato, S.: Dual scaling and its historical development. Math. Sci. **190**, 76–83 (1979).
- Nishisato, S.: Dual scaling and its variants. In: Traub, R.E. (ed.) New Directions in Testing and Measurement, pp. 1–12. Josey Bass, San Francisco (1979).
- Nishisato, S., Sheu, W.J.: Piecewise method of reciprocal averages for dual scaling of multiple choice data. Psychometrika **45**, 467–478 (1980).
- Nishisato, S.: Dual scaling of successive categories data. Jpn. Psychol. Res. **22**, 134–143 (1980).
- Nishisato, S.: Mathematical expositions of dual scaling. In: Chaubey, Y.P., Dwivedi, T.D. (eds.) Topics in Applied Statistics, pp. 629–640. Concordia University Press, Montreal (1981).
- Nishisato, S., Sheu, W.J.: A note on dual scaling of successive categories data. Psychometrika **49**, 493–500 (1984).
- Nishisato, S.: Dual scaling by reciprocal medians. Estratto Dagli Atti della XXXII Riunione Scientifica, Sorrento, Italy, pp. 141–147 (1984).
- Nishisato, S.: Forced classification: A simple application of a quantification method. Psychometrika **49**, 25–36 (1984).
- Nishisato, S.: Generalized forced classification for quantifying categorical data. In: Diday, E. (ed.) Data Analysis and Informatics, IV, pp. 351–362. Elsevier Science Publishers B. V., North Holland, Amsterdam (1986).
- Nishisato, S.: Classification with a variety of categorical data. In: Gaul, W., Schader, M. (eds.) Classification as a Tool of Research, pp. 353–359. Elsevier Science Publishers B. V., North Holland, Amsterdam (1986).
- Weingarden, P., Nishisato, S.: Can a method of rank ordering reproduce paired comparison? An analysis by dual scaling (correspondence analysis). Can. J. Market. Res. **5**, 11–18 (1987).

- Nishisato, S.: Robust techniques for quantifying categorical data. In: MacNeil, I.B., Umphrey, G.J. (eds.) Foundations of Statistical Inference, pp. 209–217. D. Reidel Publishing Company, Dordrecht, The Netherlands (1987).
- Nishisato, S.: Dual scaling: its development and comparisons with other quantification methods. In: Proceedings of the Annual Meeting of the German Society of Operations Research, Berlin, pp. 376–389 (1988).
- Nishisato, S.: Assessing quality of joint graphical display in correspondence analysis and dual scaling. In: Diday, E., Escoufier, Y., Lebart, L., Page, J., Schektman, Y., Tommasone, R. (eds.) Data Analysis and Informatics, V., pp. 409–416. North Holland, Amsterdam (1988).
- Nishisato, S.: Market segmentation by dual scaling through generalized forced classification. In: Gaul, W., Schader, M. (eds.) Data, Expert Knowledge and Decisions, pp. 268–278. Springer, Berlin (1988).
- Nishisato, S.: Forced classification procedure of dual scaling: its mathematical properties. In: Bock, H.H. (ed.) Classification and Related Methods, pp. 523–532. North Holland, Amsterdam (1988).
- Nishisato, S., Gaul, W.: Marketing data analysis by dual scaling. Int. J. Res. Market. **5**, 151–170 (1989).
- Nishisato, S., Lawrence, D.R.: Dual scaling of multiway data matrices: Several variants. In: Coppi, R., Bolasco, S. (eds.) Multiway Data Analysis, pp. 317–326. North Holland, Amsterdam (1989).
- Nishisato, S., Gaul, W.: An approach to marketing data analysis: The forced classification procedure of dual scaling. Journal of Marketing Research **27**, 354–360 (1990).
- Nishisato, S.: Dual scaling of designed experiments. In: Schader, M., Gaul, W. (eds.) Knowledge, Data and Computer-Assisted Decisions, NATO ASI Series F: Computers and Systems Science Vol. 61, pp. 115–125. Springer, Berlin (1990).
- Nishisato, S.: Standardizing multidimensional space for dual scaling. Proceedings of the 20th Annual Meeting of the German Operations Research Society, pp. 584–591. Hohenheim University, Germany (1991).
- Yamada, F., Nishisato, S.: Several mathematical properties of dual scaling as applied to item category data. Japanese Journal of Behaviormetrics **20**, 56–63 (1993) (in Japanese).
- Nishisato, S.: On quantifying different types of categorical data. Psychometrika **58**, 617–629 (1993).
- Bean, G., Nishisato, S., Rector, N.A.: The psychometric properties of the Competency Interview Schedule. Canadian Journal of Psychiatry **39**, 368–376 (1994).
- Nishisato, S.: Graphical representation of quantified categorical data: Its inherent problems. Journal of Statistical Planning and Inference **43**, 121–132 (1995).
- Nishisato, S.: Optimization and data structure: Seven faces of dual scaling. Annals of Operations Research **55**, 345–359 (1995).
- Nishisato, S., Ahn, H., When not to analyze data: Decision making on missing responses in dual scaling. Annals of Operations Research **55**, 361–378 (1995).
- Nishisato, S.: An overview and recent developments in dual scaling. In: Gaul, W., Pfeifer, D. (eds.) From Data to Knowledge, pp. 73–85. Springer, Berlin (1995).

- Beans, G., Nishisato, S., Rector, N.A., Glancy, G.: The assessment of competence to make a treatment decision: An empirical approach. Canadian Journal of Psychiatry **41**, 85–92 (1996).
- Nishisato, S.: What is quantification? A point of view. Festschrift for Professor Yoshio Sugiyama, 187–192 (1996).
- Nishisato, S.: Gleaning in the field of dual scaling. Psychometrika **61**, 559–599 (1996). (Presidential address of the Psychometric Society).
- Nishisato, S.: Graphing is believing: Interpretable graphs for dual scaling. In: Blasius, J., Greenacre, M.J. (eds.) Visualization of Categorical Data, pp. 185–196. Academic Press, London (1997).
- Nishisato, S.: Exploring multidimensional quantification space. In: Hayashi, C., Yajima, K., Bock, H.-H., Ohsumi, N., Tanaka, Y., Baba, Y. (eds.) Data Science, Classification and Related Methods: Proceedings of the Fifth Conference of the International Federation of Classification Societies (IFCS-96), pp. 441–451. Springer, Kobe, Japan (1998).
- Nishisato, S.: Data types and information: Beyond the current practice of data analysis. In: Decker, R., Gaul, W. (eds.) Classification and Information Processing at the Turn of the Millennium, pp. 40–51. Springer, Heidelberg (1999).
- Nishisato, S., Baba, Y.: On contingency, projection and forced classification of dual scaling. Behaviormetrika bf 26, 207–219 (1999).

(Retirement, June 2000)

- Nishisato, S.: Le *dual scaling* et ses applications. In: Moreau, J., Doudin, P.-A., Cazes, P. (eds.) L'Analyse des Correspondances et les Techniques Connexes: Approaches Nouvelles pour l'Analyse Statistique des Données, pp. 9– 31. Springer, Berlin (2000) (in French).
- Nishisato, S.: A characterization of ordinal data. In: Gaul, W., Opitz, O., Schader, M. (eds.) Data Analysis: Scientific Modeling and Practical Applications, pp. 285–298. Springer, Heidelberg (2000) (*Festschrift* for Professor Dr. Hans-Hermann Bock).
- Nishisato, S., Hemsworth, D.: Quantification of ordinal variables: A critical inquiry into polychoric and canonical correlation. In: Baba, Y., Hayter, A. J., Kanefuji, K., Kuriki, S. (eds.) Recent Advances in Statistical Research and Data Analysis, pp. 49–84. Springer, Tokyo (2002).
- Nishisato, S.: Measurement and multivariate analysis. In: Nishisato, S., Baba, Y., Bozdogan, H., Kanefuji, K. (eds.) Measurement and Multivariate Analysis, pp. 25–36. Springer, Tokyo (2002).
- Nishisato, S.: Data analysis of rank order data by dual scaling. In: Yanai, H., Okada, A., Shigemasu, K., Takagi, H., Iwasaki, M. (eds.) Handbook for Concrete Examples of Multivariate Analysis, pp. 614–623. Asakura Shoten, Tokyo (2002) (in Japanese).
- Nishisato, S.: Structural differences between continuous and discrete variates from the dual scaling point of view, and suggestions for an integrated approach for analysis. Japanese Journal of Sensory Evaluation **6**, 89–94 (2002) (in Japanese).

- Nishisato, S.: Analysis of qualitative data: dual scaling. In: Institute of Industrial Technology, Department of Human Welfare Medical Technology (ed.) Handbook of Human Measurement, pp. 398–402. Asakura Shoten, Tokyo (2003).
- Nishisato, S., Clavel, J.G.: A note on between-set distances in dual scaling and correspondence analysis. Behaviormetrika **30**, 87–98 (2003).
- Nishisato, S.: Total information in multivariate data from dual scaling perspectives. Alberta J. Educ. Res. **49** 244–251 (2003) (Special Fall Issue in Honor of Ross E. Traub, XLIX).
- Nishisato, S.: Geometric perspectives of dual scaling for assessment of information in data. In: Yanai, H., Okada, A., Shigemasu, K., Kano, Y., Meulman, J. (eds.) New Developments in Psychometrics, pp. 453–462. Springer, Tokyo (2003).
- Nishisato, S.: New look at and suggestions for integrated multidimensional data analysis from the dual scaling point of view. Japanese Journal of Sensory Evaluation **8**, 4–10 (2004).
- Nishisato, S.: Dual scaling. In: Lewis-Beck, M., Bryman, A.E., Liao, T.F. (eds.) The Sage Encyclopedia of Social Science Research Methods, vol. 1, pp. 285–288. Sage Publication, Thousand Oaks, California (2004).
- Nishisato, S.: Dual scaling. In: Kaplan, D. (ed.) The Sage Handbook of Quantitative Methodology for the Social Sciences, pp. 3–24. Sage Publication, Thousand Oaks, California (2004).
- Nishisato, S.: Correspondence analysis and dual scaling. In: Kempf-Leonard, K. (ed.) Encyclopedia for Social Measurement, pp. 531–536. Elsevier, San Diego, California (2005).
- Nishisato, S.: New framework for multidimensional data analysis. In: Weihs, C., Gaul, W. (eds.) Classification—the Ubiquitous Challenge, pp. 280–287. Springer, Heidelberg (2005).
- Nishisato, S.: Empirical approach as a scientific framework for data analysis. In: Decker, R., Schmidt-ThiÃ©me, L., Baier, D. (eds.) Data Analysis and Decision Support, pp. 108–116. Springer, Heidelberg (2005).
- Nishisato, S.: On the scaling of ordinal measurement: A dual scaling perspective. In: Maydeu-Olivares, A., McArdle, J.J. (eds.) Contemporary Psychometrics: A Festschrift for Roderick P. McDonald, pp. 479–508. Lawrence Erlbaum Associates (2005).
- Nishisato, S.: Correlational structure of multiple-choice data as viewed from dual scaling. In: Greenacre, M.J., Blasius, J. (eds.) Multiple Correspondence Analysis and Related Methods, pp. 161–178. Chapman & Hall/CRC, London (2006).
- Clavel, J.G., Nishisato, S.: Joint analysis of within-set and between-set distances. In: Shigemasu, K., Okada, A., Imaizumi, T., Hoshino, T. (eds.) New Trends in Psychometrics, pp. 41–50. Universal Academy Press, Tokyo (2008).
- Nishisato, S., Clavel, J.G. Interpreting data in reduced space: A case of what is not what in multidimensional data analysis. In: Shigemasu, K., Okada, A., Imaizumi, T., Hoshino, T. (eds.) New Trends in Psychometrics, pp. 357–366. Universal Academy Press, Tokyo (2008).
- Nishisato, S., Clavel, J.G.: Total information analysis: Comprehensive dual scaling. Behaviormetrika **37**, 15–32 (2010).

- Nishisato, S.: Generating optimal data through regression of measurement onto data. Theory and Applications of Data Analysis, 1–10 (2011) (in Japanese).
- Nishisato, S.: Quantification theory: Reminiscence and a step forward. In: Gaul, W., Geyer-Schultz, A., Schmidt-Thiéme, L., Kunze, J. (eds.) Challenges at the Interface of Data Analysis, Computer Science and Optimization, pp. 109–119. Springer, Berlin (2012).
- Clavel, J.G., Nishisato, S.: Reduced versus complete space configurations in total information analysis. In: Gaul, W., Geyer-Schultz, A., Schmidt-Thiéme, L., Kunze, J. G. (eds.) Challenges at the Interface of Data Analysis, Computer Science and Optimization, pp. 91–99. Springer, Berlin (2012).
- Nishisato, S.: Applications of mathematics to behavioural sciences: relations between exploratory data analysis and measurement. Jpn. J. Behav. **41**, 89–102 (2014) (in Japanese).
- Nishisato, S.: Structural representation of categorical data and cluster analysis through filters. In: Gaul, W., Geyer-Schultz, A., Baba, Y., Okada, A. (eds.) German-Japanese Interchange of Data Analysis Results, pp. 81–90. Springer, Cham (2014).
- Nishisato, S.: Dual scaling: Revisit to *Gleaning of the Field*. Theory Appl. Data Anal. **5**, 1–9 (2016) (in Japanese).
- Nishisato, S.: Multidimensional joint graphical display of symmetric analysis: back to the fundamentals. In: van der Ark, L.A., Bolt, D.M., Wang, W.C., Douglas, J.A., Wiberg, M. (eds.) Quantitative Psychology Research, pp. 291–298. Springer, Cham (2016).
- Nishisato, S.: Outcries of dual scaling: the key is duality. In: van der Ark, L.A., Wiberg, M., Culpepper, S.A., Douglas, J.A., Wang, W.-C. (eds.) Quantitative Psychology, Springer Proceedings in Mathematics & Statistics, pp. 105–115. Springer, Cham (2017).
- Nishisato, S.: Reminiscence: quantification theory and graphs. Theory Appl. Data Anal. **8**, 47–57 (2019) (in Japanese).
- Nishisato, S.: From joint graphical display to bi-modal clustering: [1] a giant leap in quantification theory. In: Imaizumi, T., Okada, A., Miyamoto, S., Sakaori, F., Yamamoto, Y., Vichi, M. (eds.) Advanced Studies in Classification and Data Science, pp. 157–168. Springer, Singapore (2020).
- Clavel, J.G., Nishisato, S.: From joint graphical display to bi-modal clustering: [2] cluster analysis. In: Imaizumi, T., Okada, A., Miyamoto, S., Sakaori, F., Yamamoto, Y., Vichi, M. (eds.) Advanced Studies in Classification and Data Science, pp. 131–143. Springer, Singapore (2020).
- Nishisato, S.: Quantification theory: categories, variables and mode of analysis. In: Imaizumi, T., Nakayama, A., Yokoyama, S. (eds.) Advanced Studies in Behaviormetrics and Data Science: Essays in Honor of Akinori Okada, pp. 253–264. Springer, Singapore (2020).
- Nishisato, S.: Gratitude: a life relived. In: Beh, E.J., Lombardo, R., Clavel, J.G. (eds.) Analysis of Categorical Data from Historical Perspectives: Essays in Honor of Shizuhiko Nishisato, pp. 3–25. Springer, Singapore (2023).

- Nishisato, S.: Propositions for quantification theory. In: Okada, A., Shigemasu, K., Yoshino, R., Yokoyama, S. (eds.) Facets of Behaviormetrics: The 50th Anniversary of the Behaviormetric Society, pp. 173–191. Springer, Singapore (2023).

Nishisato's Psychometric World

Pieter M. Kroonenberg

1 Introduction

This chapter contains photographs rather than words to show Prof Shizuhiko Nishisato's psychometric world. This contribution is made up of photographs of him and his colleagues as they appeared in front of the lens of my camera. Obviously not all of them are present, as at the time neither they nor I knew that they had a role to play in the *Festschrift* for Nishi's 88th birthday. Given selfies came only in vogue after 2012, I had to include my own existence captured by other, to me unknown, photographers. I am afraid this mostly precludes giving proper acknowledgements (Fig. 1).

Fig. 1 Nishisato at the European Psychometric Conference, 1995; Leiden

P. M. Kroonenberg (✉)
Faculty of Social and Behavioural Sciences, Leiden University, Leiden, The Netherlands
e-mail: p.m.kroonenberg@fsw.leidenuniv.nl

E. J. Beh et al. (eds.), *Analysis of Categorical Data from Historical Perspectives*, Behaviormetrics: Quantitative Approaches to Human Behavior 17,
https://doi.org/10.1007/978-981-99-5329-5_2

I have tried to contact the persons displayed, but not all have responded to my email. Those that did, were all in favour of their portrayal, and some have kindly send me a photograph of themselves. The symbol † indicates that the person is deceased, as far as I know; when known, the year is provided.

2 Nishisato: The Man Himself

(a) Nishisato at IMPS2005 (Willem Heiser in the background)

(b) Nishisato at IMPS2007

(c) Nishisato at IMPS2015

(d) Nishisato at IMPS2015

Fig. 2 Nishisato at conferences: IMPS2005, IMPS2007 and IMPS2015

(a) Nishisato at IFCS2017

(b) Nishisato at IFCS2017

Fig. 3 Nishisato at conferences: IFCS2017

3 Nishisato and Colleagues

(a) With William Stout, Haruo Yanai, Ivo Molenaar

(b) With David Thissen, William Stout, Haruo Yanai

Fig. 4 Nishisato and friends: IMPS2001

(a) With Wim van der Linden

(b) With Hiroshi Ikeda

Fig. 5 Nishisato and friends: IMPS2007

(a) With Terry Ackerman; Susan Rees at the back

(b) With Larry Hubert

(c) With Sy Miin Chow and Sophia Rabe Hesketh

Fig. 6 Nishisato and friends: IMPS2015

(a) With Helga Gaul; IFCS2017

(b) With Yasumasa Baba, 2013; Photo courtesy of Nishisato

(c) With Michel van de Velden, Rosaria Lombardo, Eric Beh and Pieter Kroonenberg; IFCS2017

Fig. 7 Nishisato and friends, IFCS2017

4 Japan

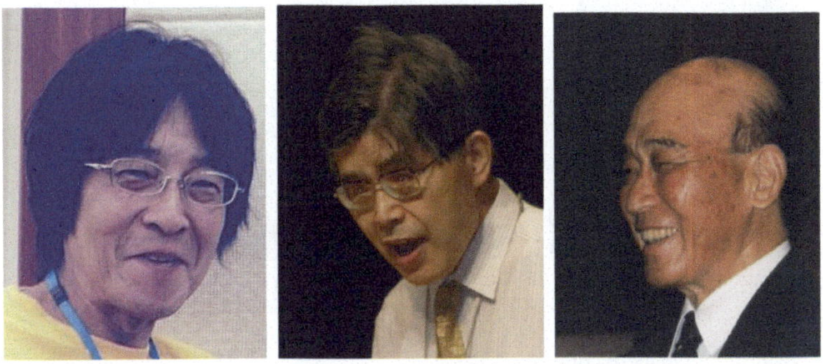

(a) Kohei Adachi, 2015; Osaka

(b) Haruo Yanai, 2007; † 2013

(c) Chikio Hayashi, 1998; † 2002

Fig. 8 Nishisato's Japanese colleagues: Part I

(a) Kentaro Hayashi, 2007; Mānoa, Hawaii

(b) Tadashi Imaizumi, 2017; Tama, Tokyo

(c) Yukio Inukai, 1997; Nagoya

Fig. 9 Nishisato's Japanese colleagues: Part II

(a) Akinori Okada, 2005; Rikkyo, Tama, Tokyo

(b) Shuichi Iwatsubo, 1997

(c) Yutaka Kano, 2007; Osaka

Fig. 10 Nishisato's Japanese colleagues: Part III

(a) Kazuo Shigemasu, 2007; Tokyo

(b) Tatsuo Otsu, 2007; NCUEE, Komaba

(c) Takashi Murakami, 2016; Nagoya, Chukyo

Fig. 11 Nishisato's Japanese colleagues: Part IV

5 North America

(d) Terry Ackerman, 2015; UNCG, Greensboro, USA

(e) Darrell Bock, 2008; thesis supervisor; † 2021

(f) Peter Bentler, 2017; UCLA, USA

Fig. 12 Nishisato's North-American colleagues: Part I

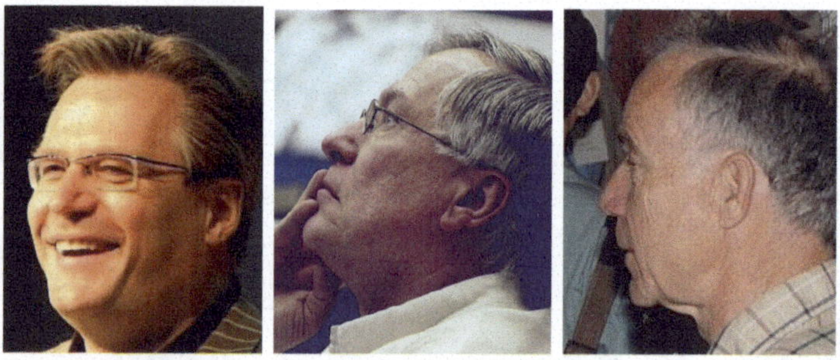

(a) Ulf Böckenholt, 2006; Kellog, Evanston, USA

(b) Robert Cudeck, 2006; Columbus, USA

(c) Norman Cliff, 2017; USC, USA

Fig. 13 Nishisato's North-American colleagues: Part II

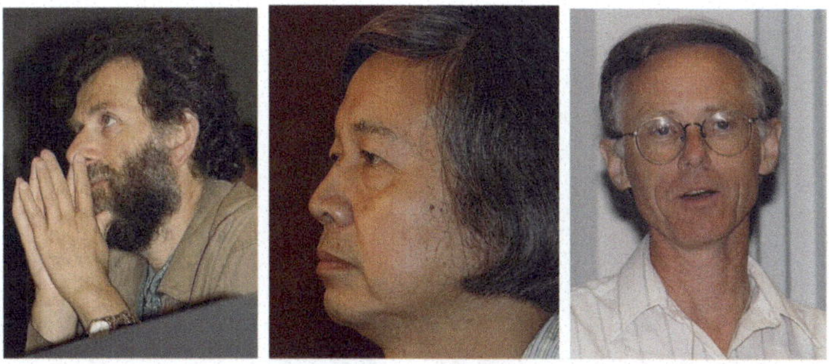

(a) Vartan Choulakian, 2015; Moncton, Canada (b) Yoshio Takane, 2007; Victoria, Canada (c) Jim Ramsay, 1995; Montreal, Canada

Fig. 14 Nishisato's Canadian colleagues: Part III

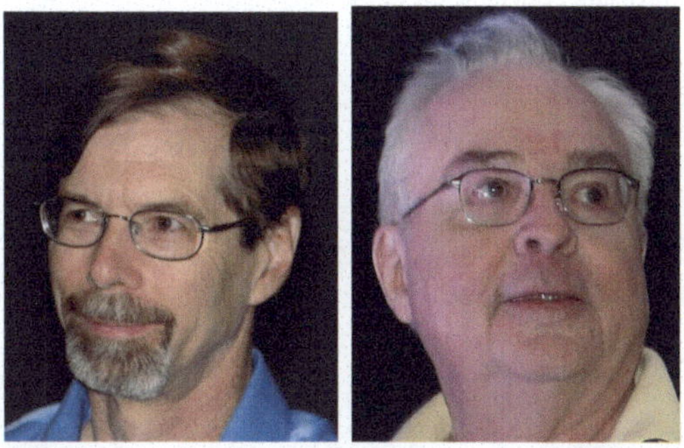

(a) David Thissen, 2006; Chapel Hill, USA (b) Larry Hubert, 2016; Urbana-Champaign, USA

Fig. 15 Nishisato's North-American colleagues: Part IV

6 Europe

(a) Wolfgang Gaul, 2017; Bonn, Germany (b) Gerhard Fischer, 2017; Vienna, Austria (c) Hans-Hermann Bock, 2010; Aachen, Germany

Fig. 16 Nishisato's European colleagues: Germany, Austria

(a) Brigitte Le Roux, 2015; Paris

(b) Gilbert Saporta, 2011; Paris

(c) Ludovic Lebart, 2015; Paris

Fig. 17 Nishisato's European colleagues: France

(a) Jan de Leeuw, 2011; UCLA, USA

(b) Willem Heiser, 2017; Leiden

(c) Jacqueline Meulman, 2019; Leiden

Fig. 18 Nishisato's European colleagues: Leiden, The Netherlands

(a) Pieter Kroonenberg, 2012; Leiden

(b) Wim van der Linden, 2016; Twente

(c) Peter van der Heijden, 2006; Utrecht

Fig. 19 Nishisato's European colleagues: The Netherlands

(a) Ivo Molenaar, 1995; † 2018
(b) Jos ten Berge, 2005
(c) Henk Kiers, 2006

Fig. 20 Nishisato's European colleagues. Groningen: The Netherlands

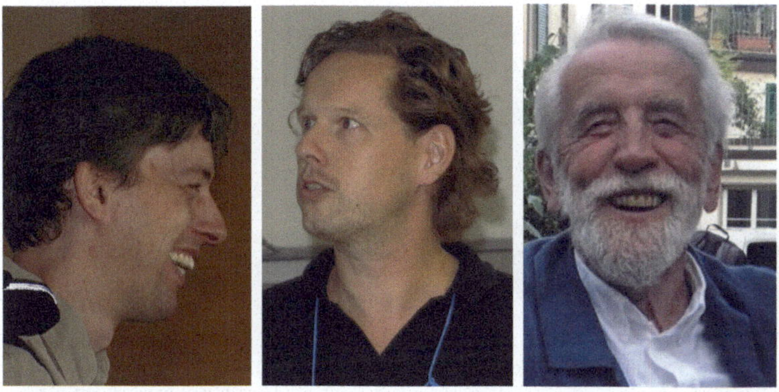

(a) Patrick Groenen, 2005; Rotterdam
(b) Michel van de Velden, 2017; Rotterdam
(c) Helmut Vorkauf, 2021; Switzerland; Courtesy of HV

Fig. 21 Nishisato's colleagues: The Netherlands

(d) John Gower, 1999; † 2019; UK
(e) Frank Critchley, 2019; Milton Keynes, UK
(f) David Hand, 2010; Imperial College London, UK

Fig. 22 Nishisato's European colleagues: Great Britain

(a) Karl Jöreskog, 2008; Uppsala, Sweden

(b) Michael Greenacre 1999; Barcelona, Spain

(c) Eeke van der Burg, 2008; Leiden † 2019

Fig. 23 Nishisato's European colleagues: Sweden, Spain, Switzerland

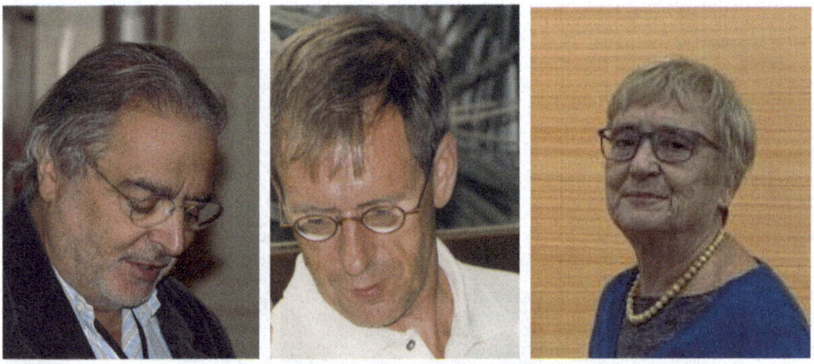

(d) Michel Tenenhaus, 2011; Paris, France

(e) Jan van Rijckevorsel, 2017; Amsterdam

(f) Ineke Stoop, 2021; Courtesy of Ineke Stoop Den Haag, The Netherlands

Fig. 24 Nishisato's European colleagues: France, The Netherlands

7 South Africa

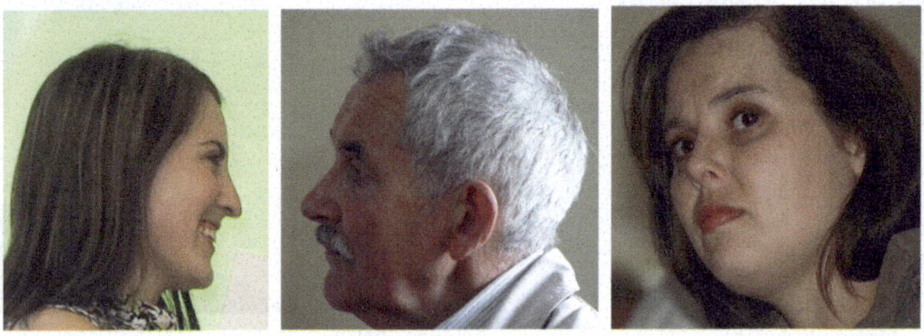

(g) Johané Nienkemper 2017; Stellenbosch

(h) Niel Le Roux, 2010; Stellenbosch

(i) Sugnet Lubbe, 2010; Stellenbosch

Fig. 25 Nishisato's South African colleagues

8 Nishisato with Many Friends

(a) Top row: Caussinus, Rouanet, Zárraga, Nishisato, Pagés, Galindo, Saporta, Friendly, Heiser, Kroonenberg, Lewi, Gower. Bottom row: Takane, Cuadras, Lauro, Ter Braak, Lebart

Fig. 26 Nishisato and friends: CARME2003

9 The Distinguished Editors

(a) Eric J. Beh, 2013; Newcastle, Australia

(b) Rosaria Lombardo, 2013; Capua, Italy

(c) Jose G. Clavel, 2016; Murcia, Spain

Fig. 27 Editors of this volume

My Recollections of People in the World of Data Science

Shuichi Iwatsubo

1 Prologue

I felt very honoured to be asked to contribute to the *Festschrift* of Professor Emeritus Shizuhiko Nishisato. At the same time, however, I hesitated before I could write anything. Why? Well, Nishisato has absorbed himself in the world of multivariate methods for categorical data and has contributed greatly to these areas; but my main interest changed a long time ago from developing data analysis utilised to clarify universal human behaviour to investigating *individual* human personality. Then I recalled how lucky I have been to know Nishisato, who is so open-minded, kind, and generous, a man who loves people. I have also had the fortune to get to know many attractive scientists, not only in Japan but also overseas, directly and indirectly, including those Nishisato has known very well. I felt that this might be a valuable opportunity to inform scientists living abroad about some aspects of the past activities of Japanese data scientists, which will most likely be largely unknown to them. And I have to admit that I have personally felt great pleasure to be a member of this world. So, I decided to take the plunge…

I would like to share my humble but precious memories of the great scientists I met, mainly, during the 1960s, '70s, and '80s however fragmental they may be. I firmly believe that a happy family originates from the warmest humanity, so I would also like to mention their families that I have had a chance to get to know very well. I will be happy if my memories can offer some previously unfamiliar episodes featuring distinguished scientists of multivariate categorical data analysis and the data sciences.

S. Iwatsubo (✉)
Research Division, The National Center for University Entrance Examinations, Tokyo, Japan
e-mail: iwatsubo@waseda.jp

E. J. Beh et al. (eds.), *Analysis of Categorical Data from Historical Perspectives*, Behaviormetrics: Quantitative Approaches to Human Behavior 17,
https://doi.org/10.1007/978-981-99-5329-5_3

2 Chikio Hayashi's 'Type III' Method

The history of science sometimes reveals the interesting fact that similar methods are sometimes discovered and developed independently and almost simultaneously. It was in the spring of 1968 that I learned about one of Chikio Hayashi's quantification methods called 'Type III'. It was later recognised to be similar to the 'Dual Scaling' developed by Shizuhiko Nishisato (Nishisato 1975, 1980), the 'Correspondence Analysis' developed by J.P. Benzécri (Benzécri 1973), and the 'Reciprocal Averaging Method' described by M. O. Hill (Hill 1973).

In 1968, I was a member of the research section of the Electro-Technical Laboratory (ETL) in Tokyo. Hirohiko Nishimura, my senior colleague, sent me to Dentsu, Japan's largest commercial advertising company. The idea was to learn Hayashi's Methods that were utilised there and to train in the techniques of manipulating the FORTRAN computer programming language.

At Dentsu's Marketing Research Division at that time, a computerised method was under construction to predict TV audience ratings based on categorical data characterised from a new planned TV programme. Hayashi's 'Type I' Method, which formally corresponds to the multi-correlation method for categorical data, had been adopted. Meiko Sugiyama of NHK, Japanese semi-public broadcasting company, had already been struggling energetically to predict the audience ratings of radio and TV programmes.

I was handed a copy of the small textbook prepared for seminar attendants that had been written by Toshio Uematsu, who was then a member of the Institute of Statistical Mathematics (ISM) in Tokyo. Produced using a kanji-character typewriter, it was designed to make Hayashi's Methods easy to understand. I immediately got absorbed in them, especially 'Type III'.

The following is a brief sketch of 'Type III' with the data to which the method was applied for the first time: 'Thirty Persons by ten Designs of Cans' binary data are given, where '1' of the binary data represents a person likes one design and '0' where that person who does not. Hayashi put x_i to design i and y_j to person j. Then $\{X|x_i, i = 1, \ldots, 30\}$ and $\{Y|y_j, j = 1, \ldots, 10\}$ are adjusted so that the correlation coefficient between X and Y is maximised. The result leads to the simultaneous construction of one-dimensional scaling with regard to persons and designs.

Soon after I returned to ETL, Nishimura and I started to give both words and sentences numerals, determined by applying 'Type III' to 'words by sentences' binary data. Sixty-one sentences were selected from four areas: software and hardware papers in computer sciences and two American literary works by John Steinbeck and Saul Bellow. Forty-one words were selected according to the running occurrence frequency order occurring in sixty-one sentences. The considerable results of applying 'Type III' to 61 by 41 binary data led to our contribution to a scientific magazine (Nishimura and Iwatsubo 1970), the percept of which was sent to Chikio Hayashi at ISM. That launched our close communication.

3 The Behaviormetric Society (BS)

The Behaviormetric Society (BS) was established in Japan in 1973. Its members, who belonged to a broad range of scientific fields, were methodologically very interested in statistical methods, especially such multivariate methods as factor analysis, MDS, Hayashi's Methods, etc. Chikio Hayashi was the first President and Haruo Yanai was the Secretary. Yanai was supported by several young members, one of whom, Kumiko Maruyama, enthusiastically persuaded many people to participate in this new interdisciplinary society.

I helped Yanai to dispatch scientific journals, newsletters, and so on to members. After our busy work was over, we often enjoyed a break together. As soon as we sat down for a cup of coffee, Yanai would ask me to listen about how he had obtained his recent new results, most of which were theorems and lemmas concerning mainly projector, generalised inverse of matrices.

> Can you spare me a few minutes? Five minutes will be alright. No, no! Just three minutes will be enough! Please listen carefully!

Then he would vividly and cheerfully continue for half an hour, or sometimes more than one hour, to tell me how to infer new propositions and deduce a lot of lemmas. He would do this by writing down hard-to-decipher symbols and formulae very quickly on any small sheet of paper available!

Yoshio Takane was one of the students who was taught multivariate statistical analysis by Yanai at the University of Tokyo. I believe that Takane was encouraged a great deal by Yanai to study multivariate methods. In fact, I feel it would be no exaggeration to say that his professional life was basically determined by meeting Yanai. In 1986, Takane was the 1st prize winner of the Hayashi Chikio Award (Achievement Award) from the Behaviormetric Society as a great methodological contributor to behaviormetric research. It was Yanai who had earnestly recommended him.

4 University of California, San Diego

In the summer of 1975, J. D. Carroll and Taro Indow organised a US-Japan Seminar on 'Theory, Methods and Applications of Multidimensional Scaling and Related Techniques'. Sponsored by the National Science Foundation, U.S.A. and the Japan Society for the Promotion of Science, it was held at the University of California, San Diego (UCSD), from the 19th to the 23rd of August. I remember that was soon after the end of the Vietnam War.

The participants from Japan were Taro Indow, Chikio Hayashi, Masaaki Yoshida, Meiko Sugiyama, Akio Kameoka, Keiko Matsushima (now Keiko Watanabe), Akinori Okada, and myself. Participants from the U.S.A. were J. D. Carroll, J. J. Chang, J. B. Kruskal, Myron Wish, Norman Cliff, James C. Lingoes, R. N. Shepard, W. S. Torgerson, L. R. Tucker, and other well-known researchers. Jean-Marie Bouroche was the only participant from France. It was thanks to him that four years later we got the opportunity to be in contact with French scientists and to meet J. P. Benzécri.

When we arrived at UCSD on the 19th of August, three hippie-style North Carolina gentlemen appeared. They were Forest Young, Jan de Leeuw and Yoshio Takane. The young Takane was long-haired and wearing sunglasses. According to the notes I wrote at the time, I invited him to my room in Tioga Hall of UCSD one evening and we kept talking until 1 a.m. It was the day before he left for the University of North Carolina, Chapel Hill, where he was to spend two years. The only thing I remember now from all that talking was that Takane had enjoyed seeing the movie *Jaws*, which was then enjoying great popularity in the U.S.A!

In September 1980, J. B. Kruskal came to Japan as an invited speaker at the Annual Meeting of the Behaviormetric Society held at the Hiyoshi Campus of Keio University. He gave us a lecture on 'Analysis of Data by Geometric and Multilinear Methods' with splendid interpretation by Takane, who was then at McGill University in Canada.

5 ISI Session, New Delhi

In 1977, Haruo Yanai and I attended the 41st Session of the International Statistical Institute (ISI) in New Delhi. We shared a room at a hotel in New Delhi, very near Old Delhi. Another important purpose for Yanai to visit to India was to meet C. R. Rao, who had recently retired from his position as Director of the Indian Statistical Institute to concentrate on his research. After the ISI Session, Yanai remained in New Delhi to contact Rao. I remember that one day after returning from a discussion Yanai was in his room making great efforts to solve problems proposed by Rao. I suppose it meant that he had to carefully prepare answers ready for the next discussion a few days later! Their discussions culminated in the following statistical papers: Rao and Yanai (1979, 1985).

After returning to Japan, Yanai told me that the Japanese Ambassador to India had invited him as a guest to both the Christmas and New Year parties at the Japanese Embassy, and in lotteries at them he had won a large traditional Indian wooden table (Christmas) and a TV set (New Year)! The Embassy staff and their families had apparently been very envious! The generous Yanai donated his prizes to Embassy staff.

I can never forget Yanai's deep kindness to me during the visit I made to Calcutta despite his busy days with Rao. Thanks to him, then the Director of the Indian Statistical Institute, Gopinath Kallianpur, invited me as a guest researcher. In Calcutta, I called on Mrs Mahalanobis, who was a friend of my parents-in-law. She led me to the spacious drawing room where R. A. Fisher once enjoyed talks with her late husband, P. C. Mahalanobis. I presented the gentle old lady with a Kabuki Theatre calendar which featured woodblock prints of famous scenes from Kabuki performances. She turned the calendar sheets one by one and asked me the meaning of every scene. In particular, she wanted to know why most of the characters looked angry. I discovered afterwards that Kabuki actors pause to pose dramatically for a few seconds to emphasise their character's actions, especially when punishing the wicked, and naturally their eyes are widely opened, which makes them look very angry. I felt sad that the combination of my poor English and my lack of knowledge about a traditional art of my country prevented me from giving Mrs Mahalanobis a satisfactory answer.

In 1979, on his way back from the 42nd ISI Session held in Manila, Yanai consulted with B. N. Mukherjee, one of his co-authors, about the publication of their book *The Foundation of Multivariate Analysis*. It was published in 1982 and received a high evaluation (Takeuchi et al. 1982).

6 International Symposium at Versailles

I have to admit that, at first, we all supposed that 'J. P. Benzécri' was not actually a personal name but a collective one just like 'Nicolas Bourbaki'. We later realised that we had been mistaken! Many bright young scientists actively collaborated with Professor Benzécri as honourable members of the 'Banzécri Clan'. One of them, Ludovic Lebart, told us that Benzécri had written his two-volume *L'analyse des Données* in old-style French. We were familiar with Benzécri's legendary refusal to travel by air, which prevented him from attending international meetings abroad. Apparently, he spent most his time meditating on data analysis at his mysterious lodge 200 km's from Paris, so the fact that Chikio Hayashi managed to realise a meeting with him in Paris in October 1979 was a rare achievement indeed (Fig. 1).

The 2nd International Symposium on Data Analysis and Informatics (ISDAI) was held at Versailles, France, in the autumn of 1979. The Japanese attendants

Fig. 1 Sketch by the author on 22 Oct. 1979

were Chikio Hayashi, Setsuko Takakura (a good French speaker), Meiko Sugiyama, Noboru Ohsumi, Fumi Hayashi (no relation with Chikio Hayashi, but his very reliable research collaborator), and myself. I had the honour of shaking hands with Benzécri at his meeting with Hayashi. The delicate softness of the genius' hand was unforgettable. (In strong contrast, Gilbert Saporta's handshake was so strong it left my right hand numb; I felt it embodied the dynamism of young French scientists!) Besides Lebart and Saporta, we could also meet Edwin Diday, Yves Escoufier, Michel Jambu, Alain Morineau, Maurice Roux, Jean-Pierre Nakache, and other active French scientists.

It was also my great pleasure to meet Jean-Marie Bouroche again. He invited us to visit his apartment in Paris, and we met his three beloved young daughters. When he attended the 46th ISI Session held in Tokyo in 1987, I welcomed him and his wife to my house. My father, who had learnt to speak French in his youth, sang *La Marseillaise* accompanied by my mother on the piano. Mr and Mrs Bouroche were so pleased to hear their national anthem, they sang an old French song in beautiful harmony to thank my parents.

Ever since that time, Ludovic Lebart has been extending kindness to Japanese data scientists. A deep friendship of trust between Ludovic and Noboru Ohsumi began and has continued right up to the present. Ohsumi later spent one year at ENST, Ludovic's institute, as a visiting scientist. In 1994, Ohsumi and Yasumasa Baba were the co-authors of a Japanese book (Ohsumi et al. 1994) which was basically a translation of the book written by Lebart, Morineau, and Warwick (Lebart et al. 1984) and with additional contents. It was warmly welcomed in Japan. In the Preface, Ohsumi expressed his gratitude to Kinji Mizuno (ISM), who had offered his survey data for the book, and to Haruo Yanai for his detailed comments on the manuscript.

Ludovic loves *The Little Prince* by Antoine de Saint-Exupéry and collects the versions published in other countries, so when he came to Japan he made sure to find a Japanese edition! He is also very interested in Japanese culture. He once told me how much he enjoyed reading *I am a Cat* written by Soseki Natsume, one of Japan's greatest novelists, which is often compared in Japan to E. T. A. Hoffmann's *Lebensansichten des Katers Murr*.

In October 1985, Haruo Yanai participated in the 5th International Symposium on Data Analysis and Informatics at Versailles as an invited speaker. He told me that his book *The Foundation of Multivariate Analysis* might be of interest to the French organisers of the Symposium. I also attended the Symposium on my way to the UK. We shared a room at a small hotel near Pont Mirabeau on the Seine, the bridge which features in *Le Pont Mirabeau*, the famous love poem by Guillaume Apollinaire. One evening we were standing on the bridge in the autumn twilight gazing at the salmon-pink sky. Yanai suddenly asked me to give him some advice. He had been offered a professorship from The National Centre for University Entrance Examinations (DNC) in Tokyo, to which I belonged. I expressed my opinion that he would probably rather have more freedom at his university than at DNC. Then next spring in the UK I heard from a DNC colleague that Yanai had in fact joined DNC! He was such a great figure who contributed to both activities for DNC and the development of data analysis. In July 2007, he presided over the International

Psychometric Meeting held in Japan. After retiring from DNC, he belonged to St. Luke's International University, a nursing university in Tokyo and made great efforts to establish a system of common entrance examinations for nurses.

Haruo Yanai sadly passed away in December 2013 at the age of only seventy-three. A condolatory telegram from C. R. Rao expressing deep sadness was read at the funeral. Yanai was indeed a highly distinguished scientist who was loved by so many people both in Japan and abroad. For me he will remain forever just as if he were my kind elder brother.

7 John C. Gower

From the autumn of 1985 to the summer of 1986, I was a visiting scientist at the Statistics Department of Rothamsted Experimental Station. I owed that happy spell in the UK to the kindness of Akinori Okada who willingly took over as Secretary of the Behaviourmetric Society, to which I had succeeded from Kinji Mizuno.

I stayed at the Rothamsted Manor House, built of sturdy English oak in the seventeenth century, which offered accommodation to visitors. Staying there were many students and researchers not only from the UK but from all over the world. I made friends with a doctorate student from Germany who studied entomology. During World War II, a treaty was forged between Japan and Germany to fight against the UK, but now, in the UK, I was able to forge a private friendship between Japan and Germany, a friendship that has continued right up to the present. That made me think how wonderful peace is. My German friend was a pious evangelist and he introduced me to his British evangelist friends, including Keith Goulding, who was in the RES Soils Division.

Needless to say, the English 'tea break' took place every morning and afternoon. I envy British people being able to enjoy those tea breaks, although I have known some people (even British gentlemen) call it an 'infamous' custom… Well, I believe that most of the best aspects of British culture might have originated from it!

It was during those breaks that I met Pete Digby, Alan Todd, Simon Harding, Gavin Ross, Peter Lane, Rodger Payne, and other fine people. At one of them, Gavin Ross told me that it was very difficult in the UK to find a mug with an ape on it. Well, I decided to surprise him if I could, so I searched for one every weekend, sometimes wandering all around London. But it was all in vain… Gavin sympathised with me when I told him, and he kindly presented me with a mug he happened to have acquired with a monkey hanging from a tree branch by its tail. It is now one of my favourites in my collection of eleven mug cups decorated with various figures of apes.

What led me to go to RES was my interest in John Gower's general coefficients of similarity (Gower 1971) rather than his principal coordinates analysis. John had just succeeded to the position of Head of the Biomathematics Division from John Nelder, who had encouraged the members to complete GENSTAT. John and I were very soon on first name terms and I was also welcomed by his wife Janet and their children Sally and James. Janet was a splendid cook and the Gowers invited me to dinner at home while I was staying at the Manor House. I still clearly remember the delicious taste of the soft, juicy roast lamb I was served. James, who was then sixteen, loved tortoises, tropical fishes, and other little creatures. He told me that he was not very interested in the biology lessons he received at school, so I gave him a paperback titled *The Green Year,* the story of a boy who loves natural history, written by A. J. Cronin.

My family went to the UK in May 1986. After detailed investigation, Keith Goulding introduced us to a lovely small elementary school for my two young daughters. Their teacher, Christopher Rowlatt, of whom our girls grew very fond, was later successful as a marbling artist, and we still keep in touch.

In April 1987, John attended the 46th ISI Session held in Tokyo. Janet accompanied him, and one day my wife took her sightseeing in Tokyo. They visited the famous Senso-ji Temple in Asakusa and tasted some small Japanese cakes baked at a traditional confectionary shop while strolling along Nakamise Street. After taking lunch at Tatsumi-ya, a long-established Japanese restaurant, they went on a boat trip down the Sumida River to Hama-Rikyu Gardens. On their way, Janet saw some people eating and drinking under the cherry trees on the riverbank, even though it was quite a cold day. They had probably expected to enjoy full blossoms, but despite the cold and being disappointed that only a few flowers were in bloom they had decided to go ahead with their cherry-blossom-viewing party! Janet was amused and took some photos of them. The two ladies took a pleasant ramble around the Hama-Rikyu Gardens and then enjoyed tea and cakes at Shiseido Parlour, a popular café in Ginza. Janet really appreciated the beautiful cakes subtly decorated with a net of thin starch jelly strings and took a photo of them.

The Gowers' son James grew up to became a very successful exhibition organiser. One Sunday when I was visiting the UK in 1989, I telephoned John from St. Pancras Station in London. The Gowers were all out at a local school where an exhibition organised by James was being held, but, as luck would have it, just at that moment John happened to return home to fetch something he'd forgotten. He invited me to join them, so by chance I had the honour of participating in James's debut exhibition!

The Great Hanshin Earthquake occurred in 1995 just when James and his wife Jo were staying with us in Tokyo. Janet quickly telephoned us from the UK to confirm their safety. In 1996 James very kindly offered accommodation to our two daughters, then university students, when they visited the UK. In February James often took my elder daughter to an underground station to pay visits to London. In summer travelling by car in Spain with James and Jo still remains vividly in my younger daughter's mind thanks to many amusing events caused from their rigid rule never to reserve accommodations in advance, interesting conversations with local Spaniards who could not speak English, etc.

In the summer of 1999, I went to the UK again on business and stayed a few days with James and Jo at Hatfield near London. John and Janet had moved to Nailsworth, a town in Gloucestershire not far from Wales. James presented me with a CD of compositions by Michael Greenacre, his father's good friend, titled *You, Woman.* I must confess that the moment I heard it I was so fascinated by the music that I could hardly believe that the composer Michael Greenacre was really the same person as the distinguished South African data analyst! When I listened to the CD again later, I discovered that it was Michael himself who sang 'Cradle Song', one of the thirteen songs, so beautifully. I apologise to him for misunderstanding up to then that all the tracks were sung by Gurdeep Stephens.

John was a keen lover of natural orchids. His delight in seeing a rare one was for him just as if he found a new property in his algebraic studies of multivariate methods! He had been greatly affected by the wild flowers he saw in a wood near Selborne, made famous by Gilbert White's *The Natural History of Selborne*, where John spent his elementary school days during World War II. When he visited Japan in July 2001, we went to Lake Shirokoma located in the mountainous Nagano Prefecture. To my great surprise, on our way to the lake he quickly spotted a little orchid, *Dactylostalix ringens Reichb.f.*, blooming quietly in the dark wood. In June 2013, when my wife and I took a two-week journey around the UK, he escorted us (fellow wild orchid lovers!) to a natural grass field with fine red British wild orchids. During our stay in the UK, we were able to lodge for a few days in the same room (The Pink Room) of the Manor House as I had inhabited for six months around thirty years earlier. At that time I was also to stand before the memorial monument to Pete Digby built after his early death. It's near the former building of the Statistics Department and the Manor House (Figs. 2 and 3).

After retirement from RES, John remained very active in his studies. He stayed in the Netherlands for a few years. Janet accompanied him wherever he went. He wrote two books (Gower and Hand 1996; Gower and Dijksterhuis 2004). Everyone who acquired '*Procrustes Problem*', one of two books, from him as a complimentary copy would see on the back cover, 'I hope you enjoy some Procrustes bedside reading' and John's autograph.

In July 2001, John visited DNC and introduced to us the Open University system of certification. He also presented us a video of R. A. Fisher produced by RES. One weekend afternoon, John and I called in at the National Museum in Ueno Park

Fig. 2 Dactylostalix ringens Reichb.f.

Fig. 3 *Orchis mascula* in the UK

in Tokyo. When we came out, John stopped on the steps to look at several people lingering to look at the sunset. 'I like to see people *lingering*…' John murmured. I'm not sure why, but I was so impressed by the sound and meaning of the word 'linger' that I have made it a habit to collect any sentence in which the word appears!

John Gower, a very happy man who was dearly loved by friends all over the world, passed away in May 2019 soon after his birthday. And a very sad mail came from Sally telling us that Janet, his greatest supporter, had peacefully followed him in October. My younger daughter and her husband had been able to meet John, Janet, and their family in the summer of 2018 before the onset of the COVID-19 pandemic. It is a precious memory.

8 French Connections

On our way back home from the UK in the summer of 1986, we dropped into France. Ludovic Lebart picked us up and took us to see the night view of Paris from Montmartre. Fifteen years later, from 2001 to 2002, my younger daughter was in a graduate course at Nottingham Trent University in the UK studying art. When she took a holiday in France, Ludovic kindly welcomed her to stay with him in Paris.

During our 1986 trip, Yves Escoufier and his wife invited us to supper at their home in Montpellier near the Mediterranean Sea. My elder daughter still says that the salad served by Mrs Escoufier on that occasion was one of the two most delicious dishes she has ever tasted in her life: tomato and paprika-based bouillabaisse risotto with mussels, shrimps, octopus, other seafood, containing onions, and garlic tomatoes, mussels, harmoniously blended with white wine as a secret ingredient. (She insists that her memory is precise with a probability of seventy per cent!) By the way, her other favourite dish was served at a small restaurant in Bath, the beautiful English town known for its Roman bath and 'Beau' Nash: sliced duck filet, lightly browned, served with roasted duck with orange juice, Grand Marnier, white wine, and orange zest in a brown sauce.

In March 1987, the Japanese-French Scientific Seminar on 'Recent Developments in Clustering and Data Analysis' was held at the Institute of Statistical Mathematics (ISM) in Tokyo. The French participants were Edwin Diday, Michel Jambu, Yves-Max Schektman, Alain Morineau, Maurice Roux, Isräel C. Lerman, Yves Escoufier, Brigitte Escofier, Ludovic Lebart, Guy Der Megreditchian, and Jean-Pierre Nakache. I enjoyed the beautiful flowing French of Megreditchian's lengthy address at the welcome party. After returning home, he kept sending me his papers. When I heard of his death after a battle with cancer, I really missed him and understood why he had earnestly continued sending his papers to a lot of scientists: he must have been writing papers in defiance of his physical condition.

9 Kinji Mizuno

Kinji Mizuno was one of Chikio Hayashi's great research supporters from his twenties (Iwatsubo 2018). After belonging to the Institute of Behavioural Sciences (IBS) in Tokyo and the University of Nagoya, he moved to ISM in 1973. He took over from Haruo Yanai as Secretary of the BS, which was then headed by Chikio Hayashi as President. I sometimes went to his office at ISM and helped him prepare materials for delivery to BS members. I was soon moved by his sincere personality, and my respect for him grew day by day. I would go so far as to say I have never seen such a gentleman like him who devoted himself so much to public welfare.

Mizuno contributed a great deal to the nationwide survey of the Japanese national character conducted by ISM every five years, especially in 1978 (Res. Committee for the Study of the Japanese National Character 1982). Each survey has been based on face-to-face interviews regarding 50 items. To facilitate comparisons, the same questions have been included for a long time. Between 3000 and 6000 Japanese nationals aged twenty and over was selected at random by a stratified three-stage probability sampling method based on voter lists. Mizuno struggled to maintain the reliability of the survey but the number of respondents who rejected the interviews increased year after year, which caused him great stress.

In March 1981, I accompanied Mizuno when he visited Atami, the well-known Japanese resort, to survey the awareness and attitude of citizens regarding the widely anticipated Tokai Earthquake. I had kept asking him to give me a chance to participate in his surveying activities and on that occasion he allowed me to join him. Following his instructions, I interviewed some citizens. It taught me how important, and also how difficult, it is to collect reliable data. My experiences in Atami considerably influenced my attitude towards research. I sometimes said to myself, 'Of course it's very important to develop data analytic methods mathematically with computer programmes. But aren't good methods, including a set of questionnaires and interviewing to collect reliable data, indispensable and important as a major premise?'

Mizuno's study was developed into disaster preparedness education for school children (Iwatsubo 2002). In his elementary school days, he had read a story in a Japanese language textbook about an old village headman who helped to save his villagers from a tsunami disaster. One day, he felt a strong earthquake and noticed that the sea waves were changing. He immediately set fire to sheaves of the precious new rice that had just been harvested near his house on the hill. The villagers all stopped working and rushed up the hill, thinking that the headman's house was on fire. Just as they reached it, a terrific tsunami surged in and swallowed up the fields they had just been working in.

Titled *Burning the Rice Sheaves (Inamura no Hi)*, the story had been written by an elementary school teacher, Tsunezo Nakai, with reference to Lafcadio Hearn's *A Living God*. In fact, the story was based on an actual earthquake and tsunami that occurred in 1854. The village headman was a real person named Go-ryo Hamaguchi. Nakai's words were so simple, clear, and vivid that not only Mizuno but also most of his fellow pupils were impressed, and the message 'Whenever you feel a strong earthquake near the seashore, run to higher ground as soon as possible' was engraved forever on their young minds. Mizuno made great efforts to make it possible for Tsunezo Nakai to be awarded the Japanese government's prize for contributing to disaster prevention in 1987.

Hyon-Jun Rho from South Korea, who came to ISM from 1987 to 1988 and belonged to Mizuno's research section, is still very grateful for the deep kindness shown to him by Mizuno. He studied Hayashi's methods and introduced them to data scientists in South Korea. He also made a large contribution to the development of quality control techniques in his country. After retirement from his university, he started writing books on modern politics and history. We used to meet every time he came to Japan and enjoyed many happy moments. Since the end of 2019, it is a great shame that the COVID-19 pandemic prevented us from meeting.

It was my great pleasure to hear that Mizuno was going to move from ISM to DNC in 1991. Sadly, however, he passed away in 1999; eight years after his move. At the memorial gathering for him in 2000, Chikio Hayashi presented an hour-long speech to express how much he missed the indispensable partner in his studies. Hayashi greatly appreciated the fact that Mizuno had shown us how quickly he acquired essentials from data, and he promised to establish his own data science as soon as possible, meeting Mizuno's expectations. His book for Data Science published in 2001 (Hayashi 2001), was dedicated to two late young comrades, one of whom was Kinji Mizuno.

We never imagined that Hayashi would be following Mizuno only two and a half years after that memorial party, leaving a wealth of works on the data sciences. Ryozo Yoshino, the editor of the Bulletin of Data Analysis of Japanese Classification Society, provided us with many invaluable words he had heard from Hayashi in 2001 (Takahashi 2021). In 2021 Kumiko Maruyama published a biography of Chikio Hayashi (Maruyama 2021).

Both Hayashi and Mizuno retained their great passion for inquiring into human behaviour, never losing a fresh interest in human beings. In other words, I would say it's very clear that they both loved people. Mizuno highly evaluated two younger scientists, Takashi Murakami and Ei-ichiro Nojima, to whom he taught computer programming techniques in their undergraduate days. Since retiring from his university, Murakami has continued to enjoy his studies, sometimes writing scientific papers for journals. Nojima, who became the head of the Waseda University School of Human Sciences, kindly made efforts for me to move from DNC to Waseda University.

10 Waseda University and Walking the Hakone Ekiden Course

I am rather afraid that the following section may be too personal and drifting some distance away from Nishisato. However, I beg for your forgiveness to let me include it as a personal example of someone who moved away from developing categorical data analysis.

In April 2005, I moved from DNC to Waseda University. I feel it was appropriate for me to leave DNC since my interest had been gradually changing from 'education *before* entrance examinations' to 'education *after* entrance examinations'. I started to present lectures on elementary statistics which in fact continued for eight years (later I lectured to graduate students on multivariate methods for categorical data). There were nearly 500 students every year, many of whom, unfortunately, were less than excited about studying mathematics!

I kept doing my best to prepare materials for my lectures, giving a small test after every lecture and encouraging their application of statistical analysis to their own data collection efforts. This took me some distance from my BS duties as I was very short of *multivariate* ability.

Despite all my efforts, I must say now that my teaching was not successful. I set a minimum requirement to be learnt by students: a deep understanding of *variance* as a fundamental source of information. I'm not sure that even my minimum requirement was met. Eventually, most students did not seem to be interested in statistics and data analysis.

I did, however, get to know some students who sympathised with my desperate efforts, and we started communicating frequently. One of them was a runner in the Hakone Ekiden marathon race held every New Year which features twenty university teams of ten runners. The race consists of five stages run between Otemachi in Central Tokyo and Hakone Mountain on the first day and the same five stages in reverse back to the finish line in Ohtemachi on the second day. The total distance of the ten courses is around 207 km's. As a kind of challenge to encourage the Waseda University team, in 2011 my wife proposed that between September and December every year we should walk all ten stages, at a pace of one stage a day. We finally ended our annual walking mission in 2020, meaning we had walked a total of 2070 km's, which is about one twentieth of the way around the Earth. Unfortunately, despite all our efforts, the Waseda University team did not win the race in any of those ten years!

Around 2005, I suppose my interest was gradually moving away from the development of categorical data analysis. Soon after I knew about Hayashi's 'Type III' Method, I was interested in the method applying to three-way categorical data. Naturally I got a method that gives optimal numerals (x_i, y_j, z_k) to a point (i, j, k) of three-way binary data by means of the maximisation of multiple correlation coefficients. It is easily generalised to the case of $n(n > 3)$-way binary data (Iwatsubo 1978) and led to the methods by the maximisation of canonical correlation coefficient as anyone may conclude. I think that the investigation of linear relationships latent in multi-way binary data is reduced to the inquiry into the properties of canonical correlation coefficient. I sensed that might be the reason why John Gower had concentrated on writing a book on canonical analysis until just before he passed away.

The vast world of non-linear relationships latent in categorical data is opened before us. I once proposed the method for three-way binary data in terms of the maximisation of the third correlation coefficient, inspired by Kei Takeuchi's paper (Takeuchi 1974). I expected that the method might detect the tendency to gradual changes of people's sense of value inversely with the lapse of time by applying to some cohort categorical data. Tadashi Yoshizawa generalised my method into the analysis of multiple contingency tables (Yoshizawa 1988).

Those very busy days at Waseda University seemed to make me, day by day, more and more remote from the development of multivariate methods for categorical data.

11 Shizuhiko Nishisato

Although Shizuhiko Nishisato had been a familiar name since he published the book on applied psychometric scaling in 1975, it was unfortunate that I had few chances to see and talk with him either in Japan or abroad. I was there when he presented a lecture in the autumn of 1986 for the 6th International Symposium on Data Analysis and Informatics (ISDAI) at Versailles, but I was so busy preparing for my session and another meeting that I had little time to talk with him. Due to a previous appointment, I had also lost a good opportunity to have the honour of being one of the commentators at an annual meeting of BS, which Yasumasa Baba, the organiser of a special session for Nishisato, had invited me to accept.

So, it was a great pleasure that I could have a long talk with Nishisato when he came to Japan in June 2019 to receive the Award from Annual Meeting of Japanese Classification Society and present a lecture. It reconfirmed for me his sincere personality—open-minded, kind and generous—but also impressed on me the importance of friendship. Since he returned to Canada, we have continued communicating by e-mail.

I was also very happy and surprised to get to know Nishisato's grandson, Lincoln Dugas-Nishisato. At the age of nine, Lincoln wrote a science fiction book which was published thanks to the warm support of his family and his teachers, with a translation into Japanese by his beloved grandfather. The book tells how the grown-up Lincoln, a physicist, is carried by a time-machine to meet great figures from the past. The humble but kind suggestions, encouragement and help Lincoln gives them help to trigger their great achievements. For example, in the Netherlands he meets Anne Frank and suggests she should buy a diary on her birthday, which will surely be read by countless people all over the world. Lincoln deeply laments that he was unable to save her as historical facts can never be altered.

At the age of twelve, Lincoln organised a cooking class designed for children with disabilities, and donations to a rehabilitation hospital are requested. Not only children but also many adults enjoyed making cakes and learning new recipes at his kitchen or via Zoom following his instructions. He loves cooking using various recipes of the places where he has enjoyed travelling with his parents. Ever since he was five, whenever he sees people who are unhappy, he has never hesitated to volunteer to help them. It seems to me that Lincoln shares that 'unbearable pity for the suffering of mankind' which was one of the three passions that Bertrand Russell said governed his life.

There is a café in Tokyo called 'Avatar Robot Café DAWN' where the bedridden or housebound can serve and communicate with guests in the café through the robots developed by my young friend Kentaro Yoshifuji, who studied engineering at Waseda University. Each robot is manipulated by a PC operated by a disabled person far from the café. To assist people with much more serious disabilities, Yoshifuji has recently been facing the challenge of developing a system for operating a PC using only brain waves.

It gives me feelings of both considerable relief and considerable hope that, in spite of all the anxieties filling the world today, we never fail to see such young people as Lincoln and Yoshifuji, who love human beings and offer themselves to improve the happiness of mankind.

12 Postscript

Every autumn for several years, I have stood in front of the graves of Chikio Hayashi, Kinji Mizuno, and Haruo Yanai. Those of Hayashi and Yanai are located not so far apart, so I can visit them on the same day. While offering sincere gratitude to all three of them, I also apologise for not having fulfilled my duties for the BS as they had expected me to do, and for not repaying their great kindness. I beg their forgiveness… Silence. They seem to be complaining to me, but have finally dared to permit me at least as being a humble seeker of the answer to that everlastingly difficult but fascinating question: 'What is a human being?' Well, so I believe…

I cannot help but notice once again that in this contribution to the Shizuhiko Nishisato *Festschrift*, I have talked too much about myself rather than about him.

It is, however, certain that I could never have met and known such splendid people as I have been talking about here if I had not been interested in the multivariate methods for categorical data and related topics which Shizuhiko Nishisato has loved throughout his life. I will be extremely happy if, just as a record of some episodes in the lives of great data scientists, my memories might be permitted to slip into the *Festschrift* of a man I highly respect.

Acknowledgements The author would like to thank the editors of the *Festschrift* for their very kind comments and suggestions and Mr Stuart Varnam-Atkin and Mrs Fumiha Suzuki for improving the readability of the article.

References

Benzécri, J.P.: L'Analyse des Données, Tome1: La Taxinomie, Tome2: L'analyse des Correspondences. Dunod, Paris (1973)
Hill, M.O.: Reciprocal averaging: an eigenvector method of ordination. J. Ecol. **61**, 237–249 (1973)
Gower, J.:C.: A general coefficient of similarity and some of its properties. Biometrics **27**, 857–871 (1971).
Gower, J.C., Hand, D.J.: Biplots. Chapman & Hall, London (1996)
Gower, J.C., Dijksterhuis G.B.: Procrustes Problems. Oxford University Press (2004)
Hayashi, C.: Data Science. Asakura Shoten, Tokyo (2001). (in Japanese)
Iwatsubo, S.: Two classification techniques of 3-way data—quantification by means of correlation ratio and three-dimensional correlation coefficient. Jpn. J. Behav. **2**, 54–65 (1974). (in Japanese)
Iwatsubo, S.: An optimal scoring method for detecting clusters and interrelations from multi-way qualitative data. Behaviormetrika **5**, 1–22 (1978)
Iwatsubo, S.: Contribution of Behaviormetrics to human society—learning from researches by the late Professor Kinji Mizuno on disaster preparedness education for school children. Jpn. J. Behav. **29**, 55–60 (2002). (in Japanese)
Iwatsubo, S.: Kinji Mizuno and behaviormetrics: research collaborations with Chikio Hayashi. Jpn. J. Behav. **45**, 85–94 (2018). (in Japanese)
Lebart, L., Morineau, A., Warwick, K.M.: Multivariate Descriptive Statistical Analysis and Related Techniques for Large Matrices. Wiley, New York (1984)
Maruyama, K.: Chikio Hayashi—The Man Who Liked Data: The Life and Work of a Data Science Founder. SINFONICA (2021)
Nishimura, H., Iwatsubo, S.: An approach to computational semantics of natural languages. Inf. Process. Jpn. **10**, 127–134 (1970). (in Japanese)
Nishisato, S.: Applications of Psychological Scaling: Analysis of Qualitative Data and Interpretation. Seishin Shobo Press, Tokyo (1975). (in Japanese)
Nishisato, S.: Analysis of Categorical Data, Dual Scaling and Its Application. University of Toronto Press, Toronto (1980)
Ohsumi, N., Lebart, L., Morineau, A., Warwick, K.M., Baba, Y.: Multivariate Descriptive Statistical Analysis. JUSE Press Ltd., Tokyo (1994). (in Japanese)
Rao, C.R., Yanai, H.: General definition and decomposition of projectors and some applications to statistical problems. J. Stat. Plan. Inference **3**, 1–17 (1979)
Rao, C.R., Yanai, H.: Generalized inverses of partitioned matrices useful in statistical applications. Linear Algebra Appl. **70**, 105–113 (1985)
Research Committee for the Study of the Japanese National Character: A Study of the Japanese National Character—Part VI—Sixth Nationwide Survey, Idemitsu Shoten, Tokyo (1982) (in Japanese; outline in English)

Takahashi, M.: Open interview with Dr. Chikio Hayashi on collaborative research. Bull. Data Anal. Jpn. Classif. Soc. **10**, 1–28 (2021)

Takeuchi, K.: A test for multivariate normality. Behaviormetrika **1**, 59–64 (1974)

Takeuchi, K., Yanai, H., Mukherjee, B.N.: The Foundation of Multivariate Analysis. Wiley Eastern, New Delhi (1982)

Yoshizawa, T.: Singular value decomposition of multiarray data and its applications. In: Hayashi, C., Diday, E., Jambu, M., Ohsumi, N. (eds.) Recent Developments in Clustering and Data Analysis, pp. 241–257. Academic, London (1988)

On Association and Scaling Issues

A Straightforward Approach to Chi-Squared Analysis of Associations in Contingency Tables

Boris Mirkin

1 Introduction

A two-way contingency table, or cross-classification, is a type of data relating two sets of categories, usually being mutually exclusive values of two nominal or ordinal features. This data structure has attracted considerable attention from researchers for the analysis of interrelations between the features.

A number of loglinear models were proposed, as were low-rank approximations of the tables such as dual scaling, and common-sense considerations; for the latest descriptions see, for example, Agresti (2019), Bland (2020), Goodman (1991) and Nishisato (1994). Yet Pearson's chi-squared independence test remains the most popular approach to analysing contingency tables. There is an issue inherent to this approach, though: it gives a global assessment of whether the hypothesis of "global" independence between features should be rejected or not. Whenever the independence hypothesis is rejected, an open issue remains of investigation of those associations between categories that cause the rejection. Sharpe (2015) puts the issue as a blunt question:

> Chi-square test is statistically significant: Now what?

He proceeds to review the main approaches to the analysis of associations between individual categories. They all establish a fact of statistical dependence but fail to evaluate that quantitatively. The only exception is the so-called odds-ratio in 2×2 contingency tables. This is a case at which both features have only two categories each. Therefore, one may compare one category of a feature with respect

B. Mirkin (✉)
Department of Data Analysis and Artificial Intelligence, NRU HSE Moscow, Moscow, Russian Federation
e-mail: bmirkin@hse.ru; mirkin@dcs.bbk.ac.uk

School of Computing and Mathematical Sciences, Birkbeck University of London, London, UK

E. J. Beh et al. (eds.), *Analysis of Categorical Data from Historical Perspectives*, Behaviormetrics: Quantitative Approaches to Human Behavior 17,
https://doi.org/10.1007/978-981-99-5329-5_4

to the other feature by comparing respecting probabilities in the categories. This, however, is not directly applicable to larger contingency tables. Therefore, in larger contingency tables, the researcher can move on to a heuristic analysis of differences between observed probabilities and those corresponding to the independence case, standardised by the analogues of their standard deviations (standardised residuals).

This author argues that, in larger contingency tables, one should compare probabilities at one category not with those at another category but rather with the average probabilities at the entire dataset. There is nothing new in this proposal. In fact, that value—the change of the probability of a category when a category of the other feature becomes known—was proposed at the very dawn of the era of statistics research by its founding father, Adolphe Quetelet (1796–1874); see Quetelet (1832) and Mirkin (2001). Currently, there is not much interest in the Quetelet index as is among statistics researchers. For example, it is mentioned, in passing, by Goodman (1991, Eq. (2.2.3)) before moving on to more interesting subjects. Perhaps, the only exceptions from the rule are Greenacre (2009) and Beh and Lombardo (2014).

This author has discovered that there is an interest in the Quetelet index as is. It relates to the Pearson's chi-squared statistic. In fact, the values of the Quetelet index averaged over the bivariate probabilities total to the phi-squared, the Pearson's chi-squared related to the number of elements. This shows that the Pearson's chi-squared has an operational meaning. The index value is proportional to the average change of probability of a category of one feature when a category of the other feature becomes known. Moreover, the averaging formula represents a decomposition of the chi-squared statistic in contributions by individual pairs of feature categories. This allows for both capturing important contributions and assigning them operational meaning. This is a novel tool for the analysis of contingency tables.

The remainder of this paper is structured as follows. Section 2 describes the conventional concepts of Pearson's chi-squared statistic and standardised residuals in their relation to the former using an example from Sayassatov and Cho (2020). Section 3 introduces the concept of the Quetelet index and relates it to the Pearson's chi-squared statistic using the very same example to point out the strongest associations together with their quantitative values. Section 4 provides three more examples from the literature to illustrate the action of Quetelet indexes. Section 5 provides some final remarks.

2 Pearson Chi-Squared Index and Association Patterns

2.1 Statistical Independence and Pearson's Statistic

The contingency table is a conventional way of representing bivariate distributions. Given a set of objects I, and a set of categories over I indexed by symbols $k = 1, 2, \ldots, K$ and of categories over I indexed by $l = 1, 2, \ldots, L$, a contingency table $\mathbf{T}$ is defined as a $K \times L$ matrix, the (k, l)th entry of which is the number N_{kl} of objects from I falling in category k and category l simultaneously, that is, the frequency of (k, l) pair. Any reasonable analysis of contingency tables involves a nonoverlapping of the categories constraint: no object may fall in two categories $k1$, $k2$ such that $k1 \neq k2$, $k1, k2 = 1, 2, \ldots, K$, nor in two categories $l1$, $l2$ such that $l1 \neq l2$, $l1, l2 = 1, 2, \ldots, L$. This, basically, means that the categories $k = 1, 2, \ldots, K$ belong to one nominal feature over set I, and $l = 1, 2, \ldots, L$, to another. This is assumed further on in this text.

Then the category frequencies, frequently referred to as marginal frequencies are defined so that:

$$N_{k+} = \sum_{l=1}^{L} N_{kl}, \quad N_{+l} = \sum_{k=1}^{K} N_{kl}, \quad \sum_{k=1}^{K} N_{k+} = \sum_{l=1}^{L} N_{+l} = N,$$

where N_{k+} and N_{+l}, is the marginal frequency for row category k, and column category l, respectively, and N is the total number of objects in I. By dividing these by N, one arrives at similar equations for the relative frequencies (empirical probabilities):

$$p_{k+} = \sum_{l=1}^{L} p_{kl}, \quad p_{+l} = \sum_{k=1}^{K} p_{kl}, \quad \sum_{k=1}^{K} p_{k+} = \sum_{l=1}^{L} p_{+l} = 1. \tag{1}$$

Two features represented by the categories are referred to as statistically independent if the equations:

$$p_{kl} = p_{k+} p_{+l} \tag{2}$$

hold for all pairs (k, l).

The most popular tool for the analysis of associations via a contingency table is what is called Pearson's chi-squared statistic, frequently referred to as Pearson's index in the computer sciences. This index measures the summary deviation of the observed frequencies from the statistical independence.

Given a category pair (k, l), its deviation from the statistical independence is convenient to measure with what is referred to as the standardised residual:

$$s(k,l) = \frac{p_{kl} - p_{k+} p_{+l}}{\sqrt{p_{k+} p_{+l}}}. \tag{3}$$

This is the difference between the real and "ideal" frequencies moderated by their size in the denominator.

The Pearson's chi-squared index is defined as N times the sum-of-squares of these values:

$$X^2 = N \sum_{k=1}^{K} \sum_{l=1}^{L} s(k,l)^2 = N \sum_{k=1}^{K} \sum_{l=1}^{L} \frac{(p_{kl} - p_{k+}p_{+l})^2}{p_{k+}p_{+l}}. \tag{4}$$

The presence of factor N is justified by a theorem proven by Pearson (1904) (see also Pearson (1948)): If a contingency table is based on a sample of objects randomly and independently drawn from a population in which the statistical independence holds (so that all deviations are due to just randomness in the sampling), then the probabilistic distribution of X^2 converges, at N tending to infinity, to the chi-square distribution with $(K - 1)(L - 1)$ degrees of freedom. K. Pearson defined the probabilistic chi-square distribution (with p degrees of freedom) as a distribution of the sum-of-squares of p independent random variables, each distributed according to the standard Gaussian $N(0, 1)$ distribution. This leads to a simple universal criterion for testing the hypothesis of statistical independence between the features (see any statistics textbook or statistical distribution tables). Still, one should not overestimate the universality of this criterion: first, the sample size N should not be too small; second, no zero entries N_{kl} are permitted in the table when the marginal probabilities are not zero. As all concerned know getting around this latter commandment requires some fantasy and rigour in modelling more suitable candidates for the zeros; see, for example, Agresti (2019) and Devore (1995).

Pearson's theorem focuses on the testing of the statistical independence hypothesis and implies no instructions for the analysis of statistical 'dependence' or 'association' in a case at which a chi-squared value is greater than an accepted independence threshold. Therefore, in such situations, researchers use various heuristics for capturing associations behind the failure of the independence test. A most popular heuristic comes from observation of the standardised residuals defined by (3): positive associations correspond to positive values in (3), and negative associations, to negative values in (3). The larger the absolute value of a standardised residual, the greater the association; see Sharpe (2015) and Sayassatov and Cho (2020).

2.2 Example

To see, how this may work, let us use an example from Sayassatov and Cho (2020). This paper analyses the association between two features that the authors evaluate of their sample of 40 students. The authors accept a classification of learning styles from Mumford and Honey (1986). According to this view, there are four different approaches people take to learning new information. Their labels are listed below together with the main characteristics of them, in parentheses:

- Activists (Learn by doing and happy to jump),
- Reflectors (Learn through observation and reflecting on results),
- Theorists (Like to understand the theory behind action),
- Pragmatists (Need to be able to see how they apply their learning to the real world).

The second feature under study is the student's preferences among artificial "Internet of Things" devices (IoT). There were four artificial devices under consideration defined by Sayassatov and Cho (2020) and described as follows:

- D1: Smart Organised Backpack. This device has certain sensors to help students not to lose their college belongings.
- D2: Smart Voice Recorder for Group Discussions. This device helps students at group meetings or discussions.
- D3: Smart Headset for Concentration. This device helps students to be more concentrated at individual studies.
- D4: Smart Education Storage Ring. By wearing this device, students keep all their education related data in its memory card.

The authors cross-classified the students according to their learning style and preferred artificial IoT device; see Table 1. The value of X^2 for this table is 30.498, which leads to the rejection of the independence hypothesis with a confidence level greater than 99.9%.

Sayassatov and Cho (2020) then engage in an investigation of meaning behind this value. They first turn to a natural version of X^2, the so-called phi-squared which is X^2 without the factor N:

$$\varphi^2 = \sum_{k=1}^{K} \sum_{l=1}^{L} \frac{(p_{kl} - p_{k+}p_{+l})^2}{p_{k+}p_{+l}}. \tag{5}$$

The values that (5) can take lie within the interval between 0 (at statistical independence) and the minimum of $K - 1$ and $L - 1$. The latter value is reached for a contingency table in which every row k (at $K \geq L$) has just one non-zero entry, in a column $l(k)$. In this case, the pattern of association between the features can be expressed as a purely logical implication rule $k \Rightarrow l(k)$ $(k = 1, 2, \ldots, K)$ from k to $l(k)$. Unfortunately, for Table 1, neither value of φ^2, nor the value of its derivative, called Cramér's V, can provide any information on the association pattern between the learning style and an IoT model of preference (Sayassatov and Cho 2020).

Therefore, the authors turn to an analysis of the residuals (3) as presented on the left in Table 2.

Table 1 Cross-classification of learning styles and preferred artificial IoT devices from Sayassatov and Cho (2020)

Device preferred	Learning style				Total
	Activist	Reflector	Theorist	Pragmatist	
D1	8	2	1	1	12
D2	1	6	1	1	9
D3	1	1	5	2	9
D4	1	2	1	6	10
Total	11	11	8	10	40

Table 2 Standardised residuals for the data in Table 1 (left part) and the Quetelet index values (part on the right)

Device preferred	Learning style				Learning style			
	Activist	Reflector	Theorist	Pragmatist	Activist	Reflector	Theorist	Pragmatist
	Standardised residuals				*Quetelet index values*			
D1	**0.4091**	– 0.1132	– 0.1429	– 0.1826	1.4242	– 0.3939	– 0.5833	– 0.6667
D2	– 0.1482	**0.3543**	– 0.0943	– 0.1318	– 0.5960	1.4242	– 0.4444	– 0.5556
D3	– 0.1482	– 0.1482	**0.3771**	– 0.0264	– 0.5960	– 0.5960	1.7778	– 0.1111
D4	– 0.1669	– 0.0715	– 0.1118	**0.3500**	– 0.6364	– 0.2727	– 0.5000	1.4000

One can see that all the standardised residuals here are negative, except for those on the diagonal (highlighted in bold) which shows an exceptionally clear-cut pattern of associations. Each learning style one-to-one corresponds to a specific IoT device preferred: Activist to D1, Reflector to D2, Theorist to D3, and Pragmatist to D4. There is no quantitative evaluation of the degree of association, though (Sayassatov and Cho 2020).

3 Quetelet Indexes for a Comprehensive Analysis of Associations

In fact, there is a similar normalised difference expression, both related to the chi-squared statistic and having a very clear meaning. This is what we refer to as Quetelet index, due to Quetelet, the founding father of statistics; see Quetelet (1832) in our paper Mirkin (2001).

The Quetelet index is defined as the relative difference between (empirical) conditional probability $P(l/k) = p_{kl}/p_{k+}$ that category l occurs under condition k and the (empirical) probability that category l occurs at all, $P(l) = p_{+l} = N_{+l}/N$ (Mirkin 2001):

$$q(l/k) = \frac{P(l/k) - P(l)}{P(l)}. \tag{6}$$

That is, the Quetelet index expresses association between categories $k = 1, 2, \ldots, K$ and $l = 1, 2, \ldots, L$ as the relative change in the probability of l when k is taken into account; see also an earlier description in Lebart and Mirkin (1993).

With a little algebra, one can derive simpler expressions:

$$q(l/k) = \frac{p_{kl} - p_{k+}p_{+l}}{p_{k+}p_{+l}} = \frac{p_{kl}}{p_{k+}p_{+l}} - 1, \tag{6'}$$

which do not differ that much from the standardised residuals in (3). The difference in semantics, however, is huge: Quetelet indexes in (6) and (6′) have a clear-cut quantitative interpretation as the relative probability changes, whereas the standardised residuals have no operational meaning.

It may seem somewhat odd that $q(l/k) = q(k/l)$, the change in the probability is a same in both directions, l under condition k and k under condition l, as follows from the right part of (6′). This means one may use symmetric notation $q(k, l)$ for asymmetric $q(k/l)$ and $q(l/k)$. In this author's view, this symmetry can be considered as a mathematical expression of the idea that static data, by themselves, can give no information of casual dependencies—those must be derived from beyond the table.

Quetelet indexes for Table 1 are presented on the right of Table 2. Obviously, the patterns of plus/minus signs in the left and right parts of Table 2 coincide because the numerators in (3) and (6′) coincide. However, in contrast to the entries in the table of standardised residuals, these entries are meaningful. One can see that attending to 'Pragmatist' learning style increases the probability of choosing D4 IoT device by 140%, and attending to 'Theorist' learning style increases the probability of choosing D3 IoT device by 178%.

Now one can take an averaged Quetelet index:

$$Q = \sum_{k=1}^{K}\sum_{l=1}^{L} p_{kl}q(l, k) = \sum_{k=1}^{K}\sum_{l=1}^{L} p_{kl}\left(\frac{p_{kl}}{p_{k+}p_{+l}} - 1\right) = \sum_{k=1}^{K}\sum_{l=1}^{L} \frac{p_{kl}^2}{p_{k+}p_{+l}} - 1, \quad (7)$$

and interpret it as the average change in the probability of a random object to fall into k-category when its l-category becomes known.

It is well-known, though, that the expression in (7) on the right is equal to the phi-squared of (5), so that $Q = \varphi^2$. Therefore, the chi-squared statistic does have an operational meaning. Its structural part, $\varphi^2 = X^2/N$, is the average change in the probability of a random object to fall into k-category when its l-category becomes known. One should draw attention to this claim. It contradicts conventional claims that the chi-squared is but a statistical criterion for testing the statistical independence.

Moreover, the left-side of (7) gives a meaningful decomposition of Pearson's chi-squared statistic in the sum of relative Quetelet indexes, $p_{kl}q(k, l)$. The relative Quetelet index takes into account both the probability p_{kl} of (k, l) and its prognostic power $q(k, l)$. Although the pattern of $\pm$ signs does not change at the set of relative Quetelet indexes, as can be seen in the right part of Table 3, the values do change, so that the total, in this case $Q = \varphi^2 = 0.7625$, is always non-negative. This value shows that, for any category, information of a category of the other feature increases, on average, the category's probability by 76.25%. Moreover, all within-row and within-column sums are positive as well (proven in Mirkin 2001). The within-row sums in the column "Total" of Table 3 shows what part of the total relative probability change, 76.25%, comes from each row category. For example, 23.39% are provided by the preference for D1 device.

The decomposition (7) shows that the total probability change is the difference between the sum of its positive entries, 0.9307 in Table 3, and the sum of its negative

Table 3 Quetelet index values (part on the left) and the relative Quetelet index values (on the right)

Device preferred	Learning style				Learning style				Total
	Activist	Reflector	Theorist	Pragmatist	Activist	Reflector	Theorist	Pragmatist	
	Quetelet index values				*Relative Quetelet index values*				
D1	1.4242	– 0.3939	– 0.5833	– 0.6667	0.2848	– 0.0197	– 0.0146	– 0.0167	0.2339
D2	– 0.5960	1.4242	– 0.4444	– 0.5556	– 0.0149	0.2136	– 0.0111	– 0.0139	0.1737
D3	– 0.5960	– 0.5960	1.7778	– 0.1111	– 0.0149	– 0.0149	0.2222	– 0.0056	0.1869
D4	– 0.6364	– 0.2727	– 0.5000	1.4000	– 0.0159	– 0.0136	– 0.0125	0.2100	0.1680

entries, 0.1682 in Table 3. The former (0.9307) reflects the positive associations between categories and the latter (0.1682), the negative associations. Distinguishing between positive and negative category-to-category associations may be a subject in the area of data engineering, but this author has nothing to say about it as yet.

4 More Examples

4.1 Sleeping Pills Action

Consider an example from Nishisato (1989) that involves a contingency table—see Table 4—summarising the answers of 140 individuals to the following two questions:

- Q.1: "How do you feel about taking sleeping pills?" with a range of answers: strongly for, for, neutral, against, strongly against;
- Q.2: "Do you sleep well every night?" with answers of either (1) never, (2) rarely, (3) sometimes, (4) often, (5) always.

Nishisato (1989) analyses this dataset with respect to the dual scaling. That is, he assigns categories with quantitative values so that the correlation between these quantified features becomes maximum. We apply the Quetelet index approach to the table, so that Table 5 summarises the Quetelet pairwise indexes.

Table 5 shows that opinions on usage of sleeping pills strongly correlate with the level of sleep disorders. Those always sleeping well are 175.86% more likely than the average to be strongly against sleeping pills, whereas those never sleeping well are more likely, 177.78%, than the average to be strongly for them. Medium sleeping disorders bring forward more moderate probability changes such as, say, 45.83% increase for the pair Q1.sometimes/Q2.against.

Table 6 presents the relative Quetelet values for the partition of the phi-squared for this dataset. What is interesting about this is that the value $Q = \varphi^2$ here is just 0.5581, which is far smaller than the maximum possible value of 4 (for 5 × 5 contingency tables), less than 14% of the maximum value. A similar value for Table 1, 0.7625,

Table 4 Cross-classification of the answers to Questions 1 and 2 below by 140 respondents

Q2	Q1					Total
	1. Never	2. Rarely	3. Sometimes	4. Often	5. Always	
Strongly for	15	8	3	2	0	28
For	5	17	4	0	2	28
Neutral	6	13	4	3	2	28
Against	0	7	7	5	9	28
Strongly ag.	1	2	6	3	16	28
Total	27	47	24	13	29	140

Table 5 Quetelet index values for data in Table 4 (those greater than 0.35 are highlighted in bold)

Q2	Q1				
	1. Never	2. Rarely	3. Sometimes	4. Often	5. Always
Strongly for	**1.7778**	– 0.1489	– 0.3750	– 0.2308	– 1.0000
For	– 0.0741	**0.8085**	– 0.1667	– 1.0000	– 0.6552
Neutral	0.1111	**0.3830**	– 0.1667	0.1538	– 0.6552
Against	– 1.0000	– 0.2553	**0.4583**	**0.9231**	**0.5517**
Strongly ag.	– 0.8148	– 0.7872	0.2500	0.1538	**1.7586**

Table 6 Relative Quetelet index values for data in Table 4 (those highlighted are greater than 0.05)

Q2	Q1					Total	Total, %
	1. Never	2. Rarely	3. Some	4. Often	5. Always		
Strongly for	**0.1905**	– 0.0085	– 0.0080	– 0.0033	0	0.1706	30.6
For	– 0.0026	**0.0982**	– 0.0048	0	– 0.0094	0.0814	14.6
Neutral	0.0048	0.0356	– 0.0048	0.0033	– 0.0094	0.0295	5.3
Against	0	– 0.0128	0.0229	0.0330	0.0355	0.0786	14.1
Strongly ag.	– 0.0058	– 0.0112	0.0107	0.0033	**0.2010**	0.1979	35.5

is just about 25% of the maximum value. That means that the association between two features here is rather weak, according to the holistic estimate, whereas there is a clear-cut association between sleeping disorders and opinions on pills described above.

Another interesting feature of the decomposition of $Q = \varphi^2$ according to Table 6 is that zeros in it are just zeros, meaning no contribution to $Q = \varphi^2$—and that is all. This drastically contrasts the treatment of zeros according to formula (5) for φ^2. Indeed, (5) is to test the hypothesis that $p_{kl} = p_{k+}p_{+l}$. Since p_{k+} and p_{+l} are positive, the value of p_{kl} must be positive, too!. This implies that "continuity correction" of zero counts in contingency tables is needed in conventional research projects; see, for example, Devore (1995).

4.2 Voting Preferences

Table 7 presents voting preferences of USA citizens as related to their incomes according to a survey undertaken by the Pew Research Centre in 2014. They classified household income in 4 groups: (1) Less than \$30,000, (2) More than \$30,000 but less than \$50,000, (3) More than \$50,000 but less than \$100,000, and (4) \$100,000 or more. Voter party affiliation is defined as either R (Republican or leaning to

Table 7 Contingency table income-party voting (left part) and its Quetelet index values (part on the right)

Income	Party			Total	Party		
	R	U	D		R	U	D
	Respondent counts				*Quetelet index values*		
1	2388	2034	4423	8845	**– 0.3034**	**0.4629**	0.0985
2	2286	938	2696	5920	– 0.0037	0.0079	0.0004
3	3885	1126	3712	8723	**0.1491**	– 0.1788	– 0.0652
4	3258	695	3049	7002	**0.2005**	– 0.3686	– 0.0435

Table 8 Relative Quetelet index values for data in Table 7

Income	Party		
	R	U	D
1	– 0.0238	0.0309	0.0143
2	– 0.0003	0.0002	0
3	0.019	– 0.0066	– 0.0079
4	0.0214	– 0.0084	– 0.0043

Republican) or U (Undecided) or D (Democrat or leaning to Democrat). Although the Quetelet index values do not reach the highs seen in Table 5, one can easily see that the richer people tend to lean to that identify as being Republican, 14.91% and 20.05% in group (3) and group (4), respectively, over the proportions in the entire population. On the other hand, the group of poor (1) do not like Republicans at all (– 30.34% with respect to the proportion on the total sample), while remaining mostly Undecided (+ 46.29%).

The relative Quetelet index values for Table 7 are summarised in Table 8. The sum of positive entries in this table is 0.0859, and its negative entries sum to – 0.0513. The total is $Q = \varphi^2 = 0.0345$, which is just above 1% of its maximum value 3. However, this is quite enough to warrant rejection of the independence hypothesis with 99.9% confidence according to the Pearson's chi-squared independence test because of a large number of respondents.

4.3 Marital Status Versus Medical Treatment

Table 9 presents a contingency data from DeViva (2014) who compared the treatment status of military veterans and their marital status (Married or Not). The treatment categories are: (1) never seen for therapy, (2) seen but not completing therapy, and (3) completed therapy.

Table 9 Contingency table for treatment versus marital status (left part) and its Quetelet index values (on the right, those positive highlighted in bold)

Marital status	Treatment				Treatment		
	Never seen	Seen, didn't complete	Completed	Total	Never seen	Seen, didn't complete	Completed
	Counts				*Quetelet index values*		
Not married	57	53	11	121	**0.1650**	– 0.057	– 0.3068
Married	17	32	13	62	– 0.3219	**0.1112**	**0.5988**

Table 10 Relative Quetelet index values (on the left) and Quetelet values (on the right)

Marital status	Treatment			Total	Treatment		
	Never seen	Seen, didn't complete	Completed		Never seen	Seen, didn't complete	Completed
	Relative Quetelet index values				*Quetelet index values*		
Not married	0.0514	– 0.0165	– 0.0184	0.0164	0.1650	– 0.0570	– 0.3068
Married	– 0.0299	0.0194	0.0425	0.0321	– 0.3219	0.1112	0.5988
Total	0.0215	0.0029	0.0241	0.0485			

The Quetelet index values on the right of Table 9 show that being married highly increases the chance of getting their treatment completed—by 59.88%—whereas being not married decreases that chance by 30.68% and increases the chance of never seeing a doctor by 16.5%. This explanation gives a clear picture of the associations in the data, in contrast to the analyses given in Sharpe (2015) where only some general claims of statistical dependence are made.

The relative Quetelet index values are provided on the left of Table 10. The total increase of category probabilities by the positive Quetelet index values, 0.1134, is drastically reduced by the negative total, – 0.0648, leading to the summary $Q = \varphi^2$ value of 0.0485 which is so small, that the resulting $X^2 = N\varphi^2 = 8.8774$ it is not enough to reject the independence hypothesis at 99% confidence level (critical chi-squared value is 9.210 at 2 degrees of freedom), yet quite enough at the 95% confidence level (critical value 5.991). Once again we see that the global independence testing turns a blind eye to the local dependencies clearly visible when using the Quetelet index values.

4.4 *Aspirin and Heart Attacks*

Table 11 is a contingency table from Agresti (2019, p. 30), on the relation between the usage of aspirin and Myocardial infarction in a medical survey.

Table 11 Cross-classification of Aspirin/Placebo use and having or not Myocardial infarction (on the left) and its Quetelet and relative Quetelet values on the right

Aspirin use	Myocardial infarction						
	Yes	No	Total	Yes	No	Yes	No
	Counts			*Quetelet index*		*Relative Quetelet*	
Placebo	189	10,845	11,034	**0.2903**	– 0.0039	0.0025	– 0.0019
Aspirin	104	10,933	11,037	– 0.2902	0.0039	– 0.0014	0.0019
Total	293	21,778	22,071			0.0011	0.0000

In spite of a rather small value of $\varphi^2 = 0.0011$ here, the chi-squared is $X^2 = 25.0139$ which, at 1 degree of freedom, leads to the rejection of the independence hypothesis at more than 99.9% confidence level (the critical value is 10.828). To follow a conventional advice, one should take a look at the odds-ratio here; see Bland (2020). The odds-ratio is 1.83 meaning that the estimated odds of Myocardial infarction are 83% higher for the Placebo group than for the aspirin group (Agresti 2019). The odds-ratio counterposes the two groups, whereas the Quetelet index compares any group rates with the grand mean. One can see that use of the Placebo increases the risk of the illness by 29.03% in comparison to the average risk.

5 Conclusion

In contrast to conventional wisdom that Pearson's chi-squared statistic is a criterion of statistical independence, rather than a measure of association, this paper demonstrates that the Pearson's chi-squared indeed is a measure of association between nominal or ordinal features if the scaling N factor is removed. Its normalised version, the phi-squared, is the average change of the probability of a category of a feature when a category of the other feature becomes known. Associations between individual categories are captured with indexes introduced by the celebrated Belgian statistician Adolphe Quetelet quite early, not later than 1832. Even at smaller values of the phi-squared for the total association, contributions of individual category pairs can be significant.

References

Agresti, A.: An Introduction to Categorical Data Analysis, 3rd edn. Wiley, Hoboken, NJ (2019)

Beh, E.J., Lombardo, R.: Correspondence Analysis: Theory, Practice and New Strategies. Wiley, Chichester (2014)

Bland, M.: Introduction to Statistics for Research: Proportions, Chi-squared Tests and Odds Ratios (2020). Available at: https://www-users.york.ac.uk/~mb55/yh_stats/chiodds.htm#yates. Accessed 14 April 2023

DeViva, J.C.: Treatment utilization among OEF/OIF veterans referred for psychotherapy for PTSD. Psychol. Serv. **11**, 179–184 (2014)

Devore, J.L.: Probability and Statistics for Engineering and the Sciences, 4th edn. Duxbury Press (1995)

Goodman, L.A.: Measures, models, and graphical displays in the analysis of cross-classified data. J. Am. Stat. Assoc. **86**, 1085–1111 (1991)

Greenacre, M.: Power transformations in correspondence analysis. Comput. Stat. Data Anal. **53**, 3107–3116 (2009)

Lebart, L., Mirkin, B.G.: Correspondence analysis and classification. In: Cuadras, C.M., Rao, C.R. (eds.) Multivariate Analysis: Future Directions 2, pp. 341–357. North-Holland, Amsterdam, The Netherlands (1993)

Mirkin, B.: Eleven ways to look at the chi-squared coefficient for contingency tables. Am. Stat. **55**, 111–120 (2001)

Mumford, A., Honey, P.: Developing skills for matrix management. Ind. Commer. Train. **18**(5), 2–7 (1986)

Nishisato, S.: Dual scaling: its development and comparisons with other quantification methods. In: Pressmar, D., Jager, K.E., Krallman, H., Shellhass, H., Steitferdt, L. (eds.) Operations Research Proceedings 1988, pp. 376–389. Springer, Berlin (1989)

Nishisato, S.: Elements of Dual Scaling: An Introduction to Practical Data Analysis. Lawrence Elbaum Associates, Hillsdale, NJ (1994)

Pearson, K.: On the theory of contingency and its relation to association and normal correlation. Draper's Memoirs, Biometric Series **1**, 46 pp. (1904)

Pearson, K.: Early Statistical Papers. University Press (1948)

Pew Research Center (PRC): Party Affiliation by Household Income, Available at: https://www.pewresearch.org/religion/religious-landscape-study/compare/party-affiliation/by/income-distribution. Accessed 14 Apr 2023

Quetelet, A.: Sur la Possibilité de Mesurer l'Influence des Causes qui Modifient les élémens Sociaux: Lettre à m. Villermé de l'Institut de France (No. 27294). M. Hayez (1832)

Sayassatov, D., Cho, N.: The analysis of association between learning styles and a model of IoT-based education: chi-square test for association. J. Inf. Technol. Appl. Manag. **27**(3), 19–36 (2020)

Sharpe, D.: Chi-square test is statistically significant: now what? Pract. Assess. Res. Eval. **20**, Article 8, 10 pp. (2015)

Contrasts for Neyman's Modified Chi-Square Statistic in One-Way Contingency Tables

Yoshio Takane and Sébastien Loisel

1 Introduction

This and its companion paper (Loisel and Takane 2022) were initially conceived as a paper dealing with "A theory of contrasts for Pearson's chi-square statistic" for multiple comparisons in the analyses of contingency tables (Lancaster 1949; Loisel and Takane 2016; Lombardo et al. 2020; Takane and Jung 2009). While we were working on this topic, we gradually came to realise that Pearson's statistic was not ideal for use in multiple comparisons, because in this statistic, mean and variance-covariance structures assumed on observed frequencies (proportions) are closely connected to each other. This means that if parts of mean structure are rejected, the corresponding parts of variance-covariance structure are also rejected. To illustrate, let there be C response categories, and let $\mathbf{p}_C$ denote the C-component vector of their true probabilities. Let $\hat{\mathbf{p}}_C$ denote the observed counterpart of $\mathbf{p}_C$. Define $\mathbf{D}_C = \text{diag}(\mathbf{p}_C)$, where the diag operator turns a vector into a diagonal matrix. Then, Pearson's statistic can be stated as:

$$X^2_{\text{Total}} = n(\hat{\mathbf{p}}_C - \mathbf{p}_C)'\mathbf{D}_C^{-1}(\hat{\mathbf{p}}_C - \mathbf{p}_C), \tag{1}$$

where n indicates the total sample size (the number of independently replicated observations) to calculate $\hat{\mathbf{p}}_C$. This statistic is known to follow an asymptotic chi-square distribution with $C - 1$ df (degrees of freedom) when the prescribed $\mathbf{p}_C$ is correct. Note that $n\mathbf{D}_C^{-1}$ is a g-inverse (generalised inverse) of the variance-covariance matrix of $\hat{\mathbf{p}}_C$, denoted by $\boldsymbol{\Sigma}_C/n$, where:

$$\boldsymbol{\Sigma}_C = \mathbf{D}_C - \mathbf{p}_C\mathbf{p}_C'. \tag{2}$$

Y. Takane (✉)
Department of Psychology, University of Victoria, Victoria, BC, Canada
e-mail: yoshio.takane@mcgill.ca

S. Loisel
Department of Mathematics, Heriot-Watt University, Edinburgh, Scotland
e-mail: sloisel@gmail.com

E. J. Beh et al. (eds.), *Analysis of Categorical Data from Historical Perspectives*, Behaviormetrics: Quantitative Approaches to Human Behavior 17,
https://doi.org/10.1007/978-981-99-5329-5_5

Notice that $\mathbf{\Sigma}_C$ is completely determined by $\mathbf{p}_C$. This means that if hypotheses about $\mathbf{p}_C$ are rejected, the corresponding parts of $\mathbf{\Sigma}_C$ are also rejected. As a consequence, any subsequent tests that assume $\mathbf{\Sigma}_C$ is correct become invalidated. We need a statistic in which mean structure and variance-covariance structure can be specified independently. A statistic that satisfies this requirement and that immediately comes to our mind is Neyman's modified chi-square statistic (Neyman 1949). In this paper, we develop a theory of contrasts for Neyman's statistic to be used for multiple comparisons in one-way tables. We refer to Loisel and Takane (2022) for a similar theory in two-way and higher-order contingency tables.

The plan of this paper is as follows. In the following section (Sect. 2), we introduce Neyman's modified chi-square statistic, and show how the variance-covariance structure assumed of $\hat{\mathbf{p}}_C$ is free from its mean structure $\mathbf{p}_C$. Like Pearson's statistic, Neyman's statistic (Y^2_{Total}) is a global index of overall discrepancies between observed and prescribed mean vectors. In Sect. 3, we introduce the notion of contrasts useful in multiple comparisons. Contrasts capture specific aspects of the overall discrepancies. We also introduce orthogonal contrasts, and how to generate them when a set of non-orthogonal contrast vectors are given. Orthogonal contrasts partition the overall discrepancies into non-overlapping components, each of which represents a unique aspect of the overall discrepancies. In Sect. 4, we show the existence of a contrast, denoted by $\mathbf{v}_{\max}$, which captures the entire variation in Y^2_{Total}. The existence of such a contrast justifies Schéffe's type of post-hoc tests by Goodman (1964). Section 5 deals with a special case in which homogeneous cell probabilities are postulated for $\mathbf{p}_C$. This special case is important because it is deemed to cover a majority of applications in the analysis of one-way tables. In Sect. 6, we briefly touch on the subject of statistical issues by presenting results from small Monte-Carlo studies examining statistical properties of Neyman's statistic. Section 7 concludes the main topic of the paper. Throughout this paper, a numerical example is provided to illustrate the computations involved. An additional section, Sect. 8, briefly discusses Nishisato's influences on our work in the past.

2 Neyman's Modified Chi-Square Statistic

Consider a one-way table of observed cell probabilities. It could also be a one-way marginal table derived from a higher-order contingency table or a slice of a conditional probability table at a particular level of a conditioning variable. Neyman's modified chi-square statistic is stated as:

$$Y^2_{\text{Total}} = n(\hat{\mathbf{p}}_C - \mathbf{p}_C)'\hat{\mathbf{D}}_C^{-1}(\hat{\mathbf{p}}_C - \mathbf{p}_C), \tag{3}$$

where $\hat{\mathbf{D}}_C = \text{diag}(\hat{\mathbf{p}}_C)$ is the observed counterpart to $\mathbf{D}_C$. Here it is tacitly assumed that $\hat{\mathbf{D}}_C$ is nonsingular. That is, there are no empty cells in the table. This statistic is known to follow asymptotically the same distribution as Pearson's statistic. The difference is that in Pearson's statistic, $\mathbf{D}_C^{-1}$ is used as the weight matrix, while in Neyman's statistic, its sample estimate, $\hat{\mathbf{D}}_C^{-1}$ is used. An important point is that $\mathbf{D}_C^{-1}$ is completely determined by $\mathbf{p}_C$, while $\hat{\mathbf{D}}_C^{-1}$ is not, although the latter is expected to approach the former as the sample increases indefinitely ($\hat{\mathbf{D}}_C^{-1}$ is a consistent estimate of its population counterpart, $\mathbf{D}_C^{-1}$). Note that $n\hat{\mathbf{D}}_C^{-1}$ is a g-inverse of the variance-covariance matrix of $\mathbf{p}_C$, namely $\hat{\boldsymbol{\Sigma}}_C/n$, where:

$$\hat{\boldsymbol{\Sigma}}_C = \hat{\mathbf{D}}_C - \hat{\mathbf{p}}_C\hat{\mathbf{p}}_C', \tag{4}$$

(Compare (4) with (2)). Provided that $\hat{\mathbf{D}}_C^{-1}$ exists, it can be easily verified that (3) is invariant over the choice of a g-inverse of $\hat{\boldsymbol{\Sigma}}_C/n$ with $n\hat{\mathbf{D}}_C^{-1}$ being just a special case.

Note 1. While we have never seen a formal proof of this invariance in Neyman's statistic, Puntanen et al. (2011, p. 120) shows a similar invariance in Pearson's statistic. The proof for Neyman's statistic should not be difficult, following a similar line of proof for Pearson's statistic by Puntanen et al. (2011).

For later use, it is convenient to rewrite (3) as:

$$Y^2_{\text{Total}} = n(a-1), \tag{5}$$

where

$$a = \mathbf{p}_C'\hat{\mathbf{D}}_C^{-1}\mathbf{p}_C. \tag{6}$$

This follows trivially from (3) by simply expanding its terms.

An illustrative example: Assume that the following frequency table is observed: $\mathbf{F} = \begin{pmatrix}40 & 50 & 10\end{pmatrix}'$ with $C = 3$ and $n = 100$. The corresponding table of observed proportions is given by $\hat{\mathbf{p}}_C = \begin{pmatrix}0.4 & 0.5 & 0.1\end{pmatrix}'$. It is postulated that the true probabilities of the three cells in the table are $\mathbf{p}_C = \begin{pmatrix}0.5 & 0.2 & 0.3\end{pmatrix}'$. The value of Y^2_{Total} is found to be 60.5000, and $a = 1.6050$. This data set will be used in the following discussion to exemplify various aspects of multiple comparisons in one-way tables.

3 Contrasts

Let $\mathbf{v}$ be a C-component nonzero vector such that:

$$\mathbf{v} \in \text{Ker}(\mathbf{p}_C') \tag{7}$$

where $\text{Ker}(\mathbf{p}_C')$ indicates the null space of $\mathbf{p}_C'$. That is, the space spanned by vectors $\mathbf{x}$ such that $\mathbf{p}_C'\mathbf{x} = 0$. A linear function of $\mathbf{p}_C$ of the form:

$$\phi(\mathbf{v}) = \mathbf{v}'\mathbf{p}_C \tag{8}$$

is called a population contrast associated with the contrast (weight) vector $\mathbf{v}$. An analogous function:

$$\hat{\phi}(\mathbf{v}) = \mathbf{v}'\hat{\mathbf{p}}_C \tag{9}$$

with $\mathbf{p}_C$ in (8) replaced by its observed counterpart $\hat{\mathbf{p}}_C$ is called a sample contrast. The size of the effect due to a contrast is measured by:

$$Y^2_{\phi(\mathbf{v})} \equiv n(\mathbf{v}'\hat{\mathbf{p}}_C)^2/\mathbf{v}'\hat{\mathbf{\Sigma}}_C\mathbf{v}. \tag{10}$$

Obviously, the size of the effect of a contrast is invariant over the transformation of $\mathbf{v}$ of the form $d\mathbf{v}$ for any nonzero scalar d.

Note 2: We typically require (7) on $\mathbf{v}$. This, however, is not an absolute necessity. For example, we may wish to test $p_1 = 0.5$. For that, we define $\mathbf{w} = \begin{pmatrix} 1 & 0 & 0 \end{pmatrix}'$ and test the hypothesis that $\mathbf{w}'\mathbf{p}_C = 0.5$. Obviously, this $\mathbf{w}$ does not satisfy (7), as indicated by the fact that $\mathbf{w}'\mathbf{p}_C \neq 0$. This $\mathbf{w}$ is still admissible if its effect size is measured by:

$$Y^2_{\phi(\mathbf{w})} = n(\mathbf{w}'(\hat{\mathbf{p}}_C - \mathbf{p}_C))^2/\mathbf{w}'\hat{\mathbf{\Sigma}}_C\mathbf{w}, \tag{11}$$

which generalises (10). Is it then just a matter of convenience to require (7)? No, because the word "contrast" implies comparing two quantities. How do we compare? By taking a difference between the two and checking if the difference is significantly different from zero. The difference of zero is typically postulated as the null hypothesis to be tested. Also, note that the above $\mathbf{w}$ can always be turned into an equivalent $\mathbf{v}$ that satisfies (7) by the following transformation:

$$\mathbf{v} = (\mathbf{I}_C - \mathbf{1}_C\mathbf{p}'_C)\mathbf{w} = (\mathbf{I}_C - \mathbf{1}_C(\mathbf{1}'_C\mathbf{D}_C\mathbf{1}_C)^{-1}\mathbf{1}'_C\mathbf{D}_C)\mathbf{w}. \tag{12}$$

If we apply this transformation to the above $\mathbf{w}$, we obtain $\mathbf{v} \propto \begin{pmatrix} 1 & -1 & -1 \end{pmatrix}'$, and $Y^2_{\phi(\mathbf{w})} = Y^2_{\phi(\mathbf{v})}$. The hypothesis to be tested is also turned into $\mathbf{v}'\mathbf{p}_C = 0$, which is equivalent to (7). It may sound a bit surprising at a first glance to find that $p_1 = 0.5$ and $p_1 - p_2 - p_3 = 0$ represent the same hypotheses, but this makes perfect sense because if $p_1 = 0.5$, $p_2 + p_3 = 0.5$, so that $p_1 - p_2 - p_3 = 0$. That $Y^2_{\phi(\mathbf{w})} = Y^2_{\phi(\mathbf{v})}$ can be easily verified.

A pair of contrasts are said to be $\hat{\mathbf{\Sigma}}_C$-orthogonal (or simply orthogonal) if and only if $\mathbf{v}'_i\hat{\mathbf{\Sigma}}_C\mathbf{v}_j = 0$, where $\mathbf{v}_i$ and $\mathbf{v}_j$ $(i \neq j)$ are two contrast vectors. The effect sizes of the two orthogonal contrasts evaluated separately by (10) add up to the size of the joint effects of the two contrasts obtained by:

$$Y^2_{\phi(\mathbf{V})} = n\hat{\mathbf{p}}'_C\mathbf{V}(\mathbf{V}'\hat{\mathbf{\Sigma}}_C\mathbf{V})^{-1}\mathbf{V}'\hat{\mathbf{p}}_C, \tag{13}$$

where $\mathbf{V} = [\mathbf{v}_i, \mathbf{v}_j]$.

A set of K contrasts are said to be orthogonal if every pair of contrasts in the set are mutually orthogonal. When the contrasts are orthogonal, the size of their joint effects can be obtained by adding the effect sizes of the contrasts calculated separately. The size of the joint effects of more than one contrasts can generally (whether they are orthogonal or not) be calculated by (13), where $\mathbf{V}$ is redefined as $\mathbf{V} = [\mathbf{v}_1, \ldots, \mathbf{v}_K]$. If $K = C - 1$, $Y^2_{\phi(\mathbf{V})} = Y^2_{\text{Total}}$.

Let $\mathbf{U} = [\mathbf{u}_1, \ldots, \mathbf{u}_K]$ $(K \leq C - 1)$ be a matrix of non-orthogonal but linearly independent vectors. These vectors can be successively orthogonalised by the following procedure. We assume that $\mathbf{U}$ already satisfies the condition (7). If not, we simply apply (12) to $\mathbf{U}$ to satisfy the condition.

Step 1. Set $\mathbf{v}_1 = \mathbf{u}_1$ and set $\mathbf{V} = \mathbf{v}_1$.
Step 2. For $k = 2, \ldots, K$, set $\mathbf{v}_k = (\mathbf{I}_C - \mathbf{V}(\mathbf{V}'\hat{\mathbf{\Sigma}}_C\mathbf{V})^{-1}\mathbf{V}'\hat{\mathbf{\Sigma}}_C)\mathbf{u}_k$, and append $\mathbf{V}$ by $\mathbf{v}_k$.

In the end $\mathbf{V}$ contains a set of orthogonalised contrast vectors. Obviously, the above sequential process will produce different results if the columns of $\mathbf{U}$ are arranged differently. In general, $\mathbf{v}_k$ indicates the effect of $\mathbf{u}_k$ eliminating all previous effects, $\mathbf{u}_1$ through $\mathbf{u}_{k-1}$, and ignoring all subsequent effects, $\mathbf{u}_{k+1}$ through $\mathbf{u}_K$.

An example continued: Consider two contrasts defined by $\mathbf{v}_1 = \begin{pmatrix} 1 & -1 & -1 \end{pmatrix}'$ and $\mathbf{v}_2 = \begin{pmatrix} 0 & 3 & -2 \end{pmatrix}'$. The first one is the same as the one discussed in Note 2 above. The second one concerns the ratio of p_2 to p_3 is 2/3, i.e. $p_2/p_3 = 2/3$, which turns into $3p_2 = 2p_3$ or $3p_2 - 2p_3 = 0$. We find that $Y^2_{\phi(\mathbf{v}_1)} = 4.1667$. We also find $Y^2_{\phi(\mathbf{v}_2)} = 52.6480$. These two Y^2 values do not add up to $Y^2_{\text{Total}} = 60.5000$ because the two contrasts are not orthogonal. If $\mathbf{v}_2$ is orthogonalised with respect to $\mathbf{v}_1$ using the procedure given above, $\mathbf{v}_2$ eliminating $\mathbf{v}_1$, is given by $\mathbf{v}_2|\mathbf{v}_1 \propto \begin{pmatrix} 0.4333 & 0.7667 & -1.2333 \end{pmatrix}'$, and its Y^2 value by $Y^2_{\phi(\mathbf{v}_2|\mathbf{v}_1)} = 56.3333$, so that $Y^2_{\phi(\mathbf{v}_1)} + Y^2_{\phi(\mathbf{v}_2|\mathbf{v}_1)} = 60.5000 = Y^2_{\text{Total}}$, as expected. If, on the other hand, $\mathbf{v}_1$ is orthogonalised with respect to $\mathbf{v}_2$, $\mathbf{v}_1$ eliminating $\mathbf{v}_2$ is given by $\mathbf{v}_1|\mathbf{v}_2 \propto \begin{pmatrix} 4.0000 & -0.1121 & 6.5919 \end{pmatrix}'$ with the Y^2 value of $Y^2_{\phi(\mathbf{v}_1|\mathbf{v}_2)} = 7.8521$, so that $Y^2_{\phi(\mathbf{v}_2)} + Y^2_{\phi(\mathbf{v}_1|\mathbf{v}_2)} = 60.5000 = Y^2_{\text{Total}}$, as expected. So here we have two sets of two mutually orthogonal contrast vectors, $[\mathbf{v}_1, \mathbf{v}_2|\mathbf{v}_1]$ and $[\mathbf{v}_1|\mathbf{v}_2, \mathbf{v}_2]$.

4 The Contrast That Captures the Whole Variations in Y^2_{Total}

In Loisel and Takane (2022), it was shown that there exists a contrast that captures the entire interaction effects in two-way contingency tables. An analogous contrast that captures the entire between-cell effects in one-way tables also exists, and can be defined in a similar manner, namely:

$$\mathbf{v}_{\max} = \mathbf{S}(\mathbf{S}'\hat{\mathbf{\Sigma}}_C\mathbf{S})^{-1}\mathbf{S}'\hat{\mathbf{p}}_C, \tag{14}$$

where $\mathbf{S}$ is a matrix of linearly independent bases vectors spanning $\text{Ker}(\mathbf{p}'_C)$, and $\hat{\mathbf{\Sigma}}_C$ is as defined in (4). We want to show that:

$$Y^2_{\phi(\mathbf{v}_{\max})} \equiv n\mathbf{v}'_{\max}\hat{\mathbf{p}}_C = Y^2_{\text{Total}}. \tag{15}$$

The proof of (15) is much more difficult than the analogous proof in Loisel and Takane (2022), since:

$$\text{Sp}(\hat{\boldsymbol{\Sigma}}_C) \supset \text{Sp}(\mathbf{S}), \tag{16}$$

(where $\supset$ indicates the space on the lefthand side of $\supset$ includes the space on the righthand side) does not necessarily hold (unless $\mathbf{p}_C \in \text{Sp}(\mathbf{1}_C)$), and consequently Khatri's (1966) extended theorem:

$$\mathbf{S}(\mathbf{S}'\hat{\boldsymbol{\Sigma}}_C\mathbf{S})^{-1}\mathbf{S}' = \hat{\boldsymbol{\Sigma}}_C^+ - \hat{\boldsymbol{\Sigma}}_C^+\mathbf{p}_C(\mathbf{p}'_C\hat{\boldsymbol{\Sigma}}_C^+\mathbf{p}_C)^{-1}\mathbf{p}'_C\hat{\boldsymbol{\Sigma}}_C^+, \tag{17}$$

where $\hat{\boldsymbol{\Sigma}}_C^+$ is the Moore-Penrose (MP) inverse of $\hat{\boldsymbol{\Sigma}}_C$, does not necessarily hold. As a result, we need to take a somewhat different route to prove (15).

Using (4), we can rewrite $n\mathbf{v}'_{\max}\hat{\mathbf{p}}_C$ as:

$$\begin{aligned} n\mathbf{v}'_{\max}\hat{\mathbf{p}}_C &= n\hat{\mathbf{p}}'_C\mathbf{S}(\mathbf{S}'\hat{\mathbf{D}}_C\mathbf{S} - \mathbf{S}'\hat{\mathbf{p}}_C\hat{\mathbf{p}}'_C\mathbf{S})^{-1}\mathbf{S}'\hat{\mathbf{p}}_C \\ &= n\hat{\mathbf{p}}'_C\mathbf{S}[(\mathbf{S}'\hat{\mathbf{D}}_C\mathbf{S})^{-1} - (\mathbf{S}'\hat{\mathbf{D}}_C\mathbf{S})^{-1}\mathbf{S}'\hat{\mathbf{p}}'_C \\ &\qquad \times (1 - \hat{\mathbf{p}}'_C\mathbf{S}(\mathbf{S}'\hat{\mathbf{D}}_C\mathbf{S})^{-1}\mathbf{S}'\hat{\mathbf{p}}_C)^{-1}\hat{\mathbf{p}}'_C\mathbf{S}(\mathbf{S}'\hat{\mathbf{D}}_C\mathbf{S})^{-1}]\mathbf{S}'\hat{\mathbf{p}}_C \\ &= n(b + b^2/(1-b)) = nb/(1-b), \end{aligned} \tag{18}$$

where

$$b = \hat{\mathbf{p}}'_C\mathbf{S}(\mathbf{S}'\hat{\mathbf{D}}_C\mathbf{S})^{-1}\mathbf{S}'\hat{\mathbf{p}}_C \equiv \mathbf{v}^{*'}\hat{\mathbf{p}}_C, \tag{19}$$

and $\mathbf{v}^* = \mathbf{S}(\mathbf{S}'\hat{\mathbf{D}}_C\mathbf{S})^{-1}\mathbf{S})^{-1}\mathbf{S}'\hat{\mathbf{p}}_C$. The second equality in (18) holds; see, for example, Rao (1973, p. 33, Complements and Problems 2.8). Using Helmert-Khatri's original lemma (Khatri 1966; Takane 2016), we can rewrite (19) as:

$$b = \hat{\mathbf{p}}'_C(\hat{\mathbf{D}}_C^{-1} - \hat{\mathbf{D}}_C^{-1}\mathbf{p}_C(\mathbf{p}'_C\hat{\mathbf{D}}_C^{-1}\mathbf{p}_C)^{-1}\mathbf{p}'_C\hat{\mathbf{D}}_C^{-1})\hat{\mathbf{p}}_C = 1 - 1/a, \tag{20}$$

where a is as given in (6). Note that whereas (17) does not necessarily hold, a more restricted version (which we call Helmert-Khatri's original lemma):

$$\mathbf{S}(\mathbf{S}'\hat{\mathbf{D}}_C\mathbf{S})^{-1}\mathbf{S}' = \hat{\mathbf{D}}_C^{-1} - \hat{\mathbf{D}}_C^{-1}\mathbf{p}_C(\mathbf{p}'_C\hat{\mathbf{D}}_C^{-1}\mathbf{p}_C)^{-1}\mathbf{p}'_C\hat{\mathbf{D}}_C^{-1} \tag{21}$$

holds. Compare (21) and (17). While $\hat{\mathbf{D}}_C$ is assumed nonsingular, $\hat{\boldsymbol{\Sigma}}_C$ is bound to be singular. Putting the expression of b given in (20) into (18), we obtain:

$$Y^2_{\phi(\mathbf{v}_{\max})} = n\mathbf{v}'_{\max}\hat{\mathbf{p}}_C = n\frac{(a-1)/a}{1-(a-1)/a} = n(a-1) = Y^2_{\text{Total}}, \tag{22}$$

as anticipated. This implies that the asymptotic distribution of $Y^2_{\phi(\mathbf{v}_{\max})}$ is the same as that of Y^2_{Total} if the hypothesised mean structure $\mathbf{p}_C$ is correct. It in turn implies that no matter how many comparisons are made, the joint α level, the probability of making a Type 1 error in at least one of the tests performed, never exceeds a prescribed α level if each test is performed with the same critical value as in the test of Y^2_{Total}, i.e. the critical value of chi-square with $C-1$ df and a prescribed α level. This is because if $n\mathbf{v}'_{\max}\hat{\mathbf{p}}_C$ is smaller than this critical value, no other contrasts can exceed the critical value (Maxwell et al. 2018). It may be noted in passing that $\mathbf{v}_{\max} = a\mathbf{v}^*$, so that $\mathbf{Y}^2_{\phi(\mathbf{v}_{\max})} = \mathbf{Y}^2_{\phi(\mathbf{v}^*)}$, where the latter is calculated by $n(\mathbf{v}^{*'}\hat{\mathbf{p}}_C)^2/\mathbf{v}^{*'}\hat{\mathbf{\Sigma}}_C\mathbf{v}^*$.

An example continued: For the example data set we are using, $\mathbf{v}_{\max}$ is found to be $\mathbf{v}_{\max} = \begin{pmatrix} 0.3550 & 1.2050 & -1.3950 \end{pmatrix}'$, and $Y^2_{\phi(\mathbf{v}_{\max})} = 60.5000 = Y^2_{\text{Total}}$, as expected. For every hypothesised pairwise ratio, $\mathbf{v}_3 = \begin{pmatrix} 2 & -5 & 0 \end{pmatrix}'$, $\mathbf{v}_4 = \begin{pmatrix} 3 & 0 & -5 \end{pmatrix}'$, and $\mathbf{v}_5 = \begin{pmatrix} 0 & 3 & -2 \end{pmatrix}'$ (this is the same as the $\mathbf{v}_2$ in the previous numerical illustration), we have $Y^2_{\phi(\mathbf{v}_3)} = 25.7806$, $Y^2_{\phi(\mathbf{v}_4)} = 8.7344$, and $Y^2_{\phi(\mathbf{v}_5)} = 52.6480$. These values are to be compared with the critical value of chi-square with 2 df (the same df as in the test of Y^2_{Total}) and a prescribed α level. With $\alpha = 0.05$ the critical value is found to be 5.9915, and with $\alpha = 0.01$ it is 9.2103. Compare these critical values with 3.8415 and 6.6349, respectively, for planned comparisons. The differences are not so large in this example, because $C = 3$ is rather small. (Note that Y^2's due to pairwise ratios do not add up to Y^2_{Total} because the corresponding contrast vectors cannot be orthogonal to each other).

5 The Special Case in Which $\mathbf{p}_C = \mathbf{1}_C/C$

We now hypothesise equal cell probabilities for $\mathbf{p}_C$, i.e.:

$$\mathbf{p}_C = \mathbf{1}_C/C. \tag{23}$$

This is a special case of the more general case treated above, but it may be more predominant in practical applications. One prominent difference this assumption makes is that now(16) holds, so that (17) holds, and so:

$$\mathbf{S}(\mathbf{S}'\hat{\mathbf{\Sigma}}_C\mathbf{S})^{-1}\mathbf{S}' = \hat{\mathbf{\Sigma}}_C^+ - \hat{\mathbf{\Sigma}}_C^+\mathbf{1}_C(\mathbf{1}'_C\hat{\mathbf{\Sigma}}_C^+\mathbf{1}_C)^{-1}\mathbf{1}'_C\hat{\mathbf{\Sigma}}_C^+ \tag{24}$$

(Khatri 1966); this is just (17) with $\mathbf{p}_C$ replaced by $\mathbf{1}_C/C$. Since $\hat{\mathbf{\Sigma}}_C^+\mathbf{1}_C = \mathbf{0}$, it follows that:

$$Y^2_{\phi(\mathbf{v}_{\max})} = n\hat{\mathbf{p}}'_C\mathbf{S}(\mathbf{S}'\hat{\mathbf{\Sigma}}_C\mathbf{S})^{-1}\mathbf{S}'\hat{\mathbf{p}}_C = n\hat{\mathbf{p}}'_C\hat{\mathbf{\Sigma}}_C^+\hat{\mathbf{p}}_C. \tag{25}$$

Since $\hat{\mathbf{\Sigma}}_C^+$ can be expressed as:

$$\hat{\mathbf{\Sigma}}_C^+ = \mathbf{Q}_C\hat{\mathbf{D}}_C^{-1}\mathbf{Q}_C, \tag{26}$$

(Tanabe and Sagae, 1992), where $\mathbf{Q}_C = \mathbf{I}_C - \mathbf{1}_C\mathbf{1}_C'/C$, we obtain:

$$n\mathbf{p}_C'\hat{\boldsymbol{\Sigma}}_C^{+}\mathbf{p}_C = n(\mathbf{1}_C'\hat{\mathbf{D}}_C^{-1}\mathbf{1}_C/C^2 - 1) = n(a^* - 1) = Y^2_{\text{Total}}, \tag{27}$$

as expected. Here, $a^* = \mathbf{1}_C'\hat{\mathbf{D}}_C^{-1}\mathbf{1}_C/C^2$ is a special form of a defined in (6) with $\mathbf{p}_C = \mathbf{1}_C/C$.

An example continued: With the same data set as before, but with the new hypothesis of $\mathbf{p}_C = \left(1/3\ 1/3\ 1/3\right)'$, we obtain $Y^2_{\text{Total}} = 61.1111$, and $\mathbf{v}_{\max} = \left(0.7778\ 0.9444\ -1.7222\right)'$. The value of $Y^2_{\phi(\mathbf{v}_{\max})}$ is 61.1111, as expected. We test all possible pairwise differences among three cells, 1 versus 2 (with the associated contrast vector of $\mathbf{v}_6 = \left(1\ -1\ 0\right)'$), 1 versus 3 (with the contrast vector of $\mathbf{v}_7 = \left(1\ 0\ -1\right)'$), and 2 versus 3 (with the contrast vector of $\mathbf{v}_8 = \left(0\ 1\ -1\right)'$). The values of Y^2 for the three contrasts are, respectively, 1.1236, 21.9572, and 36.3636. Again, these values do not add up to Y^2_{Total} because the corresponding contrast vectors are not mutually orthogonal in this case.

6 Some Statistical Concerns

So far, our discussion has mainly focused on algebraic properties of contrasts. In this section, we briefly discuss some statistical issues concerning Neyman's statistic, namely the problem of sample size needed for Neyman's statistic to achieve its asymptotic distributional properties. We remind the reader that all the tests discussed in this paper are based on a large sample theory, as we assumed throughout the analyses of the example data set. But was this really justifiable? Recall that the example data set has $n = 100$. The first question we ask is if it is considered large enough to rely on the asymptotic theory.

6.1 A Monte-Carlo Study with n = 100

A small-scale numerical experiment was conducted to address the above issue. One thousand replicated data sets were generated with $n = 100$ according to a set of prescribed cell probabilities, $\mathbf{p}_C = \left(0.5\ \ 0.2\ \ 0.3\right)'$ (the same as the example data set). For each data set, Y^2_{Total}, $Y^2_{\phi(\mathbf{v}_1)}$ and $Y^2_{\phi(\mathbf{v}_2|\mathbf{v}_1)} = Y^2_{\text{Total}} - Y^2_{\phi(\mathbf{v}_1)}$ were calculated, where $\mathbf{v}_1 = \left(1\ -1\ -1\right)'$, and $\mathbf{v}_2 = \left(0\ 3\ -2\right)'$. (These contrasts vectors were also the same as in the example data set). Note that $\mathbf{v}_2$ is not orthogonal to $\mathbf{v}_1$ in Neyman's statistic, while it is so in Pearson's statistic. (The symbol $\mathbf{v}_2|\mathbf{v}_1$ indicates the effect of $\mathbf{v}_2$ eliminating the effect of $\mathbf{v}_1$). The quantile values of these quantities were plotted against the theoretical chi-square quantile values to obtain Q-Q plots, which are presented in the first column of Fig. 1. For comparisons, analogous X^2 values

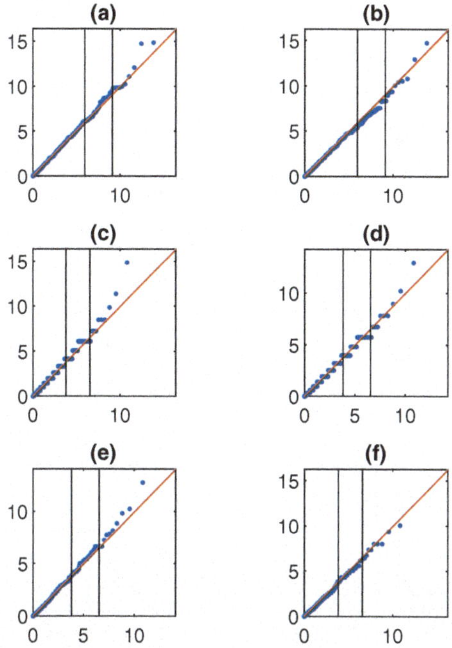

Fig. 1 Q-Q plots of Neyman's modified chi-square statistic (the first column) and Pearson's chi-square statistic (the second column) for $n = 100$ and with theoretical chi-square quantiles on the x-axis, and observed quantiles on the y-axis. **a** Y^2_{Total}, **b** X^2_{Total}, **c** $Y^2_{\phi(\mathbf{v}_1)}$, **d** $X^2_{\phi(\mathbf{v}_1)}$ **e** $Y^2_{\phi(\mathbf{v}_2|\mathbf{v}_1)}$, and **f** $X^2_{\phi(\mathbf{v}_2)}$, where $\mathbf{v}_1$ and $\mathbf{v}_2$ are given in the main text. Of two vertical lines in each plot, the left one indicates the 95% theoretical and the right one the 99% theoretical quantiles used as critical values in planned comparisons. With $n = 100$, $C = 3$, and no cell probabilities radically close to 0, observed and theoretical distributions show a fairly close match, and little differences are observed between the two statistics

were also calculated, and their Q-Q plots are displayed in the second column. Q-Q plots visually indicate how good an agreement there is between observed and theoretical distributions. In Fig. 1, agreements are good in all cases. This means that the asymptotic theory holds reasonably well for the example data set.

6.2 *A Monte-Carlo Study with n* = 50

Our next question is how much we can reduce the sample size without compromising the asymptotic theory. Another numerical experiment similar to the one above was conducted with the sample size cut down to one half of the original size (i.e. $n = 50$). Results are presented in Fig. 2. We find that agreements are still good for X^2's, while they are not as good for Y^2, particularly for quantile values beyond 95%. This means that we can still use the asymptotic theory for Neyman's statistic with a sample size

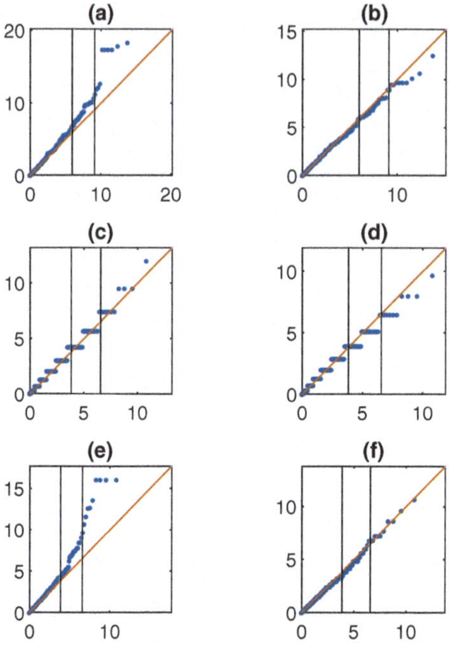

Fig. 2 Q-Q plots of the two statistics for $n = 50$, Neyman's modified chi-square in the first column, and Pearson's chi-square statistic in the second column. The basic construction of Fig. 2 is identical to that of Fig. 1. With $n = 50$, the asymptotic chi-square theory barely holds up to 95% quantiles for Neyman's statistic, while it upholds relatively well all the way for Pearson's statistic

of $n = 50$, if the test is performed at the significance level of 0.05, but not at the level of 0.01. It is understandable that Neyman's statistic needs a larger sample size to achieve its asymptotic properties than does Pearson's statistic, since the former uses $\hat{\mathbf{\Sigma}}_C/n$ as an estimator of the variance-covariance matrix of $\hat{\mathbf{p}}_C/n$, whereas the latter uses its true population value, i.e. $\mathbf{\Sigma}_C$ itself. But exactly how much lager sample is necessary is difficult to determine from this small study. More systematic studies are necessary to obtain more generalisable results.

What can we do if the asymptotic theory fails? We may ignore the asymptotic theory altogether, and we may focus on empirical distributions only, which are derived as intermediary results to construct the Q-Q plots. We can directly evaluate how rare the observed value of Y^2 is against its empirical distribution. If it is smaller than a prescribed α level, the null hypothesis is rejected. There have been vigorous attempts to minimise the number of evaluations of key statistics in simulated data (Hope 1968; Langeheine et al. 1996; Feng and McCulloch 1996) rather than deriving an entire range of empirical distributions.

When the sample size is so small that the minimum expected cell frequency is less than 5, it gets increasingly more difficult to obtain a reliable empirical distribution due to increased numbers of simulated data sets with zero frequency cells. Read and Cressie (1980, p. 75) proposed a correction formula for Y^2 for small samples. This formula is convenient because it can be used without deriving an empirical distribution by a simulation study.

7 Concluding Remarks

This paper presented a theory of contrasts for Neyman's (1949) modified chi-square statistic for one-way tables. This complements an earlier paper by the same authors (Loisel and Takane 2022) on a similar theory for the tests of (part) interaction effects in two-way and higher-order contingency tables.

The method proposed in this paper analyses the departure from hypothesised mean structures by postulating linear models on observed proportions. A comment is in order on what will happen if nonlinear transformations are applied to the observed proportions. In recent literature on correspondence analysis (CA), considerable attention has been paid to applying nonlinear transformations to the observed proportions, e.g. the log transformation (Greenacre 1984), and power transformations (Beh and Lombardo 2023; Beh et al. 2018; Greenacre 2010). A simple solution to this problem is already available due to Grizzle et al. (1969), which only involves replacing $\hat{\mathbf{\Sigma}}_C$ (n times the covariance matrix of $\hat{\mathbf{p}}_C$) in (4) by $\mathbf{J}\hat{\mathbf{\Sigma}}_C\mathbf{J}'$, where $\mathbf{J} = \left[\frac{\partial \mathbf{t}(\mathbf{p}_C)}{\partial \mathbf{p}_C}|_{\mathbf{p}_C=\hat{\mathbf{p}}_C}\right]$ is the matrix of the first derivatives of the transformations $\mathbf{t}(\mathbf{p}_C)$ with respect to their arguments evaluated at $\hat{\mathbf{p}}_C$. For example, if $\mathbf{t}$ is the element-wise logarithmic transformations of $\mathbf{p}_C$, $\mathbf{J} = \hat{\mathbf{D}}_C^{-1}$, and if it is the element-wise square-root transformations of $\mathbf{p}_C$, $\mathbf{J} = (1/2)\hat{\mathbf{D}}_C^{-1/2}$. This asymptotic covariance matrix of $\mathbf{t}(\hat{\mathbf{p}}_C)$ (multiplied by n) can be justified by the delta method; see, for example, Rao (1973, pp. 385–391). Some preliminary analysis on the numerical example used earlier indicates that Grizzle et al.'s method works very well in the present context.

8 About Nishi and His Work

In what follows, "I" refers to Yoshio Takane. Nishi is exactly ten years older than I. So he was already a full-fledged psychometrician when I started my career. Since then, I learned a lot from him. Indeed, I wrote book reviews (Takane 1982 1994) on two of his monographs on dual scaling (Nishisato 1980, 1994). This means that I read at least two of his books very closely. Here, dual scaling means simple and multiple correspondence analysis (Greenacre 1984) as well as other variants of scaling methods for ranking and pair comparison data. Although Nishi's influences on my subsequent work are many and profound, I may point out the following three, optimal scaling, analysis of sorting data, and incorporations of external information, as particularly important.

When I arrived at University of North Carolina (UNC) at Chapel Hill in 1973 as a new graduate student, my first duty as a research assistant was to "nonmetrise" various linear multivariate analysis methods, such as ANOVA, regression analysis, principal component analysis (PCA). Incorporating optimal monotonic transformations into the traditional multivariate analysis techniques fitted very nicely to the idea of dual scaling, that of assigning numbers to the subjects according to their patterns of responses to item categories. This basic idea has served as a landmark for many important developments in scaling that followed in the past forty years (Takane 2005).

In 1980, I published a paper on a method of analysis of sorting data (Takane 1980). In sorting data, a group of subjects are asked to sort a set of stimuli into several groups according to their similarity. The method finds an optimal representation of stimulus points as well as the centroids of sorting clusters in a joint multidimensional space. This method can be regarded as my first and concrete contribution to the area of dual scaling. It has turned out that the method is essentially equivalent to dual scaling of multiple-choice data arranged in such a way that the rows correspond with stimuli, and the columns with sorting clusters elicited by the subjects (Nishisato 1994).

The third point concerns how to incorporate external information in dual scaling. Nishisato (1980) proposed two alternative methods. One simply takes the product of the main data matrix and the matrices of external information regarding the rows and/or columns of the data matrix. The product is then subjected to the singular value decomposition (SVD) for further analysis. The second method, on the other hand, first projects the data matrix onto the spaces spanned by the matrices of external information, which is then subjected to SVD. Nishisato seemed to favour the first method on the ground that the second method involves the SVD of a larger matrix. Takane and his collaborators (Takane and Shibayama 1991; Takane and Hunter 2001; Takane et al. 1991; see also Takane 1994) argued, on the contrary, that the second method is superior on the ground that it is scale-invariant, and that there is a simple way to get around the computational problem pointed out by Nishisato (1980).

Acknowledgements The work reported in this paper was supported by research grants from the Natural Sciences and Engineering Research Council of Canada to the first author.

References

Beh, E.J., Lombardo, R.: Correspondence analysis and the Cressie-Read family of divergence statistics. Int. Stat. Rev. (in press) (2024)

Beh, E.J., Lombardo, R., Alberti, G.: Correspondence analysis and the Freeman-Tukey statistic: a study of archaeological data. Comput. Stat. Data Anal. **128**, 73–86 (2018)

Feng, Z.D., McCulloch, C.E.: Using Bootstrap likelihood ratios in finite mixture models. J. Royal Stat. Soci. Ser. B (Methodol.) **58**, 609–617 (1996)

Goodman, L.A.: Simultaneous confidence intervals for contrasts among multinomial populations. Ann. Math. Stat. **35**, 716–725 (1964)
Greenacre, M.J.: Theory and Applications of Correspondence Analysis. Academic Press, London (1984)
Greenacre, M.J.: Power transformations in corresponding analysis. Comput. Stat. Data Anal. **53**, 3107–3116 (2009)
Greenacre, M.J.: Log-ratio analysis is a limiting case of corresponding analysis. Math. Geosci. **42**, 129–134 (2010)
Grizzle, J.E., Starmer, C.F., Koch, G.G.: Analysis of categorical data by linear models. Biometrics **25**, 489–504 (1969)
Hope, A.C.A.: A simplified Monte Carlo significance procedure. J. Royal Stat. Soc. (Ser. B) **30**, 582–598 (1968)
Khatri, C.G.: A note on a MANOVA model applied to problems in growth curves. Ann. Inst. Stat. Math. **18**, 75–86 (1966)
Khatri, C.G.: Some properties of BLUE in a linear model and canonical correlations associated with linear transformations. J. Multivar. Anal. **34**, 211–226 (1990)
Lancaster, H.O.: The derivation and partition of χ^2 in certain discrete distributions. Biometrika **36**, 117–129 (1949)
Langeheine, R., Pannekoek, J., van de Pol, F.: Bootstrapping goodness-of-fit measures in categorical data analysis. Sociol. Methods Res. **24**, 492–516 (1996)
Loisel, S., Takane, Y.: Partitions of Pearson's chi-square statistic for contingency tables: a comprehensive account. Comput. Stat. **31**, 1429–1452 (2016)
Loisel, S., Takane, Y.: An algebraic theory of contrasts for Neyman's modified chi-square statistic. Behaviormetrika **50**, 335–360 (2022)
Lombardo, R., Takane, Y., Beh, E.J.: Familywise decompositions of Pearson's chi-square statistic in complex contingency tables. Adv. Data Anal. Classif. **14**, 629–649 (2020)
Maxwell, S.E., Delaney, H.D., Kelley, K.: Designing Experiments and Analyzing Data: A Model Comparison Perspective, 3rd edn. Routledge, New York (2018)
Neyman, J.: Contribution to the theory of the χ^2 test. In: Proceedings of the First Berkeley Symposium on Mathematical Statistic and Probability, pp. 239–274. University of California Press, Berkeley, CA (1949)
Nishisato, S.: Analysis of Categorical Data: Dual Scaling and Its Applications. University of Toronto Press, Toronto (1980)
Nishisato, S.: Elements of Dual Scaling: An Introduction to Practical Data Analysis. Lawrence Erlbaum Associates, Hillsdale, NJ (1994)
Puntanen, S., Styan, G.P.H., Isotalo, J.: Matrix Tricks for Linear Statistical Models. Springer, Berlin (2011)
Rao, C.R.: Linear Statistical Inference and Its Applications. Wiley, New York (1973)
Takane, Y.: Analysis of categorizing behavior by a quantification method. Behaviormetrika **8**, 57–67 (1980)
Takane, Y.: A review of "Analysis of Categorical Data: Dual Scaling and its Applications" by S. Nishisato (University of Toronto Press, Toronto, 1980). Can. J. Stat. **10**, 67–68 (1982)
Takane, Y.: A review of "Elements of Dual Scaling: An Introduction to Practical Data Analysis" by S. Nishisato (Erlbaum, Hillsdale, NJ, 1994). Appl. Psychol. Meas. **18**, 379–382 (1994)
Takane, Y.: Optimal scaling. In: Everitt, B.S., Howell, D.C. (eds.) Encyclopedia of Statistics for Behavioral Sciences, pp. 1479–1482. Wiley, Chichester (2005)
Takane, Y.: Professor Haruo Yanai and multivariate analysis. Special Matrices **4**, 283–295 (2016)
Takane, Y., Hunter, M.A.: Constrained principal component analysis: a comprehensive theory. Appl. Algebra Eng. Commun. Comput. **12**, 391–419 (2001)
Takane, Y., Jung, S.: Tests of ignoring and eliminating in nonsymmetric correspondence analysis. Adv. Data Anal. Classif. **3**, 315–340 (2009)
Takane, Y., Shibayama, T.: Principal component analysis with external information on both subjects and variables. Psychometrika **56**, 97–120 (1991)

Takane, Y., Yanai, H., Mayekawa, S.: Relationships among several methods of linearly constrained correspondence analysis. Psychometrika **56**, 667–684 (1991)

Tanabe, K., Sagae, M.: An exact Cholesky decomposition and the generalised inverse of the variance-covariance matrix of the multinomial distributions with applications. J. Royal Stat. Soc. Ser. B (Methodol.) **54**, 211–219 (1992)

From *DUAL3* to `dualScale`: Implementing Nishisato's Dual Scaling

Jose G. Clavel and Roberto de la Banda

1 Introduction

Nishisato (1980) presented dual scaling as a technique for finding measurements through their regression on data. Dual scaling is a versatile technique that handles not only multiple-choice data but also other types of data formats for categorical data (e.g. contingency tables, rank-order data, sorting data, paired comparison data and successive categories data), all of which are used to explore the hidden structure of the association between the categorical variables. Mathematically, dual scaling is equivalent to such quantification methods as optimal scaling, Hayashi's quantification theory, correspondence analysis (CA) and homogeneity analysis. For a complete historical overview of these quantification methods see Nishisato (2007) while Beh and Lombardo (2014) gave a detailed view of scaling specifically related to CA.

To cover a variety of categorical data, Nishisato and Nishisato (1994) provided a software package called *DUAL3* in Basic for (1) the total-space quantification of multiple-choice data, sorting data, paired comparison data, rank-order data and successive categories data and (2) subspace quantification, also referred to as forced classification. Our proposal is launched in order to 'power up' their *DUAL3* into a completely new programming language, i.e. (`R` R Core Team 2013), to meet the demands of current practises of data analysis. This present paper provides a description of the this new dual scaling package.

The package `dualScale` (Clavel, Nishisato and Pita 2014) examines the kind of incidence data that was also analysed by other functions such as

- `MCA()` in the `FactoMineR` package (Husson, Josse, Le and Mazet 2014),
- `mjca()` in the `ca` package (Nenadic and Greenacre 2007),
- `mca()` in the `MASS` package (Venables and Ripley 2002),
- `dudi.acm()` in the `ade4` package (Dray and Dufour 2007),

J. G. Clavel (✉)
Department of Quantitative Methods, University of Murcia, Murcia, Spain
e-mail: jjgarvel@um.es

R. de la Banda
Universidad Nacional de Educación a Distancia (UNED), Madrid, Spain
e-mail: rdelaband2@alumno.uned.es

E. J. Beh et al. (eds.), *Analysis of Categorical Data from Historical Perspectives*, Behaviormetrics: Quantitative Approaches to Human Behavior 17,
https://doi.org/10.1007/978-981-99-5329-5_6

- `CA3variants` package which performs a whole range of correspondence analysis and multi-way components analysis techniques for nominal/ordinal contingency tables (Lombardo and Beh 2016; Lombardo, van de Velden and Beh 2023), and
- `homals()` from the package of the same name (de Leeuw and Mair 2009).

Our package is easy to use and powerful enough to handle a large data set. It is therefore particularly attractive and useful for those students, teachers and researchers who are involved in social, political, health, medical and educational studies. It can also be used to examine general opinion polls and marketing research where questionnaires are typically used to collect data and data sets are generally large. As we know, an ingredient for a popular programme is that it should be easy to follow with a clear rationale. This is exactly what the current paper aims to provide.

In Sect. 2, the `R` functions used to perform a dual scaling of a contingency table will be introduced. In Sect. 3, the functions used for multiple-choice data are presented. Special attention is given to the function `dsFC()` which is used to perform forced classification and its use is demonstrated using several examples, that will be presented with several examples. Following Nishisato (1996), the `R` package `dualScale` creates `ds` objects for the analysis of the following types of categorical data:

- Incidence data, where the elements are either the presence or absence of an attribute, and a chi-squared metric is used to reflect distance relationships. It includes:
  - contingency/frequency data
  - multiple-choice data.
- Dominance data: formed from ordinal measurements. Here the scaling is to find a multidimensional configuration of row variables and column variables such that the information in the data is best approximated in a low-dimensional space. This type of data includes:
  - paired comparison data
  - rank-order data.

Due to space constraints, we will present here only the functions related with incidence data analysis: `ds_cf()` and `ds_mc()`.

2 Dual Scaling of Contingency Tables: `ds_cf()`

Contingency or frequency tables are the most straightforward type of data to handle with dual scaling. In summary, given a table with categorical information contained in its rows and columns, dual scaling will determine weights for the rows y_i and weights for the columns x_j in such a way that the correlation between the rows and the columns is maximised (Nishisato 1994).

Included in our R package is the object name curricula that Nishisato and Nishisato (1994) used in their package. The data we analyse is from Hollingshead (1949) and examines how the youth of a small Midwestern community called Elmtown from different social classes would enrol in different curricula. There are 390 students classified into four social classes and three curricula: college preparation, general and commercial. The following gives the R object containing this data:

```
>  curricula
               s.class1 s.class2 s.class3 s.class4
collegPrep          23       40       16        2
general             11       75      107       14
commercial           1       31       60       10
```

Let f_{ij} be the frequency of responses in row i (for $i = 1, 2, \ldots, I$) and column j (for $j = 1, 2, \ldots, J$) of an $I \times J$ contingency table. Let $f_{\cdot j}$ be the sum of the responses of row i, $f_{\cdot j}$ be the sum of the responses of column j and $f_{\cdot\cdot}$ the total number of responses in the table ($f_{\cdot\cdot} = 390$ respondents in our example). The trivial solution coincides with the expected value of each frequency when the row and column variables are independent—i.e. there is statistical independence between curricula and social class. In this case, the expected values are:

```
> ds_cf(curricula)$appro0
Distribution of Order 0 Approximation
        V1      V2      V3   V4
1  7.2692 30.3231 38.0077  5.4
2 18.5769 77.4923 97.1308 13.8
3  9.1538 38.1846 47.8615  6.8
```

Dual scaling analyses the portion of the data which is free from the effect of the trivial solution, f^*_{ij}. This is reflected in what is called Distribution of order 0 Residual Matrix with elements $f_{ij} - f^*_{ij}$, where:

$$f_{ij} - f^*_{ij} = f_{ij} - \frac{f_{i\cdot} \times f_{\cdot j}}{f_{\cdot\cdot}}.$$

For the data in curricula these values can be found using the command:

```
> ds_cf(curricula)$residual0
Distribution of Order 0 Residual Matrix
        V1      V2       V3   V4
1 15.7308  9.6769 -22.0077 -3.4
2 -7.5769 -2.4923   9.8692  0.2
3 -8.1538 -7.1846  12.1385  3.2
```

Next, we extract the most dominant pattern of the association. The first solution set, including y_{i1}, x_{j1} and ρ_1 (a measure of association), minimises the sum-of-squares discrepancies reflected on the above `Order 0 Residual Matrix` or, in other words, the first solution is the one that maximally explains the variation in f^*_{ij}. Assuming that the first solution $(y_{i1},\ x_{j1},\ \rho_1)$ does not explain all of the association captured in the f^*_{ij} elements, dual scaling will analyse the unexplained portion of association by finding the second most dominant pattern, and continue until all of the association will be accounted for. In our example, only two dimensions are needed to explain the data:

```
> ds_cf(curricula)$out
Component Eigenvalue SingValue   Delta CumDelta
        1     0.1765    0.4202 99.2289  99.2289
        2     0.0014    0.0370  0.7711 100.000
```

where:

- `Eigenvalue` is the squared correlation ratio ρ_k^2 and indicates the proportion of information one can gain from the rows given our knowledge of the data in the columns, and vice versa; here $k = 1,\ 2,\ \ldots,\ K$, where $K = \min\,(I,\ J) - 1$. The sum of the squared correlation ratios is the total of variance contained in the data;
- `SingValue` is the positive square root of the eigenvalue, also call ρ_k, and indicates the amount of linear relationship between the responses weighted by the row weights y_{ik} and by the column weights x_{jk};
- `Delta` denotes the δ_k values and is the percentage of the total association explained by solution k. It also reflects its relative importance in the full set of solutions.

Following the terminology of Nishisato (1980), both normed and projected weight vectors are provided. The normed weights y_i (and x_j) are scaled in such a way that the sum-of-squares of the set of row weights by y_{ik} (and column weights x_{jk}) is equal to $f_{..}$. For example, the normed weights for the columns (the four social class in our example) for solutions one and two is:

```
> ds_cf(curricula)$norm_opt
       V1      V2
1 -2.6785  1.1522
2 -0.4133 -0.7654
3  0.7216 -0.0568
4  0.8474  3.1465
```

Since the total number of elements in each of the social classes (columns) is (35, 146, 183, 26), then for the first solution:

$$\sum_{j=1}^{4} y_{j1} f_{j1} = -2.6785 \times 35 - 0.4133 \times 146 + \cdots + 0.8474 \times 26 = 390$$

and the same is obtained for the second solution: $\sum_j y_{j2} f_{j2} = 390$. The projected weights are given by $\rho_k y_{ik}$ and $\rho_k x_{jk}$ so that the sum-of-squares of the responses weighted by them is equal to $\rho_k^2 f_{..}$ thus reflecting the importance of that solution.

3 Dual Scaling of Multiple-Choice Data: `ds_mc()`

3.1 An Overview of Dual Scaling

Multiple-choice data consist of a table of N rows (subjects) by n columns (items) of chosen response option numbers. For quantification purposes, the data are transformed into the 'N subjects'-by-'total number of response options' table of response patterns of 1's and 0's, where 1 indicates a choice of that option and 0 is a non-choice. We use m_j to indicate the number of response options of item j and m to show the total number of options of all the items. For our package, we assume that each person selects only one response option per item. Thus, the response-pattern matrix for item j is expressed as a $N \times m_j$ matrix $\mathbf{F}_j$ such that each of the N rows—for example, $(0, 1, 0, \ldots, 0)$—contains only one choice, coded as 1, out of the m_j options. The entire data matrix for the N subjects and n items is therefore expressed as an $N \times m$ matrix, which we denote by $\mathbf{F}$ so that:

$$\mathbf{F} = [\mathbf{F}_1, \mathbf{F}_2, \ldots, \mathbf{F}_n]$$

with

$$m = \sum_{j=1}^{n} m_j .$$

The task of dual scaling is to determine m option weights as the least-squares regressions on the input data $\mathbf{F}$ (Nishisato 1980), so as to optimise mathematically equivalent criteria. For example, the variance of the subjects' weighted scores being a maximum, or the average inter-item correlation is maximised. These equivalent criteria lead to a generalised eigen-equation, yielding K orthogonal components. Since the rank of $\mathbf{F}$ is $m - n + 1$, assuming that $N \gg m - n + 1$ and there exists the so-called trivial component, dual scaling typically provides $K = m - n$ components. In other words, dual scaling typically yields $m - n$ sets of option weights.

The function `ds_mc` carries out dual scaling, using the following steps:

1. Calculate the matrix $\mathbf{C} = \mathbf{F}^\top \mathbf{D}_n^{-1} \mathbf{F}$ where $\mathbf{D}_n$ is the diagonal matrix of row totals of $\mathbf{F}$ (the number of responses from individual subjects, which are equal to n when no missing responses are involved).
2. Obtain $\mathbf{A} = \mathbf{D}^{-1/2}\mathbf{C}\mathbf{D}^{-1/2} - \frac{\mathbf{D}^{1/2}\mathbf{1}\mathbf{1}^\top\mathbf{D}^{1/2}}{f_t}$. Here, f denotes the $m \times 1$ vector of columns totals of $\mathbf{F}$ so that $\mathbf{D} = \mathrm{diag}\,(\mathbf{f})$ is the diagonal matrix of column totals

of $\mathbf{F}$, and f_t is the sum of the elements of $\mathbf{F}$, which is equal to nN, provided that no missing responses are involved.

3. Carry out a singular value decomposition on $\mathbf{A}$. That is, solve $\left(\mathbf{A} - \rho^2\mathbf{I}\right)\mathbf{D}^{1/2}\mathbf{x} = \mathbf{0}$ where ρ^2 is the maximum eigenvalue and its square root, ρ, is the singular value. The singular vector $\mathbf{D}^{1/2}\mathbf{x}$ is then converted to $\mathbf{x}$, which is referred to as the normed vector of weights for the columns (response options) of $\mathbf{F}$.
4. There exist dual relations—see Nishisato (1980)—such that:

$$\rho\mathbf{y} = \mathbf{D}_n^{-1}\mathbf{F}^\top\mathbf{x}$$
$$\rho\mathbf{x} = \mathbf{D}_n^{-1}\mathbf{F}^\top\mathbf{y}$$

where $\mathbf{y}$ is the normed score vector for the rows (subjects) of $\mathbf{F}$, $\rho\mathbf{y}$ is the vector of projected scores for subjects, and $\rho\mathbf{x}$ is the vector of projected weights for the options.

5. From the above computations we obtain $m - n$ set of components: one consisting of $m - n$ non trivial eigenvalues, singular values, reliability coefficients (Cronbarch's α), and delta coefficients δ (see Nishisato (2007) for an explanation of the terms), together with weights for the options, $\mathbf{x}_j$ and $\rho_j\mathbf{x}_j$, and scores for the subjects, $\mathbf{y}_j$ and $\rho_j\mathbf{y}_j$.
6. Compute the inter-item correlation matrix, based on the optimal scores, for component k, $r_{jj'(k)}$, and the item-total correlation $r_{jt(k)}$ as expressed by:

$$r_{jt(k)} = \frac{\mathbf{x}_j^\top\mathbf{F}_j^\top\mathbf{F}\mathbf{x}}{\sqrt{\mathbf{x}_j^\top\mathbf{D}_J\mathbf{x}_j\mathbf{x}^\top\mathbf{F}^\top\mathbf{F}\mathbf{x}}}, \quad (1)$$

where $\mathbf{x}_j$ is the vector of weights for the options of item j on component k, $\mathbf{F}_j^\top$ is the $m_j \times N$ matrix of response patterns for item j, $\mathbf{D}_J$ is the diagonal matrix of the column totals of $\mathbf{F}_j^\top$. See Nishisato (1994) for a number of interesting roles that the item-total correlation can play in data analysis.

3.2 An Application: Singapore 1985

In the `R` package we are describing here, the function `ds_mc` is used to carry out ordinary dual scaling of multiple-choice data. Let us use the `singaporean` data set that is included in the package to explain the output features of this function. This data set, presented in Nishisato and Nishisato (1994), contains the responses of 23 subjects to a 4-item questionnaire on the view that adults aged 20 years and older have to Singaporean children. The data were collected at Nishisato's workshop in Singapore in 1985. The four items are:

1. How old are you?
   (1) 20–29; (2) 30–39; (3) 40 or over
2. Children today are not as disciplined as when I was a child.
   (1) agree; (2) disagree; (3) I cannot tell
3. Children today are not as fortunate as when I was a child.
   (1) agree; (2) disagree; (3) I cannot tell
4. Religion should be taught at school.
   (1) agree; (2) disagree; (3) indifferent

All four items have 3 options each, so that $m = 12$. The total number of possible components is $m - n = 12 - 4 = 8$. To obtain the m option weights (weights of the different categories) for each component we can use the `R` command `ds_mc(singaporean)`. The following output is given using the default options, and other possible outputs are controlled by using the `print()` for `dualScale`:

```
> ds_mc(singaporean)
  Component Eigenvalue SingValue   Delta CumDelta     Alpha
1         1     0.6476    0.8047 32.3780  32.3780    0.8186
2         2     0.4407    0.6638 22.0333  54.4113    0.5769
3         3     0.3170    0.5631 15.8520  70.2633    0.2819
4         4     0.2136    0.4622 10.6806  80.9439   -0.2271
5         5     0.1843    0.4293  9.2134  90.1573   -0.4756
6         6     0.1157    0.3402  5.7852  95.9425   -1.5476
7         7     0.0528    0.2297  2.6380  98.5804   -4.9847
8         8     0.0284    0.1685  1.4196 100.0000  -10.4072
```

The first part of the output contains the 8 eigenvalues, ρ_k^2, of matrix $\mathbf{A}$, which sum to:

$$\frac{m}{n} - n = \frac{12}{4} - 1 = 2\,.$$

Other outputs include the corresponding singular values ρ_k, `Alpha` (the Kurder-Richardson generalised reliability coefficient, or Cronbach's α), and values of delta (the percentage of information accounted for by component k, i.e. the percentage of the eigenvalue divided by the sum of all the eigenvalues). These statistics are obtained for all components. As Nishisato (1994) discusses, theoretically, the coefficient α is the ratio of the expected values of two positive quantities, hence it is expected to be positive. The formula we use is an approximation to this theoretical quantity, and can become negative. Besides, Nishisato (1994) has shown that α becomes negative when the corresponding eigenvalue becomes smaller than $1/n$. He suggested that we should consider only those components with non-negative values of α for interpretation. If this strategy is adopted, the above values of `Delta` and `CumDelta` can be redefined for those adopted set of components. For example, since only the first three eigenvalues of the Singaporean data are greater than $1/n = 1/4$ we would only consider the first three components. The redefined `Deltas` for the first three components are then adjusted to 46.09%, 31.35% and 22.56%, respectively, so that these three components account for 100% of the total information. The values of `CumDeltas` should then also be adjusted accordingly.

Because of the orthogonality of the components, the total number of components is synonymously referred to as dimensions. Let us define by $r^2_{jt(k)}$ the information of item j on component k. Then, the distribution of information over the k components is given as follows:

```
> ds_mc(singaporean)$info
Distribution of Information Over 8 Components
Comp    Item1  Item2  Item3  Item4   Avge
   1   0.8615 0.7022 0.3637 0.6629 0.6476
   2   0.7418 0.1904 0.0185 0.8120 0.4407
   3   0.0124 0.4541 0.7370 0.0646 0.3170
   4   0.0758 0.1967 0.5174 0.0646 0.2136
   5   0.1664 0.2640 0.2644 0.0423 0.1843
   6   0.0062 0.1298 0.0710 0.2558 0.1157
   7   0.0709 0.0392 0.0235 0.0774 0.0528
   8   0.0651 0.0237 0.0044 0.0205 0.0284
```

The last column lists the average item contribution to component k, which turns out to be equal to the eigenvalue of the component; see Nishisato (1980) for an explanation of this feature. The statistic r^2_{jt} indicates the extent to which item j is correlated with component k: the higher the value, the greater the relevance of the item to that component. In our case, Item 1 contributes the most to component 1 ($r^2_{1t(1)} = 0.8615$), and Item 4 to component 2 ($r^2_{4t(2)} = 0.8120$). Another interesting aspect of this statistic is that its column sum is equal to the number of options of the items minus 1. For the current example, this is 2 for each column; for example, $0.8615 + 0.7418 + \cdots + 0.0651 = 2$ for the first item. This is the total contribution of the item.

The next output offers k tables of inter-item correlations for the components, $r_{jj'(k)}$. These values indicate the amount of linear relationship (defined by their product-moment correlation) between two items, of which the options are optimally scaled, usually through nonlinear transformations. Note that dual scaling determines option weights so as to maximise the sum-of-squares of all the inter-item correlations. For the Singapore data, eight correlation matrices are produced but only one of them is shown below:

```
> ds_mc(singaporean)$rij[,,1]
Inter Item Correlation for Component 1:
       Item1  Item2  Item3  Item4
Item1 1.0000 0.8005 0.3856 0.7033
Item2 0.8005 1.0000 0.3303 0.4795
Item3 0.3856 0.3303 1.0000 0.3983
Item4 0.7033 0.4795 0.3983 1.0000
```

One should keep in mind a very important feature of these correlation coefficients: since each correlation coefficient is optimised for all option weights, each coefficient is dependent on what other items are involved in the data set. In other words, the correlation coefficient between items 1 and 2 of component 1 will change if another item is discarded or additional items are added to the data set. Nishisato (2007) investigated this problem, and considered the projection of one item onto the space of the other item as the basis for assessing the correlation between two items in multidimensional space. To this end, he used his forced classification procedure (Nishisato 1984) and derived his coefficient $\nu_{j,j'}$. One remarkable aspect of his derivation is that he went one step further and proved successfully that his coefficient $\nu_{j,j'}$ is, in spite of its different appearance, identical to Cramér's coefficient $V_{j,j'}$. That is:

$$V_{j,j'} = \sqrt{\frac{\chi^2_{(j,j')}}{f_t\,(p-1)}} = \nu_{j,j'}, \tag{2}$$

where p is the smaller number of options of the two items, j and j', and $\chi^2_{(j,j')}$ is the Pearson chi-squared statistic calculated from the contingency table consisting of the options of item j and item j'. For the Singaporean data, we obtain the following matrix ν or, equivalently, $\mathbf{V}$:

$$\nu = \mathbf{V} = \begin{pmatrix} 1.000 & 0.579 & 0.292 & 0.682 \\ 0.579 & 1.000 & 0.308 & 0.394 \\ 0.292 & 0.308 & 1.000 & 0.332 \\ 0.682 & 0.394 & 0.332 & 1.000 \end{pmatrix}.$$

Thus, in multidimensional space, the correlation between Item 1 and Item 4 ($\nu_{14} = 0.682$) is the highest. The result is also available with the command:

```
> ds_mc(singaporean)$Cramer
```

More specific information from `ds_mc()` can be obtained from `names()`. The output of `dualScale` is structured as a list-object. For example, the normed option weights are obtained by:

```
> ds_mc(singaporean)$norm_opt
```

while one can obtain a more general series of output by:

```
> summary(ds_mc(singaporean))
```

3.3 Visualising the Results

The function `ds_mc()` produces a `dualScale` object, which includes optimal scores for the subjects and optimal weights for the options. The scores are projected or normed weights; for an explanation of the distinction of these two ideas; see Nishisato and Clavel (2003). Lebart, Morineau and Warwick (1984) pointed out that:

> A great deal of caution is needed in interpreting the distance between a variable point and an individual point because these two points do not belong to the same space.

This leads to a perennial problem of constructing a joint graphical display of both option weights and subject scores in the same space. In our package, a measure of the row-column space discrepancy (Nishisato and Clavel 2010) is provided to assist the user and should be used as a precaution for making a direct interpretation of a symmetric graph that is constructed using the projected scores of subjects and projected weights of options. If the separation angle is large, the data points corresponding to the rows and columns should not have coordinates on a single continuum of the component.

Following the approach of Clavel and Nishisato (2012), `dualScale` provide three kinds of plots that are available through the argument `type`. They are:

- Plots for one type of elements: `type = "Sub"` for only subjects and `type = "Ite"` for only item options. They are the projected subject scores and the projected option weights.
- Plots for two types of elements called asymmetric plots: `type = "Asy1"` for joint plots of normed row weights and projected column weights (the default option) and `type = "Asy2"` for joint plots of projected row weights and normed column weights. Although these options are logically correct, the projected quantities always have a smaller norm than the other set of quantities, and this difference in norms would typically make it difficult to make between-set (rows versus columns) comparison. This might be the main reason why many researchers still use the symmetric plot in spite of the fact that the symmetric display is an erroneous representation of two sets of variates. By default, an asymmetric graph (`Asy1`) for the first and second component is plotted. For example, the asymmetric plot of the `singaporean` data set is created with the following command:

```
> plot.ds(ds_mc(singaporean))
```

When choosing other combinations of components we must indicate those components. For example:

```
> plot.ds(ds_mc(singaporean), dim1 = 2, dim2 = 4)
```

produces a plot that is constructed using the first two components and is shown in Fig. 1. The symbol ▲ is given to the subjects, labelled with an `s.` and a number.

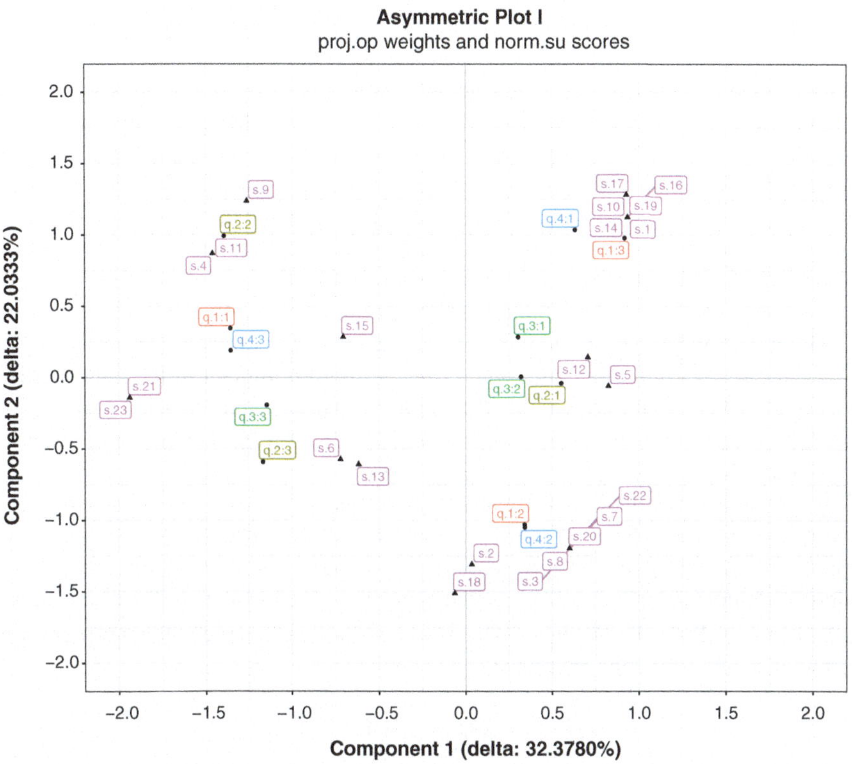

Fig. 1 Asymmetric plot for the `singaporean` data set using components 1 and 2

The symbol · is given to the options. They are labelled with a `q.` and two numbers separated by a colon. The first number indicates the item and the second number is the option within that item. For example, `q.2:1` is the label given to the first option of Item 2. To facilitate the reading of the plot, all the options of the same item are shown with the same colour. The plots also show the type of graph and the percentage of the total variance explained by each component.

For Fig. 1, the cumulative δ is 54.40%, and it shows that some distinct clusters can be identified. In the upper right quadrant are located those subjects aged 40 or over (`q.1:3`) who think that religion should be taught (`q.4:1`). In the upper left quadrant we find it dominated by those subjects age 20–29 years who are indifferent about the issue of religion at school and don't believe that the children today are less disciplined than before.

3.4 *Forced Classification Analysis:* `ds_mcf()`

Sometimes, we are interested in the analysis of a particular item of the questionnaire, rather than the entire set of items. Suppose one collects data on a father's education, mother's education, the student's enrolment or non-enrolment in kindergarten and the student's graduation from high school or drop-out half way. From the policy makers point of view, the last item (the one referring to drop-out) may be of utmost interest. If this is the case, our analysis should be focussed on this particular point with an aim to see how the response to this item is related to the responses to the first three questions of the questionnaire. This is akin to examining how to scale weights for the response options of the other items in such a way that the correlation between the item of interest with each of the remaining items is maximised. This kind of focused analysis can be carried out by forced classification of dual scaling, and it is in essence equivalent to discriminant analysis of multiple-choice data.

The function `ds_mcf()` in the `dualScale` package provides this analysis. To define the item of interest (called the *criterion item*) dominant in the data matrix, Nishisato (1984) proposed to modify the input response-pattern matrix by multiplying the criterion item, say p, by a large enough constant K, called the forcing agent. That is, modify $\mathbf{F} = \left[\mathbf{F}_1, \mathbf{F}_2, \ldots, \mathbf{F}_p, \ldots, \mathbf{F}_n\right]$ to $\mathbf{F} = \left[\mathbf{F}_1, \mathbf{F}_2, \ldots, K\mathbf{F}_p, \ldots, \mathbf{F}_n\right]$, and perform the dual scaling analysis on it. Nishisato (1984) has shown that this analysis converges to the analysis of the original response-pattern matrix $\mathbf{F}$ projected onto the subspace spanned by the columns of the criterion item, as the value of K approaches positive infinity. Thus, asymptotically, it is equivalent to the dual scaling of $\mathbf{P}_p\mathbf{F}$, where $\mathbf{P}_p = \mathbf{F}_p\left(\mathbf{F}_p^\top\mathbf{F}_p\right)^{-1}\mathbf{F}_p^\top$. The asymptotic properties of this process are captured by the following relations:

$$\lim_{K\to\infty} r_{pt}^2 = 1$$
$$\lim_{K\to\infty} \rho_p^2 = 1$$

For Item p with m_p options, the first $\left(m_p - 1\right)$ components attain the above asymptotic results. These components are called the *proper components* of forced classification, and item p is referred to as the *criterion item*. The mathematics of forced classification is presented in Nishisato (1984), and its applications and further characteristics are discussed in Nishisato (1986, 1988, 1994) and Nishisato and Baba (1999). Due to the forcing agent K, the criterion item determines the main components, meaning that option weights for non-criterion items are determined so as to maximise their correlations with the criterion item (Nishisato 2007).

The results provided by executing the package are essentially the same as those obtained with `ds_mc()` with some new features. Let us examine again the `singaporean` data set. Suppose we are interested in whether or not religion should be taught at school. Then, we should select Item 4 as the criterion and we can carry out forced classification by using the following command:

```
> ds_mcf(singaporean, crit = 4)
```

The output of this command is:

```
Call: ds_mcf(input = singaporean, crit = 4)

Type of Analysis: ds_mcf

Results:

Dual Scaling---Dual Scaling---Forced multiple-choice data analysis

Forced classification of the criterion item (type A)

 Component Eigenvalue SingValue Delta CumDelta
         1     0.3074    0.5544 63.41    63.41
         2     0.1773    0.4211 36.59   100.00
```

Since the criterion item has three options—(1) agree, (2) do not agree, (3) indifferent—the number of proper components is $m_p - 1 = 3 - 1 = 2$. Remember, too, that the asymptotic aspect of forced classification, where the eigenvalues of the modified data matrix do not provide statistics proportional to the contributions of the proper components to forced classification, needs an alternative measure of within relationships. Nishisato and Baba (1999) derived the following formula to calculate the exact correlation ratios in the subspace of the criterion variable:

$$\rho_k^2 = \frac{\sum_j r_{jt(k)}^2 - 1}{n - 1}. \tag{3}$$

The function ds_mcf() provides these adjusted correlation ratios for forced classification. The statistic δ for each of the proper components is redefined accordingly. The vectors of the projected subject scores and projected option weights, associated with these proper components have special significance to the interpretation of the forced classification outcomes. From the output this command produces, we will present only those vectors for components 1 and 2:

```
> ds_mcf(singaporean, crit = 4)$proj_opt_a
       V1      V2
1  -0.5500  0.1214
2   0.0791 -0.3286
3   0.4484  0.3011
4   0.1744 -0.0096
5  -0.4354  0.2688
6  -0.3709 -0.1631
7   0.3841  0.1206
8   0.0597 -0.0561
9  -0.4096  0.0960
10  0.4964  0.4364
11  0.1594 -0.5112
12 -0.9011  0.1849
> ds_mcf(singaporean, crit = 4)$proj_sub_a
```

```
        V1      V2
1   0.5561  0.3356
2  -0.1113 -0.3022
3   0.2552 -0.5623
4  -0.6601  0.4169
...
...
22  0.2552 -0.5623
23 -0.7590  0.0542
```

These option weights of non-criterion items are optimal in the sense that they produce maximally discriminative scores of subjects who chose different options of the criterion item, and these scores of subjects produce maximally discriminative option weights for the criterion item.

The remaining output produced using the `ds_mcf` function is similar to the output already seen from `ds_mc()`. The `Distribution of Information` of the components consists of the squares of the product-moment correlation between item j and the total scores (in the first n columns) and the average of these statistics in the last column:

```
Distribution of Information Over 8 Components:
  Item1 Item2 Item3 Item4  Avge
1 0.506 0.227 0.189 1.000 0.481
2 0.419 0.080 0.034 1.000 0.383
3 0.225 0.559 0.662 0.000 0.361
4 0.217 0.769 0.070 0.000 0.264
5 0.169 0.036 0.652 0.000 0.214
6 0.225 0.210 0.302 0.000 0.184
7 0.151 0.079 0.088 0.000 0.079
8 0.087 0.041 0.004 0.000 0.033
```

As expected, the criterion item (Item 4) has a perfect squared correlation with the first two components and a zero correlation with the other components. In other words, the criterion item is accounted for by the first two proper components and the remaining components do not contain any information about the criterion item. Considering that the sum of the two proper eigenvalues is 0.307 + 0.177, and that Item 4 contributes to the total proper subspace by 2, we can say that about the 25% of Item 4 (that is $(0.484/2) \times 100 = 24.2\%$) can be explained by first, second and fourth items.

Now, let us look at the inter-item correlation, using the function:

```
> ds_mcf(singaporean, crit = 4)$rij_a
Inter-Item Correlation for Component 1:
       [,1]   [,2]   [,3]   [,4]
[1,] 1.0000 0.7769 0.3169 0.7115
```

```
[2,] 0.7769 1.0000 0.2607 0.4763
[3,] 0.3169 0.2607 1.0000 0.4344
[4,] 0.7115 0.4763 0.4344 1.0000

Inter-Item Correlation for Component 2:
          [,1]    [,2]    [,3]   [,4]
[1,]  1.0000  0.1288 -0.0578 0.6472
[2,]  0.1288  1.0000 -0.2082 0.2820
[3,] -0.0578 -0.2082  1.0000 0.1831
[4,]  0.6472  0.2820  0.1831 1.0000
```

In forced classification, the criterion variable (Item 4) is perfectly correlated with the total score, and the item-total correlation is identical to the item-criterion correlation. As we can see, the correlation coefficient of the other three items with the criterion item in Component 1—that is, 0.7115, 0.4763 and 0.4344—are equal to the square roots of the corresponding values in the `Distribution of Information` table above: $0.7115 = \sqrt{0.506}$, $0.4763 = \sqrt{0.277}$ and $0.4344 = \sqrt{0.189}$. The magnitude of the correlation is a clear reference for which item will be better predicted by the criterion item. In our case, Item 1—how old are you?—has the highest correlations with the first two components (i.e. 0.7115 and 0.6472).

Using the plots already presented in Sect. 3.3, Fig. 2 is obtained from the output of a forced classification analysis where Item 4 is the criterion. See the location of the items options in this case. The figure shows the plot of the options using their weight of the proper components as coordinates. As Item 4 is the criterion item, its options are located in the vertices of the triangle that define the projected space and all the non-criterion options are projected onto the space of the criterion item. Option 3 of Item 1 (i.e. `q.1:3`, the age group 40 or over) is close to `q.4:1` (agree that religion should be taught). Similarly, `q.1:1` is close to `q.4:3` and `q.1:2` is close to `q.4:2`. The points in the middle (i.e. `q.3:2` and `q.2:1`) do not contribute greatly to the interpretation of the results.

3.4.1 Eliminating Versus Ignoring the Effects of the Criterion Item

The first two components of our previous example are relevant to the criterion item, and asymptotically those proper components are dual scaling results obtained from the data projected onto the subspace of the criterion item. That is, dual scaling is performed on the matrix $\mathbf{P}_p\mathbf{F}$. The remaining components can tell us how other non-criterion items behave in the absence of the influence of the criterion item, and it corresponds to dual scaling of the complimentary space of the criterion item. That is, dual scaling is performed on the matrix $(\mathbf{I} - \mathbf{P}_p)\,\mathbf{F}$. Since the analysis of complimentary space is as important as the analysis of the criterion item subspace, a convincing explanation is in order.

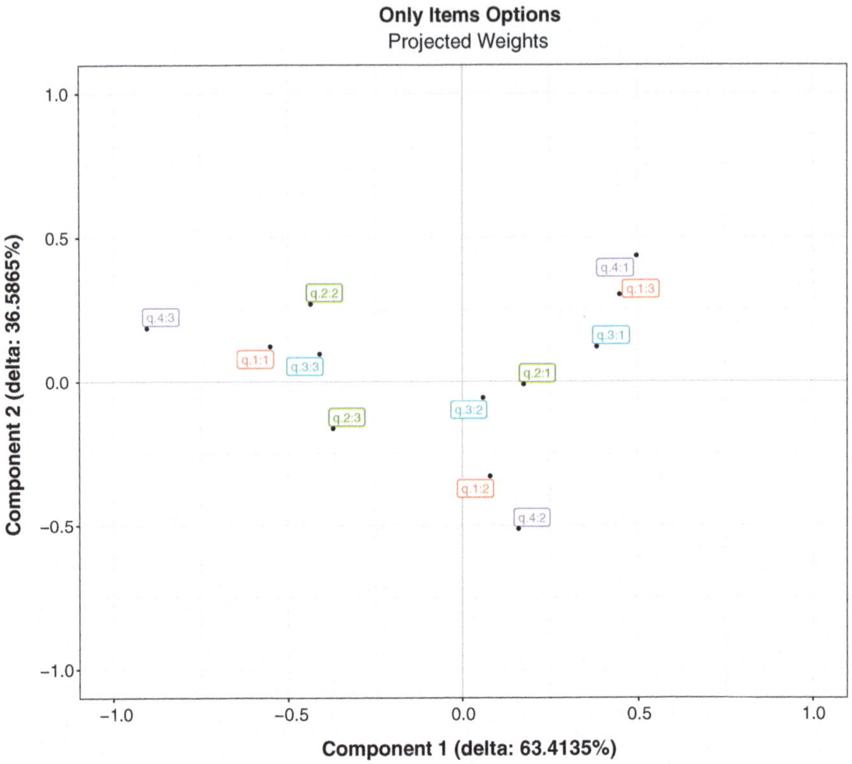

Fig. 2 Plot for forced classification of `singaporean` data when item 4 is the criterion

Suppose that health survey data were collected from several municipal hospitals where many senior patients are involved. Do we not wonder then if the dominance of senior patients somehow affects the outcome of survey results? One can then decompose the data into the analysis of the subspace for senior patients and the analysis of the complimentary space of senior patients. The former reflects how senior patients data affect the results of the survey, and the latter tells us what happens if we remove the contributions of senior patients from the analysis. In practise, there are many cases in which some variables are not taken into consideration for control, and the current two-way analysis of subspace and complimentary space can be effectively used to investigate if some uncontrolled variables have any substantial effects on the outcome of data analysis. To this end, we recommend the following strategy:

1. Suppose that age is the variable of concern. If so, carry out a forced classification and retain the results from the complimentary space. That is, we perform a dual scaling on the matrix $\left(\mathbf{I} - \mathbf{P}_p\right) \mathbf{F}$. Let us call this analysis: 'Eliminating the effect of the criterion item.'

2. Remove the age item from the data set, and subject the remainder of data to ds_mc(). Let us call this analysis as: 'Ignoring the effect of the criterion item.'

Both analyses yield the same numbers of components. When we compare the corresponding eigenvalues with the squared item-total correlations, we would expect that those values associated with 'ignoring' are larger than those from 'eliminating,' the reason being that the former may capture the hidden influences of the criterion item on results by not controlling its effects. This is a useful application of forced classification since it can be used to identify some hidden variables that influence the data analysis. To show this, let us look at a numerical example. Suppose we analyse the singaporean data once again but carry out forced classification using the age (Item 1) as the criterion. That is:

```
> ds_mcf(singaporean, crit = 1)
```

We are interested in the information distribution pertaining to the non-proper forced classification components, that is, in rows 3–8 of the following table:

```
Distribution of Information Over 8 Components:
  q.1   q.2   q.3   q.4  Avge
1   1 0.645 0.157 0.498 0.575
2   1 0.024 0.017 0.428 0.367
3   0 0.583 0.643 0.053 0.320
4   0 0.210 0.659 0.146 0.254
5   0 0.150 0.376 0.404 0.233
6   0 0.144 0.071 0.254 0.117
7   0 0.178 0.038 0.087 0.076
8   0 0.066 0.039 0.129 0.059
```

By discarding the contributions of the proper components from Table 1, we now need new mean values of the squared correlation coefficients. That is, we now must divide the sums of the item-total correlation coefficients by 3, not 4. This revised table of information distribution should be compared with the corresponding table obtained from the second data set, that is, the data set without the age item. This can be easily obtained with the command:

```
ds_mc(singaporean[-1,])$out
```

or alternatively, using the function ds_mcf()

```
> ds_mcf(singaporean, crit = 1)$out_b
```

In our case, the new data set will consist of three items; Item 2: Children today are not as disciplined as when I was a child, Item 3: Children today are not as fortunate

Table 1 Non-proper components of dsFC versus dsMC components ignoring criterion item

dsFC results					dsMC results				
Compon.	q.2	q.3	q.4	Average	Compon.	q.2	q.3	q.4	Average
3rd	0.583	0.643	0.053	0.426	1st	0.598	0.537	0.682	0.605
4th	0.210	0.659	0.146	0.338	2nd	0.779	0.276	0.257	0.437
5th	0.150	0.376	0.404	0.310	3rd	0.016	0.574	0.601	0.397
6th	0.144	0.071	0.254	0.156	4th	0.309	0.363	0.072	0.248
7th	0.178	0.038	0.087	0.101	5th	0.128	0.164	0.216	0.169
8th	0.066	0.039	0.129	0.078	6th	0.169	0.086	0.173	0.143

as when I was a child, and Item 4: Religion should be taught at school, each with 3 options. Thus the number of possible components, without the criterion item, is now $(m - n) = (3 \times 3) - 3 = 6$. Using the `ds_mc()` function produces the following output:

```
Component Eigenvalue SingValue  Alpha  Delta CumDelta
        1      0.606     0.778  0.674 30.275   30.275
        2      0.437     0.661  0.357 21.872   52.147
        3      0.397     0.630  0.240 19.849   71.996
        4      0.248     0.498 -0.516 12.399   84.395
        5      0.169     0.411 -1.454  8.464   92.859
        6      0.143     0.378 -2.001  7.141  100.000
```

Let us now present the information distributions from the two analyses, one from the complimentary space analysis of forced classification (that is, by eliminating the effects of the criterion item) and the other from dual scaling of the data set reduced by dropping the criterion item from the data set; see Table 1. Notice that the eigenvalues, indicated by `Avge`, from dual scaling are always larger than those from forced classification because the former values contain the contributions of the criterion item as a hidden contamination variable, while forced classification results on the latter were obtained by eliminating the effects of the criterion variable completely. Remember that the data set for `ds_mc()` does not have the age question, but that the data set is nonetheless under the influence of the age in a hidden way.

4 Summary

In this paper we have presented some functions of the `R` package `dualScale` for dual scaling analysis. This package contains all the features of the former commercially available software (*DUAL3*) plus various important new features, especially those related with the forced classification approach; such features include adjusted eigenvalues, analysis of complementary subspace, match-mismatch tables and more.

The three functions `ds_ct()`, `ds_mc()` and `ds_mcf()` produce a class of objects named `ds` that can be easily represented by specifically created plots by the programme using the `plot()` method for `ds` objects. Since this book is a *Festschrift* to celebrate Nishisato's career, the authors have decided to present their results using Nishisato's traditional data sets, and notation. The final goal is to include all of the findings of Nishisato fruitful career in `dualScale` thereby making it easily available for future generations.

References

Beh, E.J., Lombardo, R.: Correspondence Analysis: Theory. Practice and New Strategies. Wiley, Chichester (2014)

Clavel, J.G., Nishisato, S.: Reduced versus complete space configurations in total information analysis. In: Gaul, W.A., Geyer-Schulz, A., Schmidt-Thieme, L., Kunze, J. (eds.) Challenges at the Interface of Data Analysis, Computer Science, and Optimization, pp. 91–99. Springer-Verlag, Berlin (2012)

Clavel, J.G., Nishisato, S., Pita, A.: `dualScale`: Dual Scaling Analysis of Multiple Choice Data. Available online from CRAN.R-project.org/package=dualScale. Last accessed 1 May 2023 (2014)

de Leeuw, J., Mair, P.: Gifi methods for optimal scaling in `R`: The package `homals`. J. Stat. Softw. **31**(4), 21 (2009)

Dray, S., Dufour, A.-B.: The `ade4` Package: implementing the duality diagram for ecologists. J. Stat. Softw. **22**(4), 20 (2007)

Hollingshead, A.B.: Elmstown's Youth: The Impact of Social Classes on Adolescents. Wiley, New York (1949)

Husson, F., Josse, J., Le, S., Mazet, J.: `FactoMineR`: Multivariate Exploratory Data Analysis and Data Mining with `R`. Available online at http//CRAN.R-project.org/package=FactoMineR. Last accessed 1 May 2023 (2014)

Lebart, L., Morineau, A., Warwick, K.M.: Multivariate Descriptive Statistical Analysis. Wiley, New York (1984)

Lombardo, R., Beh, E.J.: Variants of correspondence analysis. The R Journal **8**(2), 167–184 (2016)

Lombardo, R., van de Velden, M., Beh, E.J.: Three-way correspondence analysis in `R`. The R Journal **15**(2), 237–262 (2023)

Nishisato, S.: Analysis of Categorical Data: Dual Scaling and Its Applications. University of Toronto Press, Toronto (1980)

Nishisato, S.: Forced classification: a simple application of a quantification method. Psychometrika **49**, 25–36 (1984)

Nishisato, S.: Generalized forced classification for quantifying categorical data. In: Diday, E. (ed.) Data Analysis and Informatics IV, pp. 351–362. North-Holland, Amsterdam (1986)

Nishisato, S.: Force classification procedures of dual scaling: its mathematical properties. In: Bock, H.H. (ed.) Classification and Related Methods of Data Analysis, pp. 523–532. North-Holland, Amsterdam (1988)

Nishisato, S.: Elements of Dual Scaling: An Introduction to Practical Data Analysis. Lawrence Elbaum Associates, Hillsdale, NJ (1994)

Nishisato, S.: Gleaning in the field of dual scaling. Psychometrika **61**, 559–599 (1996)

Nishisato, S., Clavel, J.G.: A note on between-set distances in dual scaling and correspondence analysis. Behaviormetrika **30**, 87–98 (2003)

Nishisato, S.: New framework for multidimensional data analysis. In: Weihs, C., Gaul, W. (eds.) Classification—The Ubiquitous Challenge, pp. 280–287. Springer-Verlag, Berlin (2004)

Nishisato, S.: Correlational structure of multiple-choice data as viewed from dual scaling. In: Greenacre, M., Blasius, J. (eds.) Multiple Correspondence Analysis and Related Methods, pp. 161–177. Chapman & Hall/CRC, Boca Raton, FL (2006)

Nishisato, S.: Multidimensional Nonlinear Descriptive Analysis. Chapman & Hall/CRC. Boca Raton, FL (2007)

Nishisato, S., Baba, Y.: On contingency, projection and forced classification of dual scaling. Behaviormetrika **26**, 207–219 (1999)

Nishisato, S., Clavel, J.G.: Interpreting data in reduced space: a case of what is not what in multidimensional data analysis. In: Shigemasu, K., Okada, A., Imaizuma, T., Hodhina, T. (eds.) New Trends in Psychometrics, pp. 357–366. Universal Academic Press, Tokyo (2008)

Nishisato, S., Clavel, J.G.: Total information analysis: comprehensive dual scaling. Behaviometrika **37**, 15–32 (2010)

Nishisato, S., Nishisato, I.: Dual Scaling in a Nutshell. MicroStats, Toronto (1994)

Nishisato, S., Yamauchi, H.: Principal components of deviation scores and standard scores. Japanese Psychol. Res. **16**, 162–170 (1974)

Nenadic, O., Greenacre, M.: Correspondence Analysis in `R`, with two- and three-dimensional graphics: the `ca` package. J. Stat. Softw. **20**(3), 13 (2007)

R Core Team: R: A Language and Environment for Statistical Computing. R Foundation for Statistical Computing, Vienna, Austria (2013)

Venables, W.N., Ripley, B.D.: Modern Applied Statistics with S, 4th edn. Springer, New York (2002)

Confounding, a Nuisance Addressed

Helmut Vorkauf

1 Introduction

I got to know Shizuhiko Nishisato in 1996 as the editor of Psychometrika, not as the editor of the nagging sort that one has to fight to get a paper published but as a friendly and resourceful helper. It is thus a great pleasure to honour such a friend by contributing to his *Festschrift*.

I happen to have a long history of problems with non-orthogonal data designs. As a teenager, still in school, I worked in an educational research institute and was assigned the task to calculate an analysis of variance with a mechanical calculating machine (no computers at that time in the 1950s), following the rules from a book by Edwards on analysis of variance (sorry, no exact recollection). As I know now, the rules were for orthogonal designs and my data were non-orthogonal survey data. When I ran the calculations using Edwards' rules and arrived at negative sums of squares for various sources of variation, my boss was ready to fire me for evident incompetence. I insisted he do some calculations himself, and he also got negative sums of squares. I was not involved in the discussions thereafter, since I was just a teenager earning some pocket money. But I kept the job.

In my professional life, problems with non-orthogonality re-surfaced often.

2 Confounding Is Not the Data's Fault, but of the Analysis with Aggregation

Recently I encountered data (Table 1) from von Kügelgen, Gresele and Schölkopf (2021) containing all confirmed COVID-19 cases in China (up to February 2020) and Italy (to March 2020).

H. Vorkauf (✉)
Bern, Switzerland
e-mail: helmut.vorkauf@gmail.com

E. J. Beh et al. (eds.), *Analysis of Categorical Data from Historical Perspectives*, Behaviormetrics: Quantitative Approaches to Human Behavior 17,
https://doi.org/10.1007/978-981-99-5329-5_7

Table 1 COVID-19 case fatality rate in China and Italy

Age	Italy		China		Case fatality rate		
	Alive	Died	Alive	Died	Italy	China	Higher in
0–9	43	0	0	0	0.000	?	?
10–19	85	0	548	1	0.000	**0.002**	China
20–29	296	0	3612	7	0.000	**0.002**	China
30–39	470	0	7582	18	0.000	**0.002**	China
40–49	890	1	8533	38	0.001	**0.004**	China
50–59	1450	3	9878	130	0.002	**0.013**	China
60–69	1434	37	8274	309	0.025	**0.036**	China
70–79	1671	114	3606	312	0.064	**0.080**	China
80–89	1330	202	1200	208	0.132	**0.148**	China
Total	7669	357	43233	1023	**0.044**	0.023	**Italy**

Unequal cell sizes led to spurious results, due to our routine practice of aggregating the data to arrive at column totals to estimate an independent variable's effect, ignoring that unequal cell sizes can lead to skewed results.

In these COVID-19 data, the erroneous result demonstrates the well-known Simpson paradox, where aggregation reverses the uniform trend of a higher fatality rate of Chinese patients in each of the age groups into an astonishing higher fatality rate of Italian patients in the aggregated total (bold-faced in the table).

The reason for this paradoxical result is the confounding variable *Age*. Chinese patients being younger than Italian patients, the many older Italian patients with a high fatality rate determine the column total and thus produce the paradox. In this example of low dimensionality, it is not difficult to identify *Age* as the responsible confounder. In an epidemiological case–control study, however, with maybe 30 potential causes of an infection it is certainly less easy to single out one of the 30 causes or one of the 435 pairs of causes as responsible for any confounding effect. The confounder might even be a triple or a higher n-tuple of causes.

It is clear that the source of the error is the summing-out of factors to arrive at marginal sums when these sums are erroneously influenced by differently skewed distributions. This problem did not exist in the early times of analysis of variance when a planned experimental design was orthogonal with equal cell frequencies. Confounding could rear its ugly head only when data started to be collected in surveys where unequal cell sizes are normal.

Alas, the change from orthogonal experimental designs to survey sampling (almost necessarily non-orthogonal) was not accompanied with a corresponding fundamental change of the method of analysis. Programs like BMDP, which I used in the late 1960s, refined the analysis to arrive at positive sums of squares, but did not eliminate confounding and its rare extreme outcomes like Simpson's paradox. Traditional analyses rely not only on the original individual cell frequencies, but also on marginal sums, ignoring Fisher's (1958) demand to use all of the data, not aggregated sub-tables:

> In inductive reasoning the whole of the data, or the available axioms, or the available observations, has to be taken into account.

3 An Approach Based on the Data, Not Marginal Sums

Searching for an analysis liberated from confounding effects, one has to look for a way to **avoid the use of aggregation** which is replacing original data with marginal sums. These sums have as their aim the isolation of variables and their effects. Could there be a way to isolate the effect of one variable on another variable without replacing the original data with partial sums and thus losing the original data?

A way was found to eliminate the association between two variables without any summing out, leaving all frequencies intact. This is done by combining the categories of two variables into one composite variable. Any association of, e.g., $Sex = [M, F]$ and $Department = [1, 2, 3]$ is eliminated by combining the values of Sex and $Department$ into the composite variable $SexDep = [M1, M2, M3, F1, F2, F3]$.

This simple operation is not new and has certainly found its uses in the past. But the approach presented here makes novel use of the **removal of interdependence** of X_i and X_j; we gave it the name ***uncoupling***.

H, the entropy of a categorical distribution with k categories, is at the basis of a proposed alternative analysis and defined by:

$$H = -\sum_{i=1}^{k} p_i \times \ln(p_i),$$

for $p_i > 0$. When $H = 0$, there is no variation since all cases are concentrated on a single category; H reaches the maximum of $\ln(k)$ for a rectangular distribution over the k categories. H, a measure of uncertainty, can readily be interpreted as a measure of variance for categorical variables.

The uncoupled composite variable $SexDep$ has less entropy than the cross-tabulation of Sex and $Department$; this loss of entropy is due to the association between Sex and $Department$. We need a way to express this loss of association as a component part of the total of the associations between all variables. We could gain a quantitative partitioning of the total of all correlation into the contribution of every pair of variables to the total correlation, i.e. an "analysis of entropy".

This total of all correlation's between all variables in a data set can be computed with the coefficient of *terseness* ζ (zeta)[1] introduced by Preuss and Vorkauf (1997).

[1] We collaborated for the publication, but the derivation of the total correlation is almost entirely the work of Lucien Preuss.

Table 2 Original three-dimensional $9 \times 2 \times 2$ table, regrouped as a two-dimensional 9×4 table by uncoupling *Country* and *Fatality*

Age	Uncoupled pair of variables			
	Italy		China	
	Alive	Died	Alive	Died
0–9	43	0	0	0
10–19	85	0	548	1
20–29	296	0	3612	7
30–39	470	0	7582	18
40–49	890	1	8533	38
50–59	1450	3	9878	130
60–69	1434	37	8274	309
70–79	1671	114	3606	312
80–89	1330	202	1200	208

ζ is a coefficient of the closeness of relations between a complete set of M variables, or a coefficient of total correlation and is defined as:

$$\zeta = 1 - \frac{\sum_{i=1}^{M} H(X_i|X_1, \ldots, X_{i-1}, X_{i+1}, \ldots, X_M)}{H(X_1, X_2, X_3, \ldots, X_M)},$$

and is valid for tables with any number of dimensions; it is normalised to 1, independent of the base of the logarithm and independent of the sample size N. Therefore, it is comparable for tables of different size and dimensionality, a quality that is highly desirable.

For the analysis of the COVID-19 data in Table 1 we first compute ζ_{Total} for the three-dimensional total table and then the reduced ζ for each of the two-dimensional sub-tables in which a different pair of variables is uncoupled (an uncoupled pair is indicated by square brackets).

1. $[country, fatality] \times age$ which eliminates the correlation of $country$ and $fatality$, shown as Table 2,
2. $[country, age] \times fatality$ which eliminates the correlation of $country$ and Age,
3. $[age, fatality] \times country$ which eliminates the correlation of age and $fatality$.

Note that the tables with a pair of variables uncoupled contain the same $9 \times 2 \times 2$ frequencies of the original table, the calculations are based on only these frequencies, with none of the marginal sums that can lead to erroneous interpretation.

The key difference is the structural interpretation, with the two-dimensional cross-tabulation of two correlated variables being replaced by the one-dimensional composite variable, thus losing the information of correlation. That is:

a	b
c	d

contains the correlation, whereas:

a	b	c	d

ignores it.

The original three-dimensional Table 1 produces a $\zeta_{\text{Total}} = 0.025893$.

We calculate the ζ of each of the three two-dimensional sub-tables and subtract from ζ_{Total} producing:

$$\Delta\zeta = \zeta_{\text{Total}} - \zeta_{\text{Subtable}},$$

which is the contribution to the total correlation of the pair of variables that are uncoupled.

1. The table uncoupling $country$ and age produces $\Delta\zeta = 0.017717$, a loss of 68%. We recognise that the dominant effect contained in the data is the overwhelming difference of the age distribution of the two countries, the cause of the confounding.
2. The table uncoupling $fatality$ and age produces $\Delta\zeta = 0.007758$, a loss of 30%; $fatality$ increases considerably with age.
3. The table uncoupling $country$ and $fatality$ produces $\Delta\zeta = 0.000146$, a negligible coefficient and a negligible loss of 1%. The disturbing Simpson's paradox has vanished as the confounder was given no chance to exert its influence.

4 Byssinosis, an Epidemiological Example

Let us now turn to a more complex data set with six variables by Higgins and Koch (1977) as shown in Table 3.

The complete $3 \times 3 \times 2 \times 2 \times 2 \times 2$ table is difficult to assess. When one tries to find the main factors leading to *byssinosis*, a lung disease caused by exposure to cotton dust, one has to take into account many interrelationships that exist between the possibly illness-inducing variables. Higgins and Koch (1977) devised a laborious χ^2-based set of rules designed to find the important factors; they concluded that *dustiness* of the workplace is the most important determinant of illness, *gender* of employee and *smoking* following next. From the content of the study, it seems curious that the *length of employment* and therefore the length of exposure to dust came in fourth place only. Could it be that some confounder has suppressed the relation between *length of employment* and *byssinosis*?[2] The $\Delta\zeta$ values summarised in Table 4 should provide an answer to this question.

[2] Higgins and Koch's division of χ^2 by degrees of freedom might also have played a role.

Table 3 Byssinosis by dustiness, employment, smoking, gender and race

Employ	Smoke	Sex	Race	Dustiness of workplace								
				Most			Medium			Least		
				No	Yes	p	No	Yes	p	No	Yes	p
< 10	Yes	M	White	37	3	0.0750	74	0	0.0000	258	2	0.0076
			Other	139	25	0.1420	88	0	0.0000	242	3	0.0122
		F	White	5	0	0.0000	93	1	0.0106	180	3	0.0164
			Other	22	2	0.0833	145	2	0.0136	260	3	0.0114
	No	M	White	16	0	0.0000	35	0	0.0000	134	0	0.0000
			Other	75	6	0.0741	47	1	0.0208	122	1	0.0081
		F	White	4	0	0.0000	54	1	0.0182	169	2	0.0117
			Other	24	1	0.0400	142	3	0.0207	301	4	0.0131
10–20	Yes	M	White	21	8	0.2758	50	1	0.0196	187	1	0.0053
			Other	30	8	0.2105	5	0	0.0000	33	0	0.0000
		F	White	0	0	??	33	1	0.0294	94	2	0.0208
			Other	0	0	??	4	0	0.0000	3	0	0.0000
	No	M	White	8	2	0.2000	16	1	0.0588	58	0	0.0000
			Other	9	1	0.1000	0	0	??	7	0	0.0000
		F	White	0	0	??	30	0	0.0000	90	1	0.0110
			Other	0	0	??	4	0	0.0000	4	0	0.0000
≥ 20	Yes	M	White	77	31	0.2870	141	1	0.0070	495	12	0.0237
			Other	31	10	0.2439	1	0	0.0000	45	0	0.0000
		F	White	1	0	0.0000	91	3	0.0319	176	3	0.0167
			Other	1	0	0.0000	0	0	??	2	0	0.0000
	No	M	White	47	5	0.0962	39	0	0.0000	182	3	0.0162
			Other	15	3	0.1667	1	0	0.0000	23	0	0.0000
		F	White	2	0	0.0000	187	3	0.0158	340	2	0.0058
			Other	0	0	??	2	0	0.0000	3	0	0.0000

From an epidemiological point of view, it is reassuring that in this analysis the order of pairs that include the dependent variable *byssinosis* is *dust*, *length of employment*, *smoking*, *gender* and *race*. This order appears more plausible for a lung disease than Higgins and Koch's (1977) order: *dust*, *gender*, *smoking*, *length of employment* and *race*.

But the largest value of $\Delta\zeta$ occurs for the uncoupling of *race* and *length of employment*; the much higher turnover of non-white employees is responsible for almost half of the terseness $\zeta = 0.0984$ of the whole table.

Table 5 shows this difference of turnover to have the effect that the clear increase of *byssinosis* with *length of employment* (and therefore exposure) seen within *race*, especially within *other race*, is reduced when *race* is summed out. This confounding has not affected the $\Delta\zeta$, however, **they are immune**.

Here, the collapsing of the table by summing out *race* was not yet an error producing a reversal of trend as in Simpson's paradox, but it is an error that led Higgins and Koch to underestimate the effect of *length of employment* on developing a *byssinosis*; the confounding was just not extreme enough to produce a rather rare Simpson's paradox.

Table 4 Byssinosis: terseness when uncoupling pairs of variables

Terseness of the full table $\zeta = 0.0984$		
$\Delta\zeta$	% Loss	Pair of uncoupled variables
0.0486	49	Race, length of employment
0.0137	14	Gender, dust
0.0102	10	Gender, smoker
0.0066	7	Race, dust
0.0060	6	Gender, length of employment
0.0057	6	**Byssinosis**, dust
0.0027	3	Smoking, length of employment
0.0027	3	Dust, length of employment
0.0026	3	Race, gender
0.0009	1	**Byssinosis**, length of employment
0.0008	1	Smoker, dust
0.0006	1	**Byssinosis**, smoker
0.0006	1	Race, smoker
0.0005	1	**Byssinosis**, gender
0.0003	0	**Byssinosis**, race

Table 5 Percentage of byssinosis: within race versus total

Years employed	White	Other	Total
< 10	1.1	3.1	2.3
10 to 19	2.8	8.3	3.7
≥ 20	3.4	9.5	3.8

For the analysis with $\Delta\zeta$, the overwhelming size of *race* and *employment* is just an effect to recognise, but we need not fear its confounding influence on the $\Delta\zeta$ of *byssinosis* and *dust*. If you continue to analyse with accepted procedures like logistic regression, you might use the analysis to guide you to appropriate steps to counter the confounding.

5 Conclusion

This is a short first presentation of a method of analysis of entropy that can provide an alternative to procedures like ANOVA, regression analysis and log-linear modelling. The method promises, first of all, relief from the perennial problem of confounding, as it is free from the practice of estimating effects from marginal sums.

It needs no distinction between methods for binary and multi-category variables; there is also no need to correct by dividing by degrees of freedom for multi-category variables.

It contains only straightforward calculation and so needs no lengthy iterations to arrive at a solution.

It is robust against sparse tables; there is no need to add a bothersome and slightly falsifying "correction" like adding 0.5 to every cell frequency.

In its simplicity and easiness, the analysis can be used by, e.g., an epidemiologist with limited statistical training when confronted with a study involving a larger number of variables. Yes, using traditional methods, he or she might get help from a statistician who can identify confounding variables by running a number of restricted logistic regression models, yet the proposed simple "analysis of entropy" could give this epidemiologist some autarchy.

6 Closing Remarks

The method is clearly oriented to $\Delta\zeta$ as the effect size, not statistical significance. I am not alone in rejecting significance as the criterion for model building.

Significance is important, however, it tells me if I can be confident that an effect can be expected to return in a repetition of my study or if it is more likely to have occurred by chance. Studying the significance of $\Delta\zeta$ needs a standard error for which I, being a psychologist with exhaustive data analysis experience and not a mathematical statistician, am unable to derive a formula. So I rely on bootstrapping to obtain standard errors, and bootstrapping rewards me with the additional freedom to adapt the bootstrap sampling to clustered or other non-srs samples.

The simplicity of uncoupling allows me, using the same procedures, to try an aimed search for confounders, such as looking for variables that both the dependent variable and a further variable depend on, a standard definition of a confounder. In my experience, the basic table of $\Delta\zeta$ has already given me the necessary information.

My program for the analysis (available on request) is written in MS Visual FoxPro. It relies on simple SQL queries, not much more, so it should be translatable into other languages that know SQL (for that purpose, the source code is also available).

7 A Reluctant Dedication

Dear Nishi, I am rather old (about your age) and tired. I tried to publish this, with different accents, in good journals. The results were negative ("nothing new") or simply neutrally negative ("already too many papers"). A less renowned journal published one such paper (Vorkauf, 2016), but it evoked only deafening silence.

So, in contributing this paper to your *Festschrift* after several futile attempts to get an idea published, I am still certain to have an important point to make on the topic of Analysis of Categorical Data, even if it did not convince some editors. I feel confident that you, as an editor, would have helped me find a more palatable way to bring it to paper.

References

Fisher, R.A.: The nature of probability. Centennial Rev. **2**, 261–274 (1958)

Higgins, J.E., Koch, G.G.: Variable selection and generalized chi-square analysis of categorical data applied to a large cross-sectional occupational health survey. Int. Stat. Rev. **45**, 51–62 (1977)

Preuss, L., Vorkauf, H.: The knowledge content of statistical data. Psychometrika **62**(1), 133–161 (1997)

von Kügelgen, J., Gresele, L., Schölkopf, B.: Simpson's paradox in COVID-19 case fatality rates: a mediation analysis of age-related causal effects. IEEE Trans. Artif. Intell. **2**(1), 18–27 (2021)

Vorkauf, H.: Uncoupling multidimensional contingency tables. Math. Stat. **4**(3), 76–80 (2016)

Correcting for Context Effects in Ratings

Michel van de Velden and Ulf Böckenholt

1 Introduction

In surveys, respondents are frequently asked to indicate their opinions and preferences on rating scales. It is well known that the responses on rating scales can be influenced by factors not related to the content of the items. In particular, when introducing the well-known range–frequency theory, Parducci (1963, 1965) and Parducci and Wedell (1986) showed in several experiments how the "rating context" can influence how respondents select the available categories for the ratings. For example, exposing respondents to a "skewed" set of objects to be rated (e.g. a few small and many large objects) affected systematically the ratings for subsequent objects, regardless of their underlying true values. These authors studied these contextual effects utilising controlled distributions of the objects' physical properties and showed that ratings can be seen as a weighted average of a range and a frequency factor. The range factor captures the objects' physical dimension range, and the frequency part captures how often each object is displayed.

Although many studies are available demonstrating the effectiveness of the range–frequency theory in accounting for contextually induced biases in observed ratings, little work is available to de-bias the ratings for these contextual factors. In this chapter, we show that the dual-scaling work on successive categories introduced in Nishisato (1980) and Nishisato and Sheu (1984) provides the foundation for such a de-biasing method. We also consider further extensions by Schoonees et al. (2015) and Takagishi et al. (2019). Specifically, we study the extent to which the individual-specific correction methods introduced in Takagishi et al. (2019) can be used to de-bias observed ratings that have been subject to range and frequency manipulations. We examine the performance of the correction methods using simulation studies that

M. van de Velden (✉)
Econometric Institute, Erasmus University Rotterdam, Rotterdam, The Netherlands
e-mail: vandevelden@ese.eur.nl

U. Böckenholt
Kellogg School of Management, Northwestern University, Evanston, IL, USA
e-mail: u-bockenholt@kellogg.northwestern.edu

E. J. Beh et al. (eds.), *Analysis of Categorical Data from Historical Perspectives*,
Behaviormetrics: Quantitative Approaches to Human Behavior 17,
https://doi.org/10.1007/978-981-99-5329-5_8

mimic previous settings of experimental studies on range–frequency theory and show that it can be used effectively to control for range and frequency effects in ratings.

In the following parts of this paper, we first review briefly the range–frequency theory (Parducci 1963, 1965), and we present the key features of the proposed bias correction for contextual effects. In a simulation study, we investigate the performance of the proposed approach and show that it can reduce effectively rating biases introduced by contextual variations.

2 Range–Frequency Theory

According to range–frequency theory, the rating of an object is affected by the distribution of other objects to be rated. In particular, the observed ratings are a compromise between range and frequency effects: The range effects amounts to distributing the ratings over the range of the objects to be rated, whereas the frequency effect concerns the tendency to evenly distribute ratings over the available rating categories. Category ratings according to the range–frequency model are a weighted average of their range and frequency values. We can formalise these concepts as follows. Let:

$$R_{ij}^{(c)} = \frac{x_{ij} - x_{\min}^{(c)}}{x_{\max}^{(c)} - x_{\min}^{(c)}} \tag{1}$$

denote the range value of individual i for object j in context c. Here x_{ij} denotes individual i's rating of object j, $x_{\min}^{(c)}$ and $x_{\max}^{(c)}$ denote, respectively, the lowest and highest ratings of all objects in context c. Hence, the range value normalises the ratings with respect to the range of the items in a context. On the other hand, the frequency value $F_{ij}^{(c)}$ of individual i for object j in context c is defined as:

$$F_{ij}^{(c)} = \frac{f_{ij}^{(c)} - 1}{N_c - 1}, \tag{2}$$

where $f_{ij}^{(c)}$ is the rank (with 1 indicating the lowest ranked item) of item j in context c and N_c is the number of items in context c. According to range–frequency theory, an individual's rating value is a weighted sum of these range and frequency values. In particular, rating $y_{ij}^{(c)}$, of an individual i, for item j in context c is:

$$y_{ij}^{(c)} = \alpha^c R_{ij}^{(c)} + \left(1 - \alpha^c\right) F_{ij}^{(c)}, \tag{3}$$

where $\alpha \in (0,\ 1)$.

Figure 1 provides ten illustrations of object distributions on a seven point rating scale, that have been considered in range–frequency studies. For example, respondents who are exposed to an object distribution with mostly “large” objects as in Context 5 tend to adjust their ratings to discriminate more among “large” than “small” objects.

3 Contextual Bias Correction

Based on the dual-scaling work on successive categories of Nishisato (1980) and Nishisato and Sheu (1984), Takagishi et al. (2019) proposed a method that corrects for response styles in rating data exhibited by different subgroups of the raters. Response styles are defined as scale usage that is independent of item content. For example, some respondents may tend to use only the extremes of the category rating scales (extreme responding), or, respondents may tend to use predominantly the middle of the scale (midpoint scaling) or only use the lower (nay-saying) or upper (yea-saying) parts of rating scales.

Rather than considering group-specific response tendencies, we implement a simplified version of the scaling method proposed by Takagishi et al. (2019) to obtain and correct for individual-specific response tendencies that could arise due to range–frequency theory. In the following, we refer to this approach as contextual bias correction (CBC).

A crucial step in CBC is the transformation of the observed rating data to successive categories data as described in Nishisato (1980) and Nishisato and Sheu (1984). That is, in addition to the rated items, we add “thresholds” that define the differences between ratings. Thus, for a r point rating scale, $r-1$ thresholds are considered. The thresholds capture the transitions from rating “1 to 2”, “2 to 3” up to “$r-1$ to r”. Next, item ratings and thresholds are jointly ranked, by sorting them from small to large. For ties, the average rank is assigned.

As an example, consider three items, A, B and C that are rated on a 5-point rating scale as 1, 4 and 5, respectively. For a 5-point rating scale, we get 4 thresholds, say $\tau_1, \ldots, \tau_4$. Jointly sorting the three item ratings and $r-1$ thresholds results in the sequence: $A < \tau_1 < \tau_2 < \tau_3 < B < \tau_4 < C$. By assigning zero to the lowest ranked item or threshold, we can code this as:

A	B	C	τ_1	τ_2	τ_3	τ_4
0	4	6	1	2	3	5

In CBC, the normalised threshold values (i.e., for the example above, the last four columns) are approximated using I-spline basis functions. That is, let f_{ij} denote the rank-ordered threshold values divided by the number of items and thresholds for individual i. Then, for $i = 1, \ldots, n$ and $j = 1, \ldots, r-1$:

$$f_{ij} \approx \phi_i^{\text{CBC}}\left(\frac{j}{r}\right),$$

where

$$\phi_i^{\text{CBC}}(x) = \sum_{s=1}^{3} \beta_{is} M_s(x),$$

and

$$\sum_{s=1}^{3} \beta_{is} = 1, \quad \beta_{is} \geq 0 \quad (s = 1, 2, 3).$$

Here, the first three I-spline basis functions are:

$$\begin{aligned} M_1(x) &= \begin{cases} \frac{2t(x-L)-(x^2-L^2)}{(t-L)^2} & (L \leq x < t) \\ 0 & (t \leq x \leq U) \end{cases} \\ M_2(x) &= \begin{cases} \frac{(x-L)^2}{(t-L)(U-L)} & (L \leq x < t) \\ \frac{(t-L)}{(U-L)} + \frac{2U(x-L)-(x^2-t^2)}{(U-t)(U-L)} & (t \leq x \leq U) \end{cases} \\ M_3(x) &= \begin{cases} 0 & (L \leq x < t) \\ \frac{(x-t)^2}{(U-t)^2} & (t \leq x \leq U) \end{cases} \end{aligned} \tag{4}$$

and $x \in [L, U]$, $t = L + 0.5(U - L)$. See, for example, Ramsay (1988) for more details on these I-splines. For convenience, and without loss of generality, we use $L = 0$ and $U = 1$. Nonnegative conditions, $\beta_{is} \geq 0$ $(s = 1, 2, 3)$, are required for ϕ_i to be a monotone-increasing function.

The individual-specific smoothing functions ϕ_i can be seen as estimates of individual-specific and context-related scale usage. Moreover, the functions can be used to estimate threshold values and derive "corrected" rating data. For more details on the model and its estimation, see Takagishi et al. (2019).

In Takagishi et al. (2019), the estimation of the response functions is combined with a cluster analysis to detect groups of individuals exhibiting a similar response behaviour. Here, rather than assuming cluster-specific preferences or response patterns, we assess whether the method can be used to identify and correct for contextual effects. We investigate the performance of this approach using a simulation study.

4 Simulation Study

We conduct several simulation studies to examine if it possible to correct for range–frequency response effects. For this purpose, we generate individual-specific continuous preferences for a set of items. These continuous preferences are discretised to map onto the rating scale by using the individual-specific "threshold" values. We consider different spacings of the threshold values to induce various contextual response effects. The frequency effect is induced by considering different spacings of the thresholds over a fixed interval. For example, increasing the intervals between the thresholds (i.e. the lower threshold are closer to each other than the higher ones) mimics discrimination at the lower end of the preference continuum. In the range–frequency framework, this corresponds to a scenario where individuals are asked to rate a set of items where most items are "small" and only a few items are "large".

For our simulation, we draw individual-specific threshold values from 10 different distributions, each corresponding to a different "context". We also consider an 11'th case where all 10 contexts are equally often present in a data set. In addition, range effects are incorporated by using the range of individual-specific preferences to translate the threshold values to a scale compatible with the underlying preferences. For the 10 context and the 11'th multi-context cases, we map the underlying preferences onto the rating scale. Finally, we apply CBC to the resulting rating data and assess how well CBC is able to recover the underlying preferences.

4.1 Study Design

In our simulation studies, we first generate the underlying true population preferences. We then add individual specific effects by drawing from a normal distribution to obtain the "true" underlying preferences. Finally, these underlying preferences are mapped onto the rating scale by considering context and individual-specific threshold values. Thus, our data-generating process follows these four steps:

1. Generate underlying preference values:
   Generate a p-dimensional mean preference vector $\boldsymbol{\mu}$ by drawing from p independent standard normal distributions.
2. Add individual-specific effects:
   For each individual, add random noise ϵ_i, from a $N(0, \sigma)$ distribution, to all elements of the mean preference vector $\boldsymbol{\mu}$ to obtain individual specific preferences $\mathbf{m}_i = \boldsymbol{\mu} + \epsilon_i \mathbf{1}$, where $\mathbf{1}$ denotes a p-dimensional vector of ones.
3. Generate individual- and context-specific threshold values:
   For each context draw $r - 1$ thresholds from a uniform distribution, where r indicates the largest rating. The intervals that we draw the $r - 1$ thresholds from, and hence their spacing, depend on the contextual specification. In Table 1, the intervals for a 7-point rating scale, for the 10 considered contexts, are given. In this table, the columns represent the threshold positions and the entries the $[r, r + 1]$ pairs. For example, for Context 2, the first threshold is drawn from the interval $[0, 1/6]$, the second threshold from the interval $[1/6, 1/3]$, etc. Note that when intervals are overlapping—as, for example, in the case of Context 1—the threshold values are drawn simultaneously and then ordered. Thus, for Context 10, we draw 6 threshold values from the interval $[1/6, 1/3]$ and then order them to reflect their different positions on the underlying preference continuum.

Table 1 Threshold intervals for considered contexts for a 7-point rating scale. Contexts in rows, threshold intervals in columns

Context	τ_1	τ_2	τ_3	τ_4	τ_5	τ_6
1	$[0, 1]$	$[0, 1]$	$[0, 1]$	$[0, 1]$	$[0, 1]$	$[0, 1]$
2	$[0, \frac{1}{6}]$	$[\frac{1}{6}, \frac{1}{3}]$	$[\frac{1}{3}, \frac{1}{2}]$	$[\frac{1}{2}, \frac{2}{3}]$	$\left[\frac{2}{3}, \frac{5}{6}\right]$	$\left[\frac{5}{6}, 1\right]$
3	$[0, \frac{1}{6}]$	$[\frac{1}{6}, \frac{1}{3}]$	$[\frac{1}{12}, \frac{11}{12}]$	$[\frac{1}{12}, \frac{11}{12}]$	$\left[\frac{2}{3}, \frac{5}{6}\right]$	$\left[\frac{5}{6}, 1\right]$
4	$[\frac{1}{3}, \frac{2}{3}]$	$[\frac{1}{3}, \frac{2}{3}]$	$[\frac{1}{3}, \frac{2}{3}]$	$[\frac{1}{3}, \frac{2}{3}]$	$[\frac{1}{3}, \frac{2}{3}]$	$[\frac{1}{3}, \frac{2}{3}]$
5	$[0, \frac{1}{6}]$	$[\frac{1}{12}, \frac{1}{4}]$	$[\frac{1}{6}, \frac{1}{3}]$	$\left[\frac{1}{4}, \frac{5}{12}\right]$	$[\frac{1}{3}, \frac{1}{2}]$	$\left[\frac{5}{12}, \frac{2}{3}\right]$
6	$[\frac{1}{3}, \frac{7}{12}]$	$[\frac{1}{2}, \frac{2}{3}]$	$[\frac{7}{12}, \frac{3}{4}]$	$\left[\frac{2}{3}, \frac{5}{6}\right]$	$[\frac{3}{4}, \frac{11}{12}]$	$\left[\frac{5}{6}, 1\right]$
7	$[0, \frac{1}{4}]$	$[0, \frac{1}{4}]$	$[0, \frac{1}{4}]$	$[\frac{1}{12}, \frac{1}{4}]$	$[\frac{1}{12}, \frac{1}{2}]$	$[\frac{1}{12}, 1]$
8	$[0, \frac{11}{12}]$	$[\frac{1}{2}, \frac{11}{12}]$	$[\frac{1}{2}, \frac{11}{12}]$	$[\frac{3}{4}, 1]$	$[\frac{3}{4}, 1]$	$[\frac{3}{4}, 1]$
9	$[0, \frac{1}{3}]$	$[0, \frac{1}{3}]$	$[0, \frac{1}{3}]$	$[0, \frac{1}{3}]$	$[0, \frac{1}{3}]$	$[0, \frac{1}{3}]$
10	$[\frac{2}{3}, 1]$	$[\frac{2}{3}, 1]$	$[\frac{2}{3}, 1]$	$[\frac{2}{3}, 1]$	$[\frac{2}{3}, 1]$	$[\frac{2}{3}, 1]$

For the multi-context case, we randomly select one of the 10 contexts for each observation so that, on average, data sets are comprised of an equal number of observations for each of the 10 single-context scenarios.

4. Transform the thresholds to a scale commensurable with the observed preference scale:
   Since the individual-specific thresholds generated in the previous step are between 0 and 1, we multiply them by the range of an individual's preferences ($\mathbf{m_i}$) and subtract the smallest preference in absolute value. We then apply an inflation factor to the resulting thresholds so that the smallest and highest preferences do not automatically receive the smallest and highest ratings.
5. Transform the preferences to discrete ratings:
   Using the thresholds, we map the preferences onto the discrete rating scale. This yields, for each context, a matrix of observed ratings

There are several factors that we control for in our set-up. In particular:

- The number of observations. We consider two cases: $n = 50$ and $n = 200$.
- The number of items. We consider three cases: $p = 10$, $p = 20$ and $p = 50$ items.
- The size of individual-specific effects, controlled through σ in step 2. We consider three cases: $\sigma = 0$, $\sigma = 2/r$ and $\sigma = 1$, corresponding, respectively, to no, medium and high individual effects.

The resulting design leads to $2 \times 3 \times 3 = 12$ matrices of underlying preferences each with 11 (the number of contexts and the multi-context case) corresponding sets of observed ratings. We consider 100 replications for each setting.

4.2 Measures

Our simulation framework allows us to generate rating data sets where the observed ratings are biased due to contextual effects. This makes it possible to study how contexts affect observed ratings, and how this impacts the recovery of the underlying preferences. Moreover, by applying CBC to the simulated data sets, we are able to study the effects of the dual-scaling bias-correction method. In particular, we consider whether and under which circumstances the correction method results in a better recovery of the underlying preferences.

To appraise the results of the correction method, we calculate for each vector of simulated ratings:

- R_o^2: The squared correlation between observed ratings and underlying preferences.
- R_c^2: The squared correlation between CBC corrected ratings and underlying preferences.
- MAD_c: The mean (over the items) of the absolute differences between the CBC corrected ratings and the normalised underlying preferences.
- MAD_o: The mean (over the items) of the absolute differences between the normalised observed ratings and the normalised underlying preferences.

For each of the 100 generated data set, we collect the means of the above measures.

4.3 Results

Note that in Fig. 1, the distributions for contexts 1, 2 and 3 are similar. There are, however, important differences between them. For context 1, all thresholds are drawn from uniform [0,1] distributions. For context 2, the thresholds are selected from equispaced intervals corresponding to the number of thresholds. For context 3, the middle two thresholds are drawn from wider intervals, whereas the lowest and highest thresholds are equivalent to those in context 2. The differences between these three contexts are primarily related to the variance of the threshold distributions. In particular, in context 2, there is very little variation in the threshold distributions and true preferences are mapped onto an equal-spaced rating scale. On average, these three contexts all lead to equispaced threshold distributions, however, as the underlying preferences are generated using normal distributions, they result in bell-shaped frequency distributions of the observed ratings.

To appraise the performance of the CBC analysis, we consider the measures described in Sect. 4.2. In particular, for each combination of the factors that are varied in the simulation study, we calculate, for each generated data set, the mean values for the measures described in Sect. 4.2. Boxplots of the results of the 100 replications can be found in Figs. 2, 3, 4 and 5.

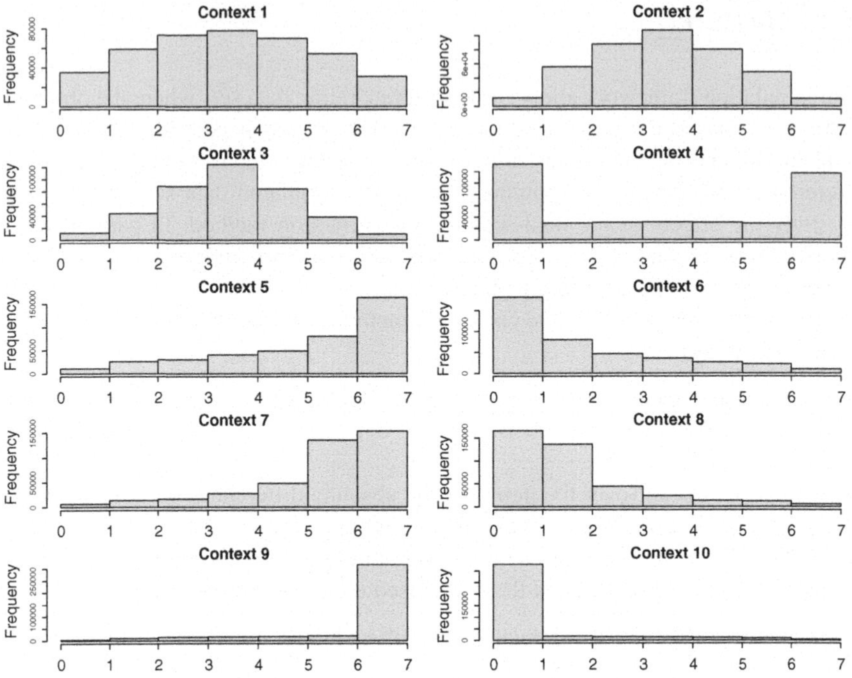

Fig. 1 Histograms of observed ratings per context

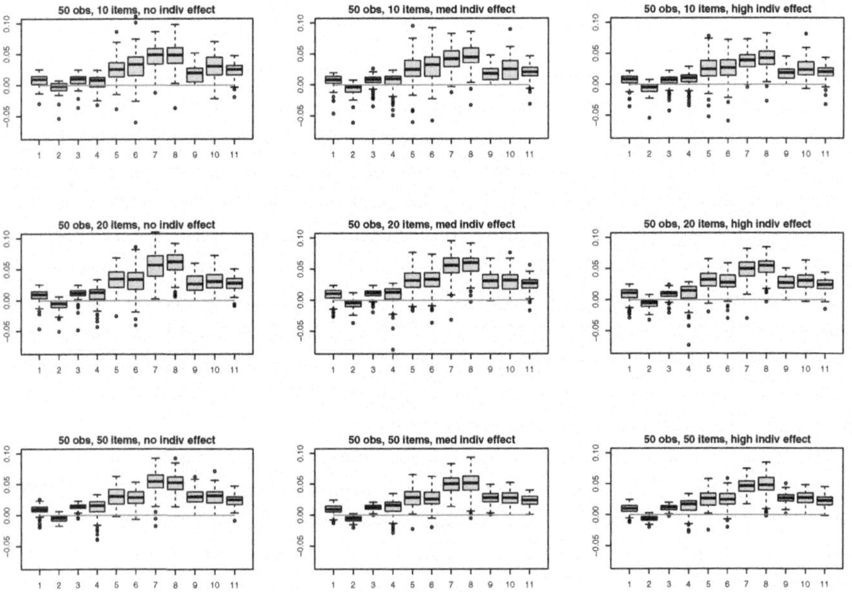

Fig. 2 Mean differences in squared correlations with underlying preferences for $n = 50$: $R_c^2 - R_o^2$ where 'c' and 'o' refer to 'corrected' and 'observed' respectively

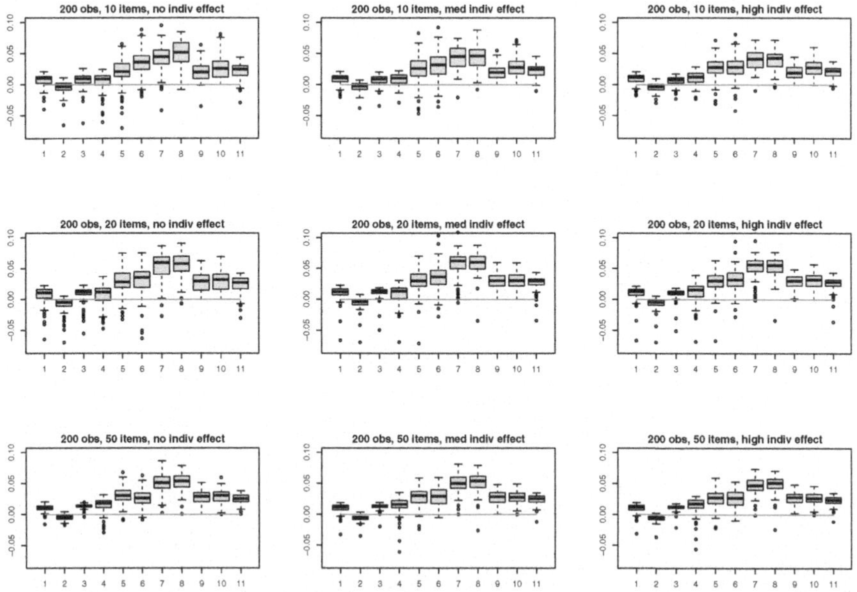

Fig. 3 Mean differences in squared correlations with underlying preferences for $n = 200$: $R_c^2 - R_o^2$ where 'c' and 'o' refer to 'corrected' and 'observed' respectively

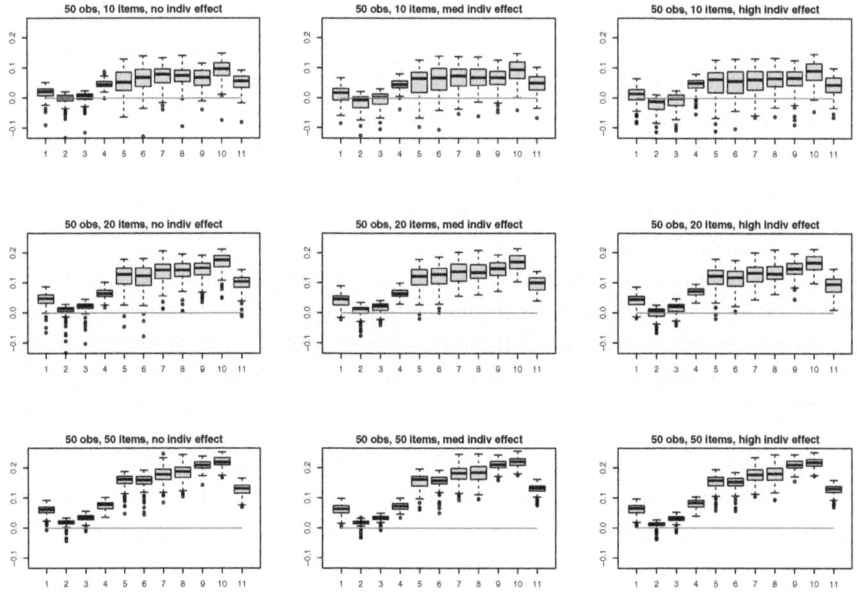

Fig. 4 Mean differences between mean absolute deviations (normalised) observed and corrected ratings for $n = 50$: $\text{MAD}_o - \text{MAD}_c$

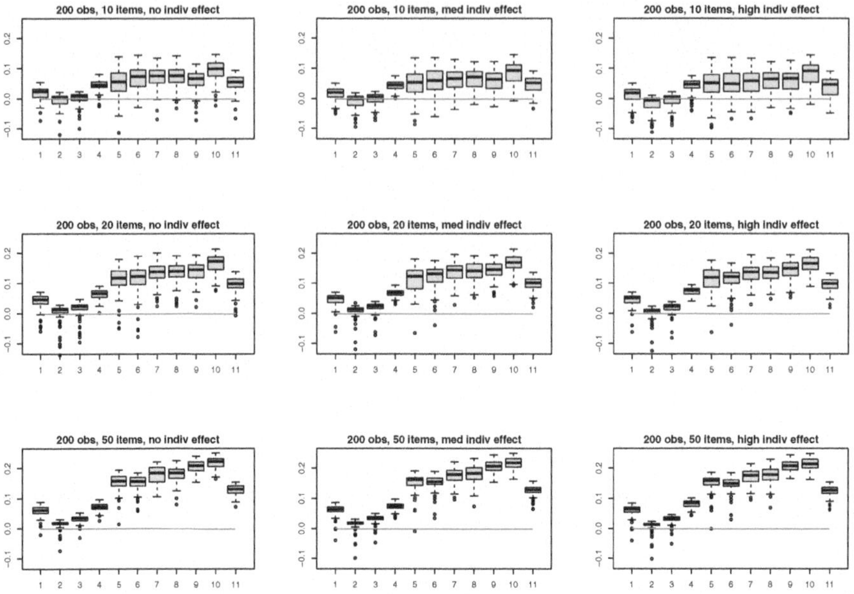

Fig. 5 Mean differences between mean absolute deviations (normalised) observed and corrected ratings for $n = 200$: $\text{MAD}_o - \text{MAD}_c$

The results across conditions paint a rather consistent picture concerning the effectiveness of the CBC corrections. In particular, with the exception of context 2 (equispaced threshold distribution), the correction leads to an improved relationship with the underlying preferences. Furthermore, the performance improves with an increasing number of items and seems to be more effective for the skewed contexts. We also note that the CBC method is beneficial in the multi-context case when all 10 contexts are considered.

For the scenario with 200 observations, 20 items and medium individual effects, Figs. 6 and 7 depict how the mean absolute differences between the observed ratings and the underlying preferences (left panels) and the CBC-corrected ratings and the underlying preferences (right panels) are related to the underlying preferences. We note the strong and systematic differences between the observed ratings and generated preferences. These differences are much reduced and less systematic for the CBC-corrected ratings demonstrating the usefulness of the CBC methodology. However, we stress that the CBC correction does not fully account for all of the contextual effects. This is especially so for scenarios with extreme skewness, where the CBC approach does not fully correct for these contextual effects.

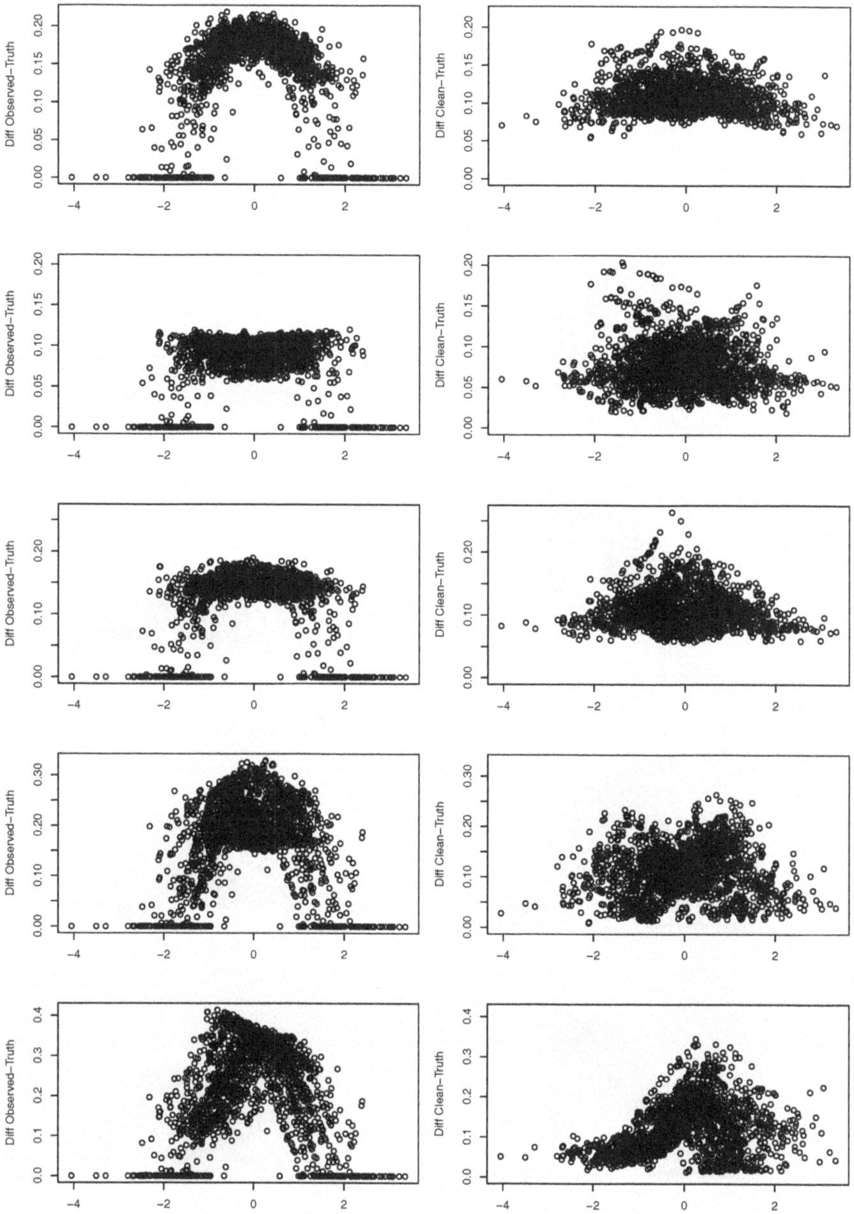

Fig. 6 Observed and corrected ratings against underlying preferences for contexts 1 (top) to 5 (bottom). The plots in the left panels depict the mean absolute differences with normalised observed ratings. The plots in the right panels show the mean absolute differences with the CBC corrected ratings

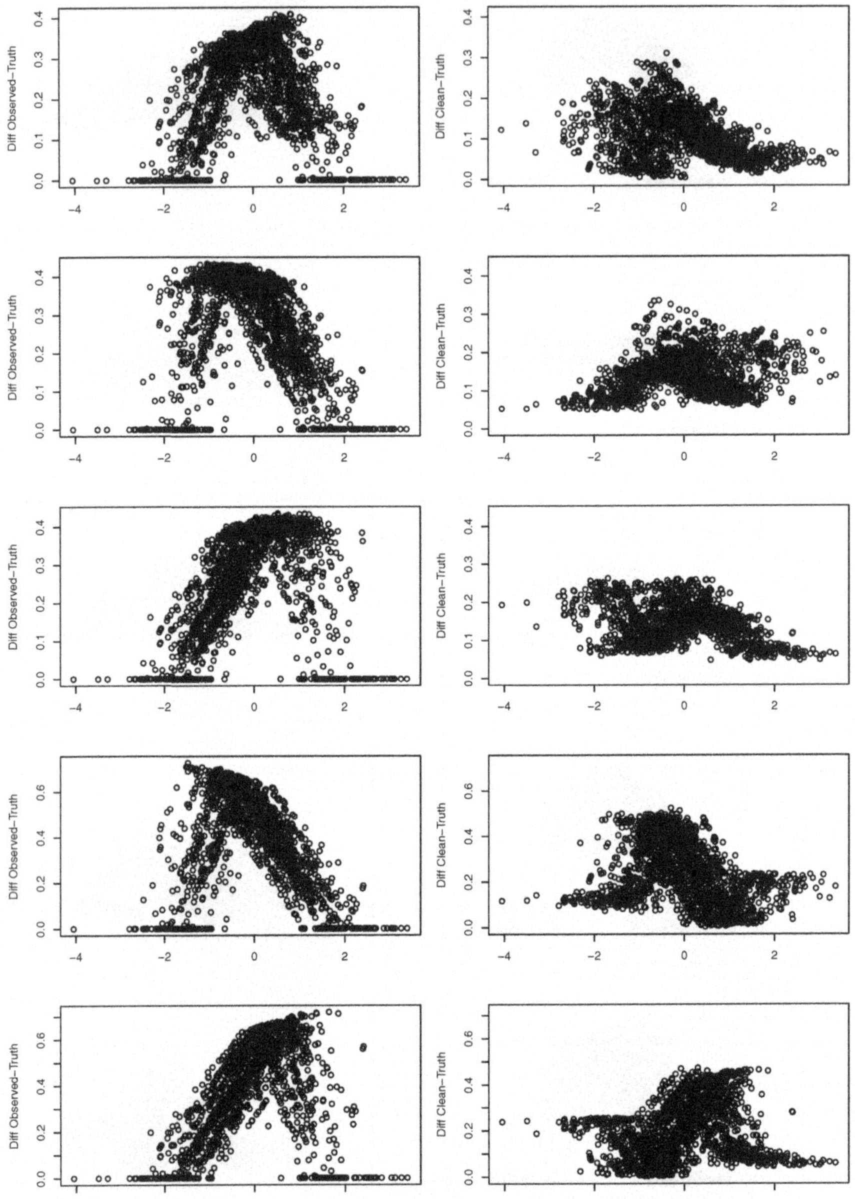

Fig. 7 Observed and corrected ratings against underlying preferences for contexts 6 (top) to 10 (bottom): the plots in the left panels depict the mean absolute differences with normalised observed ratings. The plots in the right panels show the mean absolute differences with the CBC corrected ratings

5 Conclusion

This paper studied whether the impact of contextual effects introduced by the range–frequency mechanism can be overcome by recent extensions of the dual-scaling work on successive categories of Nishisato (1980) and Nishisato and Sheu (1984). Using the approach of Schoonees et al. (2015) and Takagishi et al. (2019), we presented the results of several simulation studies that demonstrated that it is possible to recover the underlying preferences that gave rise to the ratings even when ratings are elicited from respondents who are exposed to different contexts. This is an important finding because it shows that CBC can improve the comparability of ratings from different respondents when rating differences across individuals are caused by contextual variations.

Importantly, we could demonstrate the superiority of the CBC in the case of a single contextual distribution and in the more realistic scenario of multiple contextual distributions. Thus, even when each individual is exposed to different item distributions, the observed ratings can be corrected for these item distribution differences. These corrections can prove to be useful in subsequent analyses of the ratings. Such item statistics as mean item differences or item correlations may be estimated more accurately when rating data are corrected for contextual influences.

References

Nishisato, S.: Dual scaling of successive categories data. Japan. Psychol. Res. **22**(3), 134–143 (1980)
Nishisato, S., Sheu, W.: A note on dual scaling of successive categories data. Psychometrika **49**(4), 493–500 (1984)
Parducci, A.: Range-frequency compromise in judgment. Psychol. Monogr. Gen. Appl. **77**, 1–50 (1963)
Parducci, A.: Category judgment: a range-frequency model. Psychol. Rev. **72**, 407–418 (1965)
Parducci, A., Wedell, D.H.: The category effect with rating scales: number of categories, number of stimuli, and method of presentation. J. Exp. Psychol. Hum. Percept. Perform. **12**, 496–516 (1986)
Ramsay, J.O.: Monotone regression splines in action. Stat. Sci. **3**, 425–441 (1988)
Schoonees, P.C., van de Velden, M., Groenen, P.J.F.: Constrained dual scaling for detecting response styles in categorical data. Psychometrika **80**(4), 968–994 (2015)
Takagishi, M., van de Velden, M., Yadohisa, H.: Clustering preference data in the presence of response-style bias. Br. J. Math. Stat. Psychol. **72**(3), 401–425 (2019)

Old and New Perspectives on Optimal Scaling

Hervé Abdi, Agostino Di Ciaccio, and Gilbert Saporta

1 Introduction

Qualitative variables are ubiquitous in many fields, but genetic and human sciences (especially psychology) have been some of the first disciplines to routinely incorporate qualitative variables in their practice. This importance of qualitative variables prompted the psychologist Stevens (1946) to create the now classic typology of measurement scales. In this typology, qualitative (also called categorical) variables come in two varieties:

- Nominal variables, so called because the modalities—also named levels or categories—of a nominal variable are "names." Formally, a nominal variable corresponds to a partition of a set.
- Ordinal variables (a nominal variable whose modalities are ordered); formally, an ordinal variable corresponds to a pre-order on a set.

Because most multivariate statistical methods are designed for quantitative variables (in Stevens's typology: interval and ratio scales), an obvious problem is to *optimally* transform a qualitative variable into a quantitative variable. This problem being relevant for several disciplines, similar procedures to solve it were independently developed multiple times and therefore come under different names with *scaling*, *quantification*, *coding* and *encoding* being favourites. So, a nominal or ordinal variable is *quantified*, *(en)coded*, or *scaled* when its modalities are replaced by numbers having at least the properties of an interval scale.

H. Abdi
School of Behavioral and Brain Sciences, University of Texas at Dallas, Richardson, TX, USA
e-mail: herve@utdallas.edu

A. Di Ciaccio
Dipartimento Scienze Statistiche, Sapienza Università di Roma, Rome, Italy
e-mail: agostino.diciaccio@uniroma1.it

G. Saporta (✉)
Center for Studies and Research in Computer Science and Communication (Cédric), Conservatoire National des Arts et Métiers, Paris, France
e-mail: gilbert.saporta@cnam.fr

E. J. Beh et al. (eds.), *Analysis of Categorical Data from Historical Perspectives*, Behaviormetrics: Quantitative Approaches to Human Behavior 17,
https://doi.org/10.1007/978-981-99-5329-5_9

Note that the terms *coding* and *encoding* are ambiguous because they can refer either to the transformation of a qualitative variable into a numerical variable (quantification) or to a way of representing a qualitative variable such as, for example, disjunctive coding.

The problem of transforming qualitative variables into quantitative variables has a long history. In statistics, its history goes back to the early contributions of major figures such as Hirschfeld (1935), Horst (1935), who coined the named "reciprocal averaging", Fisher (1940) and Hayashi (1950). In psychology (and of course psychometrics) early contributions of other major figures include Guttman (1941, 1944), Festinger (1947) and even Coombs (1964) in his classic work *a Theory of Data*, see also Coombs (1948). The statisticians were mostly interested in maximising the (squared) correlation between sets of variables; but the psychologists (influenced by factor analytic models) were concerned about *scaling* (i.e. estimating a *quantitative* latent variable or factor from qualitative measurements). The maximisation approach of the statisticians would lead to (*simple*) correspondence analysis, whereas the factorial approach of the psychologists would lead to *multiple* correspondence analysis; see, for details, the historical review of Lebart and Saporta (2014).

This early work matured in the 1970s and early 1980s, which were the years of the search for optimal codes (called factor scores or scaling scores) in supervised or unsupervised contexts, an endeavour where researchers such as de Leeuw (1973), Nishisato (1980), Takane (1980), Tenenhaus (1988) and Young (1976, 1978, 1981), see also Tenenhaus and Young (1985) distinguished themselves. This research was then implemented by commercial software with procedures such as PRINQUAL and TRANSREG for SAS, or CATEGORIES for SPSS.

In the next 30 years or so, after this first foray in the theory of optimal scaling, the topic did not generate much research: routine applications involved computing predictive scores, such as risk scores in banking and insurance. However, recent interest in the scaling problem was reignited by the availability of massive data sets. Nowadays, machine learning researchers and practitioners need to handle categorical data (which are ill-suited for most machine learning algorithms such as neural networks) that often have large numbers of modalities (e.g. from dozens or even hundreds of modalities, such as postal codes; for details, see, for example, Hancock and Khoshgoftaar 2020).

This new interest in qualitative data stimulated the development of several coding methods—mostly developed in the ignorance of the early work of statisticians and psychometricians. As an illustration of this trend, Di Ciaccio (2023) recently reported that the popular Python package `scikit-learn` offers 17 different methods that he categorised into three groups:

- methods where the encoding of a variable does not depend on the other variables, in particular the response (e.g. hash encoding),
- methods where the encoding only depends on the response (e.g. conditional mean), and
- *One-Hot Encoding* (OHE), which is nothing more than the usual disjunctive representation with as many indicators as modalities; see Eq. (2).

The large size of certain categorical data sets raises problems of stability and over-fitting, problems that were neglected in classical statistical applications where the number of modalities was typically small and the learning-testing methodology rarely used. Because of their different view points, the confrontation of the early approach of the statisticians and psychometricians with the newer approach from data scientists could foster a renewal of coding methods for qualitative data; for details, see Meulman et al. (2019).

The rest of the chapter is organised as follows: Sects. 2 and 3 are devoted to notations and to the mathematical structures of quantifications. Section 4 describes early works from 1935 till the 1960s. Section 5 is devoted to the "golden seventies" dominated by optimal scaling (performed with alternating least squares) and Nishisato's dual scaling. Section 6 describes how machine learning has taken over the problem of encoding, with its connection to multivariate statistics and how this can foster a re-interpretation of correspondence analysis from a nonlinear point of view.

2 Matrix Representation of Categorical Encoding and Notations

When dealing with I observations, it is often practical to represent a nominal variable as a binary *group* matrix (called a complete disjunctive coding matrix) denoted by **X** whose rows are observations and whose columns represent the modalities of the nominal variable.[1]

For example, consider a sample with $I = 5$ observations, denoted $\{S_1, \ldots, S_5\}$, and a nominal scale with $J = 3$ modalities: $\{1, 2, 3\}$ that could be, for example, {disagree, neutral, agree}, with the following answers for these five observations:

$$\mathbf{X} = [1,\ 2,\ 3,\ 1,\ 2]^{\mathrm{T}}, \tag{1}$$

then the group matrix would be equal to:

$$\mathbf{X} = \begin{matrix} S_1 \\ S_2 \\ S_3 \\ S_4 \\ S_5 \end{matrix} \begin{bmatrix} 1 & 0 & 0 \\ 0 & 1 & 0 \\ 0 & 0 & 1 \\ 1 & 0 & 0 \\ 0 & 1 & 0 \end{bmatrix} = \left[\mathbb{1}_1,\ \mathbb{1}_2,\ \mathbb{1}_3\right] \tag{2}$$

where, for example, $\mathbb{1}_1 = [1,\ 0,\ 0,\ 1,\ 0]^{\mathrm{T}}$ is the indicator variable for the first category.

[1] As noted above, and developed later on, this is a procedure rediscovered in machine learning under the name of *one hot encoding*.

In this chapter, the following notations are used:

- I is the number of units/observations $\{1, 2, \ldots, i, \ldots, I\}$,
- X is a nominal variable, namely a sequence of I modalities,
- $\mathbf{x}$ is a quantification of X (i.e. a real vector of length I),
- K is the number of nominal variables,
- J is the number of modalities of a variable, $\{1, 2, \ldots, j, \ldots, J\}$,
- J_k is the number of modalities of the kth variable (when $K > 1$),
- $\mathbf{X}$ is the disjunctive matrix (of dimensions $I \times J$) for variable X,
- L is the dimension of a vector space $\{1, 2, \ldots, \ell, \ldots, L\}$,
- $\mathbf{a}_k$ is the single category quantification of variable k (i.e. a real vector of length J_k),
- $\mathbf{A}_k$ is the category quantification array on L dimensions (of dimensions $J_k \times L$),
- $\mathbf{q}_k$ is the vector of a single quantified variable k, (a real vector of Length I),
- $\mathbf{Q}_k$ is the quantified array of variable $\mathbf{X}_k$ (of dimensions $I \times L$) for L dimensions.

3 The Structure of Quantifications

Quantifying or encoding a categorical variable can be written using simple transformations that we explicitly define in the following sections.

3.1 Categorical Encoding

Let X be a nominal variable with J unordered modalities $\{1, \ldots, j, \ldots, J\}$ and $\mathbf{x}$ a quantification of X using at most J distinct values $\{a_1, \ldots, a_j, \ldots, a_J\}$. Then, if $\mathbb{1}_j$ denotes the indicator variable of the jth category, we have:

$$\mathbf{x} = \sum_{j=1}^{J} a_j \mathbb{1}_j \,. \tag{3}$$

Quantifying X boils down to defining a linear combination (with the weights a_j called the code or scale values) of the indicator variables. When there is no constraint on the a_j weights, the set of possible quantifications $\mathbf{x}$ is a vector subspace $\mathcal{W}$ with dimension J.

Because:

$$\sum_{j=1}^{J} \mathbb{1}_j = \mathbf{1}, \tag{4}$$

(with **1** begin a commensurable vector of 1's) the set Δ of constant variables (which is a one-dimensional subspace) is included into $\mathcal{W}$. If **x** is required to have zero mean, then:

$$\mathbf{x} \in \left\{\Delta^{\perp} \cap \mathcal{W}\right\} . \tag{5}$$

Note that the encoding from (3) is redundant because the value of any $\mathbb{1}_j$ variable can be deduced from the values of the other $(J - 1)$ variables. Another possibility could be to use only $J - 1$ indicator variables as done, for example, with the dummy coding scheme used in the general linear model and logistic regression. We will not use this coding scheme here so that all modalities play the same role.

3.2 *Ordinal Encoding*

If there is a natural order between the modalities (i.e. a pre-order on the set of responses), it is natural to require that:

$$a_1 \leq a_2 \leq \cdots \leq a_J .$$

Let us consider the following reparamaterisation:

$$a_1 = b_1,\ a_2 = b_1 + b_2,\ \ldots,\ a_J = b_1 + \cdots + b_j + \cdots + b_J \text{ with } \begin{cases} b_1 \in \mathbb{R} \\ b_2, \ldots, b_J \geq 0; \end{cases} \tag{6}$$

then

$$\begin{aligned} \mathbf{x} &= \sum_{j=1}^{J} a_j \mathbb{1}_j \\ &= b_1 \mathbb{1}_1 + (b_1 + b_2)\, \mathbb{1}_2 + \cdots + (b_1 + b_2, \ldots)\, \mathbb{1}_J \\ &= b_1\, (\mathbb{1}_1 + \mathbb{1}_2 + \cdots + \mathbb{1}_J) + b_2(\mathbb{1}_2 + \cdots + \mathbb{1}_J) + \cdots + b_J \mathbb{1}_J \\ &= b_1 + b_2\, (\mathbb{1}_2 + \cdots + \mathbb{1}_J) + \cdots + b_J \mathbb{1}_J \\ &= b_1 + \sum_{j=2}^{J} b_j \mathbb{z}_j \end{aligned} \tag{7}$$

where

$$\mathbb{z}_j = \sum_{\ell=j}^{J} \mathbb{1}_\ell . \tag{8}$$

The variable **x** is thus a linear combination of $J - 1$ variables with non-negative coefficients, which is the definition of a convex polyhedral cone (see, for example,

Tenenhaus 1988), plus one unconstrained constant term. In other words, $\mathbf{x}$ belongs to the direct sum of Δ and a $(J-1)$ convex polyhedral cone, $\mathcal{C}_{J-1}$, and so:

$$\mathbf{x} \in \{\Delta \oplus \mathcal{C}_{J-1}\} . \tag{9}$$

Note: if we also require that $\mathbf{x}$ has zero mean, the constant b_1 will be negative.

3.3 Two Simple Optimal Scaling Problems

Let Y be a numerical response variable. What is the optimal way to quantify a qualitative variable X in order to best predict Y in the least-squares sense?

If X is categorical, the solution[2] is given by the projection of Y onto the subspace $\mathcal{W}$ spanned by the set of the indicator variables $\mathbb{1}_j$. In other words, the optimal solution is obtained by performing a multiple regression without the intercept of Y onto the set of the $\mathbb{1}_j$. Because the $\mathbb{1}_j$ are orthogonal, the solution is easily found: The optimal scores $\{a_j\}$ are the conditional means for each modality $\bar{y}_j$.

If X is ordinal, the solution is less straightforward because we have to project Y onto a polyhedral cone instead of a vector subspace. However, because the cone is convex (cf. (8)), the solution is unique and boils down to computing a multiple regression:

$$\widehat{Y} = b_1 + \sum_{j=2}^{J} b_j \mathbb{Z}_j, \tag{10}$$

with positivity constraints for the b_j coefficients for $j > 1$; see (7). The solution of this constrained optimisation problem can be found using some efficient numerical methods such as the *pool adjacent violators algorithm*; see, for example, Kruskal (1964), Tenenhaus (1988) and de Leeuw et al. (2009).

3.4 Crisp Coding, Fuzzy Coding, Spline Coding

Transforming a numerical variable into a qualitative variable by splitting it into classes, and then recoding this variable according to the previously mentioned principles, is a low cost way of nonlinearly transforming a numerical variable.

Coding with (3)—called here *crisp-coding*—has the disadvantage of introducing discontinuities that can loose some information from the original variable. To alleviate this problem, various kinds of fuzzy encodings can be used—a procedure equivalent to defining membership functions for neighbouring intervals. Crisp-coding and piecewise-linear encoding (which is a form of fuzzy coding) are particular cases of

[2] Called *target encoding* in machine learning.

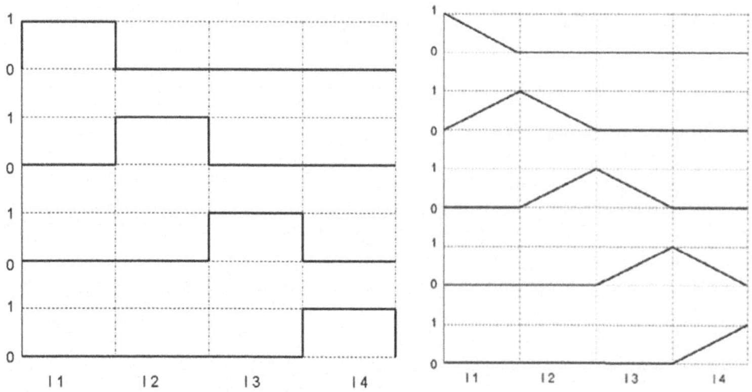

Fig. 1 Basis spline functions of degrees 0 and 1

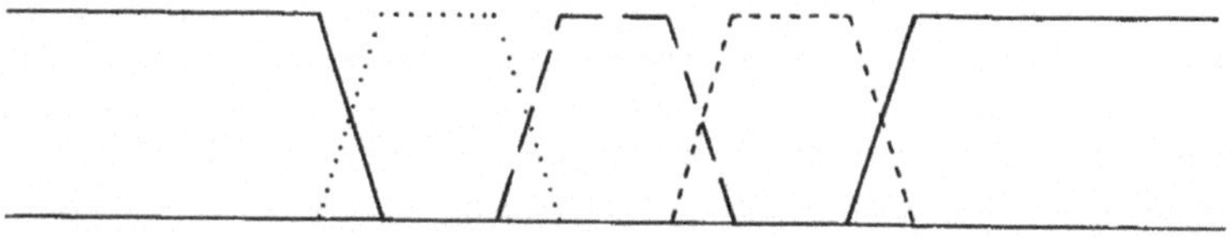

Fig. 2 Trapezoidal encoding (from Gallego 1982)

linear combinations of spline functions as illustrated in Fig. 1 that shows examples of splines of, respectively, degrees 0 and 1 associated to (discontinuous) crisp-coding and piecewise continuous linear transformations.

An additional example of spline function is suggested by Ramsay (1988) who advocates the use of monotonous spline functions. Gallego (1982), who also considers fuzzy coding, uses trapezoidal encodings as illustrated in Fig. 2.

4 Early Works

Quantifying a qualitative variable on its own makes little sense if it is not linked to a goal, such as explaining another variable. Statisticians were concerned very early on with the search for nonarbitrary quantifications by seeking to optimise specific criteria (which were, most of the time, expressed as maximising squared scalar products such as correlations). The early works were naturally concerned with the case of two categorical variables and their associated contingency table.

4.1 The Case of Bivariate Distributions

Hirschfeld (1935, p. 520)—better known under his American identity of Hartley—is apparently the first researcher to ask the following question (and to answer it):

> It is well known that the correlation theory for such a distribution gives much better results if both regressions are linear [...]. Given a discontinuous distribution p_{vq}, is it always possible to introduce [...] new values for the variates x_v, y_q, such that *both* regressions are linear?

Later on (and without reference to Hirschfeld), as summarised by Lancaster (1957, pp. 289–290):

> In 1940, Fisher considered contingency tables from the point of view of discriminant analysis. Suppose that 'scores,' i.e. arbitrary variate values, are assigned to the rows and also to the columns of a contingency table: what are the best scores to assign to the rows so that a linear function of them will best differentiate the classes determined by the columns, and vice versa. This turns out to be a problem in maximising the correlation between the scores and the required correlations are those known as 'canonical' in the sense of Hotelling (1936).

Lancaster was referring to the algorithm described by Fisher (1940, p. 426), and now considered as an early example of alternating least squares or dual scaling, applied to the (now) famous table cross-tabulating the eye and hair colours of Scottish schoolchildren (from the county of Caithness):

> ...starting with arbitrarily chosen scores for eye colour, determining from these average scores for hair colour, and using these latter to find new scores for eye colour.

This "optimal coding" algorithm converges to the solution given by the coordinates of the rows and columns along the first axis of the correspondence analysis of the contingency table.

Maung (1941, p. 200)—who was interested in the higher order encodings corresponding to the successive pairs of canonical variables—attributes to Fisher a formula giving the value of each cell in the contingency table from the margins, the canonical correlations and the successive codings. This formula—also called the RC canonical correlation model—is none other than the well-known reconstitution formula of correspondence analysis.

Williams (1952) is also a notable reference about the development of significance tests for canonical correlations.

Further details on the relationship between optimal scaling and correspondence analysis are given in Saporta (1975), Nishisato (2006, Chap. 3), Lebart and Saporta (2014) and many others, including Hill (1974), and Beh and Lombardo (2014).

4.2 Lancaster's Theorem

The search for optimal scores is unexpectedly related to the problem of transforming a given probability distribution into a normal distribution. Lancaster (1957) showed that the (squared) correlation coefficient between the two components of a bivariate normal vector cannot be increased regardless of the (nonlinear) transformations that can be applied to them.

This result inspired the following comments to Kendall and Stuart (1961, pp. 568–569):

> We may ask: What scores should be allotted to the categories in order to maximise the correlation coefficient between the two variables? Surprisingly enough, it emerges that these 'optimum' scores are closely connected with the transformation of the frequencies in the table to bivariate normal frequencies [...] And the theoretical implication of the [Lancaster's] result is clear: if we seek separate scoring systems for the two categorised variables such as to maximise their correlation, we are basically trying to produce a bivariate normal distribution by operations upon the margins of the table.

4.3 Quantifying More Than Two Attributes: Guttman, Hayashi

Guttman (1941), in a famous paper, referred to the method of reciprocal averaging (as described by Horst 1935) and proposed to simultaneously quantify K categorical variables in such a way that they are as similar as possible and that their means are as dispersed as possible. The rationale behind this criterion was that such an approach would be optimal when the K variables, collected in a multiple choice questionnaire, measured more or less the same construct (as in a factor analysis model with only one latent variable). When the total variance is fixed, this amounts to maximising the measure of internal consistency as described below.

Let $\mathbf{X} = [\mathbf{X}_1 | \dots |\mathbf{X}_k| \dots |\mathbf{X}_K]$ be the super-matrix of all K disjunctive matrices, $\mathbf{a}_k$ the category quantification vector of variable X_k, $\mathbf{a}$ the super-vector concatenating all category quantifications, $\mathbf{z}_k = \mathbf{X}_k\mathbf{a}_k$ the corresponding vector of object scores and:

$$\bar{\mathbf{z}} = \frac{1}{K}\sum_{k=1}^{K}\mathbf{z}_k = \frac{1}{K}\mathbf{X}\mathbf{a}, \tag{11}$$

the vector of average object scores.

Guttman (1941) showed that the scores, which maximise the variance of $\bar{\mathbf{z}}$ under a scaling constraint for $\mathbf{a}$, are given by the coordinates of the modalities of the K variables along the first axis of what will later be called multiple correspondence analysis (MCA). On this occasion, Guttman coined the term "*chi-square* metric" now routinely associated with correspondence analysis.

Independently, Hayashi (1950) developed an approach similar to Guttman's under the name of Type III quantification. Three other types of quantification using (or not) an external response variable were also developed by Hayashi. Tanaka (1979), and Takeuchi et al. (1982, Chap. 8) are useful references for the Japanese contributions. A bit later Slater (1960) proposed a method to analyse personal preference data that represents these data in a multi-dimensional space where observations and stimuli can be represented simultaneously and, as noted by Nishisato (1978, p. 263), his approach was "essentially the same as Guttman's, but the close relationship between them was apparently left unnoticed."

5 The Golden Seventies

The 1970s were a particularly fertile period for the development of optimal scaling and the journal *Psychometrika* was the privileged venue for publishing on this topic with no less than 145 articles appearing between 1968 and 1982 using the keywords "Optimal Scaling" (799 using the same keywords without dates and 199 using only the keywords "Dual Scaling"). It is therefore impossible to be exhaustive.

5.1 The Alternating Least Squares (ALS) Approach for Optimal Scaling

In his 1981 Presidential Address to the Psychometric Society's Spring Meeting, Young (1981) returned at length to his work carried out in collaboration with, on one hand de Leeuw and Takane and with, on the other hand, Tenenhaus. He reflected that these collaborations constituted an important new stream because:

> Optimal scaling is a data analysis technique which assigns numerical values to observation categories in a way which maximises the relation between the observations and the data analysis model while respecting the measurement character of the data (Young 1981, p. 358).

A large number of algorithms were then developed using the alternating least squares (ALS) approach, which consists in separating the parameters of the problem into two sets:

1. the *model* parameters, and
2. the *data* parameters (the codings).

The optimisation then proceeds by obtaining the least squares estimates of the model parameters while assuming that the data parameters are constant. One then switches to the other set: obtaining the least squares estimates of the data parameters given the model parameters and so on until convergence. Even though convergence to a local optimum is guaranteed, convergence to a global optimum is *not* guaranteed because convergence depends upon the initial values (i.e. there are multiple local optima where the search could converge). Note that the ALS approach can also be applied to regression or predictive type problems which are now called *supervised* approaches, whereas the pioneers were not particularly interested in these methods.

MORALS-type algorithms (Young et al. 1976) make it possible to carry out multiple regressions by transforming both a response Y and the predictors $X_1, \ldots, X_k$,

..., X_K with monotonic or nonmonotonic optimal transformations according to the nature of the variables by using successions of projections on vector subspaces or cones. Denoting by ψ and φ_1, ..., φ_K the transformations of the original variables, the optimisation problem is the following:

$$\max_{\psi,\varphi_1,\varphi_2,\ldots,\varphi_K} R^2 \left[\psi(Y);\, \varphi_1(X_1),\, \varphi_2(X_2),\, \ldots,\, \varphi_K(X_K)\right] . \tag{12}$$

Transformed variables are usually constrained to be standardised in order to avoid degeneracy.

The PRINQUAL (Bouroche et al. 1977) and PRINCALS (Young et al. 1978) algorithms implement a principal component analysis of K coded qualitative variables while respecting the nominal or ordinal nature of these variables. However, the optimality criterion is not as obvious as is the maximisation of the (squared) multiple correlation in multiple regression, because this is an unsupervised problem. The most commonly used criterion maximises the percentage of variance explained by the first L principal components $C_1, \ldots, C_L$; the default value is $L = 2$ in the PRINQUAL procedure of SAS because two-dimensional displays are the ones most frequently used. Formally, the maximisation problem can be expressed as the solution of:

$$\max_{\substack{\varphi_1,\varphi_2,\ldots,\varphi_k \\ C_1,\ldots,C_L}} \sum_{k=1}^{K} \sum_{\ell=1}^{L} r^2 \left(\varphi_k(X_k),\, C_\ell\right) . \tag{13}$$

Note that if $L = 1$, the solution for K nominal variables is identical to the solution provided by the first dimension of multiple correspondence analysis, (i.e. this is the solution of the problem from Guttman 1941). However, there is a fundamental difference between the algorithms of the PRINQUAL-type—which look for unique codings of the categorical variables—and the algorithms of the MCA and HOMALS types—which look for as many codings as the number of dimensions of the data; for more, see Tenenhaus and Young (1985) Gifi (1990).

In the late 1980s, van Buuren and Heiser (1989) developed GROUPALS, a method for optimising simultaneously a clustering of units and quantifications of categorical variables, which was taken up almost 30 years later by van de Velden et al. (2017) for their development of *cluster* correspondence analysis.

5.2 *Dual Scaling: Nishisato's Synthesis*

In the 1970s Nishisato (originally a psychologist, later turned into a psychometrician) revisits the problem of the quantification of qualitative variables (both nominal or ordinal) and integrates the two quantification traditions (i.e. statistics and psychometrics). Faced with so many names for equivalent methods, Nishisato preferred the appellation of *dual scaling*. In his early book, Nishisato (1980) presents an early synthesis of these two branches in the first chapter dedicated to the history of the "scaling" problem for qualitative variables—a review that remains one of the best sources for its origins and early efforts but that also often suggests future developments. Nishisato anchors dual scaling in the early psychometric approach of Horst (1935) and Guttman (1941), but also integrates Maung's (1941) and Fisher's contributions; that is, his "additive scoring" (Fisher 1940). Nishisato describes dual scaling as a maximisation problem as previously defined by Bock (1960) as an approach that:

> assign[s] numerical values to alternatives, or categories, so as to discriminate optimally among the objects (Bock 1960, p. 1).

From this definition, Nishisato generalised and adapted the dual scaling methodology to a wider set of data types whose extension can only be compared to the, then, contemporary, French developments. For the specific problem of quantifying a set of nominal variables, Nishisato uses the super matrix approach described in (11) and derives from there the equations and properties of multiple correspondence analysis.

5.3 A Success Story: Credit Scoring

Credit scoring techniques are used to check if a loan applicant is worthy of credit. Using historical data on whether or not debtors have correctly repaid their instalments, the problem reduces for numerical predictors to an application of a supervised classification method such as discriminant analysis or logistic regression.

However, for individual applicants, most of the predictors are categorical variables such as gender, marital and employment status. Scoring methods assign a score to each modality of a variable so that the addition of these partial scores best separates the two groups. Because the quantification of each predictor is equivalent to defining a linear combination of the indicators of its modalities, the optimal solution is obtained from a discriminant analysis using the columns of the associated disjunctive table as predictors:

$$\mathbf{X} = [\mathbf{X}_1 | \dots | \mathbf{X}_k | \dots | \mathbf{X}_K] \ . \tag{14}$$

Because $\mathbf{X}$ is not of full rank, Bouroche et al. (1977) proposed to replace it by the P best components $\mathbf{z}_p$ of the multiple correspondence analysis of $\mathbf{X}$. Here "best components" means the components that best predict the target, instead of the ones with the largest eigenvalues. Fisher's linear discriminant function is then computed as and redecomposed as a linear combination of all indicator variables which gives the optimal scores—a procedure similar to "principal component regression" for qualitative instead of quantitative variables. The previous method known as DISQUAL (see Niang and Saporta 2006, for a detailed illustration of DISQUAL) as well as logistic regression (which eliminates an indicator in each $\mathbf{X}_k$) are routinely used by banks, insurance companies and so on: Optimal coding has become transparent!

The interest of scores compared to black box approaches is to lead to easily interpretable decision rules—a feature now socially required.

6 Machine Learning and Variable Encoding

In the machine learning terminology, the modality quantification (or encoding) can be obtained by "embedding" the modalities in a low-dimensional space. For neural networks, a well-known embedding is called word-embedding; see, for example, Bengio et al. (2003). Embedding in Natural Language Processing (NLP, which is the set of techniques that use machine learning to analyse textual data) is a vector representation of the words in such a way that words which frequently appear in similar contexts are close to each other. It is possible to use the same approach for representing modalities in a vector space, in order to use models that require numerical data. Using neural networks, interesting connections appear with the optimal scaling methods described in the previous paragraphs. One of the advantages of the approach showed here is the ability to analyse categorical variables with hundreds of modalities, as long as the number of observations is adequate.

It is convenient to distinguish the supervised case, in which we need to predict a quantitative target Y, from the unsupervised case, in which we do not have a target variable. In the supervised case, quantification is only a tool for applying the model to qualitative data and generally has no interest in itself: The best quantification is the one that best predicts the target. By contrast, in the unsupervised case, the interest is precisely in the quantifications of the modalities: here the embedding of the modalities, and eventually of the units, should best represent the information present in the data.

6.1 *Traditional Encoding Methods*

In addition to the approaches described in the previous paragraphs, other methods have been proposed to encode categorical variables; for details, see the review by Hancock and Khoshgoftaar (2020). These are simple and popular methods because they can be used for qualitative data with both classical models and machine learning algorithms. These methods either:

1. only use the target,
2. consider the target and other variables, or
3. do not consider any other data than the variable to be quantified.

In the latter case (i.e. ignoring the data), a criterion is chosen that does not use other data and the result is usually a single numeric variable. This way, there is no risk of over-fitting, but the encodings obtained cannot be unambiguously interpreted. Such methods include: The *label encoder*—which assigns a different integer to each modality—and the *ordinal encoder*—which constrains the assignments to respect the natural modality order. The *hash encoder* uses a hash function to embed the J modalities of a variable into a small number of dimensions, but multiple values can be represented by the same hash value—an effect known as a *collision*. Because this encoder is extremely efficient, it is sometimes used with big data sets when the number of modalities of some variables is very high. But, in these cases, it is not possible to perform a reverse lookup to determine what the input was and so the quantifications provided by collision could be meaningless.

There are many methods that use the target to obtain a numerical coding of the modalities in such a way that the availability of other explanatory variables does not influence the coding. The result of such a procedure can be either:

1. a single numeric variable for regression tasks (whose dimensionality would be the same as the dimensionality of the original data), or
2. multiple numerical variables that can then be used for classification.

Applying target-based encoding often produces *data leakage*—a problem leading to over-fitting and poor predictive performance. To correctly work, this method needs large amounts of data, a small number of categorical variables and the same target distribution in training and test data sets. To overcome data leakage, it has been suggested to add noise, or to use cross-validation techniques, or other forms of regularisation. The *simple target encoder*—a popular method for regression tasks—belongs to this group. This method assigns the conditional mean target value to each modality of the explanatory variable.

For classification tasks, where the target is also categorical, the explanatory categorical variable is encoded with J new variables (where J is the number of classes of the target). These variables contain the relative conditional frequencies of each class given the modality of the categorical variable.

Other methods in this approach are based on the *contrast* between some modalities and other modalities of the variable; these methods are called *contrast encoders* (an approach often used in the general linear model framework for testing specific predictions). For example, the *Helmert encoder* requires a quantitative target and ordered levels of the categorical variable; this encoder generates a set of contrasts where each modality is compared in turn to all the subsequent ones. This method is also routinely used in multiple regression and analysis of variance.

A favourite method to analyse qualitative variables is the, previously mentioned, *one hot encoding* which assigns one indicator matrix to each variable. Note that OHE differs from *dummy coding* that excludes one modality of the variable (to avoid multicollinearity). But, when applying machine learning models it is necessary to include all the modalities, otherwise the omitted modality disappears—a standard problem (called "the dummy variable trap") in multiple regression when using dummy coding; see, for example, Darlington and Hayes (2017).

In fact, one hot encoding is not a real quantification method, but just a binary transformation of the original data. Using OHE makes it possible to take into account the other explanatory variables because the quantifications are obtained as parameters of a model. The main drawback of OHE follows from the tendency of indicator variables to cause over-fitting. Moreover, if a variable has many modalities, OHE generates a large number of new features and a sparse array in which the new indicator variables are perfectly independent—an unrealistic assumption. OHE is used in the optimal scaling approach (see MORALS in Sect. 5) but is also widely used in machine learning.

6.2 Nonlinear Encoding in the Supervised Case

In the supervised case, modality quantifications are generally just a tool for applying a predictive model. The best quantification will therefore be the one that gives the best predictions for the model used.

As shown in Sect. 5.1, MORALS makes it possible to perform a multiple regression considering optimal transformations of the variables. Let $\mathbf{X}_k$ (of dimensions $I \times J_k$) be the indicator matrix of variable k, and Y a numerical response variable. If we have K categorical explanatory variables, MORALS defines the *residual sum of squares* (RSS) as:

$$\text{RSS} = \left\| \mathbf{y} - \sum_{k=1}^{K} \beta_k \mathbf{X}_k \mathbf{a}_k \right\|^2 = \left\| \mathbf{y} - \sum_{k=1}^{K} \beta_k \mathbf{q}_k \right\|^2, \tag{15}$$

where $\mathbf{q}_k = \mathbf{X}_k \mathbf{a}_k$ is the vector of the quantified kth variable, $\mathbf{a}_k$ is the vector with the (single) quantification of the modalities of the kth variable, with the centering and normalisation constraints:

$$\mathbf{1}^{\mathrm{T}} \mathbf{q}_k = 0, \quad \frac{1}{I} \mathbf{q}_k^{\mathrm{T}} \mathbf{q}_k = 1, \quad k = 1,\, 2,\, \ldots,\, K\,. \tag{16}$$

The algorithm then defines the following optimisation problem, solved by an alternating least squares algorithm:

$$\min_{\substack{\mathbf{a}_1, \mathbf{a}_2, \ldots, \mathbf{a}_K \\ \beta_1, \beta_1, \ldots, \beta_K}} \left\| \mathbf{y} - \sum_{k=1}^{K} \beta_k \mathbf{X}_k \mathbf{a}_k \right\|^2. \tag{17}$$

With only explanatory nominal variables—unless a different normalisation of the parameters is used—MORALS essentially corresponds to a linear regression with OHS. This approach is likely to over-fit data sets with few observations or when variables have many modalities. It is also possible to obtain multiple quantifications by creating copies of the variables; see, for example, Gifi (1990). However, this approach would increase the number of free parameters and having more parameters to fit the data would worsen the over-fitting problems of MORALS.

In machine learning, and specifically for neural networks, OHE encoding is often used to analyse categorical variables. All the dummies of all the variables, put together, constitute the input of the network. However, this method is not an optimal choice because it greatly increases the size of the dataset by adding orthogonal binary variables.

A different and more adequate strategy (proposed by Di Ciaccio 2020) is described below. Let L be the chosen dimensionality of the embedding space. To explicitly introduce the quantification of modalities in a neural network, it is possible to define an architecture which provides a distinct input for each categorical variable. Each input will be of the OHS type and will be followed by a "dense layer" (the classical

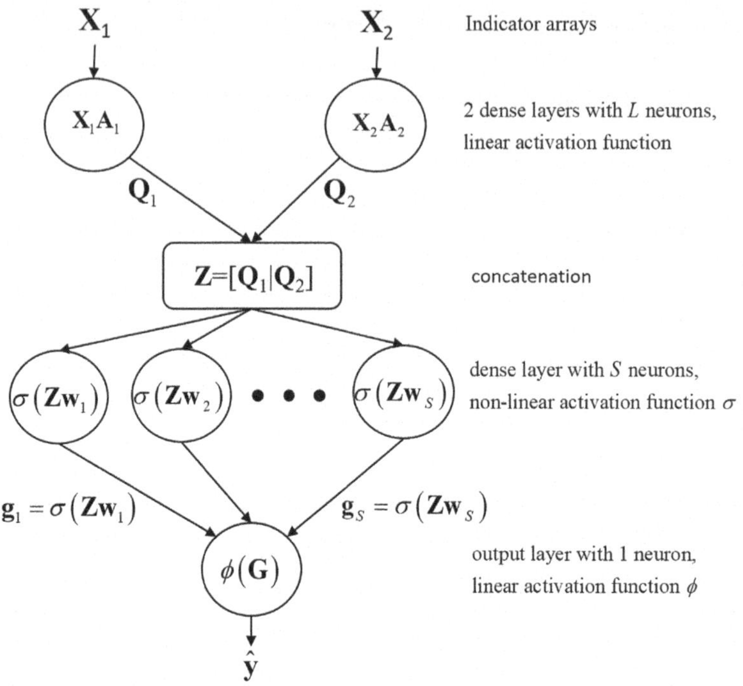

Fig. 3 Supervised neural network for two nominal explicative variables

fully connected layer) with L neurons without bias and with a linear activation function. Layers and activation functions are the basic elements of a neural network; for definition of these terms see, for example, Abdi et al. (1999), or Bengio et al. (2003). The output of this step is an array $\mathbf{Q}_k$ (of dimensions $I \times L$) for each variable, which gives the L-dimensional quantification of $\mathbf{X}_k$, while the modality quantifications are given by $\mathbf{A}_k$. In the next layer, the outputs, coming from all the variables, must be concatenated. At this point, we can add the classical layers of a neural network, for example, one dense layer with S neurons and activation function σ (usually nonlinear, chosen by the researcher), and one output dense layer with only one neuron and a linear activation function φ (if Y is quantitative). The final network architecture is shown in Fig. 3. The corresponding neural network can be defined as:

$$\widehat{\mathbf{y}} = \beta_0 + \sum_{s=1}^{S} \beta_s \sigma \left(\sum_{k=1}^{K} \sum_{\ell=1}^{L} \mathbf{X}_k \mathbf{a}_{k\ell} w_{k\ell s} + w_{0s} \right). \tag{18}$$

Conversely, in the classical OHE encoding:

$$\widehat{\mathbf{y}} = \beta_0 + \sum_{s=1}^{S} \beta_s \sigma \left(\sum_{k=1}^{K} \mathbf{X}_k \mathbf{w}_{ks} + w_{0s} \right). \tag{19}$$

The function σ is the activation function of the dense layer with S neurons and is usually nonlinear. The embedding dimension is given by L, while S is the number of neurons which determines the adaptive capacity of the network. In (18), $\mathbf{X}_k\mathbf{a}_{k\ell}$ is equal to $\mathbf{q}_{k\ell}$, which is the ℓth column of $\mathbf{Q}_k$.

A relevant difference between the two expressions is the different number of parameters. If the qualitative variables have more than two modalities and if $L = 2$, there are fewer parameters in (18). Even if the variables have many modalities (for example, 100 or 200), the embedding of (18) makes it possible to perform the analysis without difficulty because it involves a smaller number of parameters. Di Ciaccio (2023) showed how this approach—compared to OHS or *target encoding*—leads, with neural networks, to much better predictions. Other works that consider a comparison between different techniques in the supervised approach are, for example, Di Ciaccio (2020), and Potdar et al. (2017).

6.3 Nonlinear Encoding in the Unsupervised Case

In the unsupervised case, the quantifications can be the true goal of the analysis and must therefore highlight the information present in the data. The modalities can be represented in a vector space obtaining multiple quantifications, as in the case for HOMALS and MCA.

With HOMALS or MCA, the modalities are "optimally" encoded by using the eigenvectors with the largest eigenvalues of the cross-product matrix. In MCA, the problem is solved analytically, while in HOMALS, the problem is solved numerically. This numerical variant offers great flexibility in machine learning. The MCA/HOMALS approaches are linear methods that give a map where both units and variables are represented in a low L-dimensional Euclidean space in such a way that an observed unit is relatively close to the modalities that characterise it and away from the modalities that do not. In this representation, the modality embeddings are the centres of gravity of the units that share the same modality.

Let $\mathbf{Z}$ (of dimensions $I \times L$) be the score matrix (the observations coordinates on the vector space), $\mathbf{X}_k$ (of dimensions $I \times J_k$) the indicator matrix of variable k, $\mathbf{A}_k$ (of dimensions $J_k \times L$) the multiple quantification of the modalities, and $\mathbf{U}_k$ the unitary matrix (of dimensions $L \times L$). The HOMALS loss finds the object scores $\mathbf{Z}$ and the quantifications $\mathbf{A}_k$ so that:

$$\min_{\substack{\mathbf{A}_1,\mathbf{A}_2,\ldots,\mathbf{A}_K \\ \mathbf{Z}}} \text{LOSS} = \sum_{k=1}^{K} \| \mathbf{Z} - \mathbf{X}_k\mathbf{A}_k \|^2, \tag{20}$$

with the centring and normalisation constraints $\mathbf{u}^{\mathrm{T}}\mathbf{Z} = \mathbf{0}$, $\mathbf{Z}^{\mathrm{T}}\mathbf{Z} = I\mathbf{U}$, to avoid the trivial solutions: $\mathbf{Z} = \mathbf{0}$, $\mathbf{A}_k = \mathbf{0}$. The LOSS function in (20) can be written as:

$$\text{LOSS} = \sum_{k=1}^{K} \| \mathbf{Z} - \mathbf{X}_k \mathbf{A}_K \|^2 = \sum_{k=1}^{K} \| \mathbf{X}_k - \mathbf{Z}\mathbf{A}_k^+ \|^2$$
$$= \sum_{k=1}^{K} \| \mathbf{X}_k - \widehat{\mathbf{X}}_k \|^2 = \sum_{i=1}^{I} \sum_{k=1}^{K} \sum_{j=1}^{J_k} \left(x_{ikj} - \widehat{x}_{ikj} \right)^2, \tag{21}$$

where $\widehat{\mathbf{X}}_k$ is the best "reconstruction" of $\mathbf{X}_k$ and $\mathbf{A}_k^+$ the Moore–Penrose inverse of $\mathbf{A}_K$. Considering that, to minimise this loss, $\mathbf{Z}$ has to be the mean of the K matrices $\mathbf{X}_k \mathbf{A}_K$, the modality quantifications $\mathbf{A}_K$ are the only parameters to estimate.

The previous expression suggests an alternative formulation as an auto-encoder neural network. An auto-encoder (also called an auto-associator) associates a pattern to itself, often as a way of de-noising a signal; an auto-encoder can also be seen as a nonlinear version of principal component analysis; for more, see Bengio et al. (2003). Within our framework, an auto-encoder is a particular neural network able to minimise the LOSS:

$$\min_{\sigma,\varphi} L\left(\mathbf{X}, \sigma\left(\varphi\left(\mathbf{X}\right)\right)\right), \tag{22}$$

where φ and σ introduce some constraints in the reconstruction of $\mathbf{X}$ and the LOSS penalises the difference between $\mathbf{X}$ and $\widehat{\mathbf{X}}$. Using the residual sum of squares (RSS), (22) becomes:

$$\min_{\sigma,\varphi} \| \mathbf{X} - \sigma(\varphi(\mathbf{X})) \|^2, \tag{23}$$

where φ maps the indicator array $\mathbf{X}$ to an L-dimensional latent space (the bottleneck), σ maps this representation to the output, which is the same as the input. Considering only linear φ, σ, and a low embedding of dimension L, the architecture of the corresponding auto-encoder for only two nominal variables is shown in Fig. 4. This neural network includes only dense layers (also called standard or fully connected layers).

The first layer is composed by two dense sublayers with L neurons for each variable and linear activation function. The output layer has two dense sublayers with as many neurons as the number of modalities of the corresponding variable and a linear activation function. The auto-encoder produces the modality quantification $\mathbf{A}_1$ and $\mathbf{A}_2$ on L dimensions (usually $L = 2$ or 3). The score matrix, $\mathbf{Z}$, is the mean of the quantified variables $\mathbf{Q}_1$ and $\mathbf{Q}_2$ on L dimensions. To obtain the same results as HOMALS, the score matrix, $\mathbf{Z}$, needs to be orthonormalised and column centred. Of course, actually performing all these computations would not make sense, because with much less effort we can use the elegant analytical solution provided by MCA or the alternating least squares algorithm of HOMALS.

The neural network architecture shown in Fig. 4 highlights two constraints:

1. the weights of the output layer are the inverse weights of the first layer, and
2. for all layers, the activation function is linear.

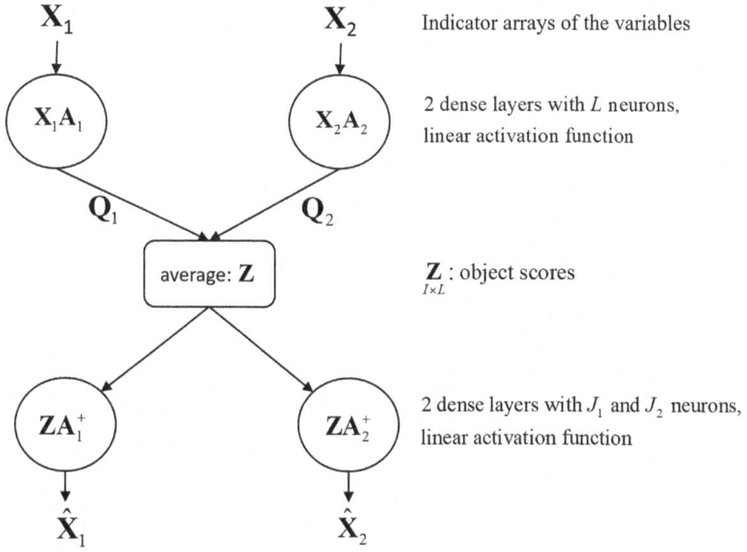

Fig. 4 An auto-encoder that reproduces HOMALS

Moreover, the LOSS function of HOMALS is based on the classical RSS, which may not be the best choice to compare $\widehat{\mathbf{X}}_k$ to $\mathbf{X}_k$. It is possible to extend the previous approach by eliminating these two constraints and introducing a better LOSS function. The new architecture of the auto-encoder for only two nominal variables and dimension L is shown in Fig. 5.

Note that in the output layer there is a new parameter matrix $\mathbf{W}_k$ (of dimensions $L \times J_k$) and the activation function is now *Softmax* (see Bengio et al. 2003)—the same function as used in multinomial logistic regression. Specifically, Softmax is a function, denoted $\sigma : \mathbb{R}^J \rightarrow (0, 1)^J$, defined as:

$$\sigma(\mathbf{v})_j = \frac{e^{v_j}}{\sum_{m=1}^{J} e^{v_m}}, \quad j = 1, \ldots, J, \ \mathbf{v} = (v_1, v_2, \ldots, v_J)^{\mathrm{T}}. \tag{24}$$

This way, $\widehat{\mathbf{X}}_k$ contains, for each unit, the estimated probability of assuming the different modalities of variable k. Then, the categorical cross-entropy $H\left(\mathbf{X}_k, \widehat{\mathbf{X}}_k\right)$ (also called logistic LOSS) is more appropriate to compare the reconstructed array to the indicator array $\mathbf{X}_k$:

$$\sum_{k=1}^{K} H\left(\mathbf{X}_k, \widehat{\mathbf{X}}_k\right) = -\sum_{i=1}^{I}\sum_{k=1}^{K}\sum_{j=1}^{J_k} x_{ikj} \log \widehat{x}_{ikj} = -\sum_{k=1}^{K}\sum_{i=1}^{I} \log\left(\sigma\left(\mathbf{z}_i \mathbf{W}_k\right)\right) \mathbf{x}_{ik}^{\mathrm{T}}, \tag{25}$$

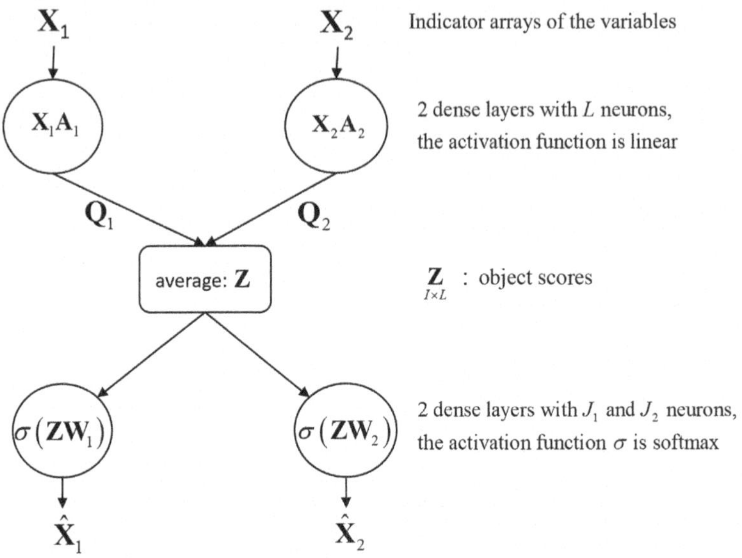

Fig. 5 Auto-encoder to extend HOMALS to nonlinear encoding

where $\mathbf{z}_i$ is the ith row vector of $\mathbf{Z}$ (with length L). Then the minimisation problem becomes:

$$\min_{\substack{\mathbf{W}_1,\mathbf{W}_k,\ldots,\mathbf{W}_K\\ \mathbf{Z}}} \sum_{k=1}^{K} H\left(\mathbf{X}_k, \sigma\left(\mathbf{Z}\mathbf{W}_k\right)\right). \tag{26}$$

Considering that, by definition, $\mathbf{Z}$ is the mean of $\mathbf{X}_k\mathbf{A}_K$, the modality quantifications $\mathbf{A}_K$ and the weights $\mathbf{W}_K$ are the parameters to estimate. The nonlinear encoding achieved in this way can be much more effective than the encoding provided by HOMALS/MCA. Note that both methods (i.e. HOMALS and its nonlinear extension) use the same OHE coding of the categorical variables as its input. However, the parameterisation is different, and the extension includes more parameters, a nonlinear transformation and a different objective function. As a simple example, consider only two categorical variables, X and Y, each with 5 modalities denoted (respectively) by (A, B, C, D, E) and (a, b, c, d, e), which, together, produce the contingency table shown in Table 1 (from Di Ciaccio 2023). The strong associations of the pairs of modalities (A, a), (B, b), (C, c), (D, d), (E, e) are evident because of the dominant cell frequencies that appear in the main diagonal of the table.

We would therefore expect a representation on two components that highlights these associations: a representation where strongly associated pairs are close to each other and equally far away from the other modalities. By applying MCA, the first four components have the same eigenvalue and are all necessary to obtain a satisfactory representation of the modalities. This is a feature of the matrix being symmetric; see Beh and Lombardo (2022). Figure 6 shows the result obtained from the first two

Table 1 A contingency table showing the association between variables X and Y

X/Y	a	b	c	d	e	Total
A	801	100	100	100	100	1201
B	100	800	100	100	100	1200
C	100	100	800	100	100	1200
D	100	100	100	800	100	1200
E	100	100	100	100	800	1200
Total	1201	1200	1200	1200	1200	6001

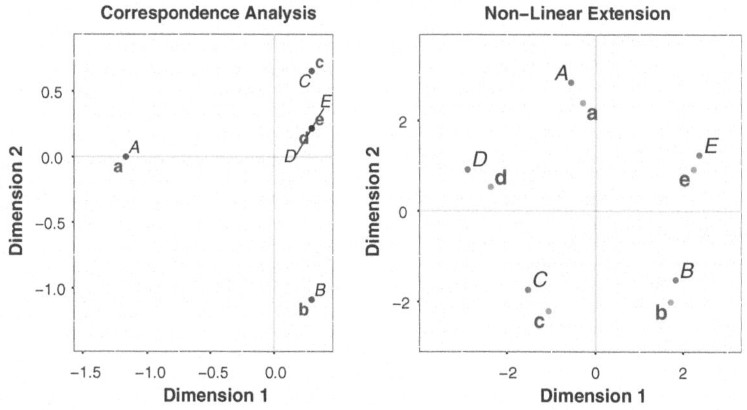

Fig. 6 Categorical encoding for CA (left) and nonlinear extension (right) on data of Table 1, first two components

components of MCA (on the left) and with the nonlinear version just described (on the right). Note how—with the presence of only one more unit for the pair (A, a)—MCA creates, on the first two dimensions, a configuration that is hard to interpret. By contrast, a nonlinear extension shows, with only two axes, a representation of the associations very consistent with the data in the table.

7 Conclusion and Perspectives: Towards a Renewal of Optimal Coding Methods

Transforming qualitative variables into numerical variables is once again a hot topic in part because the profusion of (qualitative) variables with a large number of modalities often found in big data analytics applications.

The statisticians who developed optimal scaling methods were not very concerned about the over-fitting and instability issues that could arise from the use of a large number of indicators because these statisticians often worked with low-dimensional data (they, however, developed very efficient algorithms in the linear case). The DISQUAL method is certainly a method of regularisation by projection onto a low-dimensional subspace, but this aspect remained secondary to the objective of calculating scores. Similarly, the work of Russolillo (2012) uses optimal scaling to apply PLS regression and PLS path modelling to qualitative data without really focusing on the regularising effect of projection onto the PLS components.

It is only very recently (see Meulman et al. 2019) that regularisation by Ridge, LASSO or Elastic Net has been combined with MORALS-type optimal scaling regression—a combination that opens up many new opportunities.

Largely independently, machine learning practitioners confronted with these high-dimensional problems have developed—without always being concerned with optimality or robustness—a large number of techniques, some of them arbitrary, or some of them being a rediscovery of known techniques. However, we have noticed that an approach based on neural networks leads to satisfactory results not only in supervised but also in unsupervised approaches. In the latter case, an auto-encoder network minimising the cross-entropy with the consideration of nonlinear links may give better results than the least-squares minimisation at the origin of the alternating least-squares methods.

References

Abdi, H., Valentin, D., Edelman, B.: Neural Networks. Sage, Thousand Oaks (1999)

Beh, E.J., Lombardo, R.: Correspondence Analysis: Theory, Practice and New Strategies. Wiley, Chichester (2014)

Beh, E.J., Lombardo, R.: Visualising departures from symmetry and Bowker's X^2 statistic. Symmetry **14**, 1103, 25pp (2022)

Bengio, Y., Ducharme, R., Vincent, P., Janvin, C.: A neural probabilistic language model. J. Mach. Learn. Res. **3**, 1137–1155 (2003)

Bock, R.D.: Methods and Applications of Optimal Scaling. The University of North Carolina Psychometric Laboratory Research Memorandum No. 25 (1960)

Bouroche, J.M., Saporta, G., Tenenhaus, M.: Some methods of qualitative data analysis. In: Barra, J.R. (ed.) Recent Developments in Statistics: Proceedings of the European Meeting of Statisticians, pp. 749 – 755, North-Holland, Amsterdam (1977)

Coombs, C.H.: Some hypotheses for the analysis of qualitative variables. Psychol. Rev. **55**, 167–174 (1948)

Coombs, C.H: A Theory of Data. Wiley, Chichester (1964)

Darlington, R.B., Hayes, A.F.: Regression Analysis and Linear Models. Guilford Press (2017)

de Leeuw, J.: Canonical analysis of categorical data. Doctoral Dissertation, Leiden University, Leiden, The Netherlands (1973)

de Leeuw, J., Hornik, K., Mair, P.: Isotone optimization in R: pool-adjacent-violators algorithm (PAVA) and active set methods. J. Stat. Softw. **32**(5), 1–24 (2009)

Di Ciaccio, A.: Categorical encoding for machine learning. In: Pollice, A., et al. (eds.) Book of Short Papers SIS 2020, pp. 1048–1053. Pearson, New York (2020)

Di Ciaccio, A.: Optimal coding of high-cardinality categorical data in machine learning. In: Grilli, M., Lupparelli, M., Rampichini, C., Rocco, E., Vichi, M. (eds.) Statistical Models and Methods for Data Science, pp. 39–51. Studies in Classification, Data Analysis, and Knowledge Organization. Springer, Cham (2023)

Festinger, L.: The treatment of qualitative data by scale analysis. Psychol. Bull. **44**, 149–161 (1947)
Fisher, R.A.: The precision of discriminant functions. Ann. Eugenics **10**, 422–429 (1940)
Gallego, F.J.: Codage flou en analyse des correspondances. Cahiers l'Analyse Données **7**, 413–430 (1982)
Gifi, A.: Nonlinear Multivariate Analysis. Wiley, Chichester (1990)
Guttman, L.: The quantification of a class of attributes: a theory and method of a scale construction. In: Horst, P., Wallin, P., Guttman, L. (eds.) The Prediction of Personal Adjustment, pp. 321–348. Social Research Council, New York (1941)
Guttman, L.: A basis for scaling qualitative data. Am. Sociol. Rev. **9**, 139–150 (1944)
Hancock, J.T., Khoshgoftaar, T.M.: Survey on categorical data for neural networks. J. Big Data **7**, 1–41 (2020)
Hayashi, C.: On the quantification of qualitative data from the mathematico-statistical point of view. Ann. Inst. Stat. Math. **2**, 35–47 (1950)
Hill, M.O.: Correspondence analysis: a neglected multivariate method. J. R. Stat. Soc. (Ser. C) (Appl. Stat.) **23**, 340–354 (1974)
Hirschfeld, H.O.: A connection between correlation and contingency. Math. Proc. Camb. Philos. Soc. **31**, 520–524 (1935)
Horst, P.: Measuring complex attitudes. J. Soc. Psychol. **6**, 369–374 (1935)
Hotelling, H.: Relations between two sets of variates. Biometrika, **28**(3/4), 321–377 (1936)
Kendall, M.G., Stuart, A.: The Advanced Theory of Statistics, vol. II. Charles Griffin, London (1961)
Kruskal, J.B.: Nonmetric multidimensional scaling: a numerical method. Psychometrika **29**, 115–129 (1964)
Lancaster, H.O.: Some properties of the bivariate normal distribution considered in the form of a contingency table. Biometrika **44**, 289–292 (1957)
Lebart, L., Saporta, G.: Historical elements of correspondence analysis and multiple correspondence analysis. In: Blasius, J., Greenacre, M. (eds.) Visualization and Verbalization of Data, pp. 73–86. Chapman & Hall/CRC, Boca Raton, FL (2014)
Maung, K.: Measurement of association in a contingency table with special reference to the pigmentation of hair and eye colours of Scottish school children. Ann. Eugenics **11**, 189–223 (1941)
Meulman, J.J., van der Kooij, A.J., Duisters, K.L.: ROS regression: integrating regularization with optimal scaling regression. Stat. Sci. **34**, 361–390 (2019)
Nishisato, S.: Optimal scaling of paired comparison and rank order data: an alternative to Guttman's formulation. Psychometrika **43**, 263–271 (1978)
Nishisato, S.: Analysis of Categorical Data: Dual Scaling and Its Applications. University of Toronto Press, Toronto (1980)
Nishisato, S.: Multidimensional Nonlinear Descriptive Analysis. Chapman and Hall/CRC, Boca Raton, FL (2006)
Potdar, K., Pardawala, T.S., Pai, C.D.: A comparative study of categorical variable encoding techniques for neural network classifiers. Int. J. Comput. Appl. **175**(4), 7–9 (2017)
Ramsay, J.O.: Monotone regression splines in action. Stat. Sci. **3**, 425–441 (1988)
Russolillo, G.: Non-metric partial least squares. Electron. J. Stat. **6**, 1641–1669 (2012)
Saporta, G.: Dépendance et codages de deux variables aléatoires. Rev. Stat. Appl. **23**, 4–63 (1975)
Saporta, G., Niang-Keita, N.: Correspondence analysis and classification. In: Greenacre, M., Blasius, J. (eds.) Multiple Correspondence Analysis and Related Methods, pp. 371–392. Chapman and Hall/CRC, Boca Raton, FL (2006)
Slater, P.: The analysis of personal preferences. Br. J. Stat. Psychol. **13**, 119–135 (1960)
Stevens, S.S.: On the theory of scales of measurement. Science **103**, 677–680 (1946)
Takane, Y.: Analysis of categorizing behavior by a quantification method. Behaviometrika **8**, 57–67 (1980)
Takeuchi, K., Yanai, H., Mukherjee, B.N.: The Foundations of Multivariate Analysis. Wiley Eastern, New Delhi (1982)
Tanaka, Y.: Review of the methods of quantification. Environ. Health Perspect. **32**, 113–123 (1979)

Tenenhaus, M.: Canonical analysis of two convex polyhedral cones and applications. Psychometrika **53**, 503–524 (1988)

Tenenhaus, M., Young, F.W.: An analysis and synthesis of multiple correspondence analysis, optimal scaling, dual scaling, homogeneity analysis and other methods for quantifying categorical data. Psychometrika **50**, 91–119 (1985)

van Buuren, S., Heiser, W.J.: Clustering *N* objects into *K* groups under optimal scaling of variables. Psychometrika **54**, 699–706 (1989)

van de Velden, M., D'Enza, A.I., Palumbo, F.: Cluster correspondence analysis. Psychometrika **82**, 158–185 (2017)

Williams, E.J.: Use of scores for the analysis of association in contingency tables. Biometrika **39**, 274–289 (1952)

Young, F.W.: Quantitative analysis of qualitative data. Psychometrika **46**, 357–388 (1981)

Young, F.W., de Leeuw, J., Takane, Y.: Regression with qualitative and quantitative variables: alternating least squares methods with optimal scaling features. Psychometrika **41**, 505–529 (1976)

Young, F.W., Takane, Y., de Leeuw, J.: The principal components of mixed measurement level multivariate data: an alternating least squares method with optimal scaling features. Psychometrika **43**, 279–281 (1978)

Marketing Data Analysis by the Dual Scaling Approach: An Update and a New Application

Daniel Baier and Wolfgang Gaul

1 Introduction

Dual scaling and related methods like quantification theory, correspondence analysis, or homogeneity analysis (in the following shortly summarised as the dual scaling approach) have a long history in data analysis and statistics; see Nishisato et al. (2021, pp. 5–25) for a recent review. After the first attempts by Karl Pearson and others to quantify categorical data at the beginning of the twentieth century, Shizuhiko Nishisato and others formalised and advanced this approach from the 1960s under different names. From the start, the dual scaling approach was successfully applied in various disciplines; see Malhotra et al. (2005). Also in marketing, it demonstrated its usefulness. Well-known and often cited are the early articles on applications in marketing by Franke (1985) and Hoffman and Franke (1986). They applied dual scaling for copytesting print advertisements and correspondence analysis for market structuring. Nishisato and Gaul (1988) summarised early applications of dual scaling in marketing and demonstrated its usefulness by referring to analysing complex and varied data (e.g. paired comparisons, preferences, ratings). They argued that the dual scaling approach—at least in marketing—no longer should be called the "neglected multivariate method" with a reference to Hill (1974).

However, thirty years later, at least in marketing, other methods seem to be preferred: So, Orme (2019) argues on the basis of a yearly survey among industrial users of Sawtooth Software (the market leader for conjoint analysis software) that conjoint analysis is applied more than 27,000 times a year in large-scale commer-

D. Baier (✉)
Marketing & Innovation, University of Bayreuth, Bayreuth, Germany
e-mail: daniel.baier@uni-bayreuth.de

W. Gaul
Institute of Decision Theory and Management Science, Karlsruhe Institute of Technology, Karlsruhe, Germany
e-mail: wolfgang.gaul@kit.edu

E. J. Beh et al. (eds.), *Analysis of Categorical Data from Historical Perspectives*, Behaviormetrics: Quantitative Approaches to Human Behavior 17,
https://doi.org/10.1007/978-981-99-5329-5_10

cial contexts. Baier and Brusch (2021) support these findings by analysing a large sample of conjoint analysis applications of a major European market research institute. Articles with overviews on applications of conjoint analysis (Green and Srinivasan 1978, 1990) are among the most often cited articles in marketing research journals, in contrast to the mentioned articles on applications of the dual scaling approach. Additionally, when the goal is to analyse complex and varied data—a known advantage of dual scaling—the most often applied methods according to polls among data scientists (e.g. https://www.kdnuggets.com/2016/09/poll-algorithms-used-data-scientists.html) are regression, cluster analysis, and decision trees. Visualisation is ranked fourth in this poll, but the dual scaling approach is not referred to as a solution for this task.

The paper is structured as follows: In Sect. 2, we take a closer look at applications of the dual scaling approach in marketing and the potential reasons of less usage. Section 3 shows dual scaling results of well-known paired comparisons data using recent software developments. In Sect. 4, we introduce and analyse a new dataset with preferences of a large sample of online shop customers. The paper closes with an outlook in Sect. 5.

2 Marketing Data Analysis by the Dual Scaling Approach

Collecting data and analysing them with advanced statistical methods has a long tradition in marketing (see, Ferber 1949). However, in a recent review, Wedel and Kannan (2016) argued that the spread of these methods firstly gained impact in the 1960s when developments and practical applications were published in respected journals like *Journal of Marketing* and *Journal of Marketing Research* or additionally—from the 1980s—*International Journal of Research in Marketing* and *Marketing Science*. Inspired by these articles, cluster, conjoint, correspondence, and discriminant analysis, dual scaling, logit analysis, multidimensional scaling, regression analysis, and many other methods were further developed and applied to solve real-world market segmentation, product positioning, and pricing problems.

Roberts et al. (2014) analysed the diffusion of these methods in marketing theory and practice. Based on a citation analysis and three surveys among researchers, mediators, and practitioners, they determined that some methods were outstanding in this regard. So, publications on developments and applications of conjoint analysis (e.g. Green and Srinivasan 1978, 1990) were most often cited and had the highest impact. On the other side, publications on developments and applications of the dual scaling approach—see Table 1 for an overview—also show a high number of citations. But, compared to other methods, the application numbers of the dual scaling approach are still low and corresponding publications not among the top 100 publications with outstanding impact; see Roberts et al. (2014). At least in marketing and in contrast to the expectations of Nishisato and Gaul (1988), the dual scaling approach must further be referred to as a “neglected multivariate method”.

Table 1 Overview on applications of the dual scaling approach in marketing

Marketing task	Data (sample size)	Method applied	Reference
Copytest a print ad for women's shoes	Ratings of 18 attributes w.r.t. the ad ($n = 30$)	Dual scaling of response frequencies	Franke (1985)
Structure the overnight delivery market	15 attributes describing one's shipper ($n = 252$)	Carroll-Green-Schaffer-Scaling	Carroll et al. (1986)
Structure the beverage market (cola, non-cola)	Weekly consumption of 9 brands (yes/no, $n = 34$)	Correspondence analysis with French plot	Hoffman and Franke (1986)
Position cognac brands by print ads	Paired comparisons of 10 print ads ($n = 69$)	Dual scaling of the dominance matrix	Nishisato and Gaul (1988)
Position cognac brands by print ads	Ratings of 7 attributes w.r.t. 10 print ads ($n = 69$)	Forced classification on dominance matrix	Nishisato and Gaul (1988)
Segment cosmetics markets (hypothetically)	Ratings of brands w.r.t. attributes by consumers	Forced classification on dominance matrix	Nishisato (1988)
Segment cosmetics markets (hypothetically)	Preferences of consumers w.r.t. brands	Forced classification on dominance matrix	Nishisato (1988)
Position cigarette brands by print ads	Ratings of 14 attributes w.r.t. 8 ads ($n = 126$)	Forced classification on dominance matrix	Nishisato and Gaul (1990)
Measure attribute impact on destination choice	12 paired comparisons of fictive dest. ($n = 157$)	Multiple correspondence analysis	Kaciak and Louviere (1990)
Track periodic beverage consumption ($t = 1, \ldots, 4$)	Allocation of brands to occasions ($n = 800$)	Correspondence analysis with French plot	Higgs (1991)
Improve the perceived safety of small cars	Results of 24 cars w.r.t. 4 objective tests	Multiple correspondence analysis	Hoffman and De Leeuw (1992)
Position hospitals in the eyes of referrers	Referrals of physicians ($n = 1086$) for diseases	Correspondence analysis with French plot	Javalgi et al. (1995)
Position banks in the eyes of customers	Allocation of 25 features to 10 banks ($n = 364$)	Correspondence analysis with French plot	Yavas and Shemwell (1996)
Position retailers w.r.t. purchase patterns	Allocation of motives to products ($n = 319$)	Correspondence analysis with French plot	Yavas (2001)
Position cigarette brands in the eyes of men	Allocation of 12 brands to 11 attributes ($n = 100$)	Correspondence analysis with French plot	Cholakian (2006)
Position airline brands in the eyes of customers	Allocation of brands to attributes ($n = 381$)	Correspondence analysis with French plot	Wen et al. (2008)
Sell luxury goods online and/or in-store	Reasons to shop online resp. in-store ($n = 55$)	Text mining and correspondence analysis	Liu et al. (2013)
Segment software markets w.r.t. preferences	Preferences w.r.t. 12 design attributes ($n = 128$)	Correspondence analysis with French plot	Wang (2016)
Measure effectiveness of online marketing	Usage frequencies of 5 tools ($n = 313$)	Correspondence analysis with French plot	Krizanova et al. (2019)
Find big five personality trait segments	Scores on big five personality traits ($n = 27$)	Correspondence analysis with French plot	Pitt et al. (2020)
Position plant-based meat alternatives	Evoked associations by consumers ($n = 1039$)	Correspondence analysis with French plot	Michel et al. (2021)
Improve online shops w.r.t. sustainability	Preferences w.r.t. 9 improvements ($n = 4411$)	Dual scaling of ranking data	This paper

Abbreviation w.r.t. = with respect to

The reasons for this neglect seem to be many-fold. Here, we only discuss two concerns often raised when talking to marketing researchers, mediators, and practitioners: (1) The lack of powerful software for the dual scaling approach that makes it easy to collect and analyse data as needed in a market research or consulting project as well as (2) the lack of a convincing case study where the application of the dual scaling approach leads to a measurable impact from a marketing manager's or decider's point of view.

So, concerning (1), it is often argued that conjoint analysis and structural equation modelling are very successful since powerful and dedicated software is available. Sawtooth Software offers Lighthouse for conjoint analysis, a system that handles the whole market research process from designing advanced online questionnaires, data collection, analysis, simulation, and optimisation as well as final presentation of results to marketing managers and deciders. The software is constantly improved in close collaboration with researchers, managers, and deciders. The same holds for smartPLS, a currently widespread structural equation modelling software that allows the analyst to perform acceptance analyses of new technologies in an advanced manner. For dual scaling, at a first glance, this availability of powerful and dedicated software also holds: Beh (2004), Malhotra et al. (2005) as well as Lombardo and Beh (2016) discuss a large number of procedures and packages. Shizuhiko Nishisato and Ira Nishisato developed Dual3 for dual scaling which is based on the work of Nishisato (1980). Nowadays, Dual3 is no longer available, but the R package `dualScale` by Clavel et al. (2014) helped to fill this gap. For related methods like correspondence analysis, the availability of software packages and procedures is even greater. All statistical software systems (e.g. BMBP, SAS, and SPSS) offer at least simple correspondence analysis. Moreover, Lombardo and Beh (2016) not only discuss their own powerful R package `CAvariants`, but also 12 further R packages for correspondence analysis: `ca` by Nenadic and Greenacre (2007) and `FactoMineR` by Lê et al. (2008), as well as `ade4`, `anacor`, `cabootcrs`, `CAInterprTools`, `cncaGUI`, `ExPosition`, `homals`, `MASS`, `PTAk`, and `vegan`. Most of these packages receive a yearly update and/or constant improvements; see Lombardo and Beh (2021), Greenacre et al. (2020), and Husson et al. (2020). Lombardo and Beh (2016) conclude that these packages cover broad areas from a methodological and an application-oriented point of view. However, the close collaboration between the software (and methodological) developers and the marketing research practice seems to be limited at the moment. Here, maybe, more exchange between researchers, mediators, and practitioners during conferences (e.g. COMPSTAT, ECDA, IFCS, INFORMS Marketing Science) could be a solution.

Concerning (2), dual scaling still lacks a convincing marketing case study with a convincing managerial impact from a practical point of view. At a first glance, Table 1 contains a large number of advertising testing, brand management, or market segmentation applications. Graphical displays of two modes of objects (e.g. individuals and brands or brands and attributes) are employed. However, the often used "French plot" in these applications is problematic since the coordinates of the two modes of objects come from two different subspaces. The visualisation is helpful for exploration, but it does not allow to relate inter-mode distances to market shares or profits. Consequently, modifications (e.g. modified brand positions) can not be evaluated in terms of future market shares and profits. Many recent publications propose alternative joint graphical displays and biplots included in freely downloadable, easy-to-use, and comprehensive software packages; see Lombardo and Beh (2016) for an overview and an excellent R software package. But—up to now—there is a lack of marketing applications of these packages that lead to a measurable impact from a managerial point of view.

In the next two sections, we try to show that these concerns are exaggerated. In Sect. 3, we analyse the well-known paired comparisons data from Nishisato and Gaul (1988) with a newer R package (`FactoMineR`). In Sect. 4, we apply the same analysis to a large dataset with preference data from $n = 4411$ online shop customers.

3 Analysing the Nishisato and Gaul (1988) Paired Comparisons

In their review article on marketing applications of dual scaling, Nishisato and Gaul (1988) analysed the paired comparisons data first published in Gaul and Schader (1988). Each of the 69 customers who took part in the study was asked (forced choice) to indicate for all possible pairs of presented ads which ad he/she prefers. The ten ads used in data collection were two each for the five cognac brands Remy Martin, Hennessey, Courvoisier, Bisquit, and Martell, leading to $10 \cdot (10 - 1)/2 = 45$ possible pairs of presented ads. In the following, we reanalyse this data using the R package `FactoMineR`. First, following a proposal by Torres and Greenacre (2002), we transfer them to count data as given transposed in Table 2. Each customer is represented by two rows: One row counts her/his indicated dispreferences for each ad ("−"), one row counts her/his indicated preferences for each ad ("+").

Cell $(i-, j)$ indicates how often customer i did not prefer j to another ad ($i = 1, \ldots, 69$ and $j = 1, \ldots, 10$). Cell $(i+, j)$ indicates how often customer i preferred j to another ad. Note that the sum of $(i-, j)$ and $(i+, j)$ is always nine since j is contained in nine ad pairs. The additional rows All− and All+ summarise the counts across all customers (summing up to $69 \cdot (10 - 1) = 621$ for each ad as the number of presentations of pairs with the ad contained). It can be easily seen that—across all 69 customers—ad Martell (1) was the most preferred in a paired comparison (419 times), whereas ad Hennessey (1) was the least preferred (186 times).

The organisation of the paired comparisons data as described has the advantage that the count data (without the additional rows All− and All+) now can be analysed using correspondence analysis software; for example, `FactoMineR` or `CAvariants`. Nevertheless, the same results emerge as if dual scaling and Dual3 would have been applied; see Torres and Greenacre (2002). For this analysis, Nishisato and Gaul (1988) constructed a so-called dominance matrix from the paired comparisons data with customers as rows (69 rows) and ads as columns (10 columns) where each cell counts the number of times a customer preferred the ad minus the number of times a customer dispreferred the ad. Note that dual scaling and Dual3 are able to deal with negative cells (in contrast to correspondence analysis).

Table 2 Paired comparisons data from Nishisato and Gaul (1988) arranged as count data (here: transposed presentation) by "doubling the rows" (i becomes $i-$ and $i+$ for all 69 customers) according to Torres and Greenacre (2002)

	1−	2−	3−	4−	5−	...	69−	1+	2+	3+	4+	5+	...	69+	All−	All+
Remy Martin (1)	1	5	2	7	1	...	5	8	4	7	2	8	...	4	293	328
Hennessy (1)	8	7	7	9	8	...	6	1	2	2	0	1	...	3	435	186
Courvoisier (1)	6	5	1	2	3	...	2	3	4	8	7	6	...	7	270	351
Bisquit (1)	6	8	6	5	6	...	7	3	1	3	4	3	...	2	335	286
Martell (1)	1	1	2	1	1	...	1	8	8	7	8	8	...	8	202	419
Remy Martin (2)	5	8	1	5	4	...	9	4	1	8	4	5	...	0	372	249
Hennessey (2)	4	5	7	3	8	...	3	5	4	2	6	1	...	6	342	279
Courvoisier (2)	4	4	7	3	5	...	4	5	5	2	6	4	...	5	283	338
Bisquit (2)	9	2	8	8	8	...	8	0	7	1	1	1	...	1	350	271
Martell (2)	1	0	4	2	1	...	0	8	9	5	7	8	...	9	223	398

Table 3 Summary of statistics when dual scaling is applied to the paired comparisons data from Nishisato and Gaul (1988)

	Comp. 1	Comp. 2	Comp. 3	Comp. 4	Comp. 5	Comp. 6
Eigenvalue	0.123	0.064	0.046	0.032	0.024	0.018
Singular value	0.350	0.254	0.215	0.178	0.156	0.136
Accounted for (%)	36.401	19.119	13.751	9.428	7.222	5.492
Cum. accounted for (%)	36.401	55.520	69.271	78.699	85.921	91.413
Discrepancy angle (°C)	69.494	75.293	77.566	79.730	81.023	82.179

Abbreviation Comp. = component, cum. = cumulative

Table 3 summarises the results of applying `FactoMineR` to the data in Table 2. As expected, the eigenvalues, singular values, and variances accounted for are identical to the results presented in Nishisato and Gaul (1988). Two components account for 55.520% of the inertia. The calculated high discrepancy angles for each dimension—see Nishisato et al. (2021, p. 49) for a discussion—additionally indicate that the usual joint graphical display (French plot) as given in Fig. 1 must be used with caution. Overlaying the row (customer) and the column (ad) subspaces is inaccurate. However, Nishisato et al. (2021)'s discussion of the discrepancy angles and their consequences allows the analyst to read the French plot in an appropriate way.

Taking these caveats into consideration (a topic that has always been at the core of Shizuhiko Nishisato's talks), Fig. 1 is able to provide some interesting insights. First, the graphical display of the ads is identical to the graphical display in Nishisato and Gaul (1988). Note the different sign in dimension 1 is irrelevant from an analytical point of view. The low distance between the two ad points of the same cognac brand indicates again that they are very similarly judged by the customers. Moreover, similar ads with respect to the presented motifs are near-by positioned: Courvoisier and Martell as well as Remy Martin (1) show exclusive convivial moments, whereas the others show exclusive solitude with or without a lonesome brand ambassador.

Moreover, now, in contrast to Nishisato and Gaul (1988), Fig. 1 shows one point for each customer's preferences (rows $i+$, displayed by a "+"). The point for a single customer's dispreferences (rows $i-$, displayed by a "−") is suppressed for readability reasons. Due to the dependence of the corresponding counts (the cells for "+" and "−" in Table 2 sum to the number of ads minus one for each ad), each dispreferences point ("−") would have a mirrored at the origin position to the corresponding preferences point ("+"). The many "+" in the direction of the Courvoisier and Martell ads demonstrate that many customers prefer these ads. Even if we are aware of the discrepancies of the two subspaces (for ads and customers) and take the increasing distance from the origin into account, the graphical display gives useful insights into the competition between brands. We demonstrate the advantages of this graphical display in the next section with an even larger sample of customers.

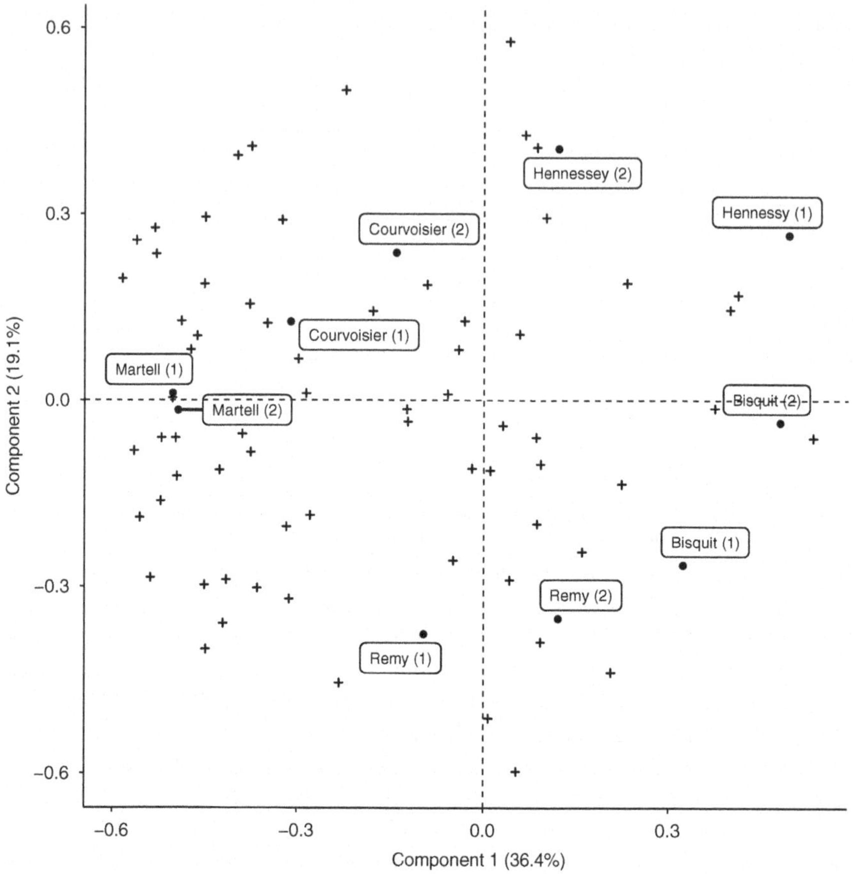

Fig. 1 Graphical display when dual scaling is applied to the paired comparisons data from Nishisato and Gaul (1988), "+" displays a customer's preferences, "−" displays a customer's dispreferences ("+" and "−" are mirrored points at the origin, "−" points are suppressed for readability reasons)

4 Analysing Sustainable Online Shop Improvement Preferences

During the last decade, sustainability has developed from a marginal to a mainstream topic in many industries; see Baier et al. (2020) and Rausch et al. (2021) for more on this development. Consumers are increasingly environmentally conscious and expect from their business partners the same. So, Rausch et al. (2021) showed in two surveys (with $n = 1770$ and $n = 1678$) that customers of a major German apparel online retailer (BAUR, www.baur.de) favour durable products, especially when they are manufactured using low-emission technologies as well as fair wages and working conditions. Baier et al. (2020) additionally found in a large survey with ADIDAS customers that consumers—on average–would accept a price increase for sustainable apparel and sportswear of about 15–20%. In spring 2021, the BAUR surveys were repeated, with a focus on potential online shop improvement options (in the following shortly: options).

A list of options was developed based on Baier et al. (2020), Rausch et al. (2021), and the references there as well as workshops with 23 BAUR customers and ten experts (the retailer's senior staff for website design, for corporate sustainability, and for customer service). Table 4 reflects the final nine options with short names (used later in this paper) and a shortened description of each option with sample images to make the options clearer for the customers (text in German).

An online questionnaire—developed and hosted using the software Qualtrics (see www.qualtrics.com)—contains these descriptions and the invitation to sort the described options according to decreasing preference ("Please indicate by ranking which of the discussed options are most important to you. Please arrange the nine options from top to bottom, starting with the most important. To do this, simply drag the respective option to the desired position while holding down the mouse button."). The questionnaire also asked for socio-demographic information and for the respondent's past sustainable and non-sustainable buying behaviour. The questionnaire was tested with a sample of ten customers. According to their understanding of the descriptions, some phrasing was slightly modified.

The questionnaire was distributed among BAUR customers via the company's June 2021 newsletter. Recipients of the newsletter were asked to participate in an improvement survey of the shop and for participating they were offered a raffle with five vouchers at 20 Euro. Within one week, $n = 4411$ completely filled out questionnaires were collected. Gender distribution (female: $n = 3502, 73.4\%$, male: $n = 900$, 18.9% male, diverse: $n = 9$) and age distribution (up to 29 years: $n = 199, 4.5\%$, 30–39 years/30s: $n = 529$, 12.0%, 40s: $n = 873$, 19.8%, 50s: $n = 1509$, 34.2%, from 60 years: $n = 1301, 29.5\%$) of the sample reflects quite well gender and age distribution of the online retailer's newsletter recipients with about 75% female and 25% male customers, most of them 45+ years old. Of course, as with many other customer surveys using newsletters for distribution, the sample is biased in so far that we expect that especially loyal and less critical customers participated in the survey. Moreover, only 2% of the newsletter recipients answered.

The collected rank-order preferences were—as in the last section—transformed into count data as given and summarised in Table 5. Cell $(i-, j)$ indicates how often customer i ranked option j less important than another option ($i = 1, \ldots, 4411$, $j = 1, \ldots, 9$). $(i+, j)$ indicates how often customer i ranked option j more important than another option. The sum of $(i-, j)$ and $(i+, j)$ is always eight (the number of options minus 1) since in each ranking the rank of option j can be compared with the rank of eight other options. Again, All− and All+ summarise the counts across all 4411 respondents. It can be easily seen that—across all respondents—product traffic light is the most preferred option (22,915 times), followed by labelled images, visible filter, and brand traffic light. Note that again, the sum of the All− and All+ counts in each row is the number of customers (here: 4411) times the number of options (here: 9) minus 1 (here: $4411 \cdot 8 = 35{,}288$). Analysing this count data (again without the rows All− and All+) using `FactoMineR` leads to the results in Table 6. Two

Table 4 Potential sustainable online shop improvement options for an online apparel retailer

Short name	Description in the questionnaire (shortened, with explanatory images)	
Product traffic light	Please imagine that there was a traffic light system in the BAUR online shop, with which the individual products were labelled with regard to their sustainability. With the help of the traffic light system, you as a customer receive a simple assessment of whether the product in question meets the three sustainability criteria (ecology, economy, social issues)	
Brand traffic light	Please imagine that there was a traffic light system in the BAUR online shop with which the brands behind the products are identified with regard to their sustainability. With the help of the traffic light system, you receive a simple assessment of whether the respective brand meets the three sustainability criteria (ecology, economy, social issues)	
Labelled images	Please imagine BAUR labelling the product images in the online shop with the word "sustainable". When shopping, you can see at first glance whether a product has been manufactured sustainably or not	
Webpages	Please imagine that BAUR would show very transparently on its website how the company is committed to more sustainability, for example, "Which specific sustainability projects are supported?", "What seals are there in the BAUR online shop?", etc. On this page you could also read general information about sustainability, for example, "Why is sustainability important?"	
Seal	Please imagine that BAUR would use different seals to label products in the online shop in order to show which criteria for sustainability a certain product met	
Visible filter	Please imagine that the BAUR online shop gave you the option of filtering products directly according to the "sustainability" criterion. The associated filter is directly visible and clickable, so you can filter for all sustainable products with just one click	

(continued)

Table 4 (continued)

<table>
<tr><th>Short name</th><th colspan="2">Description in the questionnaire (shortened, with explanatory images)</th></tr>
<tr><td>Detailed filter</td><td>Please imagine you could filter the products in the BAUR online shop according to various sustainability criteria. The filter criteria are explained in a simple and understandable way</td><td>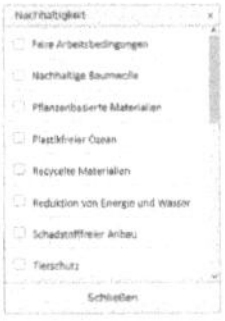
</td></tr>
<tr><td>Project filter</td><td>Please imagine you could filter the products in the BAUR online shop according to various sustainability criteria. The filter criteria would be named after specific projects that are supported. You can find more information about the projects on their own information page on baur.de</td><td>
</td></tr>
<tr><td>Product details</td><td>Please imagine that BAUR went into more detail about the sustainability of these products in the details and descriptions of the individual sustainable products in the online shop</td><td>
</td></tr>
</table>

components account for 42.011% of the inertia. Again, the high discrepancy angles for each dimension (see, Nishisato et al. 2021, p. 49, for a discussion) indicate that the joint graphical display should be used with caution.

Again, taking these caveats into consideration, Fig. 2 is able to provide some interesting insights. First, the graphical display helps to understand how similar the options were ranked by the customers. Product traffic light and brand traffic light seem to be similarly preferred, the same holds for labelled images and visible filter. Again, as in the last section, the "+" reflect single customer's preferences, whereas the "–" that reflect single customer's dispreferences are suppressed for better readability. The grouping of many "+" on the left side of the display indicates that the above mentioned options—product traffic light and brand traffic light but also labelled images and visible filter—are preferred options whereas the similar options webpages and project filter are dispreferred.

In order to analyse whether these preferences depend on the grouping of the customers (e.g. age and gender as a priori groups or groupings derived by clustering), an advantage of dual scaling is to position supplementary variables in the display. Here, the aggregate counts of the age and the gender groups were formed similar to the All− and All+ rows in Table 5, and their positions were calculated using the `FactorMineR` package. Figure 3 shows the derived graphical display. Note that here (for better interpretation), the single customers' "+" were suppressed also.

It can easily be seen that the age and gender group's preferences and dispreferences are very similar to the observed overall preferences and dispreferences in Fig. 2. All "+" are on the left side of the display, all suppressed "–" are on the right side.

Table 5 Sustainable online shop improvement preferences of $n = 4411$ customers arranged as count data (here: transposed presentation) by doubling the rows according to Torres and Greenacre 2002

	1−	2−	3−	4−	5−	…	1+	2+	3+	4+	5+	…	All−	All+
Product traffic light	5	2	2	7	8	…	3	6	6	1	0	…	12,373	22,915
Brand traffic light	1	5	5	1	6	…	7	3	3	7	2	…	15,257	20,031
Labelled images	4	0	1	4	0	…	4	8	7	4	8	…	13,346	21,942
Webpages	3	8	8	2	5	…	5	0	0	6	3	…	22,499	12,789
Seal	6	7	7	6	7	…	2	1	1	2	1	…	20,571	14,717
Visible filter	2	6	3	0	2	…	6	2	5	8	6	…	14,458	20,830
Detailed filter	7	3	4	3	3	…	1	5	4	5	5	…	18,648	16,640
Project filter	8	4	6	5	4	…	0	4	2	3	4	…	24,611	10,677
Product details	0	1	0	8	1	…	8	7	8	0	7	…	17,032	18,256

Table 6 Summary of statistics when dual scaling is applied to the sustainable online shop improvement preference data

	Comp. 1	Comp. 2	Comp. 3	Comp. 4	Comp. 5	Comp. 6
Eigenvalue	0.115	0.060	0.051	0.048	0.041	0.038
Singular value	0.394	0.246	0.226	0.219	0.202	0.194
Accounted for (%)	27.504	14.506	12.285	11.471	9.786	9.075
Cum. accounted for (%)	27.504	42.011	54.297	65.768	75.554	84.629
Discrepancy angle (°C)	70.213	75.768	76.924	77.372	78.350	78.787

Abbreviation Comp. = component, cum. = cumulative

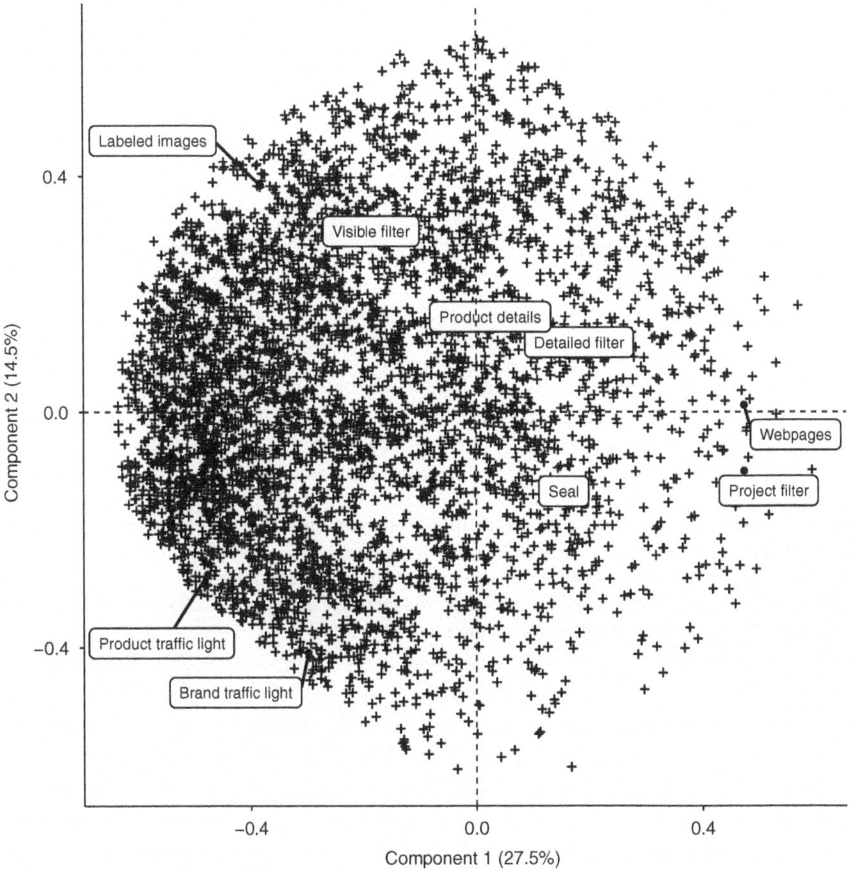

Fig. 2 Graphical display when dual scaling is applied to the sustainable online shop improvement preference data, "+" displays a customer's preferences, "−" displays a customer's dispreferences (suppressed for readability reasons)

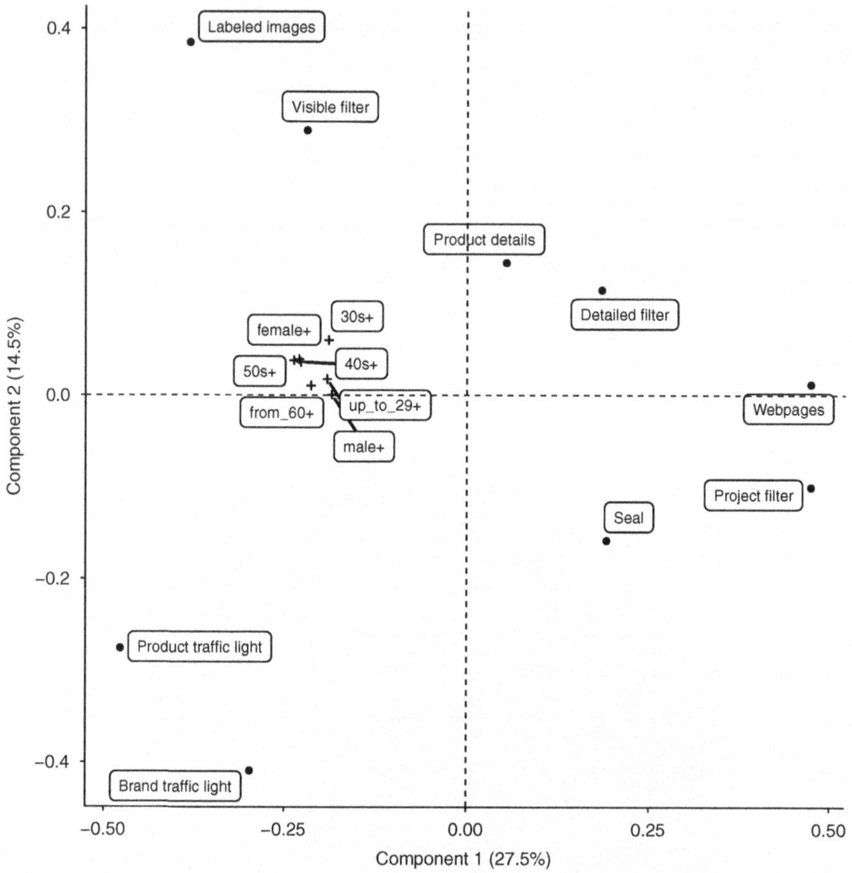

Fig. 3 Graphical display when dual scaling is applied to the sustainable online shop improvement preference data, "+" displays a customer group's preferences, "−" displays a customer group's dispreferences (suppressed for readability reasons); positions of customer groups are calculated via supplementary variables (gender groups: female and men, age groups: up to 29 years, 30 s, 40 s, 50 s and above 60 years, all other positions are the same as in Fig. 2

However, a slight indication is available that male and elder customers tend a bit more to the options in the lower part of the display (product traffic light and brand traffic light), whereas female and younger customers tend more to the options in upper part of the display (labelled images and visible filter). Further analyses have of course been conducted, but we will stop here with our discussion due to space restrictions.

5 Conclusions and Outlook

In this paper, we discussed the diffusion of the dual scaling approach in marketing and discussed a new application. The number of applications has recently considerably increased; however, for playing a major role in marketing applications, some further successful applications and convincing case studies are needed. The available software has made major progress and offers many possibilities to analyse complex and varied data (answers to open questions, associations, cross-tabulations, discrete choices, preferences, ratings) a deciding asset of dual scaling from the beginning.

At this point of reflections, the authors would like to say a big thank you to Shizuhiko Nishisato for his never-ending effort to make dual scaling popular among us and other marketing researchers. We are looking forward to even more joint work in the near future that helps to demonstrate the inspiring elegance and practical usefulness of these methods.

References

Baier, D., Brusch, M.: Conjointanalyse, 2nd edn. Springer, Germany (2021). (in German)
Baier, D., Rausch, T.M., Wagner, T.F.: The drivers of sustainable apparel and sportswear consumption: a segmented Kano perspective. Sustainability **12**(7), 2788, 21pp (2020)
Beh, E.J.: Simple correspondence analysis: a bibliographic review. Int. Stat. Rev. **72**(2), 257–284 (2004)
Carroll, J.D., Green, P.E., Schaffer, C.M.: Interpoint distance comparisons in correspondence analysis. J. Mark. Res. **23**, 271–280 (1986)
Choulakian, V.: Taxicab correspondence analysis. Psychometrika **71**(2), 333–345 (2006)
Clavel, J.G., Nishisato, S., Pita, A.: DualScale: dual scaling analysis of multiple choice data (2014). Available online on the CRAN at http://cran.nexr.com/web/packages/dualScale. Last accessed 8 May 2023
Ferber, R.: Statistical Techniques in Market Research. University of Illinois, Bureau of Economic and Business Research (1949)
Franke, G.R.: Evaluating measures through data quantification: applying dual scaling to an advertising copy test. J. Bus. Res. **13**(1), 61–69 (1985)
Gaul, W., Schader, M.: Clusterwise aggregation of relations. Appl. Stoch. Models Data Anal. **4**(4), 273–282 (1988)
Green, P.E., Srinivasan, V.: Conjoint analysis in consumer research: issues and outlook. J. Consum. Res. **5**(2), 103–123 (1978)
Green, P.E., Srinivasan, V.: Conjoint analysis in marketing: new developments with implications for research and practice. J. Mark. **54**(4), 3–19 (1990)
Greenacre, M., Nenadic, O., Friendly, M.: Package 'ca'—an R package (2020). Available online on the CRAN at https://cran.r-project.org/web/packages/ca/index.html. Last accessed 8 May 2023
Higgs, N.T.: Practical and innovative uses of correspondence analysis. J. R. Stat. Soc. Ser. D (Stat.) **40**(2), 183–194 (1991)
Hill, M.O.: Correspondence analysis: a neglected multivariate method. J. R. Stat. Soc. Ser. C (Appl. Stat.) **23**(3), 340–354 (1974)
Hoffman, D.L., de Leeuw, J.: Interpreting multiple correspondence analysis as a multidimensional scaling method. Mark. Lett. **3**(3), 259–272 (1992)

Hoffman, D.L., Franke, G.R.: Correspondence analysis: graphical representation of categorical data in marketing research. J. Mark. Res. **23**(3), 213–227 (1986)
Husson, F., Josse, J., Lê, S., Mazet, J.: Package 'FactoMineR'—an R package (2020). Available online on the CRAN at https://cran.r-project.org/web/packages/FactoMineR/. Last accessed 8 May 2023
Javalgi, R.G., Benoy, J.W., Gombeski, W.R.: Positioning your service to target key buying influences: the case of referring physicians and hospitals. J. Serv. Mark. **9**(5), 42–52 (1995)
Kaciak, E., Louviere, J.: Multiple correspondence analysis of multiple choice experiment data. J. Mark. Res. **27**(4), 455–465 (1990)
Krizanova, A., Lăzăroiu, G., Gajanova, L., Kliestikova, J., Nadanyiova, M., Moravcikova, D.: The effectiveness of marketing communication and importance of its evaluation in an online environment. Sustainability **11**(24), 7106, 19pp (2019)
Lê, S., Josse, J., Husson, F.: FactoMineR: an R package for multivariate analysis. J. Stat. Softw. **25**(1), 1–18 (2008)
Liu, X., Burns, A.C., Hou, Y.: Comparing online and in-store shopping behavior towards luxury goods. Int. J. Retail Distrib. Manag. **41**(11/12), 885–900 (2013)
Lombardo, R., Beh, E.J.: Variants of simple correspondence analysis. R J. **8**(2), 167–184 (2016)
Lombardo, R., Beh, E.J.: Package 'CAvariants'—an R package (2021). Available online on the CRAN at https://cran.r-project.org/web/packages/CAvariants. Last accessed 8 May 2023
Malhotra, N.K., Rush, C.B., Uslay, C.: Correspondence analysis: methodological perspectives, issues, and applications. Rev. Mark. Res. **1**, 285–316 (2005)
Michel, F., Hartmann, C., Siegrist, M.: Consumers' associations, perceptions and acceptance of meat and plant-based meat alternatives. Food Qual. Prefer. **87**, 104063, 10pp (2021)
Nenadic, O., Greenacre, M.: Correspondence analysis in R, with two-and three-dimensional graphics: the CA package. J. Stat. Softw. **20**(3), 1–13 (2007)
Nishisato, S.: Analysis of Categorical Data: Dual Scaling and Its Applications. University of Toronto Press, Toronto (1980)
Nishisato, S.: Forced classification: a simple application of a quantification method. Psychometrika **49**(1), 25–36 (1984)
Nishisato, S.: Market segmentation by dual scaling through generalized forced classification. In: Gaul, W.A., Schader, M. (eds.) Data, Expert Knowledge and Decisions, pp. 268–278. Springer, Berlin (1988)
Nishisato, S., Gaul, W.: Marketing data analysis by dual scaling. Int. J. Res. Mark. **5**(3), 151–170 (1988)
Nishisato, S., Gaul, W.: An approach to marketing data analysis: the forced classification procedure of dual scaling. J. Mark. Res. **27**(3), 354–360 (1990)
Nishisato, S., Beh, E.J., Lombardo, R., Clavel, J.G.: Modern Quantification Theory. Springer, Singapore (2021)
Orme, B.K.: Getting Started with Conjoint Analysis, 3rd edn. Research Publishers (2019)
Pitt, C.S., Bal, A.S., Plangger, K.: New approaches to psychographic consumer segmentation: exploring fine art collectors using artificial intelligence, automated text analysis and correspondence analysis. Eur. J. Mark. **54**(2), 305–326 (2020)
Rausch, T.M., Baier, D., Wening, S.: Does sustainability really matter to consumers? Assessing the importance of online shop and apparel product attributes. J. Retail. Consum. Serv. **63**, 102681, 16pp (2021)
Roberts, J.H., Kayande, U., Stremersch, S.: From academic research to marketing practice: exploring the marketing science value chain. Int. J. Res. Mark. **31**(2), 144–146 (2014)
Torres, A., Greenacre, M.: Dual scaling and correspondence analysis of preferences, paired comparisons and ratings. Int. J. Res. Mark. **19**(4), 401–405 (2002)
Wang, C.-H.: A novel approach to conduct the importance-satisfaction analysis for acquiring typical user groups in business-intelligence systems. Comput. Hum. Behav. **54**, 673–681 (2016)
Wedel, M., Kannan, P.: Marketing analytics for data-rich environments. J. Mark. **80**(6), 97–121 (2016)

Wen, C-H. Lai, S-C. Yeh, W-Y.: Segmentation and positioning analysis for international air travel market: Taipei-to-Tokyo route. Transp. Res. Rec. **2052**(1), 46–53 (2008)

Yavas, U.: Patronage motives and product purchase patterns: a correspondence analysis. Mark. Intell. Plann. **19**(2), 97–102 (2001)

Yavas, U., Shemwell, D.J.: Bank image: exposition and illustration of correspondence analysis. Int. J. Bank Mark. **14**(1), 15–21 (1996)

Power Transformations and Reciprocal Averaging

Eric J. Beh, Rosaria Lombardo, and Ting-Wu Wang

1 Personal Reflections and Outline

Professor Shizuhiko Nishisato has dedicated his career to the development and dissemination of ideas concerned with many areas of Statistics. Much of it has focused specifically on a vast array of issues concerned with quantification theory and categorical data analysis. We, therefore, consider it an honour and a privilege to not only be asked to contribute to this special collection of papers designed to celebrate his career but to also (in the case of the first two authors) to edit it. Our humble addition to this collection will focus on a variation of a key area of research that has garnered much of Nishisato's attention throughout his career. However, before we discuss more on the nature of this variation, we feel it is appropriate to view through a wide-field lens the contributions he has made.

Nishisato's work in quantification theory has been predominantly on quantifying, for largely categorical data, scores that help to reflect the association between the variables as well as understanding how specific categories compare. In Chap. 2 of his book titled *Multidimensional Nonlinear Descriptive Analysis* (Nishisato 2007), Nishisato outlines a variety of different ways in which quantification theory can

E. J. Beh (✉)
National Institute for Applied Statistics Research, Australia (NIASRA), University of Wollongong, Wollongong, NSW, Australia
e-mail: ericb@uow.edu.au

Centre for Multi-Dimensional Data Visualisation (MuViSu), Stellenbosch University, Stellenbosch, South Africa

R. Lombardo
Department of Economics, University of Campania Luigi Vanvitelli, Capua, Italy
e-mail: rosaria.lombardo@unicampania.it

T.-W. Wang
School of Information and Physical Sciences, University of Newcastle, Newcastle, NSW, Australia
e-mail: ting-wu.wang@uon.edu.au

E. J. Beh et al. (eds.), *Analysis of Categorical Data from Historical Perspectives*, Behaviormetrics: Quantitative Approaches to Human Behavior 17,
https://doi.org/10.1007/978-981-99-5329-5_11

be approached. These include using the method of reciprocal averaging (which is our favoured option), through a one-way ANOVA framework, by maximising the bivariate correlation between the variables (a technique related to canonical correlation analysis) and by approaching the method through geometric means. He also discusses a least squares approach that involves minimising the sum-of-squares of the difference between an observed cell frequency and its estimated value obtained from an association model. Interestingly, he starts the chapter (Sect. 2.1) with the question:

> Is Likert-Type Scoring Appropriate?

Such a question is certainly relevant when viewed from the point of view we have taken for much of our joint and independent work concerning ordinal variables. Indeed, Likert-type scores (we've referred to them as *natural scores* in the past—see, for example, Beh and Lombardo (2014, 2021))—are used as a basis for the construction of orthogonal polynomials when defining the structure of ordinal categories. While Nishisato does not immediately answer the question, he does state more recently (Nishisato et al. 2021, p. 39) that:

> . . . the Likert scale is nowadays useful only as a coding method, and it no longer serves as a scoring method.

For the range of contributions Nishisato has given to quantification theory, much of his energy has been dedicated to developing *dual scaling*—a term he coined in 1976 but proposed to the scientific community in his book titled *Analysis of Categorical Data: Dual Scaling and its Applications* (Nishisato 1980a). One may refer to Nishisato (2007, Sect. 3.3.4) for more information on the genesis of the term. While papers in this *Festschrift* provide a comprehensive discussion of his career and its many highlights, his work on dual scaling spans an extensive array of publications that specifically deal with this area of research, including the books *Elements of Dual Scaling* (Nishisato 1994), *Dual Scaling in a Nutshell* (Nishisato and Nishisato 1994) and, of course, *Analysis of Categorical Data: Dual Scaling and its Applications* (Nishisato 1980a). He has also examined the role of dual scaling on ordered, and partially ordered, categories—see Nishisato (1980b, 2000), Nishisato and Arri (1975), Nishisato and Wen-Jenn (1984) and Nishisato and Inukai (1972)—which has been a topic that we (Beh and Lombardo) have independently, and jointly, focused much of our attention to since the late 1990s. More recently, Nishisato has written extensively about some of the pit-falls inherit in using the scores obtained from quantification theory to visualise the relationship between variables (as correspondence analysis does for categorical variables); see Nishisato (1988a, 1995) and Nishisato and Clavel (2003, 2010) for an array of discussions on this matter. While we have been greatly influenced by Nishisato's early work on the analysis of ordinal categorical variables—especially of Nishisato and Arri (1975)—it is through his concerns raised on the topic of visualisation that cements our collaboration and friendship. The Nishisato/Clavel and Beh/Lombardo teams may be on opposite sides of the visualisation spectrum, something we were open about in the Preface of Nishisato et al. (2021), but this only helped to strengthen our friendship and mutual respect of each other's work. However, to avoid any potential "rift" (we say with a smile on our face) we refrain from discussing the role of data visualisation in quantification theory. Instead, we shall describe in this paper the variation to dual scaling that we alluded to in the first paragraph. This variation involves investigating the role of power transformations using an analogous technique to dual scaling—commonly referred to as *reciprocal averaging*—for two categorical variables. Such an approach is very much related to the issue of power transformations described from a correspondence analysis perspective by Michael Greenacre (Greenacre 2009, 2010) and the methods of Beh and Lombardo (2024), Cuadras and Cuadras (2006, 2015) and Cuadras et al. (2006) although we focus purely on the scaling aspects here instead of the geometric/visual elements.

To describe the role of power transformations from a reciprocal averaging perspective, this paper is divided into six further sections. Section 2 introduces the notation of a two-way contingency table that we shall be using throughout this discussion (Sect. 2.1), as well as defining the *profile* of a row and column of this table (Sect. 2.2). An overview of the traditional approach to reciprocal averaging as outlined by many, including Hill (1974) and Beh and Lombardo (2014), is also described (Sect. 2.3). Section 3 provides a discussion of the role of power transforming the elements of the row and column profile. Greenacre (2009) describes two types of transformation that can be considered and does so from the perspective of correspondence analysis. We shall be focusing our attention on the role of reciprocal averaging on his "power family 2" although one may consider Wang et al. (2023) for a related discussion on its role in a third type of power transformation. Section 4 provides the core discussion of this paper where we derive the reciprocal averaging procedure to determine a one-dimensional set of row and column scores when a power transformation is applied to the profile elements (Sect. 4.1). We also show how eigen-decomposition can be performed to obtain a multi-dimensional orthogonal set of row and column scores (Sect. 4.2). We show in Sect. 4.3 that the correlation between the set of row and column scores that is obtained is the maximum possible correlation along each dimension. Section 5 outlines the role of singular value decomposition (SVD) for obtaining the row and column scores. The practical equivalence of the scores obtained using the reciprocal averaging procedure and through the SVD of a matrix of residuals is demonstrated in Sect. 6. We study the asbestos data of Irving Selikoff (Selikoff 1981) which is described in Sect. 6.1 and then calculate the one-dimensional row and column scores using reciprocal averaging for various power transformations (Sect. 6.2). We then describe the application of SVD to obtain equivalent one-dimensional scores and, more generally, multi-dimensional scores in Sect. 6.3. Some final comments are left for Sect. 7 where we also describe how one may perform alternative reciprocal averaging methods to obtain row and column scores under a power transformation of the profile elements. Such methods advance the arithmetic averaging of the elements on which the traditional reciprocal averaging is based, but also include a modified version of *method of reciprocal medians* (Nishisato 1984) and *geometric averaging* (Clavel 2021, Chap. 8) which considers the geometric average of the profile elements. At the end of the paper we include an appendix which gives an R function `rapower.exe()` that performs the one-dimensional reciprocal averaging procedure described in Sect. 4.

2 An Overview of Reciprocal Averaging

2.1 Notation

Suppose we consider an $I \times J$ two-way contingency table, $\mathbf{N}$, where the (i, j)th cell entry has a frequency of n_{ij} for $i = 1, 2, \ldots, I$ and $j = 1, 2, \ldots, J$. Let the grand total of $\mathbf{N}$ be n and let the matrix of relative frequencies be $\mathbf{P}$ so that its (i, j)th cell entry is $p_{ij} = n_{ij}/n$ where $\sum_{i=1}^{I} \sum_{j=1}^{J} p_{ij} = 1$. Define the ith row marginal proportion by $p_{i\bullet} = \sum_{j=1}^{J} p_{ij}$ so that it is the ith element of the vector $\mathbf{r}$ and the (i, i)th element of the diagonal matrix $\mathbf{D}_I$. Similarly, define the jth column marginal proportion as $p_{\bullet j} = \sum_{i=1}^{I} p_{ij}$ so that it is the jth element of the vector $\mathbf{c}$ and the (j, j)th element of the diagonal matrix $\mathbf{D}_J$.

2.2 *Definition of a Profile*

Before we provide a broad discussion of reciprocal averaging, it is important to understand the quantities we are working with when we calculate row and column scores and the interpretation they provide when comparing the row scores or the column scores. The quantities of interest to us here are the *profile* of a chosen row or column category. The profile of the ith row category is defined as the set of relative cell frequencies of that row so that the profile takes the form:

$$\left(\frac{n_{i1}}{n_{i\bullet}}, \frac{n_{i2}}{n_{i\bullet}}, \cdots, \frac{n_{ij}}{n_{i\bullet}}, \cdots, \frac{n_{iJ}}{n_{i\bullet}}\right) = \left(\frac{p_{i1}}{p_{i\bullet}}, \frac{p_{i2}}{p_{i\bullet}}, \cdots, \frac{p_{ij}}{p_{i\bullet}}, \cdots, \frac{p_{iJ}}{p_{i\bullet}}\right).$$

Similarly, the profile of the jth column profile is:

$$\left(\frac{n_{1j}}{n_{\bullet j}}, \frac{n_{2j}}{n_{\bullet j}}, \cdots, \frac{n_{ij}}{n_{\bullet j}}, \cdots, \frac{n_{Ij}}{n_{\bullet j}}\right) = \left(\frac{p_{1j}}{p_{\bullet j}}, \frac{p_{2j}}{p_{\bullet j}}, \cdots, \frac{p_{ij}}{p_{\bullet j}}, \cdots, \frac{p_{Ij}}{p_{\bullet j}}\right).$$

If there is no association between the row and column variables, such that $p_{ij} = p_{i\bullet}p_{\bullet j}$, for all $i = 1, 2, \ldots, I$ and $j = 1, 2, \ldots, J$, then the ith row and jth column profiles simplify to:

$$\left(p_{\bullet 1}, p_{\bullet 2}, \cdots, p_{\bullet j}, \cdots, p_{\bullet J}\right)$$

and

$$(p_{1\bullet}, p_{2\bullet}, \cdots, p_{i\bullet}, \cdots, p_{I\bullet}),$$

respectively, so that the ith *centred row profile* is:

$$\left(\frac{p_{i1}}{p_{i\bullet}} - p_{\bullet 1}, \frac{p_{i2}}{p_{i\bullet}} - p_{\bullet 2}, \cdots, \frac{p_{ij}}{p_{i\bullet}} - p_{\bullet j}, \cdots, \frac{p_{iJ}}{p_{i\bullet}} - p_{\bullet J}\right).$$

Similarly, the jth *centred column profile* is:

$$\left(\frac{p_{1j}}{p_{\bullet j}} - p_{1\bullet}, \frac{p_{2j}}{p_{\bullet j}} - p_{2\bullet}, \cdots, \frac{p_{ij}}{p_{\bullet j}} - p_{i\bullet}, \cdots, \frac{p_{Ij}}{p_{\bullet j}} - p_{I\bullet}\right).$$

In both cases, if there is complete independence between the row and column categories, both sets of centred profiles will consist of zeros.

2.3 Reciprocal Averaging

Reciprocal averaging, like dual scaling and other analogous quantification methods, determines row and column scores that do two things. First, they are calculated to best discriminate between differing profiles and highlight those with a similar structure. Secondly, they are calculated to maximise the association that exists between the row and column variables. Once these scores are determined they can be used for visually exploring the nature of this association rather than relying solely on numerical summaries. Correspondence analysis is the most common approach that adapts these quantities for such a purpose and many of Nishisato's friends who have contributed to this collection have dedicated much of their career to the development of correspondence analysis and its related methods.

Before we describe the reciprocal averaging of power transformed profiles we provide a broad overview of the traditional approach to the reciprocal averaging of the profile elements. Such an overview is not new and has been described numerous times throughout the quantification literature including, for example, Hill (1974) and Hirschfeld (1935), and in various forms in many of Nishisato's publications.

Suppose we define the ith row score by a_{im} while the jth column score is denoted by b_{jm} for $m = 1, 2, \ldots, M$; the subscript m is typically included to reflect the quantity along the mth dimension of a visualisation of the association although, while we make no such comment on this, it is important to note that such dimensions are orthogonal to each other. There are many accounts given that show how reciprocal averaging can be performed to determine these scores, most of which do so by stating that the scores are subject to the following properties:

$$\mathrm{E}(a_{im}) = \sum_{i=1}^{I} p_{i\bullet} a_{im} = 0 \quad \mathrm{Var}(a_{im}) = \sum_{i=1}^{I} p_{i\bullet} a_{im}^2 = 1 \tag{1}$$

$$\mathrm{E}\left(b_{jm}\right) = \sum_{j=1}^{J} p_{\bullet j} b_{jm} = 0 \qquad \mathrm{Var}\left(b_{jm}\right) = \sum_{j=1}^{J} p_{\bullet j} b_{jm}^2 = 1\,. \tag{2}$$

We refrain from referring to (1) and (2) as the *constraints* of a_{im} and b_{jm}, as has often been done in the quantification literature. This is largely because Gower (1989, p. 222) states that properties of the type defined by (1) and (2) are "conveniences that should not be regarded as constraints".

Reciprocal averaging involves determining the row score, a_{im} and column score b_{jm}, by considering the weighted (arithmetic) mean of the elements of their centred profiles so that:

$$\begin{aligned} \lambda_m a_{im} &= \left(\frac{p_{i1}}{p_{i\bullet}} - p_{\bullet 1}\right) b_{1m} + \cdots + \left(\frac{p_{iJ}}{p_{i\bullet}} - p_{\bullet J}\right) b_{Jm} \\ &= \sum_{j=1}^{J} \left(\frac{p_{ij}}{p_{i\bullet}} - p_{\bullet j}\right) b_{jm} \end{aligned} \tag{3}$$

and

$$\begin{aligned} \lambda_m b_{jm} &= \left(\frac{p_{1j}}{p_{\bullet j}} - p_{1\bullet}\right) a_{1m} + \cdots + \left(\frac{p_{Ij}}{p_{\bullet j}} - p_{I\bullet}\right) a_{Im} \\ &= \sum_{i=1}^{I} \left(\frac{p_{ij}}{p_{\bullet j}} - p_{i\bullet}\right) a_{im}\,. \end{aligned} \tag{4}$$

The λ_m term in these two equations is the maximum (positive) correlation between the mth set of row and column scores so that:

$$\lambda_m = \sum_{i=1}^{I} \sum_{j=1}^{J} p_{ij} a_{im} b_{jm}\,.$$

Equations (3) and (4) can be expressed in matrix notation by:

$$\lambda_m \mathbf{a}_m = \left(\mathbf{D}_I^{-1} \mathbf{P} - \mathbf{1}_I \mathbf{c}^T\right) \mathbf{b}_m$$

and

$$\lambda_m \mathbf{b}_m = \left(\mathbf{D}_J^{-1} \mathbf{P}^T - \mathbf{1}_J \mathbf{r}^T\right) \mathbf{a}_m\,,$$

where

$$\mathbf{a}_m = \left(a_{1m}, \cdots, a_{im}, \cdots, a_{Im}\right)^T$$

and

$$\mathbf{b}_m = \left(b_{1m}, \cdots, b_{jm}, \cdots, b_{Jm}\right)^T.$$

The solution to $\mathbf{a}_m$ and $\mathbf{b}_m$ can be obtained by solving the eigen-decomposition equation:

$$\left(\mathbf{Z}\mathbf{Z}^T - \lambda_m^2 \mathbf{I}_I\right)\left(\mathbf{D}_I^{1/2}\mathbf{a}_m\right) = \mathbf{0}_I, \quad (5)$$

where

$$\mathbf{Z} = \mathbf{D}_I^{-1/2}\left(\mathbf{P} - \mathbf{r}\mathbf{c}^T\right)\mathbf{D}_I^{-1/2} \quad (6)$$

and $\mathbf{I}_I$ is an $I \times I$ identity matrix. Note that the (i, j)th element of $\mathbf{Z}$ is Pearson's standardised residual:

$$Z_{ij} = \frac{p_{ij} - p_{i\bullet}p_{\bullet j}}{\sqrt{p_{i\bullet}p_{\bullet j}}}$$

so that $\sqrt{n}Z_{ij}$ is asymptotically standard normally distributed. By denoting $\tilde{\mathbf{a}}_m = \mathbf{D}_I^{1/2}\mathbf{a}_m$ then (5) becomes:

$$\left(\mathbf{Z}\mathbf{Z}^T - \lambda_m^2 \mathbf{I}_I\right)\tilde{\mathbf{a}}_m = \mathbf{0}_I,$$

so that $\tilde{\mathbf{a}}_m$ is the mth eigen-vector of $\mathbf{Z}\mathbf{Z}^T$ and λ_m^2 is the mth largest eigen-value of this matrix. A similar derivation obtains the eigen-decomposition equation:

$$\left(\mathbf{Z}^T\mathbf{Z} - \lambda_m^2 \mathbf{I}_I\right)\tilde{\mathbf{b}}_m = \mathbf{0}_I,$$

where $\tilde{\mathbf{b}}_m = \mathbf{D}_J^{1/2}\mathbf{a}_m$ is the mth eigen-vector of $\mathbf{Z}^T\mathbf{Z}$.

3 Linear Transformations and Reciprocal Averaging

In Sect. 2 we outlined that the traditional approach to reciprocal averaging involves the weighted (arithmetic) mean of the elements of the centred row and column profiles. There are situations where considering a power transformation of these elements is warrented. For example, it may involve a square root power to help stabilise the variance of the cell frequencies when overdispersion is present in the data; this may be arise due to the underlying assumption that cell frequencies are Poisson random variables. One may also wish to determine the limiting value of such a transformation as the power approaches zero so that a (natural) logarithmic transformation is considered. It might be that chi-squared distributed measures of association other than Pearson's statistic are used. Such situations were considered by Beh and Lombardo (2024) although earlier discussions on the role of power transformation involving contingency tables has been a topic of much discussion. In particular, one may consider the interconnected issues described by Cuadras and Cuadras (2006, 2015), Cuadras et al. (2006), and Greenacre (2009, 2010) who provide a discussion of the role of power transformations in the context of correspondence analysis. For more general, and earlier, discussions of the power transformation for the contingency table the interested reader is directed to Anscombe (1953, pp. 229–230), Bishop et al. (2007, Example 14.6–3) and McCullagh and Nelder (1984, p. 38).

We shall be considering a purely numeric account of the role of power transformations by examining its role in the context of reciprocal averaging and canonical correlation analysis. There are two types of transformations that can be considered and involve:

1. transforming only p_{ij} so that we have p_{ij}^{δ} where the row and column marginal totals are defined as:

$$p_{i\bullet}(\delta) = \sum_{j=1}^{J} p_{ij}^{\delta} \quad \text{and} \quad p_{\bullet j}(\delta) = \sum_{i=1}^{I} p_{ij}^{\delta},$$

2. defining the transformation of the elements of the profiles such that:

$$\left(\frac{p_{ij}}{p_{i\bullet}}\right)^{\delta} \quad \text{and} \quad \left(\frac{p_{ij}}{p_{\bullet j}}\right)^{\delta}.$$

Greenacre (2009) also considered transformations related to these and referred to the first type as the "power family 1" transformation and the second type as the "power family 2" transformation. A transformation somewhat related to the first type is examined in the context of reciprocal averaging by Wang et al. (2023) and so we shall confine our attention in this paper to the second type of transformation. In doing so, we define $\mathbf{P}^{\delta}$ to be the matrix of p_{ij}^{δ} elements. Similarly, $p_{i\bullet}^{\delta}$ is ith element of the vector $\mathbf{r}^{\delta}$ and the (i, i)th element of $\mathbf{D}_I^{\delta}$ while the $p_{\bullet j}^{\delta}$ is the jth element of the vector $\mathbf{c}^{\delta}$ and the (j, j)th element of $\mathbf{D}_J^{\delta}$.

4 The Reciprocal Averaging Procedure

4.1 The Setup

Suppose we denote $a_{im}(\delta)$ as the ith row score and $b_{jm}(\delta)$ to be the jth column score for a given value of δ. Then, the reciprocal averaging of the power transformed profile elements involves solving $a_{im}(\delta)$ and $b_{jm}(\delta)$ so that:

$$\begin{aligned}\lambda_m(\delta)\, a_{im}(\delta) &= \left(\left(\frac{p_{i1}}{p_{i\bullet}}\right)^{\delta} - p_{\bullet 1}^{\delta}\right) b_{1m}(\delta) + \cdots + \left(\left(\frac{p_{iJ}}{p_{i\bullet}}\right)^{\delta} - p_{\bullet J}^{\delta}\right) b_{Jm}(\delta) \\ &= \sum_{j=1}^{J}\left(\left(\frac{p_{ij}}{p_{i\bullet}}\right)^{\delta} - p_{\bullet j}^{\delta}\right) b_{jm}(\delta) \qquad (7)\end{aligned}$$

and

$$\lambda_m(\delta)\, b_{jm}(\delta) = \left(\left(\frac{p_{1j}}{p_{\bullet j}}\right)^{\delta} - p_{1\bullet}^{\delta}\right) a_{1m}(\delta) + \cdots + \left(\left(\frac{p_{Ij}}{p_{\bullet j}}\right)^{\delta} - p_{I\bullet}^{\delta}\right) a_{Im}(\delta)$$
$$= \sum_{i=1}^{I} \left(\left(\frac{p_{ij}}{p_{\bullet j}}\right)^{\delta} - p_{i\bullet}^{\delta}\right) a_{im}(\delta)\ . \qquad (8)$$

Here

$$\sum_{i=1}^{I} p_{i\bullet}^{\delta} a_{im}(\delta) = 0, \qquad \sum_{i=1}^{I} p_{i\bullet}^{\delta} a_{im}^{2}(\delta) = 1 \qquad (9)$$

and

$$\sum_{j=1}^{J} p_{\bullet j}^{\delta} b_{jm}(\delta) = 0 \qquad \sum_{j=1}^{J} p_{\bullet j}^{\delta} b_{jm}^{2}(\delta) = 1, \qquad (10)$$

so that

$$\lambda_m(\delta) = \sum_{i=1}^{I}\sum_{j=1}^{J} p_{ij}^{\delta} a_{im}(\delta)\, b_{jm}(\delta) \qquad (11)$$

is the correlation between the set of row scores $\mathbf{a}_m(\delta) = (a_{1m}(\delta), \ldots, a_{Im}(\delta))^T$ and column scores, $\mathbf{b}_m(\delta) = (b_{1m}(\delta), \ldots, b_{Jm}(\delta))^T$—we confirm in Sect. 4.3 that, given the chosen value of δ, $\lambda_m(\delta)$ is the maximum possible correlation between $\mathbf{a}_m(\delta)$ and $\mathbf{b}_m(\delta)$. We note that when $\delta = 1$, the above equations are equivalent to those described in Sect. 2.3. That is $a_{im}(1) \equiv a_{im}$, $b_{im}(1) \equiv b_{im}$ and $\lambda_m(1) \equiv \lambda_m$.

Equations (7) and (8) may be expressed in matrix form since they are elements of:

$$\lambda_m(\delta)\, \mathbf{a}_m(\delta) = \left(\mathbf{D}_I^{-\delta}\mathbf{P}^{\delta} - \mathbf{1}_I\left(\mathbf{c}^T\right)^{\delta}\right) \mathbf{b}_m(\delta) \qquad (12)$$

and

$$\lambda_m(\delta)\, \mathbf{b}_m(\delta) = \left(\mathbf{D}_J^{-\delta}\left(\mathbf{P}^T\right)^{\delta} - \mathbf{1}_J\left(\mathbf{r}^T\right)^{\delta}\right) \mathbf{a}_m(\delta)\ , \qquad (13)$$

respectively. Properties (9) and (10) can then be expressed as:

$$\left(\mathbf{r}^{\delta}\right)^T \mathbf{a}_m(\delta) = 0 \quad \text{and} \quad \left(\mathbf{c}^{\delta}\right)^T \mathbf{b}_m(\delta) = 0 \qquad (14)$$

and

$$\mathbf{a}_m(\delta)^T\, \mathbf{D}_I^{\delta} \mathbf{a}_m(\delta) = 1 \quad \text{and} \quad \mathbf{b}_m(\delta)^T\, \mathbf{D}_J^{\delta} \mathbf{b}_m(\delta) = 1\ , \qquad (15)$$

respectively. We now define $\mathbf{A}_{\delta}$ to be the $I \times M$ column matrix containing the vector of row scores, $\mathbf{a}_m(\delta)$, and $\mathbf{B}_{\delta}$ to be the $J \times M$ column matrix containing the column scores $\mathbf{b}_m(\delta)$. Thus, these properties can be defined as:

$$\mathbf{A}_\delta^T \mathbf{D}_I^\delta \mathbf{A}_\delta = \mathbf{I}_M \quad \text{and} \quad \mathbf{B}_\delta^T \mathbf{D}_J^\delta \mathbf{B}_\delta = \mathbf{I}_M, \tag{16}$$

where $\mathbf{I}_M$ is an $M \times M$ identity matrix. In the appendix of this paper we outline an R function called `rapower.exe()` that performs the reciprocal averaging procedure described here for $m = 1$.

4.2 The Eigen-Decomposition Solution

Suppose we pre-multiply both sides of (12) by $\lambda_m(\delta)\,\mathbf{D}_I^{\delta/2}$ so that we have:

$$\begin{aligned} &\lambda_m^2(\delta)\left(\mathbf{D}_I^{\delta/2}\mathbf{a}_m(\delta)\right) \\ &\quad = \lambda_m(\delta)\,\mathbf{D}_I^{\delta/2}\left(\mathbf{D}_I^{-\delta}\mathbf{P}^\delta - \mathbf{1}_I\left(\mathbf{c}^\delta\right)^T\right)\mathbf{b}_m(\delta) \\ &\quad = \left[\mathbf{D}_I^{-\delta/2}\left(\mathbf{P}^\delta - \left(\mathbf{r}\mathbf{c}^T\right)^\delta\right)\mathbf{D}_J^{-\delta/2}\right]\left(\lambda_m(\delta)\,\mathbf{D}_J^{\delta/2}\mathbf{b}_m(\delta)\right). \end{aligned} \tag{17}$$

Now, pre-multiplying both sides of (13) by $\mathbf{D}_J^{\delta/2}$ gives us:

$$\begin{aligned} &\lambda_m(\delta)\,\mathbf{D}_J^{\delta/2}\mathbf{b}_m(\delta) \\ &\quad = \mathbf{D}_J^{\delta/2}\left(\mathbf{D}_J^{-\delta}\left(\mathbf{P}^T\right)^\delta - \mathbf{1}_J\left(\mathbf{r}^T\right)^\delta\right)\mathbf{a}_m(\delta) \\ &\quad = \left[\mathbf{D}_J^{-\delta/2}\left(\mathbf{P}^\delta - \left(\mathbf{r}\mathbf{c}^T\right)^\delta\right)^T\mathbf{D}_I^{-\delta/2}\right]\left(\mathbf{D}_I^{\delta/2}\mathbf{a}_m(\delta)\right). \end{aligned} \tag{18}$$

We shall show in Sect. 4.3 that, for both (17) and (18), $\lambda_m(\delta)$ is defined by (11) and is the maximum (positive) correlation between $\mathbf{a}_m(\delta)$ and $\mathbf{b}_m(\delta)$ for a fixed value of δ.

To simplify (17) and (18), we let:

$$\mathbf{Z}_\delta = \mathbf{D}_I^{-\delta/2}\left(\mathbf{P}^\delta - \left(\mathbf{r}\mathbf{c}^T\right)^\delta\right)\mathbf{D}_J^{-\delta/2} \tag{19}$$

be the matrix of standardised residuals after a power transformation has been applied to the profile elements. For example, when $\delta = 1$, (19) simplifies to (6). Therefore, (17) becomes:

$$\lambda_m^2(\delta)\left(\mathbf{D}_I^{\delta/2}\mathbf{a}_m(\delta)\right) = \mathbf{Z}_\delta\left(\lambda_m(\delta)\,\mathbf{D}_J^{\delta/2}\mathbf{b}_m(\delta)\right) \tag{20}$$

and (18) is:

$$\lambda_m(\delta)\,\mathbf{D}_J^{\delta/2}\mathbf{b}_m(\delta) = \mathbf{Z}_\delta^T\left(\mathbf{D}_I^{\delta/2}\mathbf{a}_m(\delta)\right). \tag{21}$$

Substituting (21) into (20) gives us:

$$\lambda_m^2(\delta)\left(\mathbf{D}_I^{\delta/2}\mathbf{a}_m(\delta)\right) = \mathbf{Z}_\delta\mathbf{Z}_\delta^T\left(\mathbf{D}_I^{\delta/2}\mathbf{a}_m(\delta)\right)$$

which can be expressed as the eigen-decomposition equation:

$$\left(\mathbf{Z}_\delta\mathbf{Z}_\delta^T - \lambda_m^2(\delta)\mathbf{I}_I\right)\left(\mathbf{D}_I^{\delta/2}\mathbf{a}_m(\delta)\right) = \mathbf{0}_I\,. \tag{22}$$

Suppose we now denote:

$$\tilde{\mathbf{a}}_m(\delta) = \mathbf{D}_I^{\delta/2}\mathbf{a}_m(\delta)$$

then (22) becomes:

$$\left(\mathbf{Z}_\delta\mathbf{Z}_\delta^T - \lambda_m^2(\delta)\mathbf{I}_I\right)\tilde{\mathbf{a}}_m(\delta) = \mathbf{0}_I\,. \tag{23}$$

Therefore, $\tilde{\mathbf{a}}_m(\delta)$ can be determined from the eigen-decomposition of $\mathbf{Z}_\delta\mathbf{Z}_\delta^T$ and is the mth eigen-vector of the matrix, while $\lambda_m^2(\delta)$ is its mth largest eigen-value. Thus the set of row scores, $\mathbf{a}_m(\delta)$ can be determined by performing an eigen-decomposition of $\mathbf{Z}_\delta\mathbf{Z}_\delta^T$ and calculating $\mathbf{D}_I^{-\delta/2}\tilde{\mathbf{a}}_m(\delta)$.

By following a similar derivation we also get:

$$\left(\mathbf{Z}_\delta^T\mathbf{Z}_\delta - \lambda_m^2(\delta)\mathbf{I}_I\right)\tilde{\mathbf{b}}_m(\delta) = \mathbf{0}_I, \tag{24}$$

where

$$\tilde{\mathbf{b}}_m(\delta) = \mathbf{D}_J^{\delta/2}\mathbf{b}_m(\delta)$$

is the mth eigen-vector of $\mathbf{Z}_\delta^T\mathbf{Z}_\delta$ so that $\lambda_m^2(\delta)$ is also the mth largest eigen-value of this matrix. Thus, the vector $\mathbf{b}_m(\delta)$ can be determined from the eigen-decomposition of $\mathbf{Z}_\delta^T\mathbf{Z}_\delta$ by pre-multiplying $\tilde{\mathbf{b}}_m(\delta)$ by $\mathbf{D}_J^{-\delta/2}$.

Given the property that $\mathbf{A}_\delta$ and $\mathbf{B}_\delta$ fulfil—see (16)—the properties met by the column matrices containing $\tilde{\mathbf{a}}_m(\delta)$ and $\tilde{\mathbf{b}}_m(\delta)$—defined by $\tilde{\mathbf{A}}_\delta$ and $\tilde{\mathbf{B}}_\delta$, respectively—are:

$$\mathbf{A}_\delta^T\mathbf{D}_I^\delta\mathbf{A}_\delta = \left(\mathbf{D}_I^{-\delta/2}\tilde{\mathbf{A}}_\delta\right)^T\mathbf{D}_I^\delta\left(\mathbf{D}_I^{-\delta/2}\tilde{\mathbf{A}}_\delta\right) = \tilde{\mathbf{A}}_\delta^T\tilde{\mathbf{A}}_\delta = \mathbf{I}_M \tag{25}$$

and

$$\mathbf{B}_\delta^T\mathbf{D}_J^\delta\mathbf{B}_\delta = \left(\mathbf{D}_J^{-\delta/2}\tilde{\mathbf{B}}_\delta\right)^T\mathbf{D}_J^\delta\left(\mathbf{D}_J^{-\delta/2}\tilde{\mathbf{B}}_\delta\right) = \tilde{\mathbf{B}}_\delta^T\tilde{\mathbf{B}}_\delta = \mathbf{I}_M\,. \tag{26}$$

4.3 A Canonical Correlation Solution

At the heart of reciprocal averaging/dual scaling is the idea that the scores are determined so that one obtains the maximum (positive) correlation that exists between them while ensuring that the scores also maximise any differences that exist between its categories of the row variable and the column variable. We can show that the correlation obtained from the reciprocal averaging procedure described in Sect. 4.2 is the maximum possible correlation between $\mathbf{a}_m(\delta)$ and $\mathbf{b}_m(\delta)$.

Following on from our above discussion, we define the correlation between $\mathbf{a}_m(\delta)$ and $\mathbf{b}_m(\delta)$ by:

$$\begin{aligned}
\lambda_m(\delta) &= \mathrm{Corr}\left(\mathbf{a}_m(\delta), \mathbf{b}_m(\delta)\right)\\
&= \frac{\left(\mathbf{a}_m(\delta) - \mathrm{E}\left(\mathbf{a}_m(\delta)\right)\right)^T}{\sqrt{\mathrm{Var}\left(\mathbf{a}_m(\delta) - \mathrm{E}\left(\mathbf{a}_m(\delta)\right)\right)}}\mathbf{P}^\delta \frac{\left(\mathbf{b}_m(\delta) - \mathrm{E}\left(\mathbf{b}_m(\delta)\right)\right)}{\sqrt{\mathrm{Var}\left(\mathbf{b}_m(\delta) - \mathrm{E}\left(\mathbf{b}_m(\delta)\right)\right)}}\\
&= \frac{\left(\mathbf{a}_m(\delta) - \mathrm{E}\left(\mathbf{a}_m(\delta)\right)\right)^T}{\sqrt{\left(\mathbf{a}_m(\delta) - \mathrm{E}\left(\mathbf{a}_m(\delta)\right)\right)^T \mathbf{D}_I^\delta \left(\mathbf{a}_m(\delta) - \mathrm{E}\left(\mathbf{a}_m(\delta)\right)\right)}}\mathbf{P}^\delta \times\\
&\qquad \frac{\left(\mathbf{b}_m(\delta) - \mathrm{E}\left(\mathbf{b}_m(\delta)\right)\right)^T}{\sqrt{\left(\mathbf{b}_m(\delta) - \mathrm{E}\left(\mathbf{b}_m(\delta)\right)\right)^T \mathbf{D}_J^\delta \left(\mathbf{b}_m(\delta) - \mathrm{E}\left(\mathbf{b}_m(\delta)\right)\right)}}.
\end{aligned}$$

At this stage, there is no need to impose any property that $\mathbf{a}_m(\delta)$ and $\mathbf{b}_m(\delta)$ must abide. Although to help simplify the derivations we shall let $\mathbf{a}_m^*(\delta) = \mathbf{a}_m(\delta) - \mathrm{E}\left(\mathbf{a}_m(\delta)\right)$ and $\mathbf{b}_m^*(\delta) = \mathbf{b}_m(\delta) - \mathrm{E}\left(\mathbf{b}_m(\delta)\right)$. Therefore:

$$\lambda_m(\delta) = \frac{\mathbf{a}_m^*(\delta)^T \mathbf{P}^\delta \mathbf{b}_m^*(\delta)}{\sqrt{\left(\mathbf{a}_m^*(\delta)^T \mathbf{D}_I^\delta \mathbf{a}_m^*(\delta)\right)\left(\mathbf{b}_m^*(\delta)^T \mathbf{D}_J^\delta \mathbf{b}_m^*(\delta)\right)}}.$$

Squaring this correlation gives:

$$\lambda_m(\delta)^2 = \left(\mathbf{a}_m^*(\delta)^T \mathbf{D}_I^\delta \mathbf{a}_m^*(\delta)\right)^{-1}\left(\mathbf{a}_m^*(\delta)^T \mathbf{P}^\delta \mathbf{b}_m^*(\delta)\right)^2\left(\mathbf{b}_m^*(\delta)^T \mathbf{D}_J^\delta \mathbf{b}_m^*(\delta)\right)^{-1}. \quad (27)$$

To maximise this squared correlation we begin by first differentiating it with respect to $\mathbf{a}_m(\delta)$. Doing so is done by noting that:

$$\frac{\partial \lambda_m(\delta)^2}{\partial \mathbf{a}_m(\delta)} = \frac{\partial \lambda_m(\delta)^2}{\partial \mathbf{a}_m^*(\delta)}\frac{\partial \mathbf{a}_m^*(\delta)}{\partial \mathbf{a}_m(\delta)} = \frac{\partial \lambda_m(\delta)^2}{\partial \mathbf{a}_m^*(\delta)}.$$

Therefore:

$$\begin{aligned}\frac{\partial \lambda_m(\delta)^2}{\partial \mathbf{a}_m(\delta)} &= 2\left(\mathbf{a}_m(\delta)^T \mathbf{D}_I^\delta \mathbf{a}_m(\delta)\right)^{-1}\left(\mathbf{a}_m(\delta)^T \mathbf{P}^\delta \mathbf{b}_m(\delta)\right)\left(\mathbf{P}^\delta \mathbf{a}_m(\delta)\right) \\ &\quad \times \left(\mathbf{b}_m(\delta)^T \mathbf{D}_J^\delta \mathbf{b}_m(\delta)\right)^{-1} - 2\left(\mathbf{a}_m(\delta)^T \mathbf{D}_I^\delta \mathbf{a}_m(\delta)\right)^{-2}\left(\mathbf{D}_I^\delta \mathbf{a}_m(\delta)\right) \\ &\quad \times \left(\mathbf{a}_m(\delta)^T \mathbf{P}^\delta \mathbf{b}_m(\delta)\right)^2\left(\mathbf{b}_m(\delta)^T \mathbf{D}_J^\delta \mathbf{b}_m(\delta)\right)^{-1} \\ &= 0\,. \end{aligned} \tag{28}$$

Similarly, differentiating (27) with respect to $\mathbf{b}_m(\delta)$ leads to:

$$\begin{aligned}\frac{\partial \lambda_m(\delta)^2}{\partial \mathbf{b}_m(\delta)} &= 2\left(\mathbf{b}_m(\delta)^T \mathbf{D}_J^\delta \mathbf{b}_m(\delta)\right)^{-1}\left(\mathbf{b}_m(\delta)^T \mathbf{P}^\delta \mathbf{b}_m(\delta)\right)\left(\left(\mathbf{P}^\delta\right)^T \mathbf{a}_m(\delta)\right) \\ &\quad \times \left(\mathbf{a}_m(\delta)^T \mathbf{D}_I^\delta \mathbf{a}_m(\delta)\right)^{-1} - 2\left(\mathbf{b}_m(\delta)^T \mathbf{D}_J^\delta \mathbf{b}_m(\delta)\right)^{-2}\left(\mathbf{D}_J^\delta \mathbf{b}_m(\delta)\right) \\ &\quad \times \left(\mathbf{b}_m(\delta)^T \left(\mathbf{P}^\delta\right)^T \mathbf{a}_m(\delta)\right)^2\left(\mathbf{a}_m(\delta)^T \mathbf{D}_I^\delta \mathbf{a}_m(\delta)\right)^{-1} \\ &= 0\,. \end{aligned} \tag{29}$$

Suppose we now let $\mathbf{a}_m(\delta)$ and $\mathbf{b}_m(\delta)$ be subject to (14) and (15). Then (28) and (29) simplify to:

$$\frac{\partial \lambda_m(\delta)^2}{\partial \mathbf{a}_m(\delta)} = 2\lambda_m(\delta)\left(\mathbf{P}^\delta \mathbf{b}_m(\delta)\right) - 2\lambda_m(\delta)^2\left(\mathbf{D}_I^\delta \mathbf{a}_m(\delta)\right) = 0 \tag{30}$$

$$\frac{\partial \lambda_m(\delta)^2}{\partial \mathbf{b}_m(\delta)} = 2\lambda_m(\delta)\left(\mathbf{a}_m(\delta)^T \mathbf{P}^\delta\right) - 2\lambda_m(\delta)^2\left(\mathbf{b}_m(\delta)^T \mathbf{D}_J^\delta\right) = 0\,, \tag{31}$$

respectively, while (27) simplifies to:

$$\lambda_m(\delta)^2 = \left(\mathbf{a}_m(\delta)^T \mathbf{P}^\delta \mathbf{b}_m(\delta)\right)^2,$$

so that taking the square root of both sides gives elements that are equivalent to (11). We can verify that this is indeed the maximum (squared) correlation between the set of row scores $\mathbf{a}_m(\delta)$ and column scores $\mathbf{b}_m(\delta)$ for the chosen value of δ since, for all $\delta \neq 0$:

$$\frac{\partial^2 \lambda_m(\delta)^2}{\partial \mathbf{a}_m(\delta)^2} = -2\lambda_m(\delta)^2 \mathbf{D}_I^\delta < 0,$$

$$\frac{\partial^2 \lambda_m(\delta)^2}{\partial \mathbf{b}_m(\delta)^2} = -2\lambda_m(\delta)^2 \mathbf{D}_J^\delta < 0\,.$$

We can verify that the solution to $\mathbf{a}_m(\delta)$ and $\mathbf{b}_m(\delta)$ are just (12) and (13). To show this, note that (30) and (31) reduce to:

$$\begin{aligned}\mathbf{P}^\delta \mathbf{b}_m(\delta) &= \lambda_m(\delta)\left(\mathbf{D}_I^\delta \mathbf{a}_m(\delta)\right) \\ \mathbf{a}_m(\delta)^T \mathbf{P}^\delta &= \lambda_m(\delta)\left(\mathbf{b}_m(\delta)^T \mathbf{D}_J^\delta\right)\end{aligned}$$

for $\lambda_m(\delta) \neq 0$ which can be alternatively, and equivalently, expressed as:

$$\lambda_m(\delta)\,\mathbf{a}_m(\delta) = \left(\mathbf{D}_I^{-\delta}\mathbf{P}^{\delta}\right)\mathbf{b}_m(\delta) = \left(\mathbf{D}_I^{-\delta}\mathbf{P}^{\delta} - \mathbf{1}_I\left(\mathbf{c}^{\delta}\right)^T\right)\mathbf{b}_m(\delta)$$
$$\lambda_m(\delta)\,\mathbf{b}_m(\delta) = \left(\mathbf{D}_J^{-\delta}\left(\mathbf{P}^T\right)^{\delta}\right)\mathbf{a}_m(\delta) = \left(\mathbf{D}_J^{-\delta}\left(\mathbf{P}^{\delta}\right)^T - \mathbf{1}_J\left(\mathbf{r}^{\delta}\right)^T\right)\mathbf{a}_m(\delta)\,,$$

since $\left(\mathbf{r}^T\right)^{\delta}\mathbf{a}_m(\delta) = 0$ and $\left(\mathbf{c}^T\right)^{\delta}\mathbf{b}_m(\delta) = 0$ – see (14). These results are just those of (12) and (13), respectively. Therefore, canonical correlation analysis yields row and column scores, $\mathbf{a}_m(\delta)$ and $\mathbf{b}_m(\delta)$, respectively, that are identical to those obtained via reciprocal averaging with $\lambda_m(\delta)$ being the maximum possible (positive) correlation between $\mathbf{a}_m(\delta)$ and $\mathbf{b}_m(\delta)$.

5 The Solution Using Singular Value Decomposition

Rather than performing two eigen-decompositions to determine the set of row scores $\mathbf{a}_m(\delta)$ and the set of column scores $\mathbf{b}_m(\delta)$—as (23) and (24) do—we can instead apply a singular value decomposition (SVD) to the matrix of residuals $\mathbf{Z}_\delta$ defined by (19). By doing so, we have:

$$\mathbf{Z}_\delta = \mathbf{D}_I^{-\delta/2}\left(\mathbf{P}^{\delta} - \left(\mathbf{r}\mathbf{c}^T\right)^{\delta}\right)\mathbf{D}_J^{-\delta/2} = \tilde{\mathbf{A}}_\delta\mathbf{D}_\delta\tilde{\mathbf{B}}_\delta^T\,, \tag{32}$$

where $\tilde{\mathbf{A}}_\delta$ is subject to (25), $\tilde{\mathbf{B}}_\delta$ is subject to (26) and $\mathbf{D}_\delta$ is the diagonal matrix where the $(m,\ m)$th element is $\lambda_m(\delta)$, the mth singular value of $\tilde{\mathbf{Z}}_\delta$. The advantage of considering (32) is that the properties underlying $\tilde{\mathbf{A}}_\delta$ and $\tilde{\mathbf{B}}_\delta$ are those adopted by the `svd()` function in R and so it is (32) that is central to the calculations performed in Sect. 6.3.

The matrix form of the row and column scores that motivated our discussion—$\mathbf{A}_\delta$ and $\mathbf{B}_\delta$—can be found from the SVD of $\mathbf{Z}_\delta$, by rescaling $\tilde{\mathbf{A}}_\delta$ and $\tilde{\mathbf{B}}_\delta$ so that:

$$\mathbf{A}_\delta = \mathbf{D}_I^{-\delta/2}\tilde{\mathbf{A}}_\delta \quad \text{and} \quad \mathbf{B}_\delta = \mathbf{D}_J^{-\delta/2}\tilde{\mathbf{B}}_\delta\,. \tag{33}$$

Recall that these matrices have the property given by (16).

6 Application: Selikoff's Asbestos Data

6.1 The Data

Consider the 5×4 contingency table of Table 1 that comes from a study undertaken in 1963 and whose findings were not published until 1981 (Selikoff 1981). Irvin Selikoff was a chest physician in New York. With his team, Selikoff examined 1117 New York construction workers that were exposed to asbestos fibres. They established that there

Table 1 Selikoff's data for studying the link between years of exposure to asbestos fibres and severity of asbestosis

	Asbestosis grade diagnosed				
Occupational exposure (years)	None	Grade 1	Grade 2	Grade 3	Total
0–9	310	36	0	0	346
10–19	212	158	9	0	379
20–29	21	35	17	4	77
30–39	25	102	49	18	194
40+	7	35	51	28	121
Total	575	366	126	50	1117

was a link between the number of years of occupational exposure to asbestos fibres and the severity of asbestosis that the worker was diagnosed with; from a preliminary analysis of the data in Table 1, Selikoff posited the "20-year rule" (Selikoff 1981, p. 948) stating that "it was only after the 20-year point that most reontgenograms became abnormal". The impact of being exposed to asbestos fibres on an person's health has since been felt internationally with many countries banning the production and importation of products containing asbestos fibres; see Beh and Smith (2011), Tran et al. (2012) and Beh and Lombardo (2014, Sect. 1.4) for a discussion of this issue from a categorical data analysis perspective and the references mentioned within for additional global contexts.

Table 1 cross-classifies 5 different lengths of time that a worker was exposed to asbestos (in intervals of 10 years) and four grades (of severity) of asbestosis that the workers were diagnosed with; this data also appears in Table 1 of Selikoff (1981) with a reorganisation of the categories. A chi-squared test of independence of Table 1 gives a Pearson statistic of 648.81 with $(5-1)(4-1) = 12$ degrees of freedom. Thus, there exists a statistically significant association between the years of exposure to asbestos and the diagnosed level of asbestosis since the p-value of each of these test statistics is less than 0.001.

One could perform a correspondence analysis to visually identify the nature of the association that exists between the variables. Although this was comprehensively done by Beh and Smith (2011) and Beh and Lombardo (2014). Therefore, we shall confine our attention to the calculation of the singular vectors and singular values for various values of δ.

6.2 *Reciprocal Averaging*

Suppose we consider for the moment determining the set of uni-dimensional row scores:

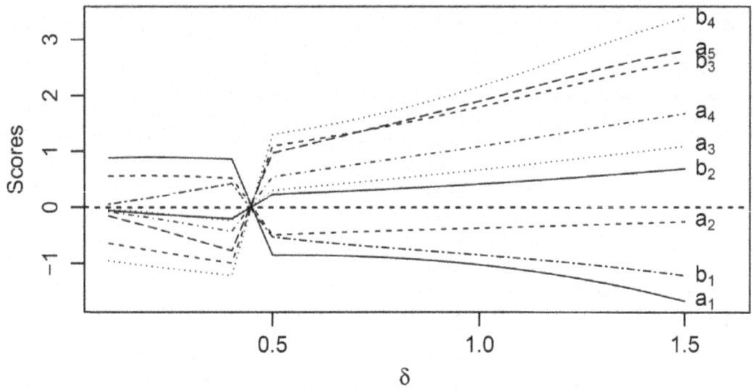

Fig. 1 Plot of the row and column scores versus $\delta \in [0.1,\ 1.5]$ for Table 1

$$\mathbf{a}_1(\delta) = (a_1(\delta), \cdots, a_5(\delta))^T,$$

and column scores:

$$\mathbf{b}_1(\delta) = (b_1(\delta), \cdots, b_4(\delta))^T.$$

We shall confine our attention to $\delta \in [0.1,\ 1.5]$ where the initial set of row scores that were used were (1, 2, 3, 4, 5) while (1, 2, 3, 4) were used as the set of initial column scores. The initial value of λ is set to 1.

Figure 1 shows the changes in the set of row scores, $\mathbf{a}_1(\delta)$, and changes in the set of column scores, $\mathbf{b}_1(\delta)$, for $\delta \in [0.1,\ 1.5]$. It shows that, irrespective of the value of δ, the association that exists between specific rows and columns remains unchanged. For example, $a_1(\delta)$ (for the row category *0 - 9 years*) is always associated with $b_1(\delta)$ (for the column category *None*). Similarly, the longest years of exposure to asbestos fibres, $a_5(\delta)$ (for the row category *40+ years*) is always associated with the most severe case of asbestosis, $b_4(\delta)$ (for the column category *Grade 3*). However, there is a change in the sign of the scores around $\delta = 0.4$.

To highlight the changes in the correlation between these row and column scores as δ shifts from 0.1 to 1.5, Fig. 2 shows $\lambda_1(\delta)$ versus δ. It shows that the maximum correlation of $\lambda_1(\delta) = 0.91755$ is achieved $\delta = 0.506$. Thus, while we know that any given value of δ ensures the correlation between the row and column scores is maximised for that value of δ, for Table 1, it is a square root transformation of profiles that produces a near maximum possible correlation between them. When $\delta = 0.506$ the row and column scores are:

$$\mathbf{a}_1(\delta = 0.506) = (-0.84997,\ -0.48866,\ 0.30848,\ 0.55218,\ 0.98048)^T$$

and

$$\mathbf{b}_1(\delta = 0.506) = (-0.53114,\ 0.23426,\ 1.10893,\ 1.31583)^T,$$

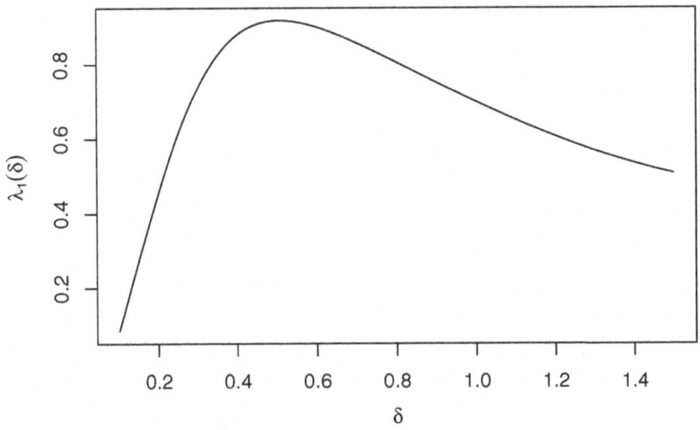

Fig. 2 Plot of $\delta \in [0.1,\ 1.5]$ versus $\lambda_1(\delta)$ for Table 1

respectively. For comparisons, a unitary transformation ($\delta = 1$) produces row and column scores:

$$\mathbf{a}_1(\delta = 1) = (-1.02368,\ -0.36766,\ 0.66890,\ 1.09353,\ 1.89988)^T$$

and

$$\mathbf{b}_1(\delta = 1) = (-0.84693,\ 0.41606,\ 1.79902,\ 2.16057)^T\ ,$$

respectively. These scores are equivalent to those obtained using the traditional reciprocal averaging method and the correlation between them is $\lambda_1(1) = 0.69940$. For these two values of δ (and for others that can be considered), $\mathbf{a}_1(\delta)^T\,\mathbf{D}_I^\delta \mathbf{a}_1(\delta) = 1$ and $\mathbf{b}_1(\delta)^T\,\mathbf{D}_J^\delta \mathbf{b}_1(\delta) = 1$. Table 2 gives the row scores, column scores, and their correlation (to five decimal places) for Table 1. We have selected values of δ ranging from 0.1 to 1.5 at increments of 0.2. The number of iterations for convergence to five decimal places to occur is also given.

6.3 SVD Solution

The solutions to $\mathbf{a}_1(\delta)$ and $\mathbf{b}_1(\delta)$ in Sect. 6.2 are only one-dimensional but can be generalised to M dimensions. To discuss these solutions we first consider the matrix of residuals, $\mathbf{Z}_\delta$, defined by (19), which are summarised in Table 3 for $\delta = 1, 0.5$ and 1.3. Since $M = \min(5, 4) - 1 = 3$ we produce the 5×3 matrix of row scores, $\mathbf{A}_\delta$. The elements of this matrix are summarised in Table 4 for $\delta = 1,\ 0.5$ and 1.3. Similarly, the 4×3 matrix of column scores, $\mathbf{B}_\delta$, are summarised in Table 5 for these δ values. These scores are calculated by first applying the `svd()` function in R to the matrix of elements in Table 3. The `svd()` function produces the the matrices

Table 2 Row scores, column scores and correlation from the reciprocal averaging for Table 1 when $\delta \in [0.1,\ 1.5]$

	δ							
Score	0.1	0.3	0.5	0.7	0.9	1.1	1.3	1.5
$a_1(\delta)$	0.88436	0.88524	−0.85037	−0.86900	−0.95510	−1.10872	−1.34407	−1.67244
$a_2(\delta)$	0.55889	0.55418	−0.49058	−0.43428	−0.38887	−0.34791	−0.30275	−0.25517
$a_3(\delta)$	−0.03714	−0.15982	0.30424	0.44444	0.59089	0.74958	0.92243	1.09741
$a_4(\delta)$	−0.07647	−0.30268	0.54535	0.76408	0.98108	1.20809	1.44709	1.67839
$a_5(\delta)$	−0.15030	−0.55994	0.96909	1.33702	1.70770	2.09875	2.48137	2.79853
$b_1(\delta)$	0.05989	0.30163	−0.52581	−0.67306	−0.78959	−0.90696	−1.04524	−1.21570
$b_2(\delta)$	−0.05475	−0.15831	0.23231	0.29773	0.37279	0.46255	0.56988	0.68624
$b_3(\delta)$	−0.63975	−0.88138	1.10218	1.34617	1.63711	1.96954	2.31247	2.60518
$b_4(\delta)$	−0.95244	−1.13114	1.30939	1.57347	1.94098	2.40186	2.91314	3.38770
$\lambda_1(\delta)$	0.08457	0.73476	0.91743	0.85718	0.75147	0.65094	0.56847	0.50784
Iterations	8	8	7	6	4	4	7	8

$\tilde{\mathbf{A}}_\delta$ and $\tilde{\mathbf{B}}_\delta$ respectively. To ensure that they have the property $\mathbf{A}_\delta^T\mathbf{D}_I^\delta\mathbf{A}_\delta = \mathbf{I}_M$ and $\mathbf{A}_\delta^T\mathbf{D}_I^\delta\mathbf{A}_\delta = \mathbf{I}_M$ we pre-multiply $\tilde{\mathbf{A}}_\delta$ by $\mathbf{D}_I^{-\delta/2}$ and $\tilde{\mathbf{B}}_\delta$ by $\mathbf{D}_J^{-\delta/2}$ thereby calculating $\mathbf{A}_\delta$ and $\mathbf{B}_\delta$; see (33).

When $\delta = 1$, the first dimensional solution to the row and column scores summarised in Tables 4 and 5 is equivalent to that obtained using the traditional reciprocal averaging procedure–see the scores $\mathbf{a}_1$ ($\delta = 1$) and $\mathbf{b}_1$ ($\delta = 1$) in the previous section. The scores calculated using reciprocal averaging procedure and the `svd()` function are accurate to at least the fourth decimal place.

When $\delta = 0.5$ and 1.3, the row and column scores summarised in Table 2 are exactly the same, to four or five decimal places, with the scores calculated using the `svd()` function; see the first column of Tables 4 and 5. A similar level of accuracy can be obtained for other values of δ.

Suppose we now turn our attention to calculating the correlation along the first dimension of the row and column scores obtained from the `svd()` function. This can be achieved using the correlation defined by (11). Table 6 summarises these correlation values and the absolute difference with those obtained from the reciprocal averaging procedure. We can see that using the `svd()` function in R produces correlation values that are accurate to at least the fifth decimal place.

7 Discussion

Profile transformations are not new to the analysis of contingency tables. One can consider Beh and Lombardo (2024), Cuadras and Cuadras (2006), Cuadras et al. (2006) and Greenacre (2009, 2010), especially since their discussion is in terms of the correspondence analysis of a two-way table. However, very little appears to be available (at least, that we are aware of) that examines the issue of power transforma-

Table 3 Residuals, $\mathbf{Z}_\delta$, using (19) when $\delta = 1$, $\delta = 0.5$ (in parentheses), $\delta = 1.3$ (in brackets)

Occupational exposure (years)	Asbestosis grade diagnosed			
	None	Grade 1	Grade 2	Grade 3
	0.296	−0.217	−0.187	−0.118
0–9	(0.202)	(−0.246)	(−0.432)	(−0.343)
	[0.320]	[−0.175]	[−0.113]	[−0.062]
	0.036	0.091	−0.154	−0.123
10–19	(0.027)	(0.074)	(−0.239)	(−0.351)
	[0.037]	[0.088]	[−0.104]	[−0.066]
	−0.089	0.058	0.084	0.009
20–29	(−0.118)	(0.069)	(0.118)	(0.018)
	[−0.064]	[0.045]	[0.059]	[0.005]
	−0.224	0.144	0.173	0.095
30–39	(−0.273)	(0.130)	(0.186)	(0.131)
	[−0.174]	[0.132]	[0.144]	[0.067]
	−0.210	−0.222	0.302	0.290
40+	(−0.323)	(−0.026)	(0.310)	(0.336)
	[−0.144]	[−0.017]	[0.260]	[0.234]

Table 4 Row scores, $\mathbf{A}_\delta$, of Table 1 when $\delta = 1$, $\delta = 0.5$ (in parentheses), $\delta = 1.3$ [in brackets]

Occupational exposure (years)	Dimension		
	1	2	3
	−1.02290	0.94851	−0.38985
0–9	(−0.85037)	(0.32697)	(0.90201)
	[−1.34416]	[1.19850]	[−1.01459]
	−0.36841	−0.91680	0.86632
10–19	(−0.49058)	(−1.00021)	(0.18633)
	[−0.30269]	[−0.88181]	[−1.30718]
	0.66843	−0.58167	−2.72979
20–29	(0.30424)	(−0.60267)	(0.11576)
	[0.92246]	[−0.51343]	[−0.75616]
	1.09292	−0.73852	−0.58441
30–39	(0.54535)	(−0.75422)	(0.45356)
	[1.44713]	[−0.61736]	[−1.69930]
	1.90128	1.71357	1.07537
40+	(0.96909)	(0.27742)	(1.15320)
	[2.48117]	[2.84114]	[−0.87060]

Table 5 Column scores, $\mathbf{B}_\delta$, of Table 1 when $\delta = 1$, $\delta = 0.5$ (in parentheses), $\delta = 1.3$ [in brackets]

Asbestos grade diagnosed	Dimension 1	2	3
	−0.84676	0.47171	−0.05555
None	(−0.52581)	(0.56963)	(−0.87972)
	[−1.04526]	[0.59196]	[−0.95817]
	0.41559	−1.33969	0.29055
Grade 1	(0.23231)	(−0.89489)	(−0.79743)
	[0.56992]	[−1.32742]	[−1.46238]
	1.79933	0.87879	−1.96347
Grade 2	(1.10218)	(−0.00212)	(−0.46582)
	[2.31241]	[1.82513]	[−1.11975]
	2.16133	2.16732	3.45997
Grade 3	(1.30939)	(1.20810)	(−0.19278)
	[2.91301]	[3.71911]	[−1.46567]

Table 6 Maximum correlation of the row and column scores from the SVD of $\mathbf{R}_\delta$ for Table 1 when $\delta \in [0, 1, 1.5]$

Correlation	δ 0.1	0.3	0.5	0.7	0.9	1.1	1.3	1.5
$\lambda_1(\delta)$	0.08456	0.73475	0.91742	0.85718	0.75146	0.65095	0.56848	0.50786
Abs Diff	0.00001	0.00001	0.00001	<0.00001	0.00001	0.00001	0.00001	0.00002

tions from a scaling perspective. Hopefully, this paper fills that void by discussing it in terms of reciprocal averaging and canonical correlation analysis, and the application of Irving Selikoff's asbestos data in Sect. 6.1. We have methodologically, and practically, shown that the calculation of the row and column scores can be found using reciprocal averaging or, for a multi-dimensional solution, from the SVD of the matrix $\mathbf{Z}_\delta$ defined by (19). While we can also show how such scores and their correlation relate to the Cressie-Read family of divergence statistics (Cressie and Read 1984) further work can be undertaken to demonstrate its practical benefits. The links that exist between this family and correspondence analysis were established by Beh and Lombardo (2024).

While we do focus on power transformations of the profile elements from the perspective of reciprocal averaging it is important to keep in mind that reciprocal averaging involves the arithmetic averaging of the transformed elements of the row and column profiles. One may also consider other strategies for finding the centre of the profiles. These include performing reciprocal averaging on the median of the profile elements, or even a geometric or harmonic averaging of the elements.

Considering reciprocal averaging in the median case when examining untransformed elements was discussed by Nishisato (1984) but can be expanded to the transformed case by considering the following two equations:

$$\lambda_m(\delta)\,a_{im}(\delta) = \text{Mdn}_j\left[\left(\left(\frac{p_{ij}}{p_{i\bullet}}\right)^{\delta} - p_{\bullet j}^{\delta}\right) b_{jm}(\delta)\right] \tag{34}$$

and

$$\lambda_m(\delta)\,b_{jm}(\delta) = \text{Mdn}_i\left[\left(\left(\frac{p_{ij}}{p_{\bullet j}}\right)^{\delta} - p_{i\bullet}^{\delta}\right) a_{im}(\delta)\right]. \tag{35}$$

Here, "Mdn_j" is the median of the I elements of the power transformed elements of the centred row profile, while "Mdn_i" is the median of the J elements of the power transformed centred column profile. Note that when $\delta = 1$, (34) and (35) simplify to Eqs. (5) and (6), respectively, of Nishisato (1984), a technique he referred to as the *method of reciprocal medians* (or simply MRM) and was proposed, in part, to "mitigate the problem of extreme weights" [p. 143].

Rather than determining the row and column scores by considering the weighted arithmetic mean of the elements of the profiles, or their median, Nishisato et al. (2021, Chap. 8) proposed a few different ways in which geometric averaging could be performed. These methods, referred to as the methods of *geometric averaging*, were designed for the untransformed case but can be easily amended when considering the transformed version of the profile elements. In this case, the row and column scores, $a_{im}(\delta)$ and $b_{jm}(\delta)$, can be determined from:

$$\lambda_m(\delta)\,a_{im}(\delta) = \left[\prod_{j=1}^{J}\left|\left(\frac{p_{ij}}{p_{i\bullet}}\right)^{\delta} - p_{\bullet j}^{\delta}\right| b_{jm}(\delta)\right]^{1/J} \tag{36}$$

and

$$\lambda_m(\delta)\,b_{jm}(\delta) = \left[\prod_{i=1}^{I}\left|\left(\frac{p_{ij}}{p_{\bullet j}}\right)^{\delta} - p_{i\bullet}^{\delta}\right| a_{im}(\delta)\right]^{1/I}. \tag{37}$$

An alternative set of geometric averaging formulae were also derived, those being:

$$\lambda_m a_{im}(\delta) = \left[\prod_{j=1}^{J}\left(\frac{p_{ij}}{p_{i\bullet}p_{\bullet j}}\right)^{\delta} b_{jm}(\delta)\right]^{1/J} \tag{38}$$

and

$$\lambda_m b_{jm}(\delta) = \left[\prod_{i=1}^{I}\left(\frac{p_{ij}}{p_{i\bullet}p_{\bullet j}}\right)^{\delta} a_{im}(\delta)\right]^{1/I}. \tag{39}$$

In both cases, there is no guarantee that the row and column scores calculated from either set of formulae will be centred at zero or have a unitary variance. However, these properties can be satisfied once convergence has been achieved. We note that when $\delta = 1$, then (36)–(37) simplify to (8.4)–(8.5) of Nishisato et al. (2021, p. 162), while (38)–(39) simplify to Eqs. (8.6)–(8.7) (Nishisato et al. 2021, p. 163). Choulakian (2023) points out that taxicab correspondence analysis (Choulakian 2006) can be performed using the framework outline in this paper by adapting (7) and (8) so that:

$$\lambda_m(\delta)\, a_{im}(\delta) = \sum_{j=1}^{J} \left(\left(\frac{p_{ij}}{p_{i\bullet}} \right)^{\delta} - p_{\bullet j}^{\delta} \right) \operatorname{sgn}\left(b_{jm}(\delta) \right)$$

and

$$\lambda_m(\delta)\, b_{jm}(\delta) = \sum_{i=1}^{I} \left(\left(\frac{p_{ij}}{p_{\bullet j}} \right)^{\delta} - p_{i\bullet}^{\delta} \right) \operatorname{sgn}\left(a_{im}(\delta) \right) .$$

Here, $\operatorname{sgn}(\bullet)$ is the coordinate-wise sign function such that $\operatorname{sgn}(x) = 1$ if $x > 1$ and $\operatorname{sgn}(x) = -1$ if $x \leq 0$. Choulakian (2023) also points out that the method of reciprocal medians (Nishisato 1984) can be performed in the context of a taxicab analysis by replacing $b_{jm}(\delta)$ with $\operatorname{sgn}\left(b_{jm}(\delta)\right)$ on the right-hand side of (34) and $a_{im}(\delta)$ with $\operatorname{sgn}(a_{im}(\delta))$ on the right-hand side of (35). Similarly, a taxicab analysis of the method of geometric averaging can be performed by making a substitution of $\operatorname{sgn}(a_{im}(\delta))$ for $a_{im}(\delta)$ in (37) and (39), and of $\operatorname{sgn}\left(b_{jm}(\delta)\right)$ for $b_{jm}(\delta)$ in (36) and (38).

One unresolved issue with these alternative "averaging" techniques is that, unlike reciprocal averaging in the classic (untransformed) or transformed case considered here, their link with eigen-decomposition and singular value decomposition has not been established. The advantage of any such links is evident in the computational simplicity that can be achieved (at least in most cases) in `R` using the `svd()` function. Although these links, and other research questions that may present themselves, will be discussed at a later date.

Appendix

We present here the `R` function `rapower.exe()` that performs the reciprocal averaging of the elements of the row and column profiles under a power transformation δ (`delta`). The input arguments of the function are:

- the contingency table, **N** which is defined as `data`,
- the power of the transformation, δ (`delta`). By default `delta = 1`,
- a logical argument `iters` that, if it is set to `TRUE` (default), prints to screen the value of each row score, column score and correlation at each iteration.

- the initial value of λ_1 (δ) (`lambda.ini`) which, by default, is set equal to 1, and
- the number of decimal places (`acc`) until convergence of the correlation is reached. By default convergence is set to five decimal places:

```
rapower.exe <- function(data, delta = 1, iters = TRUE,
                                        lambda.ini = 1, acc = 5) {

  #  This function performs a reciprocal averaging using the
  # "power family 2" transformation of Greenacre (2009)

  ################################################################
  # Some basics                                                  #
  ################################################################

  Inames <- dimnames(data)[1] # Row category names
  Jnames <- dimnames(data)[2] # Column category names

  I <- nrow(data)
  J <- ncol(data)
  n <- sum(data)
  P <- data/sum(data)
  pidot <- apply(P, 1, sum)
  pdotj <- apply(P, 2, sum)
  R <- diag(pidot, I, I)
  C <- diag(pdotj, J, J)

  ################################################################
  # The algorithm                                                #
  ################################################################

  a.ini <- c(1:nrow(data))  # Initial value of row scores
  b.ini <- c(1:ncol(data))  # Initial value of column scores

  if (iters == TRUE){
    print(round(c(0, a.ini, b.ini, lambda.ini), digits = acc))
  }
  # The first iteration of the row and columns scores, and their
  # correlation

  a.old <- a.ini/sqrt(t(a.ini)%*%R^delta%*%a.ini)[1,1]
  b.old <- (1/lambda.ini)*(solve(C^delta)%*%t(P^delta) -
                   (rep(1, times = J)%*%t(pidot^delta)))%*%a.old
  b.old <- b.old/sqrt(t(b.old)%*%C^delta%*%b.old)[1,1]
  lamb.old <- (t(a.old)%*%P^delta%*%b.old)[1,1]

  # The iterative step of the algorithm

  counter = 1

  if (iters == TRUE){
    print(round(c(counter, a.old, b.old, lamb.old), digits = acc))
  }

  repeat{
    a.new <- (1/lamb.old)*(solve(R^delta)%*%P^delta -
                    (rep(1, times = I)%*%t(pdotj^delta)))%*%b.old
    b.new <- (1/lamb.old)*(solve(C^delta)%*%t(P^delta) -
                    (rep(1, times = J)%*%t(pidot^delta)))%*%a.new

    a.new <- a.new/sqrt(t(a.new)%*%R^delta%*%a.new)[1,1]
    b.new <- b.new/sqrt(t(b.new)%*%C^delta%*%b.new)[1,1]

    lamb.new <- (t(a.new)%*%P^delta%*%b.new)[1,1]
```

```
    counter <- counter + 1

    if (iters == TRUE){
      print(round(c(counter, a.new, b.new, lamb.new), digits = acc))
    }

    lamb.comp <- abs(lamb.old - lamb.new)

    if (lamb.comp < 10^(-1*acc)) break

    a.old <- a.new
    b.old <- b.new
    lamb.old <- lamb.new
  }

  ##################################################################
  # The numerical output . . .                                     #
  ##################################################################

  dimnames(a.new) <- list(paste(Inames[[1]]), paste("row score"))
  dimnames(b.new) <- list(paste(Jnames[[1]]), paste("col score"))

  list(iterations = round(counter, acc),
       a = round(a.new, acc),
       b = round(b.new, acc),
       lamb = round(lamb.old, acc))
}
```

Therefore, when `asbestos.dat` is the R object given to the two-way contingency table of Table 1 and is defined by:

```
> asbestos.dat <- matrix(c(310, 212, 21, 25, 7, 36, 158, 35,  102,
+                  35, 0, 9, 17, 49, 51, 0, 0, 4, 18, 28), nrow = 5)
> dimnames(asbestos.dat) <- list(paste(c("0-9", "10-19", "20-29",
+                 "30-39", "40+")), paste(c("None", "Grade 1", "Grade 2",
+                  "Grade 3")))
>
> asbestos.dat
      None Grade 1 Grade 2 Grade 3
0-9    310      36       0       0
10-19  212     158       9       0
20-29   21      35      17       4
30-39   25     102      49      18
40+      7      35      51      28
>
```

The traditional reciprocal averaging approach may be performed by defining $\delta = 1$ so that:

```
> rapower.exe(selikoff.dat, iters = F)
$iterations
[1] 3


$a
       row score
0-9     -1.02368
10-19   -0.36766
20-29    0.66890
30-39    1.09353
40+      1.89988
```

```
$b
          col score
None       -0.84693
Grade 1     0.41606
Grade 2     1.79902
Grade 3     2.16057

$lamb
[1] 0.6994

>
```

Note that `$a` and `$b` are the row and column scores given by $\mathbf{a}_1$ ($\delta = 1$) and $\mathbf{b}_1$ ($\delta = 1$), respectively, in Sect. 6.2 while the correlation of 0.69940 appears as `$lamb`.

If a similar analysis is performed but with $\delta = 0.5$ then we get the row scores (`$a`), column scores (`$b`) and correlation (`$lamb`) that are summarised in the fourth column of Table 2 so that:

```
> rapower.exe(selikoff.dat, delta = 0.5, iters = F)
$iterations
[1] 7

$a
        row score
0-9      -0.85037
10-19    -0.49058
20-29     0.30424
30-39     0.54535
40+       0.96909

$b
          col score
None       -0.52581
Grade 1     0.23231
Grade 2     1.10218
Grade 3     1.30939

$lamb
[1] 0.91743

>
```

References

Anscombe, F.J.: Discussion of 'new light on the correlation coefficient and its transforms' (Hotelling, H). J. Roy Stat Soc Ser B (Methodol) **15**, 229–230 (1953)

Beh, E.J., Lombardo, R.: Correspondence Analysis: Theory. Practice and New Strategies. Wiley, Chichester (2014)

Beh, E.J., Lombardo, R.: An Introduction to Correspondence Analysis. Wiley, Chichester (2021)

Beh, E.J., Lombardo, R.: Correspondence analysis and the Cressie-Read family of divergence statistics. Int. Stat. Rev. (in press) (2024)

Beh, E.J., Lombardo, R., Alberti, G.: Correspondence analysis and the Freeman-Tukey statistic: a study of archaeological data. Comput. Stat. Data Anal. **128**, 73–86 (2018)

Beh, E.J., Smith, D.R.: Real world occupational epidemiology, Part 1: odds ratios, relative risk, and asbestos. Arch. Environ. Occup. Health **66**, 119–123 (2011)

Bishop, Y.M.M., Fienberg, S.E., Holland, P.W.: Discrete Multivariate Analysis. Springer, Berlin, (reprint of 1974 MIT Press publication) (2007)

Choulakian, V.: Taxicab correspondence analysis. Psychometrika **71**, 333–345 (2006)

Choulakian, V.: Private email communication (2023)

Cressie, N.A.C., Read, T.R.C.: Multinomial goodness-of-fit tests. J. Roy. Stat. Soc. Ser. B (Methodol.) **46**, 440–464 (1984)

Cuadras, C.M., Cuadras, D.: A parametric approach to correspondence analysis. Linear Algebra Appl. **417**, 64–74 (2006)

Cuadras, C.M., Cuadras, D.: A unified approach for the multivariate analysis of contingency tables. Open J. Stat. **5**, 223–232 (2015)

Cuadras, C.M., Cuadras, D., Greenacre, M.J.: A comparison of different methods for representing categorical data. Commun. Stat. Simul. Comput. **35**, 447–459 (2006)

Freeman, M.F., Tukey, J.W.: Transformations related to the angular and square root. Ann. Math. Stat. **21**, 607–611 (1950)

Goodman, L.A.: A single general method for the analysis of cross-classified data: reconciliation and synthesis of some methods of Pearson, Yule and Fisher, and also some methods of correspondence analysis and association analysis. J. Am. Stat. Assoc. **91**, 408–428 (1996)

Gower, J.: Generalized canonical analysis. In: Coppi, E., Bolasco, S. (eds.) Multiway Data Analysis, pp. 221–232. North Holland (1989)

Greenacre, M.: Power transformations in correspondence analysis. Comput. Stat. Data Anal. **53**, 3107–3116 (2009)

Greenacre, M.: Log-ratio analysis is a limiting case of correspondence analysis. Math. Geosci. **42**, 129–134 (2010)

Hill, M.: Correspondence analysis: a neglected multivariate technique. J. R. Stat. Soc. Ser. C (Appl. Stat.) **23**, 340–354 (1974)

Hirschfeld, H.O.: A connection between correlation and contingency. Proc Cambridge Philos. Soc. **31**, 520–524 (1935)

Kullback, S.: Information Theory and Statistics. Wiley (1959)

McCullagh, P., Nelder, J. A.: Generalized Linear Models, 2nd edn. Chapman and Hall (1984)

Mirkin, B.: Eleven ways to look at the chi-squared coefficient for contingency tables. Am. Stat. **55**, 111–120 (2001)

Neyman, J.: Contributions to the theory of the χ^2 test. Proc. Berkeley Symp. Math. Stat. Probab. **1**, 239–273 (1949)

Nishisato, S.: Analysis of Categorical Data: Dual Scaling and its Applications. University of Toronto Press, Toronto (1980)

Nishisato, S.: Dual scaling of successive categories data. Jpn. Psychol. Res. **22**, 134–143 (1980)

Nishisato, S.: Dual scaling by reciprocal medians. Atti della XXXII Riunione Sci. della Soc. Ital. Stat. 141–147 (1984)

Nishisato, S.: Assessing quality of joint graphical display in correspondence analysis with dual scaling. In: Diday, E. (ed.) Data Analysis and Informatics, V, pp. 409–416. North-Holland, Amsterdam (1988)

Nishisato, S.: Dual scaling: its development and comparisons with other quantification methods. In: Pressmar, D., Jäger, K.E., Krallmann, H., Schellhaas, H., Streitferdt, L. (eds.) Operations Research Proceedings 1988, pp. 376–389. Springer, Berlin (1988)

Nishisato, S.: Elements of Dual Scaling: An Introduction to Practical Data Analysis. Lawrence Elbaum Associates, Hillsdale, NJ (1994)

Nishisato, S.: Graphical representation of quantified categorical data: its inherent problems. J. Stat. Plann. Infer. **43**, 121–132 (1995)

Nishisato, S.: A characterization of ordinal data. In: Gaul, W., Opitz, O., Schader, M. (eds.) Data Analysis, pp. 285–298. Springer, Berlin (2000)

Nishisato, S.: Multidimensional Nonlinear Descriptive Analysis. Chapman & Hall/CRC, Boca Raton, FL (2007)
Nishisato, S., Arri, P.S.: Nonlinear programming approach to optimal scaling of partially ordered categories. Psychometrika **40**, 525–548 (1975)
Nishisato, S., Beh, E.J., Lombardo, R., Clavel, J.G.: Modern Quantification Theory: Joint Graphical Display, Biplots, and Alternatives. Springer, Singapore (2021)
Nishisato, S., Clavel, J.G.: A note on between-set distances in dual scaling and correspondence analysis. Behaviormetrika **30**, 87–98 (2003)
Nishisato, S., Clavel, J.G.: Total information analysis: comprehensive dual scaling. Behaviormetrika **37**, 15–32 (2010)
Nishisato, S., Inukai, Y.: Partially optimal scaling of items with ordered categories. Jpn. Psychol. Res. **14**, 109–119 (1972)
Nishisato, S., Nishisato, I.: Dual Scaling in a Nutshell. MicroStats, Toronto (1994)
Nishisato, S., Wen-Jenn, S.: A note on dual scaling of successive categories data. Psychometrika **49**, 493–500 (1984)
Read, T.R.C., Cressie, N.A.C.: Goodness-of-fit Statistics for Discrete Multivariate Data. Springer, Berlin (1988)
Selikoff, I.J.: Household risks with inorganic fibers. Bull. New York Acad. Med. **57**, 947–961 (1981)
Tran, D., Beh, E.J., Smith, D.R.: Real world occupational epidemiology, Part 3: An aggregate data analysis of Selikoff's "20-year rule". Arch. Environ. Occup. Health **67**, 243–248 (2012)
Wang, T.-W., Beh, E.J., Lombardo, R., Renner, I.W.: Profile transformations for reciprocal averaging and singular value decomposition. (in review) (2023)
Wilks, S.S.: The large-sample distribution of the likelihood ratio for testing composite hypotheses. Ann. Math. Stat. **9**, 60–62 (1938)

Dual Scaling of Rating Data

Michel van de Velden and Patrick J.F. Groenen

1 Introduction

The works of Nishisato has shown that dual scaling is a powerful method that can be applied to solve a wide variety of data analysis problems. Dual scaling is closely related and, for many practical purposes and applications, equivalent to correspondence analysis. The books of Nishisato (1994) and Greenacre (1984) give a detailed account of the relationships and origins of the methods. Mathematically, relationships are particularly strong; see, for example, Greenacre (1984) and van de Velden (2000a). Perhaps, the biggest difference between the methods concerns correspondence analysis' focus on geometry versus dual scaling's emphasis on the optimal scaling properties.

Although both correspondence analysis and dual scaling are often considered for analysing a contingency table, both methods can be applied to other types of data. However, with respect to the analysis of such other data types, the approaches do in fact differ. In this Chapter, we explicitly consider dual scaling and correspondence analysis of rating data.

For correspondence analysis, the analysis of rating data is explicitly treated in Greenacre (1984, Chap. 6). However, Greenacre (2017) treats the analysis of rating data in a chapter titled "Data re-coding" (Chap. 23). The new labelling of the topic is a direct result of the way the analysis of rating data is defined in the correspondence analysis literature. That is, for the analysis of rating data, the rating data are first re-coded in a specific form and subsequently the usual correspondence analysis calculations are applied to the re-coded data.

M. van de Velden (✉) · P. J.F. Groenen
Econometric Institute, Erasmus University Rotterdam, Rotterdam, The Netherlands
e-mail: vandevelden@ese.eur.nl

P. J.F. Groenen
e-mail: groenen@ese.eur.nl

E. J. Beh et al. (eds.), *Analysis of Categorical Data from Historical Perspectives*,
Behaviormetrics: Quantitative Approaches to Human Behavior 17,
https://doi.org/10.1007/978-981-99-5329-5_12

For dual scaling, the analysis of rating data is treated in the context of paired comparison and successive categories data. The proposed methods in these contexts also amount to the application of the usual dual scaling calculations (which are equivalent to the correspondence analysis calculations) to re-coded data. However, as we show in this paper, the re-coding in dual scaling and correspondence analysis is different and, consequently, properties of the solutions differ as well. Note that, a direct analysis of rating data does not appear to exist in the dual scaling literature. However, as we show in Sect. 4, we can tackle this problem by using a similar interpretation of the ratings as done in the correspondence analysis literature.

In this paper, we review the existing approaches to the analysis of rating data in dual scaling and correspondence analysis. We do so by first briefly summarising the different types of re-coding in Section's 3 and 4. Furthermore, we propose a method that allows a more direct treatment of rating data in dual scaling. Next, using the optimal scaling framework that is fundamental in the works of Nishisato, we provide insights into the theoretical differences between the methods, and we discuss the implications of these differences in practise. We illustrate the differences by means of an example data set taken from Nishisato (1994) and provide some final remarks in Sect. 7.

2 Dual Scaling

The objective of dual scaling is to find optimal scaling values or scores (or coordinates) for row categories that maximise the between row variance whilst at the same time finding scores for the column categories that maximise the between column variance. Here we only give the basic formulas needed to calculate the dual scaling solution for analysing a two-way data table $\mathbf{F}$ consisting of non-negative integers. For a complete description of the rationale and a derivation of the dual scaling solution; see Nishisato (1994).

Let $\mathbf{F}$ denote an $n \times p$ matrix consisting of non-negative entries and define diagonal matrices $\mathbf{D}_r$ and $\mathbf{D}_c$ in such a way that $\mathbf{D}_r \mathbf{1}_p = \mathbf{F}\mathbf{1}_p = \mathbf{r}$ and $\mathbf{D}_c \mathbf{1}_n = \mathbf{F}^T \mathbf{1}_n = \mathbf{c}$, where generically, $\mathbf{1}_i$ denotes an $i \times 1$ vector of ones. Consider the singular value decomposition:

$$\mathbf{D}_r^{-1/2} \left(\mathbf{F} - \frac{1}{s} \mathbf{r}\mathbf{c}^T \right) \mathbf{D}_c^{-1/2} = \mathbf{U}\mathbf{\Lambda}\mathbf{V}^T, \tag{1}$$

where $s = \mathbf{1}_n \mathbf{F} \mathbf{1}_p$, and, without loss of generality, the singular values on the diagonal of $\mathbf{\Lambda}$ are in non-increasing order. The k-dimensional optimal scaling values (i.e. the scores/coordinates) for rows and columns are $\mathbf{X} = \mathbf{D}_r^{-1/2}\mathbf{U}_k$ and $\mathbf{Y} = \mathbf{D}_c^{-1/2}\mathbf{V}_k$ respectively, where $\mathbf{U}_k$ and $\mathbf{V}_k$ correspond to the first k columns of $\mathbf{U}$ and $\mathbf{V}$.

Note that by defining $\mathbf{X}$ and $\mathbf{Y}$ in this way, they are standardised such that $\mathbf{X}^T\mathbf{D}_r\mathbf{X} = \mathbf{Y}^T\mathbf{D}_c\mathbf{Y} = \mathbf{I}_k$. In the correspondence analysis literature, the matrices $\mathbf{X}$ and $\mathbf{Y}$ are referred to as standard coordinates. Alternatively, defining $\mathbf{G} = \mathbf{D}_r^{-1/2}\mathbf{U}_k\mathbf{\Lambda}_k$ and $\mathbf{H} = \mathbf{D}_c^{-1/2}\mathbf{V}_k\mathbf{\Lambda}_k$ gives the solution in so-called principal coordinates. For more details on the different scalings and their implications on, in particular, graphical representations of results; see Greenacre (2017).

3 Correspondence Analysis of Ratings (CAr)

Let $\mathbf{R}$ denote an $n \times p$ matrix of ratings on a 1 to q scale. We use an artificial example of $n = 4$ individuals who each rate $p = 3$ objects on a $q = 5$ point rating scale, to illustrate the different data pre-processing steps required in the different variant. That is:

$$\mathbf{R} = \begin{bmatrix} 2 & 4 & 5 \\ 3 & 3 & 1 \\ 2 & 1 & 4 \\ 1 & 5 & 3 \end{bmatrix}. \tag{2}$$

Correspondence analysis is concerned with count data. The ratings can be considered as counts by considering a rating value as the number of times an object was preferred over the lowest rating number. To achieve this, we simply subtract 1 from the originals ratings. Let $\mathbf{T} = \mathbf{R} - \mathbf{1}_n\mathbf{1}_p^T$, denote the resulting matrix with values from 0 to $q - 1$. That is, if the original rating scale consists of q ratings, we first subtract 1 which leads in our toy example to:

$$\mathbf{T} = \mathbf{R} - \mathbf{1}_4\mathbf{1}_3^T = \begin{bmatrix} 2 & 4 & 5 \\ 3 & 3 & 1 \\ 2 & 1 & 4 \\ 1 & 5 & 3 \end{bmatrix} - \begin{bmatrix} 1 & 1 & 1 \\ 1 & 1 & 1 \\ 1 & 1 & 1 \\ 1 & 1 & 1 \end{bmatrix} = \begin{bmatrix} 1 & 3 & 4 \\ 2 & 2 & 0 \\ 1 & 0 & 3 \\ 0 & 4 & 2 \end{bmatrix}.$$

Thus, $\mathbf{T}$ can be interpreted as the number of scale points below a given rating, or, equivalently, as the number of times an object was considered to exceed a threshold on the original rating scale.

Mathematically, we can apply correspondence analysis to the count data in $\mathbf{T}$. However, the problem with such a procedure is that the direction of the original rating scale influences the results; reversing the scale would lead to different results. That is, if data are gathered on a scale were the lowest rating (1) corresponds to "bad" and the highest rating (q) to "good", and we decide to switch the labelling from 1 = "good" to q = "bad", the results of the analysis would change. Clearly this sensitivity to the direction of the scale is an undesirable effect.

To overcome this problem, the data are "doubled", meaning that the rating data for both directions of the rating scale are considered simultaneously. In correspondence analysis, this is done by, for each object, adding a column with the rating on the reversed scale. Consequently, instead of p columns, we obtain a matrix consisting of $2p$ columns. Let $\mathbf{S}$ denote the matrix of ratings on the reversed scale, that is, $\mathbf{S} = (q - 1)\mathbf{1}_n\mathbf{1}_p^T - \mathbf{T}$. In our running example, we get:

$$\mathbf{S} = (q-1)\mathbf{1}_n\mathbf{1}_p^T - \mathbf{T} = \begin{bmatrix} 4&4&4 \\ 4&4&4 \\ 4&4&4 \\ 4&4&4 \end{bmatrix} - \begin{bmatrix} 1&3&4 \\ 2&2&0 \\ 1&0&3 \\ 0&4&2 \end{bmatrix} = \begin{bmatrix} 3&1&0 \\ 2&2&4 \\ 3&4&1 \\ 4&0&2 \end{bmatrix}.$$

We construct a column-wise doubled matrix as $\mathbf{F}_c = [\mathbf{T} \mid \mathbf{S}]$:

$$\mathbf{F}_c = [\mathbf{T} \mid \mathbf{S}] = \left[\begin{array}{ccc|ccc} 1&3&4&3&1&0 \\ 2&2&0&2&2&4 \\ 1&0&3&3&4&1 \\ 0&4&2&4&0&2 \end{array}\right].$$

Substituting this doubled matrix $\mathbf{F}_c$ for $\mathbf{F}$ in the formulas of Sect. 2 yields the correspondence analysis solution.

The specific structure of the doubled matrix $\mathbf{F}_c$ results in structured coordinates for the columns as well. In particular, the points corresponding to the same object, with the reversed ratings, can be connected through a straight line running through the origin (however, the distances from the origin of both points differ). Greenacre (2017) uses this relationship and shows that the resulting lines can be divided into $q-1$ equal sized intervals with the endpoints corresponding to the endpoints of the rating scale. That is, rating q corresponds to the point corresponding to the original rating, and rating 1 corresponds to the point on the reversed scale. The approximated average rating value (on the original scale) can then be inferred from this plot by considering the value on this line at the origin. Furthermore, similar to the case in principal component analysis, the angles (at the origin) between the lines corresponding to the different attributes, approximate correlations between the ratings for the attributes. In fact, as shown in van de Velden (2004, pp. 103–104), the analysis of the doubled matrix $\mathbf{F}_c$ is equivalent to a principal component analysis of a particularly scaled and double centred version of the original rating data.

4 Dual Scaling of Rating Data

Dual scaling of rating data is not treated as topic of its own in Nishisato (1994). Instead, in the context of paired comparison and rank order data, Nishisato (1994) proposes two dual scaling variants that require different re-coding of the data. The first approach requires re-coding of the ratings as rankings while the second approach involves a joint ranking of objects and, unobserved, rating boundaries. To these two approaches, we add a third, more direct, re-coding that relies on an interpretation of ratings similar to the one used in correspondence analysis and described in Sect. 3. In the following subsections, we briefly discuss these three types of re-coding as well as the dual scaling analysis of them. For convenience, we have labelled these dual scaling variants DS1 up to DS3.

4.1 Converting Ratings to Rank Order Data (DS1)

For the first variant, rather than considering the observed ratings directly, for an observation i, one counts the number of times that individual i's rating for object j is rated higher than ratings for all other objects. This is equivalent to transforming the ratings to ranked (from 0 to $p - 1$) data and requires a way to deal with ties (i.e. equal ratings). For the data from our running example, that is the matrix $\mathbf{R}$ given in (2), we get:

$$\mathbf{T}^* = \begin{bmatrix} 0 & 1 & 2 \\ 1.5 & 1.5 & 0 \\ 1 & 0 & 2 \\ 0 & 2 & 1 \end{bmatrix},$$

and, on the reversed scale:

$$\mathbf{S}^* = \begin{bmatrix} 2 & 1 & 0 \\ 0.5 & 0.5 & 2 \\ 1 & 2 & 0 \\ 2 & 0 & 1 \end{bmatrix}.$$

Since the focus is now on the rank of the three objects, not their rating on a 5-point scale, this method clearly incurs a loss in information as only the direction of the difference is considered, and not the magnitude.

To analyse the resulting rank order data Nishisato (1994) proposes to construct a dominance matrix $\mathbf{E}$ consisting of the difference between the number of times an object was preferred over the other objects ($\mathbf{T}^*$) and the number of times it was not preferred over other object ($\mathbf{S}^*$). For our example we get:

$$\mathbf{E} = \mathbf{T}^* - \mathbf{S}^* = \begin{bmatrix} -2 & 0 & 2 \\ 1 & 1 & -2 \\ 0 & -2 & 2 \\ -2 & 1 & 1 \end{bmatrix}.$$

So that the sum of each row is zero. Note that the dominance matrix $\mathbf{E}$ contains positive and negative values. Moreover, as the row sums are all zero the usual dual scaling calculations, as set out in Sect. 2, cannot be applied directly. Nishisato (1994) resolves this by defining $\mathbf{D}_r = p\,(p-1)\,\mathbf{I}_n$, and $\mathbf{D}_c = n\,(p-1)\,\mathbf{I}_p$ respectively. Alternatively, as shown by van de Velden (2000b), one can apply the usual dual scaling approach to the row-wise doubled matrix $\mathbf{F}_r = \left[\mathbf{T}^{*^T} \mid \mathbf{S}^{*^T}\right]^T$ yielding, for our example data,

$$\mathbf{F}_r = \left[\begin{array}{ccc} 0 & 1 & 2 \\ 1.5 & 1.5 & 0 \\ 1 & 0 & 2 \\ 0 & 2 & 1 \\ \hline 2 & 1 & 0 \\ 0.5 & 0.5 & 2 \\ 1 & 2 & 0 \\ 2 & 0 & 1 \end{array}\right].$$

Analysing the row-wise doubled matrix yields $2n$ scores for the n rows. The scores for the first n rows corresponds to the observations (rankings) on the original scale while the scores for the second set correspond to the observations (rankings) on the reversed scale. These two sets of scores, however, are trivially related as the scores in the second set are simply -1 times those in the first set.

4.2 Converting Rating Data to Successive Category Data (DS2)

The second approach, introduced in Nishisato (1980) and further developed in Nishisato and Sheu (1984), requires the introduction of "boundaries", marking the difference between rating scale values. To each boundary we assign a rating that lies between the two values of the rating scale that the boundary represents. The observed ratings in $\mathbf{R}$ and the boundaries in $\mathbf{R}_{\text{bound}}$ are jointly ranked, resulting in so-called successive category data $\mathbf{R}_{\text{SCD}}$. Note that, in this way, in addition to the p objects, $q-1$ boundaries are added as columns to the data matrix. In our example, using 1.5 up to 4.5 as "rating" values for the boundaries, we get:

$$[\mathbf{R} \mid \mathbf{R}_{\text{bound}}] = \left[\begin{array}{ccc|cccc} 2 & 4 & 5 & 1.5 & 2.5 & 3.5 & 4.5 \\ 3 & 3 & 1 & 1.5 & 2.5 & 3.5 & 4.5 \\ 2 & 1 & 4 & 1.5 & 2.5 & 3.5 & 4.5 \\ 1 & 5 & 3 & 1.5 & 2.5 & 3.5 & 4.5 \end{array}\right] \quad \rightarrow \quad \mathbf{R}_{\text{SCD}} = \left[\begin{array}{ccccccc} 2 & 5 & 7 & 1 & 3 & 4 & 6 \\ 4.5 & 4.5 & 1 & 2 & 3 & 6 & 7 \\ 3 & 1 & 6 & 2 & 4 & 5 & 7 \\ 1 & 7 & 4 & 2 & 3 & 5 & 6 \end{array}\right].$$

The resulting $n \times (p+q-1)$ matrix of rank ordered data $\mathbf{R}_{\text{SCD}}$ can be analysed in the same way as described above, that is, using the row-wise doubled matrix. Note that it doesn't matter what the exact values are that we insert for the boundaries, as long as they are between the actual ratings.

As the boundaries are always ordered in the same way, the one-dimensional dual scaling solution for successive category data typically seems appropriate and sufficient in terms of explained variance. Moreover, as the successive category values for an individual are based on all ratings by the same individual, individual specific scale use is taken into account. For this reason, this specific coding was used by Schoonees, van de Velden and Groenen (2015) and Takagishi, van de Velden and Yadohisa (2019) to construct methods to study response style bias in questionnaires.

4.3 Converting Rating Data to Count Data (DS3)

Both re-coding methods described in the previous subsections yield individual specific rankings. Consequently, the transformed rating values are individual specific as well. This implies that if individual i assigns rating j to an object, and individual l assigns the same rating value j to that object, the re-coded values do not have to be the same for both observations. If the actual ratings are considered to be meaningful and non-individual specific, this may not be a desirable property. To overcome this problem, one can re-code and interpret the ratings as previously described in Sect. 3 in the context of correspondence analysis, that is, the rating values are re-coded to $\mathbf{T}$; the number of times an observation exceeds the boundaries on the original rating scale.

As before, we cannot apply dual scaling directly to $\mathbf{T}$ as this would lead to results that depend on the direction of the scale. Following the dual scaling approach for rank order data, we can overcome this by constructing a row-wise doubled matrix $\mathbf{F}_r = \left[\mathbf{T}^T \mid \mathbf{S}^T\right]^T$ and applying dual scaling to this matrix. From here on, we refer to this DS3 approach as dual scaling of rating data.

5 Optimal Scaling Properties

As seen in Section's 3 and 4, the difference between dual scaling and correspondence analysis of rating data amounts to a difference in doubling of the observed ratings after converting them to a 0 to $q-1$ scale. That is, correspondence analysis of ratings is defined as correspondence analysis of $\mathbf{F}_c$ whereas dual scaling of ratings is defined as dual scaling of $\mathbf{F}_r$. To better understand the implications of these differences, we briefly review the optimal scaling properties of them.

As shown by Nishisato (1978) and van de Velden (2004), the object scores obtained in an analysis of the dominance matrix $\mathbf{E}$, and, hence, the object scores in the analysis of $\mathbf{F}_r$, are equivalent to the optimal scaling values as defined and derived by Guttman (1946). As such, these values are determined "so as to best distinguish between those things judged higher and those judged lower for each individual"; see Guttman (1946). Both Guttman (1946) and Nishisato (1978) explicitly consider paired comparison data. However, crucial in the formulation of the optimal scaling framework are the matrices $\mathbf{T}$ and $\mathbf{S}$. Hence, using these matrices in the context of rating data where, respectively, the entries represent the times an object rating exceeds or does not exceed the available rating boundaries, the optimal scaling properties remain valid. That is, in dual scaling of ratings as defined in Sect. 4.3, the scale values for the objects are assigned in such a way that they best distinguish between objects.

In Guttman (1946), an optimal scaling solution for the individuals is not considered. However, we can rephrase Guttman's (1946) optimal scaling goal towards finding scale values for the observations/individuals as follows: Find scale values for individuals so as to best distinguish between individuals that judged an object

Table 1 Properties of the different variants for the analysis of rating data

Method	Intervals	Optimal scaling	Doubling
DS1	Rank order only	Objects	Implicit; row-wise
DS2	Successive categories	Objects and boundaries	Implicit: row-wise
DS3	Differences between ratings	Objects	Explicit: row-wise
CAr	Differences between ratings	Individuals	Explicit: column-wise

higher and lower, for each object. Where once again higher (lower) indicates how often a rating exceeded (did not exceed) the boundaries. This "dual" problem was considered, in the context of paired comparison data, by van de Velden (2004) who showed that the resulting optimal scaling values for individuals can be obtained by applying dual scaling/correspondence analysis to the column-wise doubled matrix $\mathbf{F}_c$. As before, interpreting the entries of $\mathbf{T}$ and $\mathbf{S}$ as the times an object rating exceeds or does not exceed the available rating boundaries, these optimal scaling properties remain valid in the analysis of $\mathbf{F}_c$. Hence, whereas dual scaling of ratings yields optimal scaling values for the objects, correspondence analysis of ratings yields optimal scaling values for the individuals.

Note that for the dual scaling analysis of rating data, the optimal scaling values for the doubled rows (i.e. individuals' ratings according to the original and reversed scales) are optimal in the usual dual scaling sense. That is, they maximise the variation between the rows of the doubled table. Similarly, for the correspondence analysis solution, the values for the doubled columns are optimal scaling values (i.e. the objects rated according to the original and reversed scales). However, when defining optimal scaling values according to the framework and rationale as presented by Guttman (1946), the typical duality associated with a dual scaling (and correspondence analysis) solution obtained using the formulas of Sect. 2, does not immediately carry over when we have rating data. That is, Guttman's optimal scaling values for individuals and objects based on rating data cannot be obtained simultaneously.

We summarised some properties of the different variants in Table 1.

In summary, the difference between the dual scaling and correspondence analysis of rating data approaches amounts to a different way of dealing with the direction of the rating scales. For the dual scaling of rating data, as introduced in Sect. 4.3, a row-wise doubling is employed. For the correspondence analysis of rating data, a column-wise doubling is used to resolve the problem. The effect of these different data pre-processing steps is that in the dual scaling analysis of rating data, the values for the objects are optimal scaling values whereas in the correspondence analysis of rating data, the coordinates for the individuals are optimal scaling values.

In order to choose one method over the other, it is important to understand these differences. Depending on the type of application and the specific research goals, a choice can be made. Dual scaling of rating data may be more appropriate when one's prime concern is a visualisation (or quantification) of a set of objects based on the observed differences in the ratings for these objects. This could be the case, when, for example, relative positions of products based on how they are perceived by a group of individuals. One the other hand, if one is more concerned with a visualisation (or quantification) of the individuals, based on differences in rating patterns for a set of objects, correspondence analysis of rating data may be better equipped to pick up on the individual differences.

6 Applications

To illustrate the differences between the dual scaling and correspondence analysis of rating approaches, we analyse an example data set from Nishisato (1994, p. 230) on perceived seriousness of crimes. In particular, we focus on the effects of the different doublings; that is, the analysis of a column-wise (CAr) and row-wise (DS3) doubled matrix of ratings.

A sample of 17 individuals indicated, on a rating scale from 1 ("somewhat serious") to 4 ("extremely serious"), the perceived seriousness of the following 8 types of crimes: Arson, burglary, counterfeiting, forgery, homicide, kidnapping, mugging and receiving stolen goods. For ease of reproducibility, we included the data here in Table 2. Note that all individuals considered "homicide" to be extremely serious. For this lack in variation, which leads to a singular $\mathbf{D}_c$ matrix in CAr, we removed this type of crime from our analyses. In the DS3 approach such a singularity does not occur and Nishisato (1994) analyses the data without removal of this object.

The two-dimensionsal dual scaling solution for the objects (crimes) can be found in Fig. 1. In accordance with the optimal scaling formulations of Guttman (1946) and Nishisato (1978) the scaling values are in so-called standard coordinates. The two-dimensional solution, which is heavily dominated by the first dimension, accounts for 89% of the variance.

Correspondence analysis of the ratings results in Fig. 2, where in accordance with Greenacre (2017, Exhibit 23.2, p. 180), the coordinates for the doubled objects are in principal coordinates, and, for each crime, we connected the points corresponding to the lower and upper ends of the scale, by axes. The CA solution accounts for 64% of the variation.

A one-to-one comparison of these two solutions for the objects is complicated due to the doubling of object points in the correspondence analysis solution. Moreover, we used standard coordinates in the dual scaling analysis, and principal coordinates for the correspondence analysis results. Still, comparing Figs. 1 and 2 immediately does show a better separation of objects (crimes) in the dual scaling approach. In Fig. 1, we see that the crimes "Counterfeiting" and "Forgery", which are somewhat similar in nature, are indeed perceived as more similar by the respondents. On the other hand, the perceptions of "Mugging", "Burglary" and "Receiving stolen goods", as indicated by the ratings, differ substantially. Note that the first dimension in this analysis is rather dominant. Moreover, this dimension appears to describe mostly the perceived seriousness of crimes from more "serious" (Arson and Kidnapping) on the left, to less "serious" (Receiving stolen goods) on the right.

Table 2 Nishisato's 1994 seriousness of crimes rating data

Individual	Arson	Burglary	Counterfeit.	Forgery	Homicide	Kidnapp.	Mugging	Rec. st. goods
1	4	2	2	2	4	3	3	1
2	4	2	2	2	4	4	3	1
3	3	2	2	2	4	3	3	1
4	4	3	2	2	4	4	4	3
5	4	3	2	2	4	4	3	2
6	4	3	3	2	4	4	3	2
7	4	1	2	2	4	4	2	1
8	4	4	2	2	4	4	3	2
9	3	2	1	2	4	4	3	1
10	4	3	3	3	4	4	3	2
11	4	2	3	3	4	4	4	1
12	4	4	3	3	4	4	4	2
13	4	3	3	2	4	4	3	1
14	4	2	2	2	4	3	3	1
15	4	2	1	1	4	4	2	1
16	3	2	2	2	4	3	3	1
17	3	2	2	2	4	4	3	2

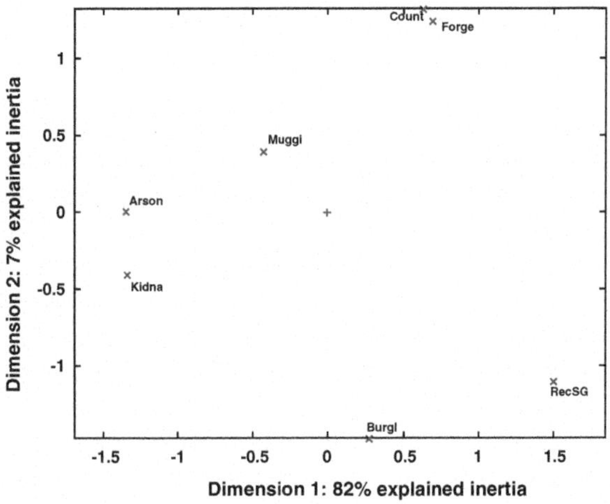

Fig. 1 Dual scaling of ratings (DS3) for the crime perception data. Optimal scaling values for objects (crimes) in standard coordinates

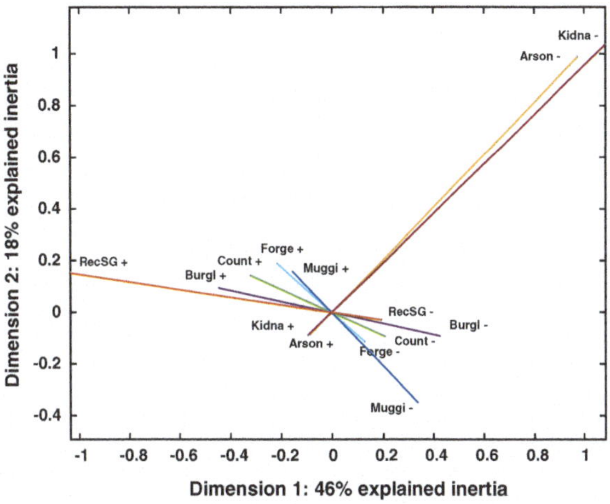

Fig. 2 Correspondence analysis of ratings (CAr) for the crime perception data. Objects (crimes) in principal coordinates

In Fig. 2, we see that the correspondence analysis approach (CAr) visualises that the ratings of "Arson" and "Kidnapping" are correlated. Furthermore, the ratings of these two crimes appear to be mostly uncorrelated to the ratings for "Forgery", "Mugging", "Burglary" and "Receiving stolen goods". Note that the endpoints of the coloured lines correspond to the end points of the scale. That is, the '−' points correspond to the lowest rating and the '+' points to the highest ratings. As the origin in a CA plot corresponds to average profiles, we can infer the approximate mean ratings for objects directly from the plot. For example, we see that both "Kidnapping" and "Arson" are rated as "extremely serious" far more often than average. Similarly, "Receiving stolen goods" tends to receive a lower (less serious) rating more often than not. For "Burglary", the results are more varied and the average rating appears to be close to the middle of the rating scale.

Figure 3, for DS3, and Fig. 4, for CAr, give, for both analyses, the corresponding solutions for the individuals. Hence, for the dual scaling solution, the scores for the individuals are in principal coordinates whereas for the CA solution they are in standard coordinates. In addition, the doubled set of "individual" scores for the dual scaling solution is ignored as these are simply the same coordinates mirrored in the origin.

Recall that the correspondence analysis solution gives optimal scaling values for the individuals. Hence, coordinates are determined in such a way that differences in the indicated rating patterns between individuals is optimally depicted. Superficially comparing Figs. 3 and 4 may not immediately expose this. However, note that for the dual scaling solution, depicted by Fig. 3, the points are not spread out along both dimensions. Instead, they are all concentrated on the negative side of the first dimension.

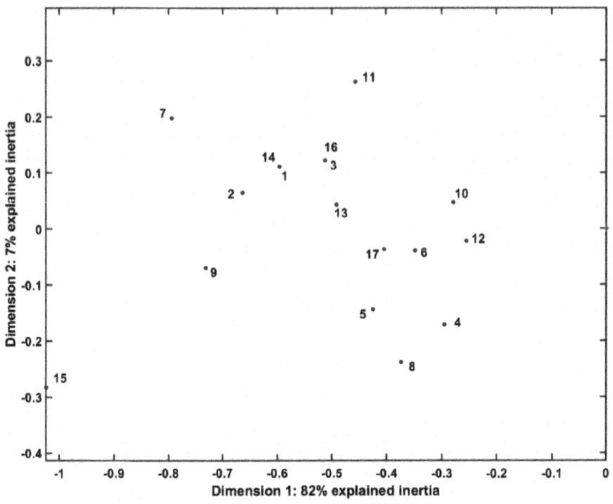

Fig. 3 Dual scaling of crime rating data. Scores for individuals in principal coordinates

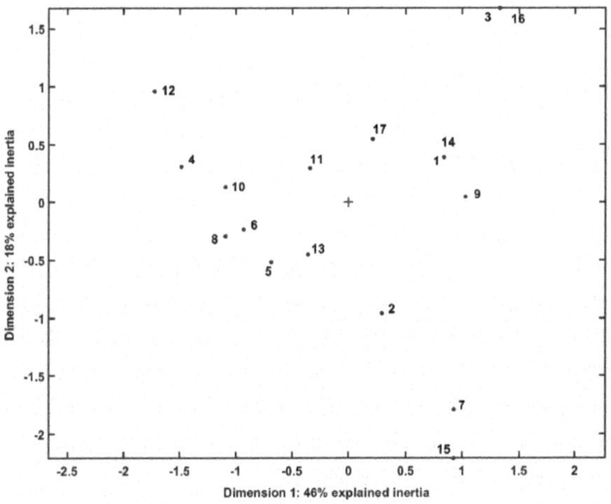

Fig. 4 Correspondence analysis of crime rating data. Optimal scaling values for individuals in standard coordinates

To better appreciate the differences in the solutions, Figs. 5 and 6 give biplots for both methods. That is, joint plots for rows and columns where projections of one set of points on the directions of other points (obtained, for example, by drawing axes from the origin through the points), can be used to reconstruct the values in the original data table; see Greenacre (1993) for more details. Note that, in both joint plots objects (crimes) are in standard, and individuals are in principal coordinates.

Interpreting the relative positions of the individuals in Fig. 5 is not so easy. For these data, differences are small and most individuals give high ratings to all crimes.

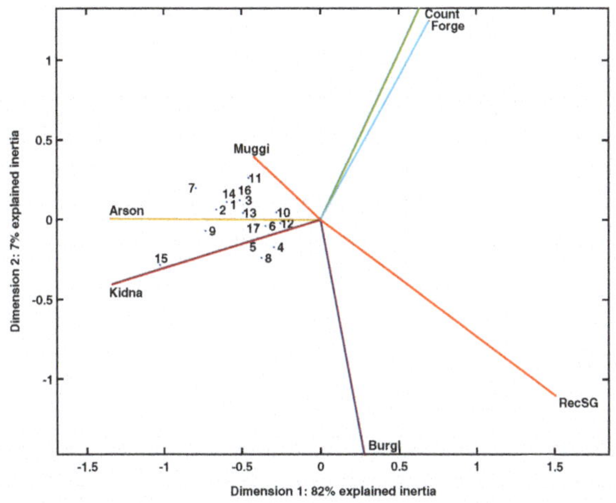

Fig. 5 Dual scaling biplot of crime rating data. Optimal scaling values for objects (crimes) in standard coordinates

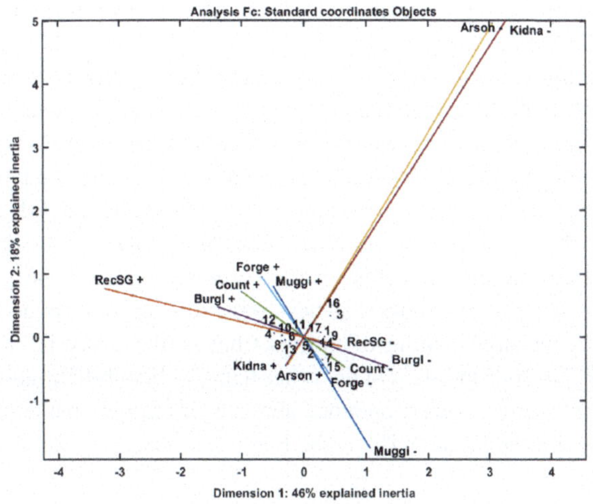

Fig. 6 Correspondence analysis biplot of crime rating data. Optimal scaling values for individuals in principal coordinates

That is, they tend to find all crimes to be *serious*. How individual 15 differentiates from the others, as its location in Figs. 3 and 5 suggests, is not clear from the plot.

The optimal scaling positions of individuals in Fig. 4 appear better separated. Moreover, the interpretation of the differences in the locations of the individuals is more straightforward. For example, individuals 15 and 7 are separated from the other points. In the biplot of Fig. 6, we see that this may be explained by both individuals giving relatively low ratings (that is, a lower rating than average) for "Mugging". Indeed, these two individuals are the only ones that assign a rating 2 to these two crimes. All others give higher ratings. In a similar way, differences in positions of other "outlying" points (e.g. 3, 16, 12, 4) can be explained by observing in what sense the corresponding rating profiles differ from the average rating profiles. For the equivalent ratings of individuals 3 and 16, we see that they differ from all other individuals with respect to their rating for "Arson" and "Kidnapping". As can be verified from the data table, they gave a rating of 3 to both of these crimes whereas all others either gave rating 4 to at least one of these crimes.

7 Conclusion

In this paper, that has been inspired by the works of Nishisato, we introduced a dual scaling of rating approach. We showed how this method relates to the correspondence analysis of rating data and that the fundamental difference between these two variants can be attributed to a difference in pre-processing of the data. In particular, the dual scaling of rating data can be described as dual scaling of a row-wise doubled matrix whereas correspondence analysis amounts to the analysis of a column-wise doubled matrix.

The dual scaling framework that has been laid out by Nishisato throughout his career offers tools to better understand the resulting differences. That is, whereas the dual scaling of rating data yields (and in fact, was defined to do so) optimal scaling values for the objects, the correspondence analysis of rating data yields optimal scaling values for the individuals. Given these rather fundamental differences, saying that one approach is better than the other, does not make much sense. A choice between these two variants depends on the research goals. If the goal is to find scale values (or: a representation) that best separates the objects according to the observed ratings, the dual scaling of ratings (that is: the analysis of the row-wise double matrix $\mathbf{F}_r$) is appropriate. On the other hand, to better distinguish individuals according to their ratings, correspondence analysis of ratings (that is: the analysis of the column-wise doubled matrix $\mathbf{F}_c$) is the better alternative.

References

Greenacre, M.J.: Theory and Applications of Correspondence Analysis. Academic Press, London (1984)

Greenacre, M.J.: Biplots in correspondence analysis. J. Appl. Stat **20**(2), 251–269 (1993)

Greenacre, M.: Correspondence Analysis in Practice, 3rd edn. Chapman & Hall/CRC Press, Boca Raton, FL (2017)

Guttman, L.: An approach for quantifying paired comparisons and rank order. Ann. Math. Stat. **17**(2), 144–163 (1946)

Nishisato, S.: Optimal scaling of paired comparison and rank order data: An alternative to Guttman's formulation. Psychometrika **43**(2), 263–271 (1978)

Nishisato, S.: Dual scaling of successive categories data. Jpn. Psychol. Res. **22**(3), 134–143 (1980)

Nishisato, S.: Elements of Dual Scaling: An Introduction to Practical Data Analysis. Lawrence Elbaum Associates, Hillsdale, NJ (1994)
Nishisato, S., Sheu, W.: A note on dual scaling of successive categories data. Psychometrika **49**(4), 493–500 (1984)
Schoonees, P.C., van de Velden, M., Groenen, P.J.F.: Constrained dual scaling for detecting response styles in categorical data. Psychometrika **80**(4), 968–994 (2015)
Takagishi, M., van de Velden, M., Yadohisa, H.: Clustering preference data in the presence of response-style bias. Br. J. Math. Stat. Psychol. **72**(3), 401–425 (2019)
van de Velden, M.: Topics in Correspondence Analysis. Thela Ph.D. Thesis (2000a)
van de Velden, M.: Dual scaling and correspondence analysis of rank order data. In: Heijmans, R.D.H., Pollock, D.S.G., Satorra, A. (eds.) Innovations in Multivariate Statistical Analysis, pp. 87–99, Springer, Berlin (2000b)
van de Velden, M.: Optimal scaling of paired comparison data. J. Classif. **21**(1), 89–109 (2004)

Whence Principal Components?

Lawrence Hubert and Susu Zhang

1 The *Journal of Educational Psychology* as a Precursor to *Psychometrika*

Before[1] the establishment of the journal *Psychometrika* in 1936, the main outlet for the publication of technical/mathematical material with a psychological bent was, somewhat surprisingly, the *Journal of Educational Psychology (JEdP). JEdP* was founded in 1910, with an opening lead article written by E. L. Thorndike (the second President of the Psychometric Society after Thurstone). By the time the 1930s arrived, *JEdP* was dominated by authors who would later become inaugural members of the Psychometric Society as well as some of its later presidents. For example, in the 1930 volume, there were quantitative articles written by the familiar names of Cureton, Dunlap, Holzinger, Spearman, Rulon, Lindquist, Edgerton, Garrett, and Carter. (We might add that in the 1930s and 40s, Jack Dunlap, one of the six founding members of the Psychometric Society, was an Editor of *JEdP* and so was responsible for all technical/quantitative submissions made for the journal). It may not be completely surprising then that Harold Hotelling, one of the leading mathematical statisticians of the 20th century, would publish his method of principal components in *JEdP* in

[1]The first author's connections with Shizuhiko Nishisato go back some forty years and are mainly editorial and through various positions and activities in the Psychometric Society. Much of this contact was in the form of being both an Associate Editor as well as a principal referee when I was the chief editor for the *Journal of Educational Statistics* (1980–1985) and *Psychometrika* (1988–1992). Nishi succeeded me as the editor of *Psychometrika* (1992–1995); I only hope that I served Nishi as well as he had served me and our profession over the years.

L. Hubert (✉)
Department of Psychology, University of Illinois, Champaign, IL, USA
e-mail: lhubert@illinois.edu

S. Zhang
Departments of Psychology and Statistics, University of Illinois, Champaign, IL, USA
e-mail: szhan105@illinois.edu

E. J. Beh et al. (eds.), *Analysis of Categorical Data from Historical Perspectives*, Behaviormetrics: Quantitative Approaches to Human Behavior 17,
https://doi.org/10.1007/978-981-99-5329-5_13

1933 (Hotelling 1933).[2] What may be more interesting historically, however, is how Hotelling came to the topic in the first place—that story is the purpose of this short essay.

2 Harold Hotelling and Truman Lee Kelley

Harold Hotelling (1895–1973) received his doctoral degree in mathematics (and economics) from Princeton in 1924. Immediately thereafter he became a Research Associate at the Stanford University Food Research Institute; from 1927 to 1931 he was an Associate Professor of Mathematics, also at Stanford. He moved to the Economics Department of Columbia University in 1931, and stayed until 1946 when he left for the University of North Carolina to found the Department of Statistics. He remained a Professor of Mathematical Statistics at North Carolina until his death. Judging from a perusal of the Harold Hotelling archives at Columbia University and those of Truman Lee Kelley at Harvard, Hotelling's work on principal components, as well as his subsequent development of canonical correlation (Hotelling 1935) also published in *JEdP*, was motivated by his association with Kelley. They overlapped as colleagues at Stanford from 1924 to 1931, where Kelley was a Professor of Education. Kelley moved in 1931 to the Harvard Graduate School of Education at exactly the same time that Hotelling moved to Columbia. As discussed below, this period of the early 1930s was a time of sustained interaction between Kelley and Hotelling that directly led to Hotelling's development of principal components and canonical correlation.

The same year that both Kelley and Hotelling left Stanford for their respective East Coast positions at Harvard and Columbia (1931) also saw the formation of the Unitary Traits Committee under E. L. Thorndike, with both Kelley and Hotelling as committee members. Several excerpts are given below from a survey that discusses the work of this group written by Karl Holzinger in the *Journal of Personality* (Holzinger 1936), entitled "Recent research on unitary mental traits":

---

[2] The technical level of Hotelling's 1933 *JEdP* article is quite high and would be unexpected in any journal devoted mainly to substantive matters. For example, Darrell Bock in his chapter, "Rethinking Thurstone," in the book, *Factor Analysis at 100* (Bock 2007) comments on Hotelling's *JEdP* article as follows (p. 42):

> Speaking of notation, I add that although Hotelling may have derived his iterative procedure for latent roots and vectors in matrix terms, in consideration of the audience, he confined his presentation to scalar algebra. Curiously, however, he introduces a notational convention from tensor calculus — namely, that when an equation is written as say, $b_i = a_{ij}$, it denotes the summation of the right-hand member with respect to the j subscript. This device is somewhat unsettling to anyone accustomed to seeing the summation sign in these equations. Surely, this is the only paper containing tensor notation in the entire psychological literature and perhaps the statistical literature.

> When Professor Spearman conceived the idea that the arrangement of a set of intercorrelations could be used to determine factors underlying a set of variables, he opened up an objective method in psychology that has been gathering momentum ever since. After the publication of *Abilities of Man*, in 1927, interest in factor theory began to spread widely throughout America, engaging the attention of such workers as Professor Truman Kelley and Professor T. V. Moore. In a book entitled *Crossroads in the Mind of Man* (1928) Professor Kelley dealt largely with group factors and new methods for their evaluation. These two volumes laid the immediate foundation for the formation of the Unitary Traits Committee in 1931.
>
> Professor E. L. Thorndike, for years a passive onlooker of methods of factorisation, now became an active promoter. Through his influence a committee was formed to study methods of factorization and apply them if possible to large bodies of data. Professor Thorndike named the committee the Unitary Traits Committee and with his characteristic symbolism, "U. T. C. for short."
>
> The Problems and Plans Committee of the American Council on Education empowered Professor Thorndike to act as chairman of this committee and secured a grant of money from the Carnegie Corporation for the purpose of preparing a plan to study unitary differential traits. The early members of this committee included Professors E. L. Thorndike, Charles Spearman, T. L. Kelley, Clark Hull, Karl Lashley, and Karl J. Holzinger. At later meetings Professors T. V. Moore, Henry Garrett, and Harold Hotelling were added to the committee.
>
> ...
>
> The sub-committees were organized as follows:
>
> 1. Mathematical theory and techniques and the improvement of methods of analysis: T. L. Kelley and Harold Hotelling.
>
> ...
>
> During the early meetings of the Unitary Traits Committee some criticism was made of existing methods of factorisation, chiefly those of Professor Kelley in *Crossroads in the Mind of Man*. Professor Kelley was already at work amending these techniques, and enlisted the aid of Professor Harold Hotelling to further this work. As a mathematical statistician Professor Hotelling was of great service to the committee. He contributed many valuable suggestions at meetings, and the factorization technique now known as the Method of Principle [sic] Components.

The remainder of the present essay can be seen as a series of interesting subtopics (or at least we hope they are) concerning the introduction of "the method of principal components" in *JEdP* (Hotelling 1933). Several of these observations result from private correspondence and material from the Unitary Traits Committee available in archives for Kelley and Hotelling at Harvard and Columbia, respectively.

3 Hotelling as a Quantitative Consultant for Psychology

For a period of time in the late 1920s and 1930s, Harold Hotelling was a favored mathematician to consult when a particularly vexing quantitative derivation task was at hand. Acknowledgments to Hotelling appeared regularly in *JEdP* in the early

1930s; others occurred in several books from around that same time.[3] For example, in Kelley's *Interpretation of Educational Measurements* (Kelley 1927), we have the footnote (p. 213):

> I am indebted to Dr. Harold Hotelling for a suggestion which readily led to the evaluation of this determinant.

Or, in Kelley's *Crossroads in the Mind of Man* (Kelley 1928), we have the following in the actual text (p. 54):

> Dr. Harold Hotelling has kindly provided the following set of necessary conditions which are more readily investigated than are the 12 sufficient equations in Formula 35.

In John Flanagan's thesis under Kelley at Harvard, *Factor Analysis in the Study of Personality* (Flanagan 1935), there is the following paragraph about Hotelling developing the method of principal components at the behest of the Unitary Traits Committee:

> This brings us directly to the last method of multiple-factor analysis which we shall consider, that of Hotelling. At the request of the Unitary Traits Committee, Hotelling attacked the problem of obtaining a serviceable solution to the problem proposed by Kelley in 1928 [in *Crossroads in the Mind of Man*], "first, a determination, having tests A, B, C, of what the independent mental traits are; and secondly an experimental construction of new tests measuring these independent traits." As we have just noted, Hotelling's least-squares conditions are identical to those in one of the solutions presented by Thurstone. Dr. Hotelling, however, has supplied a very neat iterative solution for the kth order determinant involved which makes the solution comparatively short.

The role of the Unitary Traits Committee in facilitating the development of the method of principal components is confirmed by the beginning footnote in Hotelling's paper in *JEdP* (Hotelling 1933):

> A study made in part under the auspices of the Unitary Traits Committee and the Carnegie Corporation.
>
> The author is indebted to Professor Truman L. Kelley, who was responsible for the initiation of this study and the propounding of many of the questions to which answers are here attempted; also to Professors L. L. Thurstone, Clark V. [sic; it should be L.] Hull, C. Spearman, and E. L. Thorndike, who raised some of the further questions treated.

In a four-page single-spaced letter to Kelley from Hotelling (June 2, 1932), which can be found in the Kelley archives available at the Houghton Library at Harvard, the approach that Hotelling was to take is spelled out in some detail:

> Another line of possible development in tetrad analysis (or rather factor analysis) is to take as independent factors those linear functions of a number of test scores which correspond to the principal axes of the ellipsoids of the scatter diagram.

[3] It might also be noted that Hotelling was an inaugural member of the Psychometric Society based on the membership roster published in 1936. For some unknown reason, however, he was no longer a member as of March, 1939.

Apparently, this long letter (along with some extensive handwritten notes) served as a proposal to work for the Unitary Traits Committee for two summer months in 1932 (for $800); Kelley responded to Hotelling with a letter dated June 20, 1932:

> This letter is in confirmation of our agreement that you work for the Unitary Traits Committee for a period of two months and receive therefore a total of $800.00. It is understood between us that you are to be free to meet such other obligations during this time as incidentally arise, and that we upon our part may occasionally call upon you in the future for things not involving an extended study upon your part.
>
> I am sending a copy of this letter to Dr. Thorndike, chairman of the Committee.
>
> I am returning herewith your notes, for which please accept my thanks.

Hotelling replied on June 25, 1932 (with a notation that a copy was also sent to E. L. Thorndike):

> With your letter of June 20 this will confirm our agreement that I am to work for the Unitary Traits Committee for two months this summer.
>
> Thank you for the return of my rough notes, which I hope latter to elaborate. During the past week at Syracuse I have been discussing their contents at considerable length with L. L. Thurstone, Jack Dunlap, and Ragnar Frisch. Dunlap is going to try the method of principal axes on some tests he has made of chickens. [sic?; "children"?]
>
> I hope to be at Blackey's Hotel at Gilmanton Iron Works early in July and to see you there. Meanwhile I am wrestling with some of the very beautiful and intricate mathematical problems involved.

This last letter is interesting for several reasons, particularly for the three people Hotelling mentioned that he had extensive discussions with: L. L. Thurstone, Jack Dunlap, and Ragnar Frisch. The 1932 Syracuse meeting referred to was of the American Association for the Advancement of Science and its many affiliated societies (such as the American Psychological Association). At this meeting, Thurstone presented his own principal axes solution to the problem of factor analysis. As Hotelling notes in a 1933 *JEdP* footnote:

> Since this was written Professor Thurstone has kindly sent me a pamphlet he has prepared for class use, in which he uses the same geometric interpretation as in the present section, and discusses the problem from essentially the same standpoint as that taken in [Part One]. His iterative procedure appears to have no relation to that of [Part Four]. In June, 1932, Professor Thurstone presented at the Syracuse meeting of the American Association for the Advancement of Science certain of the considerations which have served as a point of departure for this paper.

Interestingly, Thurstone abandoned his first principal axes approach because he thought it did not conform to a "true" and psychologically meaningful factor analytic model. The mention of Jack Dunlap in Hotelling's letter is also interesting, because he was to be the Editor of *JEdP* overseeing the publication of Hotelling's 1933 contribution. Ragnar Frisch, for those who might not know, was the first recipient of the Nobel Prize in Economic Sciences in 1969; he is recognised for founding the discipline of econometrics and for coining the word pair "macroeconomics/microeconomics" in 1933.

It is worth mentioning that the debate between the use of principal components and the reliance on the factor model, which rages to this day, can date back to 1935 in Thurstone's book, *The Vectors of Mind: Multiple Factor Analysis for the Isolation of Primary Traits* (Thurstone 1935). In Chapter IV, *"The Principal Axes"*, Thurstone concluded with a summary rejection of principal components as a viable approach to the factor model (p. 132):

> These considerations make it necessary to discard the method of principal axes and also Hotelling's special case of this method as solutions to the psychological factor problem.

When the first author has taught modules on principal component analysis (PCA) and factor analysis (FA) in a Multivariate Analysis class, PCA was introduced with three introductory points:

(a) PCA deals with only one set of variables without the need for categorizing the variables as being independent or dependent. There is asymmetry in the discussion of the general linear model; in PCA, however, we analyze the relationships among the variables in one set and *not* between two.
(b) As always, everything can be done computationally without the Multivariate Normal (MVN) assumption; we are just getting descriptive statistics. When significance tests and the like are desired, the MVN assumption becomes indispensable. Also, MVN gives some very nice interpretations for what the principal components are in terms of our constant density ellipsoids.
(c) Finally, it is probably best if you are doing a PCA, not to refer to these as "factors." A lot of blood and ill-will has been spilt and spread over the distinction between component analysis (which involves linear combinations of *observable* variables), and the estimation of a factor model (which involves the use of underlying latent variables or factors, and the estimation of the factor structure). We will get sloppy ourselves later, but some people really get exercised about these things.

Four introductory points were made in introducing FA:

(a) In a principal component approach, the emphasis is completely on linear combinations of the observable random variables. There is no underlying (latent) structure of the variables that I try to estimate. Statisticians generally love models and find principal components to be somewhat inelegant and nonstatistical.
(b) The issue of how many components should be extracted is always an open question. With explicit models having differing numbers of "factors," we might be able to see which of the models fits "best" through some formal statistical mechanism.
(c) Depending upon the scale of the variables used (i.e., the variances), principal components may vary and there is no direct way of relating the components obtained on the correlation matrix and the original variance-covariance matrix. With some forms of factor analysis, such as maximum likelihood (ML), it is possible to go between the results obtained from the covariance matrix and the correlations by dividing or multiplying by the standard deviations of the

variables. In other words, we can have a certain type of "scale invariance" if we choose, for example, the maximum likelihood approach.

(d) If one wishes to work with a correlation matrix and have a means of testing whether a particular model is adequate or to develop confidence intervals and the like, it is probably preferable to use the ML approach. In PCA on a correlation matrix, the results that are usable for statistical inference are limited and very strained generally (and somewhat suspect).

4 Hotelling's Power Method

At the meeting of the Unitary Traits Committee in December of 1932, several papers were read that were devoted to numerical examples of Hotelling's iterative strategy for obtaining the principal components of a correlation matrix. The procedure proposed by Hotelling would today be referred to as (a repeated use of) the power method for finding the dominant eigenvalue of a matrix. Bodewig (1956, p. 250) attributes the power method to von Mises in 1929, as published in a rather obscure German language periodical. However, because of the close date to Hotelling's own use of a power method and his not referencing von Mises (but he did so later in an *Annals of Mathematical Statistics* article in 1943 (Hotelling 1943) entitled "Some new methods in matrix calculation"), the power method itself might just as well be attributed to Hotelling. In fact, Hotelling's repeated use of the power method to find *all* the eigenvalues and eigenvectors of a matrix involves what has now become well-known as "Hotelling deflation": these are outer products of an eigenvector with itself, weighted by the eigenvalue, and subtracted from the starting matrix. We give a summary of this process taken from *Multivariate Statistical Methods* (Morrison 1967):

> Let $\mathbf{A}$ be the $p \times p$ matrix of real elements. It is not necessary that $\mathbf{A}$ be symmetric. Order the characteristic roots λ_i of $\mathbf{A}$ by their absolute values:
>
> $$|\lambda_1| > |\lambda_2| \geq \cdots \geq |\lambda_p|$$
>
> and denote their respective characteristic vectors as $\mathbf{a}_1, \ldots, \mathbf{a}_p$. Initially we shall require that only $|\lambda_1| > |\lambda_2|$. Let $\mathbf{x}_0$ be any vector of p real components, and form the sequence: $\mathbf{x}_1 = \mathbf{A}\mathbf{x}_0; \ldots \mathbf{x}_n = \mathbf{A}\mathbf{x}_{n-1} = \mathbf{A}^n\mathbf{x}_0$ of vectors. Then if the successive $\mathbf{x}_i$ are scaled in some fashion, the sequence of standardized vectors will converge to the characteristic vector $\mathbf{a}_1$. Probably the most convenient scaling is performed by dividing the elements by their maximum, with normalization to unit length merely reserved for the last, or exact, vector. Since $\mathbf{A}\mathbf{a}_1 = \lambda_1\mathbf{a}_1$ the characteristic root itself can be found by dividing any element of $\mathbf{A}\mathbf{a}_1$ by the corresponding element of $\mathbf{a}_1$. The same iterative procedure can be used to compute any distinct characteristic root of $\mathbf{A}$. To extract the second largest root and its vector we normalize the first characteristic vector $\mathbf{a}_1$ to unit length, form the $p \times p$ matrix $\lambda_1\mathbf{a}_1\mathbf{a}_1'$ and subtract it from $\mathbf{A}$ to give the residual matrix $\mathbf{A}_1 = \mathbf{A} - \lambda_1\mathbf{a}_1\mathbf{a}_1'$. [A Hotelling deflation]

In the more recent implementations of routines for finding the principal components of a covariance matrix (such as in Matlab), Hotelling's iterative procedure is not used. Instead, a Jacobi-like algorithm for finding the eigenvalues/eigenvectors of a matrix is commonly adopted [we will come back to this topic shortly]. This replacement may be due in part to the computational difficulties one might encounter with Hotelling's approach. As Bodewig (1956, p. 250) notes:

> It was R. von Mises ... who found the power method. It was a great achievement. And in many cases it gives a quick result. But it cannot be denied that in a large number of cases the convergence is extremely bad, so bad in fact that it can hardly be used at all. The convergence will be good enough only if the quotient $|\frac{\lambda_1}{\lambda_2}| > 3$. But this is only rarely the case.

We might mention that there is one prominent and current application of the power method for finding a single dominant eigenvalue/eigenvector combination—this is in Google's search engine and the use of what is called PageRank.

5 Hotelling's 1936 *Psychometrika* Paper: "Simplified calculation of principal components"

If Hotelling's seminal 1933 article in *JEdP* had appeared instead in *Psychometrika*, it would be, according to Google Scholar, the second most highly cited article in *Psychometrika* after Cronbach's (1951) survey on "coefficient alpha." The first co-editor of *Psychometrika*, Paul Horst, even relates how he tried to get something comparable for the first volume of *Psychometrika* (Horst and Stalnaker, 1986, p. 5):

> At Proctor and Gamble we had been working with the applications of the new factor analytic methods to personnel data. I had learned of a new iterative procedure that Hotelling at Columbia had developed for finding the principal axis factors of a correlation matrix, and we were using it at Proctor and Gamble. I saw Hotelling personally at Columbia during this time, to persuade him to contribute his manuscript for the maiden issue of *Psychometrika*. I asked him whether he could give us a manuscript on his new method. He at first was markedly cool to the idea and I suspected that he was not eager to conceal his production under the cover of a dubious new journal. I then told him that I very much wanted this method published in this first issue and that, if he did not feel he could do it, I would reluctantly publish the method myself and of course give him full credit. With this, he decided to provide the manuscript himself (Hotelling 1936), and we remained good friends as long as he lived.

The 1936 Hotelling paper referenced above is based on the simple idea that when the power method is applied to an integer power of a matrix (say, to $\mathbf{A}^2$) instead of to the original matrix (say, to $\mathbf{A}$), convergence will be faster. Unfortunately, such a conjecture appears generally unjustified. We give two quotes from Bodewig (1950, p. 134; 246) that make this point:

> Hotelling [in the 1943 *Annals of Mathematical Statistics* article] therefore, proposes computing the product $\mathbf{T}$ and, then to square successively: $\mathbf{T}, \mathbf{T}^2, \mathbf{T}^4, \mathbf{T}^8 \ldots$, and then to form the vector say $\mathbf{T}^{16}\mathbf{y}^{(1)}$. This method is very elegant. *Whether it is suitable, is another matter.* (emphasis added)
>
> ...
>
> Powers of Matrices: Many authors such as Kincaid, Aitken, Hammersley, and Hotelling, recommend successive squaring of $\mathbf{A}$ and iteration with $\mathbf{A}^{2m}$ on $\mathbf{v}$ instead of with $\mathbf{A}$ itself. This is done in order to speed up convergence and to save work. *But this proposal cannot be defended.* (emphasis added)

Bodewig provides a formal proof of this assertion that "this proposal cannot be defended." It is based on an elaboration of the following observation: multiplying a vector **x** by a matrix **A** and that resultant vector, **Ax**, by **A** again (i.e., **A**(**Ax**)), requires fewer operations than multiplying **A** by **A**, and then using that product matrix, $\mathbf{A}^2$, to multiply **x** (i.e., $\mathbf{A}^2\mathbf{x}$).

6 Kelley's Approach to Principal Components

In Holzinger's survey of the work completed by the Unitary Traits Committee mentioned earlier, the following short excerpt appears:

> Very recently Professor Kelley has published a volume entitled *The Essential Traits of Mental Life*(1935). In this book he has contributed a method of factorization which appears simpler than that of Hotelling, but which gives the same results. In addition to this new technique Professor Kelley makes a comparison of current methods of factorization.

In the 1936 Hotelling paper solicited by Horst, Kelley's method of obtaining principal components is explicitly commented on as follows (p. 27):

> Another method of calculating principal components has been discovered by Professor Truman L. Kelley, which involves less labor than the original iterative method, at least in the examples to which he has applied it. How it would compare with the present accelerated method is not clear, except that some experience at Columbia University has suggested that the method here set forth is the more efficient. It is possible that Kelley's method is more suitable when all the characteristic roots are desired, but not the corresponding correlations of the variates with the components. The present method seems to the computers who have tried both to be superior when the components themselves, as well as their contributions to the total variance, are to be specified. The advantage of the present method is enhanced when, as will often be the case in dealing with numerous variates, not all the characteristic roots but only a few of the largest are required.

A synopsis is given below of Kelley's method for finding the two principal components of a two-variable system, taken from his *Essential Traits of Mental Life* (1935, p. 2). He showed that by using this method iteratively for all pairs of variables, the complete set of principal components are retrieved:

> If it is desired to create two new variables, x' and y', which are completely defined by the given variables, x and y, ..., all that is necessary is to write $x' = a_1x + b_1y$; $y' = a_2x + b_2y$ and assign any values to a_1, a_2, b_1, and b_2. Solving these equations for x and y we have
>
> $$x = \frac{b_2x' - b_1y'}{a_1b_2 - a_2b_1}$$
>
> $$y = \frac{a_1y' - a_2x'}{a_1b_2 - a_2b_1}$$
>
> Of the infinite number of new sets of equivalent variables, x' and y', which can be derived by substituting different values for a_1, a_2, b_1, and b_2, that one is considered to have special merit which is a rotation of the x and y axes to the position of the major and minor axes of

the ellipse. These particular new variables, which we designate x_1 and y_1, are given by the equations

$$x_1 = x\cos\theta + y\sin\theta$$
$$y_1 = -x\sin\theta + y\cos\theta$$

where θ is the angle of rotation and is given by

$$\tan 2\theta = \frac{2p}{v_1 - v_2}.$$

Here, $p = \sigma_{12}$, $v_1 = \sigma_1^2$ and $v_2 = \sigma_2^2$. The peculiar merit of the new variables, x_1 and y_1, lies in the facts which can be immediately surmised by thinking of the elementary geometry involved.

(a) x_1 and y_1 are uncorrelated.
(b) x_1 and y_1 axes are at right angles to each other.
(c) The variance of x_1, distance from the minor axis in the direction of the major axis, is a maximum, for no other rotation of axes yields a variable with as large a variance.
(d) The variance of y_1, distance measured in the direction of the minor axis, is a minimum.

The advantage of (a), lack of correlation, need scarcely be dwelt upon, as it is the essential purpose of factorization to obtain independent measures.

The advantage of (b), orthogonality, is not quite so obvious. Though a point in two-dimensional space may be completely defined by distance from two oblique axes, nevertheless the simplicity of thought (and to create such simplicity is a basic purpose of factorization) when a point is defined in terms of perpendicular distance from two perpendicular axes, should be sufficient to commend the use of such axes.

The advantage of (c) making the variance of one of the new variables a maximum is particularly apparent when the major axis is much greater than the minor. In this case, much more about the total situation or the total field wherein variation can take place is known if variability in any other direction is known. The principle of parsimony of thought recommends a knowledge of the x_1 variable if but a single item of knowledge is available. The operation of this principle will be much more apparent when thinking of many variables, for here the variances of some of the smaller ones may be such that entire lack of knowledge of them will not be serious.

It is obvious from the geometry of the situation that there is but a single solution yielding variables with the properties mentioned. These constitute the components in the two-variable problem.[4]

It is interesting to speculate where Kelley may have come up with his approach to the calculation of principal components. He gives no explicit reference for his iterative method in the *Essential Traits of Mental Life*. In fact, he opens this text (Chapter I) as follows:

[4] As mentioned earlier, one limitation of the power iteration method is slow convergence when λ_1/λ_2 is close to 1. Kelly provided the following comment on the advantage of his method: "unlike Hotelling's method, approximate equality of variance of two components does not lead to slow convergence." (p. 9)

> A New Method of Analysis of Variables into Independent Components: Before attempting a comparison of different methods of analysis of variables into components, a new method is presented. The procedure followed is new, but the outcome is identical with that given by Hotelling's method of analysis.

One story that is at least plausible comes from a perusal of the Kelley archives at Harvard. Kelley spent a sabbatical year in the very early 1920s with Karl Pearson, who was to have a major influence on Kelley's statistical thinking. For example, in the preface to Kelley's well-received 1923 text, *Statistical Method*, there is the following acknowledgement to Karl Pearson:

> I would, however, say that my greatest inspiration has been the product of that master analyst, Karl Pearson, and that the English school entire has been most contributive.

There is also a reference in *Statistical Method* (p. 363) to Karl Pearson's paper, "On lines and planes of closest fit to systems of points in space" (Pearson 1901). As is now well-recognised, this early 1901 paper introduced "the method of principal components," although that particular terminology, introduced much later by Hotelling in 1933, was obviously not used.

The key "tan 2θ" formula in Kelley's method for finding the angle of rotation for the principal axes orientation of a two-variable system is present in Pearson (1901, p. 566). It is conceivable that Kelley could have encountered it there for the first time, but it is more likely that Kelley knew of it from his undergraduate work in mathematics at the University of Illinois in the early 1900s. Neither Pearson nor Kelley, for example, thought it necessary to include any reference for what was presumably a well-known formula in mechanics that dealt with the axes of an ellipsoid. At Illinois, Kelley did a Bachelor of Arts thesis (1909) entitled "*Graphic Evaluation of Trigonometric Functions of Complex Variables*." (A Google search on this exact title will retrieve a copy of the thesis). Kelley's trigonometric prowess as represented in his thesis is also well on display in his *Essential Traits of Mental Life*—an extensive set of trigonometric equations were derived by Kelley to make the iterative process work.

An interview done in 2006 with Darrell Bock in the *Journal of Educational and Behavioral Statistics* (Wainer and Robinson 2006) may shed some more historical light on the question of "Whence Principal Components?" The excerpts given below discuss Bock's visit to the University of Illinois in the 1950s to use the ILLIAC computer for some eigenvector/eigenvalue computations that he needed done. Note the name of the graduate student he met at Illinois, Gene Golub; Golub was soon to become a computational giant of the second half of the 20th century.

> I had heard from Charles Wrigley at Michigan State University that the new ILLIAC electronic computer at Champaign-Urbana had programs for both the one- and two-matrix eigenproblems. On his advice, I phoned Kern Dickman, who had helped Charles perform a principal component analysis on the machine, and explained my needs. He invited me to come down to Urbana and bring the matrices to be analyzed with me. By that time, I had become sufficiently proficient in using punched card equipment in the business office of the University—in particular a new electronic calculating punch that could store constants and performed cumulative multiplications as fast as the cards passed through the machine.

> I arrived in Urbana and found Kern; he took me directly to the computation center to see the ILLIAC. But there was very little to see—only a photoelectric reader of teletype tape and a box with a small slit where punched tape spewed from the machine; a few dimly revealed electronic parts could be seen behind a plate-glass window. Elsewhere in the room were teletype machines for punching numbers and letters onto paper tape, printing out the characters of an existing tape, or copying all or parts of one tape to another. My first job was to key the elements of the two covariance matrices onto tape, which in spite of my best efforts to avoid errors, took most of the afternoon.
>
> When I finished that task, Kern suggested that we should meet for dinner at his favorite watering hole in Urbana. When I arrived there I found him sitting with another person whom he introduced as Gene Golub, adding that Gene had programmed the eigenroutines for the ILLIAC. At Kern's suggestion Gene had brought along some papers for me—an introduction to programming the ILLIAC and the documentation of the eigenroutines. He said that his code was similar to that of Goldstein, who had programmed the eigen-procedures for the Maniac machine built by Metropolis at Los Alamos. It used the Jacobi iterative method, which consists of repeated orthogonal transformations of pairs of variables to reduce the elements in the off-diagonal of a real symmetric matrix to zero, all the while performing the same operation on an identity matrix. Although a given element of the matrix does not necessarily remain zero, the iterations converge to a diagonal matrix containing the eigenvalues, and the identity matrix becomes the corresponding eigenvectors.
>
> Gene told the story that Goldstein, having heard the Jacobi method described by a colleague, stopped by John von Neumann's office to ask if the method was strictly convergent. Gazing at the ceiling for about five seconds, von Neumann replied "yes, of course." Goldstein was amazed, thinking this was another of von Neuman's [sic] fabled feats of mental calculation, but as Golub and Van Loan show in their 1996 reference, *Matrix Computations*, the proof requires only a few lines of matrix expressions, which von Neumann could have easily visualized. I already knew of this method, not as Jacobi's, but as the "method of sine and cosine transformations" described by Truman Kelley in his 1935 book, *Essential Traits of Mental Life*. He presented the method as his own creation, including a proof of convergence requiring several pages of geometric argument. Considering that Jacobi had introduced the method in the middle of the 19th-century, I wondered if Kelley had heard of it from one of his fellow professors at Harvard. But I found in his 1928 book, *Crossroads in the Mind of Man*, that he had already used sine and cosine transformations in connection with Spearman's one-factor model, and I now believe that he rediscovered Jacobi's method independently.

Bock got this a little incorrect. Kelley did not "rediscover" Jacobi's method. He did not know, for example, that merely multiplying the pairwise orthogonal rotations together would give the eigenvectors directly as is done in Jacobi's method. But still, Kelley got very close by obtaining all of the eigenvalues of a correlation matrix at the end of his pairwise iterative process. Kelley generated the corresponding eigenvectors rather laboriously by keeping track of all the transformations carried out over the pairwise iterations as expressed in terms of the original variables.

7 Conclusion

So now to the opening question of "Whence principal components?" The best theoretical answer is probably Karl Pearson, given his 1901 paper mentioned earlier.[5] The numerical examples Pearson gave, however, were all extremely small and involved at most three variables. So, from a computational perspective, the answer to the question should probably be Hotelling, based upon his use of an iterative power method and the introduction of Hotelling deflation. If current computational practice is any criterion, however, Kelley could be credited with the introduction of a rudimentary Jacobi-like method. The Jacobi approach became more or less standard practice in the 1950s and 1960s. As noted by Bock in the earlier excerpts, the method had been programmed by Golub for the ILLIAC computer before Bock's visit to Illinois. From the 1970s to the present, most computer-implemented principal component computational routines (in Matlab, for instance) rely on a more basic singular value decomposition (SVD) algorithm developed by that same graduate student Bock met at Illinois in the 1950s, Gene Golub; see, for example, Golub and Reinsch (1970), "Singular value decomposition and least-squares solutions." By way of closing, it is interesting to note that the Golub-Reinsch SVD routine relies on exactly the same type of planar rotations (but now called Givens rotations) used by Kelley in his approach to computing principal components.[6]

References

Bock, R.D.: Rethinking Thurstone. In: MacCallum, R.C., Cudek, R. (eds.) Factor Analysis at 100, pp. 35–46. Erlbaum, Mahwah, N.J. (2007)

Bodewig, E.: Matrix Calculus. North-Holland Publishing, Amsterdam (1956)

Cronbach, L.J.: Coefficient alpha and the internal structure of tests. Psychometrika **16**, 297–334 (1951)

Flanagan, J.: Factor Analysis in the Study of Personality. Stanford University Press, Stanford, CA (1936)

Golub, G.H., Reinsch, C.: Singular value decomposition and least squares solutions. Numer. Math. **14**, 403–420 (1970)

Holzinger, K.: Recent research on unitary mental traits. J. Pers. **4**, 335–343 (1936)

Horst, P., Stalnaker, J.: Present at the birth. Psychometrika **51**, 3–6 (1986)

---

[5] As pointed out to us by Stephen Stigler, an approach similar to Pearson's least-squares strategy was done somewhat earlier in *The Analyst* by R. J. Adcock (**4**, 183–184 (1877); **5**, 21–22 (1878); **5**, 53–54 (1878)) and C. H. Kummell (**6**, 97–105 (1879)).

[6] It might also be noted that Hotelling in his paper introducing canonical correlations (The relations between two sets of variates, Biometrika **28**, 321–377 (1936)) relies on the same type of iterated power method for obtaining canonical correlations and canonical variates as he did in *JEdP* (1933). Kelley, in contrast, in his 1940 monograph, *Talents and Tasks: Their Conjunction in a Democracy for Wholesome Living and National Defence*, approached the canonical correlation task using planar rotations, just as he did in the *Essential Traits of Mental Life*. Also, Kelley provided a rather complete numerical example—and obviously, given the year of publication, all without any electronic computer implementation.

Hotelling, H.: Analysis of a complex of statistical variables into principal components. J. Educ. Psychol. **24**, 417–441 (1933)
Hotelling, H.: The most predictable criterion. J. Educ. Psychol. **26**, 139–142 (1935)
Hotelling, H.: Simplified calculation of principal components. Psychometrika **1**, 27–35 (1936)
Hotelling, H.: Some new methods in matrix calculation. Ann. Math. Stat. **14**, 1–34 (1943)
Kelley, T.L.: Statistical Method. Macmillian, New York, NY (1923)
Kelley, T.L.: Interpretation of Educational Measurements. World Book, Yonkers-on-Hudson, NY (1927)
Kelley, T.L.: Crossroads in the Mind of Man. Stanford University Press, Stanford, CA (1928)
Kelley, T.L.: The Essential Traits of Mental Life. Harvard University Press, Cambridge (1935)
Morrison, D.F.: Multivariate Statistical Methods. McGraw-Hill, NY (1967)
Pearson, K.: On lines and planes of closest fit to systems of points in space. Philos. Mag. **2**, 559–572 (1901)
Thurstone, L.L.: The Vectors of Mind: Multiple Factor Analysis for the Isolation of Primary Traits. University of Chicago Press, Chicago, Illinois (1935)
Wainer, H., Robinson, D.: Profiles in research: R. Darrell Bock. J. Educ. Behav. Stat. **31**, 101 – 122 (2006)

The Emergence of Joint Scales in the Social and Behavioural Sciences: Cumulative Guttman Scaling and Single-Peaked Coombs Scaling

Willem J. Heiser and Jacqueline J. Meulman

1 Introduction

In his classical textbook *Theory and Methods of Scaling*, Torgerson (1958) distinguished three groups of scaling methods. This distinction was primarily based on the consideration which part of the variability in responses of subjects (persons, judges) to stimuli (questions, tasks) was to be regarded as systematic or random. In the first group, called the *subject-centred approach*, systematic variation in the observations is attributed to individual differences in the subjects, while the stimuli are regarded as replications. In the second group, called the *stimulus-centred* or *judgement approach*, systematic variation in the observations is attributed to differences in the stimuli with respect to a designated attribute, while the subjects are regarded as replications. In the third group, called the *response approach*, systematic variation is attributed to stable differences in the subjects as well as in the stimuli. We start by briefly outlining the historical context in which the first two approaches evolved.

1.1 Galton's Subject-Centred Approach

The first group of scaling methods in Torgerson's classification originated with Francis Galton (1822–1911). As noted by Helen Walker, in her impressive dissertation on the history of educational statistics (Walker 1929), Galton's book *Hereditary Genius*

W. J. Heiser (✉)
Institute of Psychology, Leiden University, Leiden, The Netherlands
e-mail: heiser@fsw.leidenuniv.nl

J. J. Meulman
LUXs data science BV, Leiden, The Netherlands
e-mail: jmeulman@stanford.edu

Department of Statistics, Stanford University, Stanford, CA, USA

E. J. Beh et al. (eds.), *Analysis of Categorical Data from Historical Perspectives*,
Behaviormetrics: Quantitative Approaches to Human Behavior 17,
https://doi.org/10.1007/978-981-99-5329-5_14

was the reason for him to become interested in devising standardised ways of reporting individual differences in achievement and intellectual ability. She quoted Galton as follows:

> The theory of *Hereditary Genius*, though usually scouted, has been advocated by a few writers in the past as well as in modern times. But I may claim to be the first to treat the subject in a statistical manner, to arrive at numerical results, and to introduce the 'law of deviation from an average' into discussions on heredity. [...] The range of mental powers between [...] the greatest and least of English intellects, is enormous. There is a continuity of natural ability reaching from one knows not what height, and descending to one can hardly say what depth. [...] I propose in this chapter to range men according to their natural abilities, putting them into classes separated by equal degrees of merit, and to show the relative number of individuals included in the several classes. [...] The method that I shall employ for discovering all this, is an application of the very curious theoretical law of 'deviation from an average'. (Galton 1869, as quoted in Walker 1929, pp. 86–87)

The "curious theoretical law" is the normal distribution of errors, but why would a theory of errors apply to mental ability? In Chap. 3 of *Hereditary Genius*, Galton tried to show that it does. He first looked at 200 students who obtained mathematical honours at Cambridge and found that their average scores on a 'scale of merit' given by several examiners appear to follow the normal distribution, albeit that they show a longer tail in the higher end of the distribution. He then looked at another example, inspired by the inquiries of Adolphe Quetelet (1796–1874) on social and moral statistics—Galton even strongly advises readers to consult the 'very readable octavo volume' *Letters on Probabilities* (Quetelet 1849). The data were scores obtained by 73 candidates from the admissions test for the Royal Military College at Sandhurst, December 1868 (see Stigler 1992). For this data set, he found that the frequencies in ten classes compared with expected frequencies under the normal distribution according to tables published by Quetelet "accord as closely as the small number of persons examined could have led us to expect" (Galton 1869, p. 26). In this case, the tail at the lower end was shorter than expected, which he attributed to an effect of pre-selection.

Galton did not want to argue only from empirical examples. Not long after publication of *Hereditary Genius*, he tried to give some theoretical reasons why examination scores could be expected to be normally distributed (Galton 1875). He first recalled that to conform to the normal distribution, individual errors of observation were supposed to be due to the combined effect of different influences that must be all (1) independent, (2) of equal size, (3) equally likely to push the average upwards or downwards, and (4) infinitely numerous. He immediately admitted that the first three of these conditions "may occur in games of chance, but they assuredly do not occur in vital and social phenomena" (Galton 1875, p. 39). Nevertheless, he then tried to argue that, when examined more closely, they might still be approximately true at a more fundamental level. Stigler (1986) came with a stern but just assessment of this view:

> His explanation was scattered, however; and therefore incomplete. It amounted to two claims that, although true, did not get to the heart of the matter. One was a rearguing of the hypothesis of elementary errors – large influences would frequently, upon closer inspection, be seen to

> be composed of a large number of smaller influences, and hence the Laplacian conditions could be safely pushed back a stage, out of sight. The other, upon which Galton put more emphasis, was that the number of variable influences did not really need to be "infinitely numerous"; in fact, even for $n = 17$, a binomial distribution was normal for practical purposes. (Stigler 1986, pp. 274–275)

Notwithstanding the somewhat shaky defense of the assumption of normality, the 1875 paper is important for the subject-centred scaling approach, because it elaborated on what Galton called the "common statistical scale" in a letter to Editor of *Nature* (Galton 1874). First, by expressing the raw score in deviations from the mean, in standardised scores with unit defined by what was at that time called the "probable error" (before Karl Pearson conceptualised the standard deviation[1]), different distributions of all kinds could be made comparable, as well as individuals measured in different groups or on different occasions. Second, by using the cumulative frequencies of the normal distribution, a curve could be drawn, which he called the "ogive" by which scale values can be determined that partition the total frequency into one hundred equal parts—the *percentiles*. The standardised scale and the percentiles (or McCall's (1922) *T-scale* that divides individuals in ten equally frequent classes) became the mainstay of early educational and mental measurement.

The two psychometric branches that developed out of these first steps of the subject-centred approach became known as *classical test theory* and *factor analysis* on subscale scores of a test. Classical test theory introduced the concept of a systematic latent variable, on top of an error variable: "observed score = true score + error"; it concerns the reliability and validity of subject scores and various ways to establish and optimise these quality measures. A good recent source for the history of classical test theory is Clauser (2022). Factor analysis could not have been formulated before Galton's path-breaking discovery in the 1880s of the statistical concept of correlation.[2] It is concerned with the structure of correlations between scale scores, with the aim to identify (possibly overlapping) subsets of scales that are mutually highly correlated, called common factors. At the occasion of the 100th anniversary of Charles Spearman's seminal paper about the structure of intelligence in 2004, the history of factor analysis was reviewed by leading psychometricians in Cudeck and MacCallum (2007).

1.2 Fechner's Stimulus-Centred or Judgement Approach

The second group of scaling methods in Torgerson's classification started slightly earlier with Gustav Theodor Fechner (1801–1889), physicist and philosopher with important contributions to both psychology, psychometrics, and statistics (Stigler

[1] According to Yule and Filon (1936), the standard deviation was introduced in Pearson (1894, p. 75). Stigler (1986, p. 328) noted that he already mentioned it in a series of lectures from 31 January through 3 February 1893. It no longer presupposes that the variability is caused by error.

[2] But see Stigler's (1986, pp. 297–299) interesting comments about the minor role correlation played in Galton's own work.

1986, Chap. 7; Murray 2021; Heiser 2023). He is best known for *Fechner's Law*, which relates subjective sensation of stimuli to the logarithm of the amount of objective stimulation by physical excitation. As a basis, he took *Weber's Law*, which states that the perceived change in a stimulus is proportional to the initial physical value of the stimulus.

Ernest Heinrich Weber (1795–1878) started work on effects of pressure on the skin in the early 1830s and gave a matured account in *Der Tastsinn und das Gemeingefühl* (Weber 1846). According to the eminent historian of psychology Edwin Boring (1886–1968):

> immediately others began trying to establish Weber's Law for senses other than touch and for dimensions other than intensity. Weber's Law was quantitative. It was a measurement in the sense that it measured in terms of the stimulus: a sensory distance judged quantitatively. It did not, however, imply a sensory scale. (Boring 1961, pp. 241–242).

Fechner's most significant contribution to quantitative psychology was that he showed how to construct a sensory scale in his major work *Elemente der Psychophysik* (Fechner 1860). The key was to assume that on the psychological scale, all *just noticeable differences (jnds) are equal*, and then to take the *jnd as the unit of measurement*. By simple addition of *jnd*s, one could find the magnitude of a sensation above the zero point (called the *stimulus limen*).

Fechner also formulated three specific experimental designs for establishing *jnd*s: the *method of reproduction or adjustment*, the *method of minimal changes* (or *method of limits*), and the *method of constant stimuli*; see (Guilford (1936), Chaps. 2, 4, and 6) for a brief overview and Heiser (2023). In the first design, it is the subject who produces a series of stimulus adjustments to make them subjectively equal to a fixed comparison stimulus. Here, the difference between the physical value of the comparison stimulus and the *average* physical value of the reproduced stimuli is taken as the *jnd*. In the second and third designs, it is the experimenter who adjusts the intensity of a variable stimulus with respect to a constant stimulus in different ways. Then the task of the subject is merely to indicate which of the two is more intense than the other (e.g., "louder than" or "heavier than"). Here, the *jnd* is defined as the physical stimulus difference that is detected by human observers 50% of the time. It is important to notice that for the method of minimal changes and the method of constant stimuli, the actual elementary observations are *qualitative*: in a comparison of two stimuli, the subject has to declare that the *first dominates the second*, or the other way around (soon it became customary to admit judgements of equality as well). After several repetitions of the task, by the same subject and/or other subjects considered as replications, we obtain a number of relative frequencies for the "greater than" category that tend to increase as a function of the value of the comparison stimulus.

To determine the 50% point, Fechner (1860, pp. 85–93) proposed to fit a cumulative-normal distribution to these relative frequencies. It is not the place here to go into details of different ways of fitting such a curve that were developed in the next fifty years (see Urban 1907, 1910). However, it should be mentioned that Fechner's brilliant idea of fitting a nonlinear model in this context obtained an adequate name by the end of this period:

> A mathematical expression which gives the probability of a judgment as function of the comparison stimulus, is called the *psychometric function* of this judgment [...] The term psychometric function was chosen in imitation of the term biometric function, which is commonly in use for mathematical expressions which give the so-called probability of dying as function of age. (Urban 1910, p. 230).

The assumption of the normal distribution on which Fechner's sensory scale is based has become known as the *phi-gamma hypothesis* $\Phi(\gamma)$, where $\Phi(\cdot)$ denotes the normal distribution function. Here, $\gamma = h.\Delta$ is the product of the *measure of precision* in Gauss' sense h (the steepness of the curve, in Pearson's terminology $h = 1/\sigma\sqrt{2}$, where σ is the standard deviation), and the *stimulus increment* Δ corresponding to a probability of 0.5 (the inflection point of the curve). For insightful discussions of the phi-gamma hypothesis, see Boring (1917, 1924), Thurstone (1928a), and Stigler (1986, pp. 244–254).

1.3 The Response Approach: Scaling both Subjects and Stimuli

That brings us to the third type of psychological scaling, which is the main topic of this paper. Torgerson (1958) called it the *response approach*, the early history of which started in the 1940s and 1950s of the twentieth century. Its main initiators were Louis Guttman (1916–1987) and Clyde Coombs (1912–1988). The main concept of the response approach is the *joint scale*, on which both subjects and stimuli have scale values (or a rank position), and its main objective is to find these scale values by an analysis of a single data set with the responses of subjects towards a given set of stimuli. The Guttman type of joint scale will be discussed in Sect. 2, the Coombs type of joint scale in Sect. 3, and we will introduce an interesting and useful connection between the two in Sect. 4.

2 Joint Scales for Multiple Choice Data: Cumulative Guttman Scaling

Guttman's theory of a joint scale was based on the format of *multiple choice data*: responses of a group of *individuals* (alternatively called "objects") to a set of items (alternatively called "qualitative variables") with multiple response categories that are exclusive and exhaustive.[3]

---

[3] This format was developed during World War I, when mental testing and classification of almost two million army recruits necessitated group testing with efficient scoring rules, replacing the customary one-hour individual interviews by a trained psychologist. See Siegel (1992) for the

2.1 Guttman's Points of Departure

Apart from the data format that he had chosen to arrive at a joint scale, various important prerequisites and objectives for Guttman's scaling work were:

1. To consider the individuals as a sample from a *well-specified population*, and the items as a sample from a *well-designed universe of items*, which should form what Guttman called *a single class of behaviour* (Guttman 1941, p. 321).
2. To keep in mind that "a criterion for an attribute to belong in the universe is not the magnitude of the correlations of that item with other attributes known to belong in the universe. [...] It will be seen that attributes of the same type of content may have any size of intercorrelations, varying from practically zero to unity." (Guttman 1944, p. 142).
3. To regard the specific selection of categories endorsed by an individual across items as an individual's *coherent behaviour*, and the specific subgroup of individuals who checked the same category as exhibiting a *distinctive feature*, so that the entire variability of behaviour is attributed to systematic individual differences.
4. To keep away from any assumption of "a priori notions of 'units of measurement', 'interchangeability of units', 'linearity of units', 'addition of units', and the like" (Guttman 1941, p. 323), by which he distanced himself from the psychophysical tradition.
5. To work with methods that simultaneously order and/or quantify individuals and categories on the basis of one and the same data set, in which no *a priori judgment* is required whether or not one category should obtain a higher value than another.
6. To assign a single value (called *weight*) to each category and a single value (called *score*) to each individual, in order to predict responses to other items in the same universe (but outside the current sample), as well as to other individuals of the same population.

During the 1940s, Guttman designed several methods to achieve the above objectives (5) and (6) under prerequisites (1) to (4). The two best-known ones are what Guttman called his *least squares method* (Guttman 1941)—the name we also use in this chapter—and *scalogram analysis*, of which the basic principles were described in Guttman (1944, 1950a), while more detailed and practical matters can be found in Guttman (1947a, b). Let us first look at the major concepts of the least squares method.

important role of Arthur Otis, a doctoral student at Stanford University under supervision of Lewis Terman, who created the first multiple choice paper-and-pencil scale for assessing mental ability, known as the Army Group Examination Alpha.

2.2 *The Least Squares Method*

In the introductory section of Guttman (1941), he makes the following fundamental remark:

> It so happens that the "best" answer we shall derive involves rather lengthy, though simple, numerical calculations; and it can often be usefully approximated by simpler – and even intuitive – procedures. It is of little value, however, merely to say that one "weighting" system is as good as another since different weights give approximately the same numerical answer. It is of primary importance to define first a "best" answer so that one can know what it is that is being approximated, and that definition is our principal motivation in writing this paper (Guttman 1941, p. 323)

Hence, his aim was evidently to describe the *rationale* of applying already known methods of principal components analysis and reciprocal averaging—to which he referred in the bibliographical note on pp. 345–347 of his paper—when the data are qualitative. In fact, he gave three different but related criteria and proved that they lead to the same solution, after adjustments for normalisation differences. The first two have in common a new use of the *correlation ratio* to define quantifications, while the third uses the familiar *correlation coefficient* as a measure of consistency of optimal weights and scores to be found.

2.2.1 Pearson's Correlation Ratio for Nonlinear Regression

The correlation ratio η had been proposed by Pearson (1905) for the situation of a nonlinear regression of an observed quantitative variable y upon either a qualitative or a quantitative independent variable, grouped into a number of classes. The distribution of y for given class x of the independent variable was called an x-array of y's, having an average value $\bar{y}_x$. Then η^2, the square of the correlation ratio, was defined as the ratio of the variance of the means $\bar{y}_x$ of the x-arrays to the total variance of y. By what Pearson called "a *well-known property of moments*,[4] the total variance of y can be decomposed into two parts: the sum of the variance between the means and the average of the variances *within* x-arrays. Therefore, when η^2 goes to 1.0, we have perfect nonlinear correlation and no variability around the regression curve. If η^2 equals zero, the variance between the means is zero; i.e., there is no association of y's with special classes of x at all. Pearson also proved that η is always greater than r, the linear correlation coefficient, where he noted:

> except in the special case when the means of the x-arrays of y's all fall on a straight line, i.e., we have linear regression, and then the two correlation constants are equal. [...] We have now freed our treatment of correlation from any condition as to linearity of the regression. (Pearson 1905, p. 11).

[4] Particularly, it is a property of the second moment of inertia, a concept from physics that Pearson started to use a lot in the 1890s. For a historical account of Pearson's Method of Moments, see Walker (1929, Chap. III).

2.2.2 Quantification of the Category Weights

In his first application of the squared correlation ratio, Guttman introduced the important new notion to let y correspond to the parameters of his scaling problem, i.e., the *unknown quantifications of the category weights*. The independent variable x corresponded to subjects, so that an x-array contains subject-specific observations. Hence, for quantification of the weights, η^2 measures the ratio of the variance across subjects of the mean weights $\bar{y}_x$—of the categories endorsed by each subject—to the total variance of all weights. For a reason to be discussed shortly, Guttman's objective was to *minimise* the relative variability of the weights within subjects. From Pearson's decomposition of the total variance, we can be assured that this goal is in fact achieved by *maximising* η^2.

Guttman then continued to show that from this point of departure, the solution involves finding the eigenvector with the largest eigenvalue of a matrix containing all bivariate cross-tables of the qualitative variables involved in the problem, standardised with respect to the expected frequencies under the hypothesis of independence. Hence, the matrix from which we calculate the optimal category quantifications has typical elements involved in the usual chi-square statistic to test for significance of association. For this reason, Guttman remarks that although the method looks like a principal component analysis:

> There is an essential difference, however, between the present problem of quantifying a class of attributes and the problem of "factoring" a set of quantitative variates. The principal axis solution for a set of quantitative variates depends on the preliminary units of measurement of those variates. In the present problem, the question of preliminary units does not arise since we limit ourselves to considering the presence or absence of behaviour. But we [...] see that in a sense a metric has arisen out of our analysis, a metric that we shall call the "chi-square" metric. (Guttman 1941, pp. 330–331).

So, he underlines that his method satisfies prerequisites (3) and (4) discussed in the beginning of this section. We are not accounting for variance in the data, but for variability in qualitative behaviour. The chi-square metric is not assumed but follows from the aim to minimise the variance of the relevant category weights within subjects, relative to their total variance.

2.2.3 Quantification of the Subject Scores

In Guttman's second application of the squared correlation ratio, the dependent variable y corresponds to the unknown quantifications of the position of the subjects on the joint scale, i.e., the *subject scores*. The independent variable then corresponds to categories, so that an x-array contains category-specific observations that identify subgroups of subjects who share the same behaviour. Now the objective is that the values of y should be such that subjects who endorse the same category should have maximally similar scores, while subjects in different categories have maximally different scores. In this set-up, η^2 will measure the ratio of the variance across categories of the mean scores $\bar{y}_x$ to the total variance of all scores. By the same argument as before, maximising the relative variance of the mean subject scores per category implies minimising the relative variances of the scores within categories.

The solution to maximise η^2 for the scores again becomes an eigenvalue-eigenvector problem, resulting in standardised scores and in category weights that are the average of the scores of subjects belonging to or endorsing that particular category. It gives the rationale for a common statistical scale with standardised scores, with all advantages that Francis Galton had in mind, but now for purely qualitative data and without any distributional assumptions.

2.2.4 Linearising the Regression of Scores and Weights

Finally, Guttman considered a *consistency criterion* for determining the optimal category quantifications and optimal subject scores simultaneously. For this criterion, he selects all pairs of combinations of some subject i endorsing some category j. This subset of pairs is coded with ones in the binary data matrix **M**, and the other pairs are coded with zeros. The problem becomes one of finding scores $\mathbf{z} = \{z_i\}$ and weights $\mathbf{w} = \{w_j\}$ that are *maximally correlated*. Maximal correlation implies that categories endorsed by people with low scores should have similarly low values on the joint scale, while categories endorsed by people with high scores should have similarly high values on the joint scale. Guttman then showed that optimising the consistency criterion amounts to maximising a bilinear form in terms of $\mathbf{z}$ and $\mathbf{w}$ in the metric **M**, under the restrictions that the variances of $\mathbf{z}$ and of $\mathbf{w}$ are finite constants. In addition, he showed that:

1. The quantifications under the consistency criterion solution are equivalent to the two solutions based on maximising the squared correlation ratio;
2. The optimal correlation coefficient is equal to both optimal correlation ratios;
3. Therefore, due to Pearson's result quoted earlier, the regressions of the optimal category weights $\hat{w}$ on the optimal subject scores $\hat{z}$ are linear in both directions.

In sum, all three approaches *linearise the regression* between the two types of scale values on the joint scale. With these important results, we conclude our summary of the major features of Guttman's least squares technique and turn to his remark cited in the beginning of Sect. 2.2 that it could often be approximated by simpler procedures.

2.3 *Scalogram Analysis*

The basic ideas of these simpler procedures were introduced in Guttman (1944) under the name *scalogram analysis*. A scalogram is a visualisation of the joint scale, in which subjects are represented by rank scores $\mathbf{x}$ and the new concept is that an item with its response categories is required to be a *simple function* of $\mathbf{x}$. Suppose the m categories of an item V have arbitrary values $v_1, v_2, \ldots, v_m$, which are regarded to be just *labels*. Then item V is said to be a simple function of $\mathbf{x}$ if we can divide the rank scores on the scale into m consecutive intervals, for which the values within one interval are the same, while they are different from the values in the other intervals. Thus, all subjects with a score within one interval share the same unique response category. The intersection of the intervals across items yields a limited number of response profiles that may occur given the requirement of items being simple functions.

2.3.1 Definition of a Guttman Scale

Given these preliminaries, here is the explicit definition of the notion that Guttman called a *scale*, which was only implicitly playing a role in the least squares approach:

> For a given population of objects, the multivariate frequency distribution of a universe of attributes will be called a *scale* if it is possible to derive from the distribution a quantitative variable with which to characterize the objects such that each attribute is a simple function of that quantitative variable. Such a quantitative variable is called a scale variable. [...] Obviously any quantitative variable that is an increasing (or decreasing) function of a scale variable is also a scale variable [...], which is equally good at reproducing the attributes. [...] Therefore, the problem of metric is of no particular importance here for scaling. For certain problems like predicting outside variables from the universe of attributes, it may be convenient to adopt a particular metric like a least squares metric, which has convenient properties for helping analyze multiple correlations. The interesting mathematics involved here will be discussed in another paper. (Guttman 1944, pp. 140–141).

The future paper that Guttman anticipates in this quotation is most likely Guttman (1950b), in which he demonstrates that for a uniform distribution of the response profiles the least squares technique will produce optimal scores that are linear with the rank scores—i.e. they are equally spaced. A monotonically increasing function of them is obtained for a non-uniform distribution, depending on the relative frequencies of people with the same response profile. Furthermore, once the scale variable is found, there is an unambiguous meaning to the order of attribute values. One category of an attribute is higher than another if it characterises objects higher on the scale (Guttman 1944, p. 150). By ordering the categories in this way, the simple function becomes an *increasing step function* for each attribute.

2.3.2 Representations of a Guttman Scale

In a scalogram, the joint ordering of subjects and categories is represented by marking the ranked subject scores on a continuum, and then inserting *cutting points* indicating the location where one interval borders the next interval. Alternatively, items are represented as a set of parallel *bar charts*, with cutting points in proportion to the marginal frequencies, extended across all items. This construction allows reading off all possible response profiles accounted for by the joint scale. In the important special case in which each item has two categories, permuting the rows and columns of the binary data matrix **M** in the order of the scale variable will show a characteristic *parallelogram* pattern, which is often also called a scalogram (Guttman 1950b).

In joint scales of binary items, the step function has only one step, located at the cutting point between the last subject who scored in the lower category and the first subject who scored in the higher category. It is instructive to compare Guttman's step function with functions in probabilistic models that psychometricians also started to develop in the 1950s and 1960s. Such models are based on a curve that gives the probability of answering an item correctly, called a *trace line* (Lazarsfeld 1950) or *item-characteristic curve* (ICC; Lord and Novick 1968, p. 366). Initially, the most popular ICC was exactly Fechner's (and Galton's) *normal ogive* psychometric function $\Phi(\gamma)$ that we discussed in Sect. 1.2. When used as a model for the relation between ability and item responses (Lord 1952), γ is parametrised as $\gamma = a\,(\theta - b)$, where θ is the subject score, the precision parameter a is called the *discriminating power* of an item, and b is called the *item difficulty*, or more generally, *item location* parameter. In these terms, the Guttman step function is the limiting case for a growing without bound, so that items have *perfect discrimination*. Formally, we obtain $\Phi[\infty(\theta - b)] = 1$ if $\theta > b$ and $\Phi[\infty(\theta - b)] = 0$ if $\theta < b$. (Lord and Novick 1968, p. 403). If the subject score is larger than the item difficulty, we are sure the item will be answered correctly, and when it is smaller we are sure that it will not. That is why the Guttman scale is called *cumulative*: it requires some extra ability to pass the next item on the scale.

2.3.3 Advantages of Scalogram Analysis

Guttman (1944) mentioned three advantages of representing a large amount of data compactly as a joint scale:

1. It is easier to understand and remember than a large chaotic tabulation (p. 142).
2. In the ideal case, we can reconstruct the entire table from the scale scores, because that merely requires checking out which interval each score falls in for each item (p. 142–143).
3. Any outside variable can be predicted equally well from the scale scores alone as from the set of separate items (p. 150).

These properties made scalogram analysis a very popular method among social scientists. Unfortunately:

> Perfect scales are not found in practice. The degree of approximation to perfection is measured by a *coefficient of reproducibility* [...] . In practice, 85 percent perfect scales or better have been used as efficient approximations to perfect scales (Guttman 1944, p. 150).

It need not surprise us that a quite extensive literature emerged on how to find good approximate scalograms. Guttman (1947b) proposed the Cornell technique, which involved repeatedly sorting and rearranging the entries of the raw data matrix, but this soon becomes bothersome for larger number of subjects and items. Green (1956) proposed a practical and automatic method that does not involve sorting and rearranging and was based on summary statistics to estimate reproducibility. For a relatively recent overview of many other proposals and extensions of scalogram analysis, we refer to Clogg and Sawyer (1981).

3 Joint Scales for Preferential Choice Data: Single-Peaked Coombs Scaling

Coombs' theory of a joint scale was based on the format of preferential choice data: rank order responses of a group of *subjects* (treated as *individuals*) to a set of stimuli. Two major examples are (a) what Coombs (1952) called pick *k/n* data, in which subjects are asked to select k stimuli out of a set of n stimuli according to personal preference, and (b) order k/n data, in which the subjects are asked to offer their 1st, 2nd, up to k^{th} choice. When the latter set-up k is fixed equal to n, we obtain a third type of task for the subject: (c) produce a complete rank order of the stimuli. If repeatedly pairs of stimuli ($n = 2$) are presented and the subject is asked which one is preferred ($k = 1$), we get (d) the method of paired comparisons.

Coombs called this class of behaviour relative and mentioned that it requires the use of the *Method of Choice*, without indicating the origin of that group of methods. It originated as Fechner's *Wahlmethode*, which was introduced for the experimental study of aesthetics (Fechner 1871); see Guilford (1936, pp. 222–225) for its early history. What Coombs meant by *relative* is that the data do not tell us whether or not the subject actually endorses the chosen object or proposition in an absolute sense. Rather, it just indicates which one of the stimuli is psychologically closer to the subject's evaluation standard.

3.1 Coombs' Points of Departure

In terms of prerequisites and objectives, Coombs wanted to stay away from any assumption on the unit of measurement—just like Guttman. In his first paper on the so-called unfolding technique, he clearly stated his ultimate goal:

> But because we may sometimes question the meaning of the definitions and the validity of the assumptions which lead to a unit of measurement, it is our intent in this paper to develop a new type of scale not involving a unit of measurement. [...] [It] falls logically between an interval scale and an ordinal scale [...] ; on the basis of tolerable assumptions and with appropriate technique we are able to order the magnitude of the intervals between objects. We have called such a scale an ordered metric. (Coombs 1950, p. 145)

This point of departure is clearly the same as Guttman's prerequisite (4), with the added objective to establish a new kind of metric. Coombs also explicitly stated that "each stimulus has one and only one scale position for all individuals and that each individual has one and only one scale position for all stimuli" (Coombs 1950, p. 146), which corresponds to Guttman's objectives (5) and (6). However, where Guttman talks about the advantage to *predict the response to all items* from the subject score alone, Coombs phrases it with a twist:

> The unfolding technique was explicitly designed to explain preference behaviour. Existing techniques for scaling such data, as Thurstone's Law of Comparative Judgement as applied in his study of Nationality Preferences [...] [Thurstone (1928b)] are procedures [...] which best represent the preferences of the individuals in a group in some statistical sense such as least squares (see Mosteller 1951). The objective of the Unfolding Technique is to go behind the expressed preferences of individuals and to construct a model from which their preferences may be *derived* [emphasis in the original]. It is in this sense that the term explain is used (Coombs 1952, p. 56).

So Coombs, too, aims at constructing a joint scale from which all individual differences may be reproduced entirely. We have seen for the Guttman scale how to link subjects with the responses to items, but how are we going to do something like that for the Coombs joint scale, which has scale values for subjects and for stimuli (or objects), but not for responses?

3.2 *The Unfolding Mechanism*

We answer this question by examining the mechanism on which the unfolding model works. If the scale value of subject i is denoted by C_i, and the scale values of two stimuli are denoted by Q_j and Q_l, then the basic assumption is that an individual will give the response "I prefer stimulus j to stimulus l" if we have $|Q_j - C_i| < |Q_l - C_i|$. In other words, subjects will prefer the stimulus that is closer to their own scale value. In line with this principle, we expect for the case of pick k/n data that the individual selects those k stimuli that are closest to C_i (and for order k/n data, in the order of the distances from C_i). It follows that the C_i value represents a hypothetical stimulus that would perfectly represent the evaluation standard of a subject. For this reason, C_i is usually called the *ideal point* of subject i. Since preference *monotonically decreases* in both directions of the scale, with a peak at the ideal point, this property of response curves in the unfolding model is known as *single-peakedness* (see Coombs and Avrunin 1977, for the theoretical and historical background of this concept). Perhaps one of the most important consequences is that, if all preference curves are single peaked, a social or *consensus ranking* by the simple majority rule exists, and is equal to the ranking of the *median* individual on the Coombs scale (Black 1948).

3.2.1 How the Coombs Scale Limits the Number of Possible Rankings

Coombs called the preferential ordering of the stimulus objects by individual i the I scale, and the joint scale with scale values for individuals, and stimuli the J scale. He then introduced a mechanical metaphor to explain how to reconstruct the data (the I scales) from the model (the J scale), in particular:

> by imagining a hinge located on the J scale at the C_i value of the individual and folding the left side of the J scale over and merging it with the right side. The stimuli on the two sides of the individual will mesh in such a way that the quantity $|Q_j - C_i|$ will be in progressively ascending magnitude from left to right. The order of the stimuli on the folded J scale is the I scale for the individual whose C_i value coincides with the hinge.
>
> It is immediately apparent that there will be classes of individuals whose I scales will be qualitatively identical as to the order of the stimuli and that these classes will be bounded by the midpoints between pairs of stimuli on the J scale. (Coombs 1950, p. 147).

It turns out that the midpoints mentioned in this quotation play a major role in the model, so let us have a closer look at them. In the examples that follow, we use the customary convention to label the stimuli alphabetically. If there are n stimuli, there will be $\frac{1}{2}n(n-1)$ midpoints in total, which have coordinates on the J scale defined as $Q_{AB} = \frac{1}{2}(Q_A + Q_B)$, for all pairs (A,B).

Therefore, the midpoints define $\frac{1}{2}n(n-1)+1$ intervals on the J scale. For the case of complete rankings, in each of these intervals one unique I scale can be located. In case of pick k/n data, however, only a *subset of the midpoints* define feasible intervals, for a *smaller* set of I scales. Since complete rankings provide the richest information and lead to the unfolding technique, we start there and will return to the analysis of pick k/n data afterwards.

Suppose we have a J scale with four stimulus points, in the order $\{Q_A, Q_B, Q_C, Q_D\}$. Then the leftmost interval up to midpoint Q_{AB} contains the first I scale, denoted as ABCD. Going to the right, all points in the next interval will be closer to Q_B than to Q_A. Thus, passing the midpoint Q_{AB} leads to the I scale BACD; then, by moving further to the right we obtain BCAD after passing midpoint Q_{AC}, and so on. In general, the transition from one I scale to the next always involves the reversal of only one *adjacent pair* of stimuli. After having passed all six midpoints, we end up in the rightmost interval containing I scale DCBA, exactly the reverse order from where we started.

3.2.2 How Metric Information Can Be Deduced from Different Subsets of Rankings

So we see that, although there are 24 ($n!$) possible rankings of four stimuli, a perfect joint Coombs scale allows only 7 of them to be present in the data. Of course, with a different order of the stimuli on the J scale, a different set of I scales may be generated. Nevertheless, even for a given order of four stimuli on the J scale, the *subset of* I *scales* accommodated is *not unique*. In our example, the I scale in the middle interval may be either BCDA or CBAD. To understand why, consider the sequence of the three I scales in the middle: we can either have BCAD—BCDA—CBDA, or BCAD—CBAD—CBDA. In the first sequence, midpoint AD is passed first, and midpoint BC is passed next; in the second sequence, the midpoints are passed in the reverse order. An important consequence is that in the first sequence we have, in terms of coordinates, $\frac{1}{2}(Q_A + Q_D) < \frac{1}{2}(Q_B + Q_C)$. From this inequality, it follows that we must have $|(Q_D - Q_C)| < |Q_B - Q_A|$, while in the second sequence we find the reverse.

Apparently, the occurrence of the I scale BCDA in the data indicates that the interval between stimuli A and B must be greater than the interval between C and D. Conversely, the occurrence of the I scale CBAD indicates that the distance between A and B is smaller than the distance between C and D. With more than four stimuli, the variety of different sets of I scales increases rapidly and leads to a substantial amount of metric information. Coombs (1950) called a J scale with unequally sized intervals between the scale values of the stimuli a *quantitative J scale*, with an *ordered metric* measurement level.

3.3 Methods to Find Coombs Scales, Including Some Extensions and Special Cases

We are now in a position to deal with the technical problem of constructing Coombs scales, starting with early methods for finding a quantitative J scale. Next, we will indicate how the unfolding model has been extended to the multidimensional case and to probabilistic versions. The last two subsections are devoted to the analysis of *pick k/n data*.

3.3.1 Early Methods to Determine a Quantitative J Scale for a Set of Rankings

Early methods to find a quantitative J scale for a given set of I scales usually consisted of three steps. The *first step* starts by heuristically deciding on the order of the stimuli and then tries to list by trial and error the I scales from left to right, where the transition from one I scale to the next must involve only one reversal of an adjacent pair of stimuli (while keeping track of I scales that do not fit or do not occur in the data). This first step identifies the midpoints and their ordering along the scale. The *second step* involves determining metric relations between the stimulus intervals by using the order in which the midpoints change—as demonstrated earlier; it results in a partial ordering of a subset of the distances between stimulus scale values. In the *third step*, the quantitative J scale has to be derived in such a way that the stimulus intervals satisfy the metric relations found in the previous step.

A remarkable omission in the Coombs (1950) paper that introduced the unfolding technique was that Coombs completely skipped a description of the third step. One reason may have been that he encountered the difficulty that half of the subjects in his empirical example produced a different partial order of the distances than the other half. Another reason might have been that he considered this step simply to be done by trial and error for a small number of stimuli. Next, the monograph that contained an extended discussion of the unfolding model and technique (Coombs 1952) did give one empirical example of an unfolded J scale with the spacing between adjacent stimulus scale values indicated (Fig. 11 on page 82), but without any further explanation of how this result was obtained. In Coombs (1953), the unfolding of preferential choice data are embedded in his emerging theory of data, but again there is no indication on how to actually obtain the joint scale.[5]

The first method for solving the problem completely was provided by Abelson and Tukey (1959), who proposed a general *maximin criterion* for regression problems under a variety of order constraints. For the unfolding case, we consider an ordered sequence of quantitative scale values $\{Q_1, \ldots, Q_n\}$ with first differences that satisfy certain inequalities. The proposed criterion *maximises* the squared Pearson correlation r^2 between any candidate solution $\{\tilde{Q}_1, \ldots, \tilde{Q}_n\}$ and another feasible set of scale values satisfying the same inequalities, chosen to have minimal correlation with $\{\tilde{Q}_1, \ldots, \tilde{Q}_n\}$. This criterion guarantees that r^2 cannot be less than an admittedly pessimistic value between zero and one, which may be viewed as a measure of how loose or how tight the ordinal constraints on the stimulus coordinate differences determine the quantitative J scale (cf. Shepard 1966, pp. 288–292). The procedure used to actually find these maximin solutions was complete enumeration with smart heuristics, requiring computing equipment for cases with a relatively large number of stimuli.

In his wide-ranging treatise *A Theory of Data*, Coombs (1964) presented a "pencil-and-worksheet" method for finding the ordered metric scale values, called the *delta method* and developed by his colleague Frank Goode. The presentation was primarily by giving examples: one for the unfolding case of seven stimuli (o.c., pp. 96–102) and two for his ordinal method of similarities (o.c., pp. 359–362 and pp. 450–454). The delta method did not include a criterion to evaluate the quality of the solution, and all the rewritings in the worksheets were not easy to comprehend for the general reader (to put it mildly). No wonder that unidimensional ordinal unfolding never found many substantial applications, except within the limited circle of Coombs and his students (e.g. Dawes 1972, pp. 79–80). The same conclusion was reached—reluctantly—by McIver and Carmines (1981, Chap. 6).

---

[5] Interestingly, this chapter also introduced the "Method of Similarities, an adaptation of the Unfolding Technique [...] , [by which] it is possible to take a single individual subject and determine the structure of the attribute [...] as he perceived it for these stimuli. It can readily be determined whether his perception [...] satisfies a simply ordered system and what some of the metric relations are" (Coombs 1953, pp. 479–480). The first published detailed account of this method of similarities (Coombs 1954) was concerned primarily with a system of data collection procedures for finding the rank order of distances between pairs of stimuli, in several related designs. So again it turns out that, in his own words, "procedures for recovering a J space are as yet incompletely developed" (o.c., p. 193).

3.3.2 Later Methods for Extensions of Unfolding

It took 25 years after his first presentation of the ordered metric scale before Coombs could offer a practical analytical procedure in the form of the so-called ORDMET algorithm (McClelland and Coombs 1975). Meanwhile, however, other approaches to obtain single-peaked joint scales had been invented and successively improved upon that soon became more popular. We just mention the following two groups of methods.

The first approach is a special case of the *non-metric multidimensional scaling* (MDS) methods that is usually based on the least squares STRESS criterion proposed by Kruskal (1964). In the case of non-metric multidimensional unfolding, STRESS measures the average discrepancy between the best monotonically increasing transformation of each I scale and the distance between ideal and stimulus points in a Euclidean space of prespecified number of dimensions. Non-metric MDS and unfolding have generated a vast literature, with many applications in a wide spectrum of domains. For unfolding as an MDS method, we refer for more specifics and historical overviews to Heiser and Meulman (1983), Heiser and Busing (2004) and Busing (2010).

The second approach is the group of *probabilistic unidimensional unfolding* methods in the tradition of item response theory (IRT) modelling. It is important to note here that these types of methods do not attempt to model preferential choices (i.e. Coombs data), but multiple choice items (i.e. Guttman data) with binary or graded *agree-disagree responses*. In the graded case, the categories are of the ordered Likert type (Likert 1932), for example: {"strongly agree", "agree", "disagree", "strongly disagree" }. For a good example of this approach, we briefly look at the generalised graded unfolding (GGUM) model proposed by Roberts, Donoghue, and Laughlin (2000). They assume a *subjective response process*, which is single peaked in terms of the difference between the item location Q_j and the subject location C_i—the standard unfolding assumption. Then they note that for the observable categories, subjects can respond with the "disagree" response categories for either of two reasons. If $Q_j - C_i$ is negative beyond a certain threshold, then the subject will disagree with the item "from above", and if $Q_j - C_i$ is positive beyond a certain threshold, then the subject will disagree with the item "from below". The implication is that the response probabilities for the agree categories are single peaked (with different dispersions), and for the disagree categories, they are bimodal or even single dipped for items that are far away from the ideal point.

3.3.3 Parallelogram Analysis of Pick k/n Data

We now turn to the analysis of pick k/n data. Note that this type of data is a special case of rankings, because if we ask a subject to choose a subset of k stimuli out of a set of size n, the I scale so obtained is equivalent to a *tied ranking*. There will be k stimuli in the first tie block and $n - k$ stimuli in the second tie block. From the assumption of single-peakedness, it follows directly that subjects will choose k

adjacent stimuli that are all closer their ideal point than any other stimulus. Consider the case of $k = 2$ and $n = 5$, with stimulus points on the J scale in alphabetical order (Coombs 1953, pp. 496–501). Starting from the left, there are four pairs of adjacent stimuli: (AB), (BC), (CD), and (DE); in general, the number of subsets is $n - k + 1$. So the feasible I scales generated under this data collection design are the tied rankings $\{(AB), (CDE)\}$, $\{(BC), (ADE\}$, $\{(CD), (ABE)\}$, and $\{(DE), (ABC)\}$. We see that in going from one I scale to the next, the leftmost stimulus in the first tie block is dropped and replaced by the stimulus in the second tie block that is next on the J scale. So, only midpoints AC, BD, and CE are "working"—in general, there are $n - k$ midpoints to differentiate the subjects, a lot less than the $\frac{1}{2}n(n - 1)$ midpoints that are working for full rankings. When k grows with respect to n, discriminability among subjects very quickly deteriorates.

For pick k/n data, we can code a binary data matrix $\mathbf{E}$ with elements $e_{ij} = 1$ if a subject in row i has chosen the stimulus in column j, and $e_{ij} = 0$ elsewhere. Then it is not hard to see that if the rows and columns of $\mathbf{E}$ are arranged in the order of subject and stimulus points on the J scale; the rearranged data matrix will show a parallelogram pattern with k consecutive ones in each row (Coombs 1964, pp. 66–74). So an obvious procedure to analyse pick k/n data—called *parallelogram analysis*—is seeking a rearrangement of rows and columns of $\mathbf{E}$ that will yield as closely as possible a solid diagonal band from the top-left to the bottom-right. The technical problem is identical to seeking a parallelogram pattern in the matrix $\mathbf{M}$ of Guttman's scalogram analysis for binary items.

3.3.4 Mosteller's Least Squares Method

Techniques for parallelogram analysis in the early 1940s were heuristic trial-and-error procedures, which became cumbersome with growing size and error in the data (regardless how defined). We refer to Hubert (1974) for the early recognition that parallelogram analysis is formally equivalent to the *seriation problem* studied in archeology, for which theoretical results and good approximate solutions were already available. In addition, we have seen in Sect. 2.2 that the least squares technique developed by Guttman (1941) gives an optimal solution for scalogram analysis that is unique and could serve as a criterion to evaluate simpler procedures. But as first noted by Torgerson:

> More recently, Mosteller (1949) has shown that precisely the same reasoning can be applied to the nonmonotone or point item, the only difference being that, with the monotone items considered by Guttman, *each* category is included in the analysis, whereas, with the point items, only the positive category is included. Other than this, the solutions are equivalent. [...] We shall follow Mosteller's (1949) derivation mostly, rather than Guttman's, since it seems easier. (Torgerson 1958, pp. 338–339)

Mosteller's intention was to formulate a new method of scaling for attitude statements on the basis of binary agreement responses.[6] Therefore, we will in the following paragraphs call the stimuli we are dealing with in the columns of **E** *statements*. According to the summary in Torgerson (1958, pp. 339–343), the task of the subject is to check k statements for agreement, relative to the other $n - k$ statements. The subject score is defined as the average of the unknown weights for the statements selected. To find these weights, Mosteller used the same criterion as the one used by Guttman: maximising the *squared correlation ratio* η^2, i.e. the ratio of the variance of the subject scores, relative to the total variance of all weights. Next, Torgerson explained in great detail the derivation of the stationary equations for the optimal weights as a function of **E**. He also demonstrated that the procedure to obtain these optimal weights for **E** is equivalent to the procedure to obtain optimal weights for Guttman's binary data matrix **M**. However, the actual solution based on **E** is only the same as the one based on **M** if the columns of **E** are supplemented by n additional columns, with elements $\{1 - e_{ij}\}$, which represent the negative categories (disagreement) in Guttman's multiple choice format. This supplementation is called "dédoublement" in the French literature (Benzécri et al. 1973, TII A no 2, Sect. 1.4, 1.5), and "doubling" in the English literature (e.g. Benzécri 1992, pp. 390–392, pp. 513–517, or Nishisato 2007, p. 182). Hence, our conclusion must be that Mosteller (1949) had not only formulated a new scaling method for non-cumulative or non-monotone items but had also provided a procedure equivalent to correspondence analysis of binary data with equal row sums (cf. Heiser 1981, Chaps. 3 and 4) and was the originator of the concept of dédoublement as well.

The final issue is: in what sense does Mosteller's least squares method offer a good solution to parallelogram analysis? The short answer is: if there exists a permutation of the columns of **E** that yields the consecutive ones property, then the least squares method will find it. The correct order is obtained by permuting the columns of **E** in the order of the optimal weights. For a longer answer, the reader is referred to Heiser and Warrens (2008). This paper also discusses the same property for robust methods of calculating optimal weights, such as the method of reciprocal trimmed means, as proposed by Nishisato (1987).

4 A Coombs Scale of Preference Rankings Using Least Squares Guttman Scaling

Five years after his first paper about the least squares quantification of multiple choice data (Guttman 1941; see Sect. 2), Guttman published a second quantification paper, this time about the scaling of paired comparisons and rank order data (Guttman 1946).

[6] Mosteller (1949) is an internal document of Harvard university, entitled "A theory of scalogram analysis, using noncumulative types of items: a new approach to Thurstone's method of scaling attitudes". As far as we know, it is not publicly available, which is the reason that we rely on Torgerson's summary.

In the introduction, he already outlined what he wanted to achieve: "The judgments vary from person to person (and possibly within a person), and the problem is to determine a set of numerical values for the things being compared that will in some sense best represent or average the judgments of the whole population" (o.c., p. 144). As always, he wanted to avoid distributional assumptions, for example the existence of latent discriminal processes that are normally distributed, as in Thurstone's Law of Comparative Judgment (Thurstone 1927a, b). However, unlike his earlier work, which determines subject scores expressing *individual differences* in response to questions or statements, this paper was an *object-centred* type of scaling, which only aims at an *average* or *consensus* scale for the whole group of individuals.

4.1 Coding Paired Comparisons and Rankings in Multiple Choice Format

Guttman's first step was to code the paired comparison data or rank orders into the familiar format of binary items, with response categories "yes" and "no". Here, the number of items is $\frac{1}{2}n(n-1)$, and they concern the question "Did subject i prefer stimulus j over l?", with j, l ranging over all pairs of stimuli. For each subject, the response "yes" is then coded in the first category, while the response "no" is coded in the second category.

4.1.1 Recoding into a Dominance Matrix to Incorporate Stimulus Contrast Restrictions

From this point on, we will follow Nishisato (1978), who gave an alternative to Guttman's formulation that is easier to understand and leads to the same solution. Since the categories of each item specify that object j is preferred over object l or the reverse, it is natural to require that the quantification of each category is a simple linear function of the two corresponding object scale values Q_j and Q_l. If we denote the category quantifications with y_{jl_1} and y_{jl_2}, then the stimulus contrast restrictions are $y_{jl_1} = Q_j - Q_l$ and $y_{jl_1} = Q_l - Q_j$.

Nishisato then showed that maximising the correlation ratio under these restrictions leads to the same solution as Guttman (1946) and amounts to a principal components analysis of a subjects by stimuli matrix $\mathbf{S}$, called the *dominance matrix*. This matrix contains in each row the balance of how many times subject i preferred object j over the other objects, minus how often subject i preferred one of the other objects over object j. For instance, the rank order ABCDE is represented in $\mathbf{S}$ as $[4\ 2\ 0\ -2\ -4]$. When the paired comparisons contain intransitivities, the rows of the dominance matrix will contain ties.

4.1.2 Multidimensional Extension: The Vector Model of Preferences

Guttman also discussed more complicated cases, such as combinations of two things to be compared, but he did not consider a multidimensional solution. Such an extension was soon developed by other psychometricians, as explained in Nishisato (1978), and is known as the *vector model* of preferences. The most important thing to underline in the present context is that Guttman's one-dimensional solution is *not a joint scale* in the same sense as we have seen so far, because it does not give individual subjects a score from which to predict responses to the same or similar objects. Indeed, the output is a *weighted average* of the rows of **S** as the scale values of the objects, for the whole group of subjects, and the *correlations* between this weighted average and the rank orders of the subjects. These correlations give an indication of how close or far the subject rankings are from the consensus ranking.

4.2 Least Squares Guttman Scaling on the Original Coding as a First Step

Nevertheless, it is of course possible to fit a joint scale for paired comparison data *without stimulus contrast restrictions*, on the basis of standard least squares Guttman (1941) scaling (as is done in Heiser 1981, Chap. 5). Apart from subject scores, such analysis gives scale values for subgroups of subjects that prefer one object over another one, but not scale values for the objects. However, as we will now demonstrate, with the output of this standard analysis we can proceed with simple calculations to construct an unfolded Coombs scale. We were inspired by Nishisato (2000), who offered a dual scaling solution using a classical example of rankings satisfying a Coombs scale, and we are very happy to follow in his footsteps.

4.2.1 Example: Rankings Satisfying a Perfect Coombs Scale of Six Stimuli

To describe the rankings of this example, also treated in Coombs (1964, pp. 87–91), we will refer to Fig. 1. It displays the four steps used in our own analysis, to be discussed shortly. But we first describe the final result: the vertical scale on the right side of Fig. 1 showing the *stimulus points* labelled with {A,B,C,D,E,F}, and the *ideal points* of the subjects in the example. The 16 ideal points are labelled in lower case font by the rank order of the stimuli, following the ordering of the distances from the ideal point to the six stimulus points. For example, the second ideal point from the top labelled with bacdef represents the subject with ranking BACDEF, since it is closest to B, then to A, followed by C, D, E, and F.

The scale on the left of this joint scale (separated for clarity) contains the *midpoints*, which bisect the interval between two stimuli and are labelled with the corresponding pair; these pairs are labelled with a hat, to distinguish them from the labels in column (2). With six stimuli, there are $(6 \times 5)/2 = 15$ midpoints, which define 14 intermediate intervals between consecutive midpoints plus two end intervals (one beyond $\widehat{\mathsf{AB}}$ and another beyond $\widehat{\mathsf{EF}}$). Each of the intervals contains one ideal point, giving 16 ideal points in total. Two neighbouring ideal points are different only in two stimuli identifying the midpoint. For instance, when moving down on the scale, ideal point bacdef changes into an bcadef when passing midpoint $\widehat{\mathsf{AC}}$.

4.2.2 A One-Dimensional Unfolding Procedure in Four Steps

Our procedure to unfold paired comparison and rank order data on the basis of least squares Guttman scaling proceeds from left to right in Fig. 1, with the following four steps:

1. *Use standard least squares Guttman scaling* on rankings coded with paired comparison coding as described above, giving category quantifications without any restrictions. We expect that the subject points will be correctly ordered, since it is not hard to verify that the columns of the coded paired comparison matrix of an unfolding scale can be permuted into a parallelogram. If there is such a structure in data matrix **M**, then the first principal component will show it (Guttman 1950b). Category quantifications will be in the correct order for the same reason. The combined result of this step is displayed in column (1) of Fig. 1. Note that the subject scores are uniformly distributed, and the same holds for the category quantifications, which are moving averages of subsets of adjacent subject scores.
2. *Find the cutting points on the scale* that separate subjects who scored in the first category from those who scored in the second category. Recall that cutting points are characteristic for scalogram analysis and are equal to the location parameter of the item characteristic curve (ICC) of a Guttman scale, which is a step function (see Sect. 2). Warrens and Heiser (2006) studied relationships between category quantifications of the least squares method and location parameters of the ICC. They concluded that for uniformly distributed subject scores, a good approximation of the cutting point would be the sum of the category quantifications: e.g. AB = ab + ba. The result for our example is shown in column (2) of Fig. 1. As to be expected, the order remains correct, and the spacing remains uniform.
3. *Find a transformation of the cutting points* on the Guttman scale that turns them into midpoints on the Coombs scale. For paired comparisons data, a cutting point AB on the Guttman scale separates all subjects who prefer stimulus A over stimulus B from subjects who prefer the reverse. But on the Coombs scale that is exactly what a midpoint does! The only—very helpful—difference is: midpoints have a restriction that the cutting points do not have: they must satisfy the additive relation $\widehat{\mathsf{AB}} = \frac{1}{2}(\mathsf{A}+\mathsf{B})$. So we can find the stimulus scale values and the midpoints on the Coombs scale by simply fitting an additive model with n parameters to the $\frac{1}{2}n(n-1)$ cutting point values. In our example, the

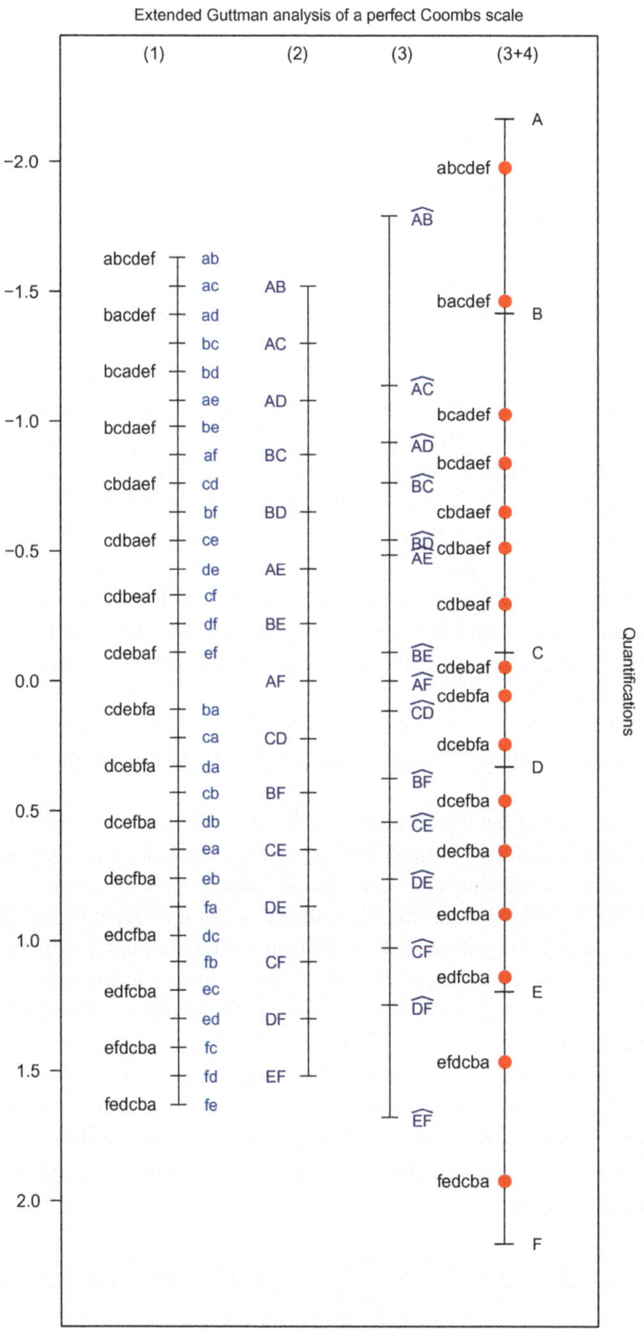

Fig. 1 Reconstruction of the joint plot for a perfect Coombs scale

fitted values of the midpoints under the additive model are shown in column (3) and the estimated stimulus scale values in column (4). Clearly, we find a *nonlinear transformation* between cutting points and midpoints, which happens to preserve the order. We used simple least squares here, because we wanted to restrict ourselves to methods available in the period in which joint scaling was conceived. But it would also be possible to fit some optimal *monotonic* transformation for the dependent variable in this additive model, as was first done by Kruskal (1965) that would *ensure* order preservation.

4. *Locate ideal points of subjects on the Coombs scale* in the middle of the interval between the two midpoints that separate them from their two neighbours. The result for our example is shown in column (4). The two extreme ideal points abcdef and fedcba are exceptions. They could have been placed anywhere beyond midpoints $\widehat{\mathsf{AB}}$ and $\widehat{\mathsf{EF}}$, but we have arbitrarily chosen the middle position between A and $\widehat{\mathsf{AB}}$ for abcdef and the middle position between F and $\widehat{\mathsf{EF}}$ for the location of fedcba. Note that by allocating 16 ideal points in intervals that follow exactly from the stimulus scale values, we can in cases of data with error simply set aside rankings that do not fit, with some measure of how close they are to the closest ideal point. That would also enable us to express the fit of the model as the percentage of rankings accounted for, or a similar measure. Note that not only stimuli and their midpoints are non-uniformly distributed, but so too are the ideal points. It testifies to the beauty of Coombs' ordered metric scale!

4.2.3 Guttman's Cutting Points Correspond to Coombs' Midpoints

The key insight driving the development of the above procedure—which to the best of our knowledge was not formulated before—is the fact that cutting points of paired comparison items on the Guttman scale divide subjects into the same two groups as the corresponding midpoints on the Coombs scale. In data with error, the first two steps could be replaced by a probabilistic IRT method that directly estimates the item location parameters in the tradition of Fechner's psychometric function (see Sect. 1); an example closely related to Guttman scaling is the two-parameter logistic model (Warrens et al. 2007).

The result of the last two steps is an ordered metric Coombs scale with non-uniform spacing. Coombs always first derived some of the metric relations with a paper-and-pencil procedure. If we denote the distance on the scale between stimuli A and B with $d(A, B)$, then for the example in Fig. 1, Coombs (1964, p. 88) found the following partial order:

$$d(C, D) < d(A, B) < \begin{pmatrix} d(D, E) \\ d(E, F) \end{pmatrix} < d(B, C) < d(D, F) < d(A, C)\,.$$

It is not difficult to verify from Fig. 1 that our inter-stimulus intervals satisfy these order relations between the distances perfectly, even though we did not impose them. We may conclude that simple and standard procedures like the one we outlined above open new avenues for revival of the one-dimensional unfolding technique.

5 Conclusions and Discussion

In the introduction, we reviewed Galton's subject-centred approach and Fechner's stimulus-centred approach to scaling. They used the same tool for constructing their scales—the normal ogive (Galton's term) and the normal psychometric function (Fechner's concept). Could it be that Galton was influenced by Fechner's work? We know for sure that Galton was familiar with *Elemente der Psychophysik*, which preceded *Hereditary Genius* by almost ten years. According to Coriale (2017), in her remarkable study of Fechner's influence on Galton:

> Although *Elemente* was not translated into English until 1966, the book made a powerful impression on scientists around the world as reviews and excerpts circulated in British, French and American periodicals during the early 1870s. [...] After learning of Fechner's ingenious experiments in James Sully's essay [entitled "Recent Experiments with the Senses", Sully (1872)] and reading *Elemente* on his own, Galton began to devise ways of making psychophysics "suitable for other applications". [...] By 1875, Francis Galton praised Fechner's book for "lay[ing] the foundations of a new science" (Coriale 2017, pp. 106–111).

As Coriale convincingly showed, Galton was especially interested in doing psychophysics himself, and in extending it to individual differences in sensory capacities of large groups of people.[7] As to methodology, it is pretty sure that there was no influence of Fechner on Galton's work on the statistical scale. We have seen that his notion to use the normal distribution to describe individual differences was inspired by Quetelet, who was the first to use the normal distribution as a model for variability in human populations. We may conclude that the subject-centred approach and the stimulus-centred approach to psychological scaling remained separated until the twenties of the twentieth century (cf. Boring 1961, p. 253).

5.1 Thurstone's Crucial Role in Preparing the Ground for the Joint Scale

It was Thurstone (1925) who started developing ideas to locate test questions in addition to test scores on the same scale, assuming normal distributions for different groups of increasing ability. He used Fechner's psychometric function to calculate

[7] He reported regularly about the results of his psychophysical experiments, for example about what he called the "auditory imagination", something that is available to us when we read silently (Galton 1893).

item locations and item discriminating power. He also initiated a lot of work on the construction of attitude scales. His first paper on this topic was called *"Attitudes can be measured"* (Thurstone 1928c), where he proposed to construct a joint scale of attitude items and person scores in two steps:

1. A first group of judges had to assess statements on the position they should have on the attitude continuum, using the classic psychophysical method of equal appearing intervals;
2. A second group of subjects had to give an agree/disagree response to all items that survived a selection procedure, and their scale scores were determined by the average of the item scale values with which they agreed.

Thurstone's follow-up was an extensive monograph on attitude scaling, where he remarked:

> "Ideally, the scale should perhaps be constructed by means of voting only. It may be possible to formulate the problem so that the scale values of the statements may be extracted from the records of actual voting. If that should be possible, the present procedure of establishing the scale values by sorting will be superseded."(Thurstone and Chave 1929).

As we have seen in Sect. 3.3.4, that was exactly what Mosteller (1949) achieved twenty years later!

5.2 The Fate of the Correlation Ratio

We have also seen that both Guttman (1941) and Mosteller (1949) used Pearson's *correlation ratio* as the criterion to be optimised for obtaining the joint scale. It might have occurred to the reader that the term correlation ratio is not current anymore in present-day statistics. Surely, this is true, but Pearson's index lives on under the name *effect size*, with the same Greek symbol η^2, or eta-squared. The urge to report measures like η^2 to show how effective interventions have been and to compare them with previously reported effects—rather than just giving F measures and p-values—has been booming for the last 25 years, and became mandatory for publication in many journals in many domains. Huberty (2002) provided an overview, rationale and history of a variety of effect size indices.

5.3 Generalisations of One-dimensional Joint Scales

A final issue to discuss is how joint scales have been generalised in the Guttman-Coombs tradition to accommodate a richer set of profiles. For the cumulative Guttman scale discussed in Sect. 2, an interesting generalisation was to use the *conjunction* of two Guttman scales (coded with only positive categories), as described by Coombs (1964, pp. 251–259). In that case, we have to deal with a special type of partial order, which is a *directed graph* called a *lattice*. A good example is the structural model for developmental processes that describes different routes from one developmental phase to another through different acquisition and deletion sequences (Coombs and Smith 1973). This model has also been recognised by students of Guttman as one of the special cases of *Partial Order Scalogram Analysis* (POSA), called the *diamond scalogram* (Shye 1985). Such generalisations of the classical scalogram deserve more attention of both theoretical and applied researchers.

For the single-peaked Coombs scale, we have mentioned in Sect. 3 that good methods for multidimensional unfolding based on Kruskal's STRESS criterion are now available. When stimuli are represented by points in two dimensions, the notion of a midpoint is replaced by the perpendicular bisector of the line segment that connects ideal points A and B (and by separating *(hyper)planes* in more dimensions). The intersection of these lines yields a set of so-called *isotonic regions*, in which all points have the same rank order of distances towards the stimulus points (Coombs 1964, pp. 140–150). Each isotonic region can contain only one ideal point with a unique rank order. So, a two-dimensional unfolding representation predicts more rankings than a Coombs scale but still a limited number. Since for the STRESS-based technique one-dimensional solutions are known to have serious problems with local minima (Hubert et al. 2002), the new procedure that we proposed in Sect. 4 could possibly help with that problem, too, by providing a good start configuration.

Acknowledgements The authors are grateful to Larry Hubert, Matthijs Warrens, and Editors for their useful comments on an earlier draft of this chapter.

References

Abelson, R.P., Tukey, J.W.: Efficient conversion of nonmetric information into metric information. In: Proceedings of the American Statistical Association Meetings, Social Statistics Section, pp. 226–230 (1959)

Benzécri, J.-P. et al.: L'Analyse des Données. II. L'Analyse des Correspondances. [Data Analysis. II. Correspondence Analysis]. Dunod, Paris (1973)

Benzécri, J.-P.: Correspondence Analysis Handbook. Marcel Dekker, New York, NY (1992)

Black, D.: On the rationale of group decision-making. J. Polit. Econ. **56**(1), 23–34 (1948)

Boring, E.G.: A chart of the psychometric function. Am. J. Psychol. **28**(4), 465–470 (1917)

Boring, E.G.: Is there a generalized psychometric function? Am. J. Psychol. **35**(1), 75–78 (1924)

Boring, E.G.: The beginning and growth of measurement in psychology. ISIS **52**(2), 238–257 (1961)

Busing, F.M.T.A.: Advances in Multidimensional Unfolding. Doctoral dissertation, Leiden University, Leiden, The Netherlands (2010)

Clauser, B.E.: A history of classical test theory. In: Clauser, B.E., Bunch, M.B. (eds.) The History of Educational Measurement: Key Advances in Theory, Policy, and Practice, pp. 157–180. Routledge (2022)

Clogg, C.C., Sawyer, D.O.: A comparison of alternative models for analyzing the scalability of response patterns. Sociol. Methodol. **12**, 240–280 (1981)

Coombs, C.H.: Psychological scaling without a unit of measurement. Psychol. Rev. **57**, 145–158 (1950)

Coombs, C.H.: A Theory of Psychological Scaling. The University of Michigan Press (1952)

Coombs, C.H.: Theory and methods of social measurement. In: Festinger, L., Katz, D. (eds.) Research Methods in the Behavioral Sciences, pp. 471–535. Dryden Press (1953)

Coombs, C.H.: A method for the study of interstimulus similarity. Psychometrika **19**(3), 183–194 (1954)

Coombs, C.H.: A Theory of Data. Wiley (1964)

Coombs, C.H., Avrunin, G.S.: Single-peaked functions and the theory of preference. Psychol. Rev. **84**(2), 216–230 (1977)

Coombs, C.H., Smith, J.E.K.: On the detection of structure in attitudes and developmental processes. Psychol. Rev. **80**(5), 337–351 (1973)

Coriale, D.: Reading through deafness: Francis Galton and the strange science of psychophysics. In: Karpenko, L., Claggett, S. (eds.) Strange Science: Investigating the Limits of Knowledge in the Victorian Age, pp. 105–124. University of Michigan Press (2017)

Cudeck, R., MacCallum, R.C. (eds.): Factor Analysis at 100. Lawrence Erlbaum, Mahwah, NJ (2007)

Dawes, R.M.: Fundamentals of Attitude Measurement. Wiley (1972)

Fechner, G.T.: Elements of Psychophysics, Volume I (translation by Adler, H.E.). Holt, Rinehart and Winston, 1966 (1860)

Fechner, G.T.: Zur Experimentellen Ästhetik [On Experimental Aesthetics]. Abhandlungen der Königliche Sächsische Gesellschaft der Wissenschaften, Math.-Physische Klasse **9**, 555–635 (1871)

Galton, F.: Hereditary Genius: An Inquiry into its Laws and Consequences. MacMillan (1869)

Galton, F.: On a proposed statistical scale. Nature **9**, 342–343 (1874)

Galton, F.: IV. Statistics by intercomparison, with remarks on the law of frequency of error. London, Edinb. Dublin Philos. Mag. J. Sci. Ser. 4 **49(322)**, 33–46 (1875)

Galton, F.: The just-perceptible difference. R. Inst. G. Br. Not. Proc. Meetings Members **14**, 13–26 (1893)

Green, B.F.: A method of scalogram analysis using summary statistics. Psychometrika **21**, 79–88 (1956)

Guilford, J.P.: Psychometric Methods. McGraw-Hill Book Company (1936)

Guttman, L.: The quantification of a class of attributes: A theory and method of scale construction. In: Horst, P., Wallin, P., Guttman, L. (eds.) The Prediction of Personal Adjustment, pp. 319–348. Social Science Research Council, New York (1941)

Guttman, L.: A basis for scaling qualitative data. Am. Sociol. Rev. **9**, 139–150 (1944)

Guttman, L.: An approach for quantifying paired comparisons and rank order. Ann. Math. Stat. **17**(2), 144–163 (1946)

Guttman, L.: On Festinger's evaluation of scale analysis. Psychol. Bull. **44**(5), 451–465 (1947a)

Guttman, L.: The Cornell technique for scale and intensity analysis. Educ. Psychol. Meas. **7**, 247–279 (1947b)

Guttman, L.: The basis for scalogram analysis. In: Stouffer, S.A., Guttman, L., Suchman, E.A., Lazarsfeld, P.F., Star, S., Clausen, J.A. (eds.) Measurement and Prediction, pp. 60–90. Princeton University Press, Princeton, NJ (1950a)

Guttman, L.: The principal components of scale analysis. In: Stouffer, S.A., Guttman, L., Suchman, E.A., Lazarsfeld, P.F., Star, S., Clausen, J.A. (eds.) Measurement and Prediction, pp. 312–361. Princeton University Press, Princeton, NJ (1950b)

Heiser, W.J.: Unfolding Analysis of Proximity Data. Unpublished doctoral dissertation, Leiden University, Leiden, The Netherlands (1981)

Heiser, W.J.: Early roots of psychometrics before Francis Galton. In: van der Ark, L.A., Emons, W.H.M., Meijer, R.R. (eds.) Essays on Contemporary Psychometrics, pp. 3–30. Springer Nature, Switzerland (2023)

Heiser, W.J., Busing, F.M.T.A.: Multidimensional scaling and unfolding of symmetric and asymmetric proximity relations. In: D. Kaplan (ed.) The SAGE Handbook of Quantitative Methodology for the Social Sciences, pp. 25–48. Sage (2004)

Heiser, W.J., Meulman. J.J.: Analyzing rectangular tables by joint and constrained multidimensional scaling. J Econ. **22(1-2)**, 139–167 (1983)
Heiser, W.J., Warrens, M.J.: On the recovery of the consecutive ones property by generalized reciprocal averaging algorithms. In: Shigemasu, K., Okada, A., Imaizumi, T., Hoshino, T. (eds.) New Trends in Psychometrics, pp. 107–110. Universal Academy (2008)
Hubert, L.J.: Problems of seriation using a subject by item response matrix. Psychol. Bull. **81**(12), 976–983 (1974)
Hubert, L.J., Arabie, P., Meulman, J.J.: Linear multidimensional scaling in the L2-Norm: Basic optimization methods using MATLAB. J. Classif. **19**, 303–328 (2002)
Huberty, C.J.: A history of effect size indices. Educ. Psychol. Meas. **62**(2), 227–240 (2002)
Kruskal, J.B.: Multidimensional scaling by optimizing goodness of fit to a nonmetric hypothesis. Psychometrika **29**(1), 1–27 (1964)
Kruskal, J.B.: Analysis of factorial experiments by estimating monotone transformations of the data. J. Roy. Stat. Soc. Ser. B (Methodol.) **27(2)**, 251-263 (1965)
Lazarsfeld, P.F.: The logical and mathematical foundation of latent structure analysis. In: Stouffer, S.A., Guttman, L., Suchman, E.A., Lazarsfeld, P.F., Star, S.A., Clausen, J.A. (eds.) Measurement and Prediction, pp. 362–412. Princeton University Press, Princeton, NJ (1950)
Likert, R.: A technique for the measurement of attitudes. Arch. Psychol. **140**, 5–53 (1932)
Lord, F.M.: A Theory of Test Scores. Psychometric Monograph No. 7, The Psychometric Society (1952)
Lord, F.M., Novick, M.R.: Statistical Theories of Mental Test Scores (with contributions by A. Birnbaum). Addison-Wesley (1968)
McCall, W.A.: How to Measure in Education. The MacMillan Company (1922)
McClelland, G.H., Coombs, C.H.: ORDMET: A general algorithm for constructing all numerical solutions to ordered metric structures. Psychometrika **40**(3), 269–290 (1975)
McIver, J.P., Carmines, E.G.: Unidimensional Scaling. Sage University Paper Series on Quantitative Applications in the Social Sciences, Series no. 07-024. Sage (1981)
Mosteller, F.: A Theory of Scalogram Analysis, Using Noncumulative Types of Items: A New Approach to Thurstone's Method of Scaling Attitudes. Report No. 9, Laboratory of Social Relations, Harvard University (1949)
Mosteller, F.: Remarks on the method of paired comparisons: I. The least squares solution assuming equal standard deviations and equal correlations. Psychometrika **16**, 3–9 (1951)
Murray, D.J.: The Creation of Scientific Psychology. Routledge (2021)
Nishisato, S.: Optimal scaling of paired comparison and rank order data: An alternative to Guttman's formulation. Psychometrika **43**(2), 263–271 (1978)
Nishisato, S.: Robust techniques for quantifying categorical data. In: MacNeill, I.B., Umphrey, G.J. (eds.) Advances in the Statistical Sciences: Foundations of Statistical Inference, pp. 209–217. Reidel (1987)
Nishisato, S.: A characterization of ordinal data. In: Gaul, W., Opitz, O., Schader, M. (eds.), Data Analysis: Scientific Modeling and Practical Applications, pp. 285–298. Springer, Berlin (2000)
Nishisato, S.: Multidimensional Nonlinear Descriptive Analysis. Chapman & Hall, London (2007)
Pearson, K.: Contributions to the mathematical theory of evolution. Philos. Trans. R. Soc. London A **185**, 71–110 (1894)
Pearson, K.: Mathematical Contributions to the Theory of Evolution, XIV: On the General Theory of Skew Correlation and Non-linear Regression. Drapers' Company Research Memoirs, Biometric series II, pp. 1–54. Dulau (1905)
Quetelet, A.: Letters Addressed to H.R.H. the Grand Duke of Saxe Coburg and Gotha, on the Theory of Probabilities as Applied to the Moral and Political Sciences, translation O.G. Downes (of original French version, 1846). Layton (1849)
Roberts, J.S., Donoghue, J.R., Laughlin, J.E.: A general item response theory model for unfolding unidimensional polytomous responses. Appl. Psychol. Meas. **24**(1), 3–32 (2000)
Shepard, R.N.: Metric structures in ordinal data. J. Math. Psychol. **3**, 287–315 (1966)

Shye, S.: Multiple Scaling: The Theory and Application of Partial Order Scalogram Analysis. North-Holland, Amsterdam (1985)
Siegel, E. J.: Arthur Sinton Otis and the American Mental Testing Movement. Unpublished doctoral dissertation, University of Miami, FL (1992)
Stigler, S.M.: The History of Statistics: The Measurement of Uncertainty Before 1900. The Belknap Press of Harvard University Press, Cambridge, MA (1986)
Stigler, S.M.: A historical view of statistical concepts in psychology and educational research. Am. J. Educ. **101**(1), 60–70 (1992)
Sully, J.: Recent experiments with the senses. Westminster Rev. **98**, 165–198 (1872)
Thurstone, L.L.: A method of scaling psychological and educational tests. J. Educ. Psychol. **16**(7), 433–451 (1925)
Thurstone, L.L.: Psychophysical analysis. Am. J. Psychol. **38**(3), 368–389 (1927a)
Thurstone, L.L.: A law of comparative judgment. Psychol. Rev. **34**(4), 273–286 (1927b)
Thurstone, L.L.: The phi-gamma hypothesis. J. Exp. Psychol. **11**(4), 293–305 (1928a)
Thurstone, L.L.: An experimental study of nationality preferences. J. Gen. Psychol. **1**(3–4), 405–425 (1928b)
Thurstone, L.L.: Attitudes can be measured. Am. J. Soc. **33**(4), 529–554 (1928c)
Thurstone, L.L., Chave, E.J.: The Measurement of Attitude: A Psychophysical Method and Some Experiments with a Scale for Measuring Attitude Toward the Church. The University of Chicago Press (1929)
Torgerson, W.S.: Theory and Methods of Scaling. Wiley (1958)
Urban, F.M.: On the method of just perceptible differences. Psychol. Rev. **14**(4), 244–253 (1907)
Urban, F.M.: The method of constant stimuli and its generalizations. Psychol. Rev. **17**(4), 229–259 (1910)
Walker, H.M.: Studies in the History of Statistical Method, with Special Reference to Certain Educational Problems. The William & Wilkins Company (1929)
Warrens, M.J., De Gruijter, D.N.M., Heiser, W.J.: A systematic comparison between classical optimal scaling and the two-parameter IRT model. Appl. Psychol. Meas. **31**(2), 106–120 (2007)
Warrens, M.J., Heiser, W.J.: Scaling unidimensional models with multiple correspondence analysis. In: Greenacre, M.J., Blasius, J. (eds.) Multiple Correspondence Analysis and Related Methods, pp. 219–235. Chapman & Hall/CRC, Boca Raton, FL (2006)
Weber, E.H.: Der Tastsinn und das Gemeingefühl. In: Wagner, R. (ed.), Handwörterbuch der Physiologie, mit Rucksicht auf physiologische Pathologie, Vol. III. part 2, pp. 481 – 588 (1846)
Yule, G.U., Filon, L.N.G.: Karl Pearson, 1857–1936. Biographical Mem. Fellows R. Soc. **2**(5), 72–110 (1936)

A Probabilistic Unfolding Distance Model with the Variability in Objects

Tadashi Imaizumi

1 Introduction

When preference data for n objects are collected from N subjects, the subjects and objects are embedded in the same multidimensional space to summarise the information in the data. We propose an embedding model in which the points representing the object are random variables.

Coombs (1950, 1964) introduced the unfolding model for analysing preference data. Multidimensional unfolding models have been used to discover hidden characteristics associated with subjects or objects in preference data. In this model, we assume the ideal point representing a hypothetical object is preferred over n objects for an individual (a subject). Carroll (1972, 1980) introduced the simple unfolding analysis, the weighted unfolding analysis, and the general unfolding analysis.

Let z_{ij}, for $j = 1, 2, \ldots, n$, denote subject i's preference for object j and is measured at least on an ordinal scale. If object j preferred to object k for subject i, then $z_{ij} \prec z_{ik}$. So, z_{ij} represents the preferential relationship among objects for subject i. Table 1 gives preference data for five objects from 4 subjects. The most preferred object for subject 1 is object 4, and that for subject 2 is object 3, and so on. Our goal is to extract the objects' and subjects' characteristics from the $N \times n$ preference matrix $\mathbf{Z} = [z_{ij}]$, for $i = 1, 2, \ldots, N$, and $j = 1, 2, \ldots, n$. We jointly attempt to embed these N subjects and n objects in the p-dimensional space. Let the N subject points $\mathbf{y}_i = [y_{it}], i = 1, 2, \ldots, N, t = 1, 2, \ldots, p$ in this p-dimensional space be represented and object points $\mathbf{x}_j = x_{jt}, j = 1, 2, \ldots, n, t = 1, 2, \ldots, p$ be also represented in this space. We call $N \times p$ matrix of N ideal points, $\mathbf{Y} = [\mathbf{y}_1, \mathbf{y}_2, \ldots, \mathbf{y}_N]^t$ subject configuration and $n \times p$ matrix $\mathbf{X} = [\mathbf{x}_1, \mathbf{x}_2, \ldots, \mathbf{x}_n]^t$ object configuration. For analysing preference data $\mathbf{Z}$, the ideal point model:

T. Imaizumi (✉)
School of Management and Information Sciences, Tama University, Tokyo, Japan
e-mail: imaizumi@tama.ac.jp

E. J. Beh et al. (eds.), *Analysis of Categorical Data from Historical Perspectives*, Behaviormetrics: Quantitative Approaches to Human Behavior 17,
https://doi.org/10.1007/978-981-99-5329-5_15

Table 1 Example of preference data

Subject	Object 1	Object 2	Object 3	Object 4	Object 5
1	5	3	4	1	2
2	2	5	1	3	4
3	4	4	2	2	2
4	3	2	4	5	1

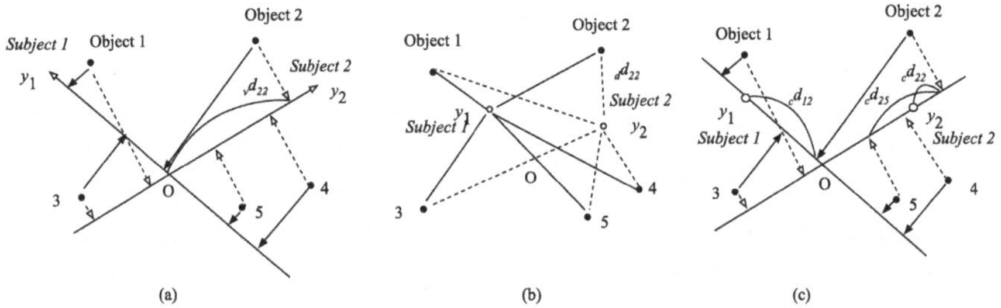

Fig. 1 (a) vector model, (b) distance model, and (c) compensatory distance model

$$d_{ij} = \sqrt{\left(\mathbf{y}_i - \mathbf{x}_j\right)^t \left(\mathbf{y}_i - \mathbf{x}_j\right)}, \tag{1}$$

has been applied. Borg and Groenen (2005) discussed the characteristics of this ideal point model. Roskam (1968) discussed two other models, the vector model and the compensatory distance model. The vector model for analysing preference data is defined as:

$$b_{ij} = \mathbf{y}_i^t \mathbf{x}_j \,.$$

Coombs (1950, 1964) discussed the compensatory distance model, and Roskam (1968) explained it explicitly, (a) there exists a unidimensional scale for each subject, and (b) all n objects are embedded as n points in multidimensional space, (c) when we represent n objects and N subjects geometrically, each point in p-dimensional space is projected onto a line that represents a subject, and the compensatory distance model is defined as:

$$_c d_{ij} = \sqrt{\left(\mathbf{y}_i - \mathbf{x}_j^t \mathbf{x}_j\right)^2} \,.$$

Figure 1 gives an illustrative representation of these three models. The vector model will be interpreted as a special case of the distance compensatory distance models. When the ideal point $||y_i||$ approaches ∞, the distance model will be expressed as:

$$d_{ij}^2 = -2\mathbf{y}_i^t \mathbf{x}_j + \mathbf{y}_i^t \mathbf{y}_i + \mathbf{x}_j^t \mathbf{x} \approx -2\mathbf{y}_i^t \mathbf{x}_j + C_{j1},$$

and for the compensatory model:

$$_c d_{ij}^2 = \mathbf{y}_i^t \mathbf{x}_j - \mathbf{x}_j \mathbf{x}_j \approx \mathbf{y}_i^t \mathbf{x}_j + C_{j2},$$

where C_{j1} and C_{j2} is a constants, respectively.

2 A Probabilistic Model

Subject i's preference for object j will include an error e_{ij}. Several probabilistic models for preference data have been proposed. Schönemann and Wang (1972) proposed the probabilistic model that the error distribution is a normal distribution. De Soete and Carroll (1983) proposed the wandering ideal point, which assumes that the ideal points are probabilistic. DeSarbo et al. (1987) proposed the model assuming the error distribution is a normal distribution. Mackay (2007) proposed a multivariate probabilistic unfolding model in which both ideal points and object points are probabilistic. As these models assume that preferences are measured on the interval scale at least, it is not easy to apply these models when the observed preference data are ordinal. So, we propose a simple model that can be applied to the ordinal preference data while estimating the error.

We propose a simple model in which the object points in p-dimensional space of a random variable such that:

$$X_{jt} = \mu_{jt} + e_j, \quad e_j \sim N\left(0, \sigma_j^2\right), \quad j = 1, 2, \ldots, n, \quad t = 1, 2, \ldots, p,$$

and distance between object j and subject i is:

$$D_{ij} = \sqrt{\sum_{t=1}^{p} \left(y_{it} - X_{jt}\right)^2}.$$

The errors are assumed to be independent of the object so that:

$$e_j \perp e_k$$

for any $i \neq j$. An error variance, σ_j^2, is assumed to be small compared with true distance so that is in Fig. 2.

Now

$$\begin{aligned} D_{ij}^2 &= \sum_{t=1}^{p} \left(y_{it} - X_{jt}\right)^2 \\ &= \sum_{t=1}^{p} \left(y_{it} - \mu_{jt} - e_j\right)^2 \\ &= \sum_{t=1}^{p} \left\{(y_{it} - \mu_{jt})\right\}^2 + pe_j^2 + 2e_j \sum_{t=1}^{p} (y_{it} - \mu_{jt}), \end{aligned} \tag{2}$$

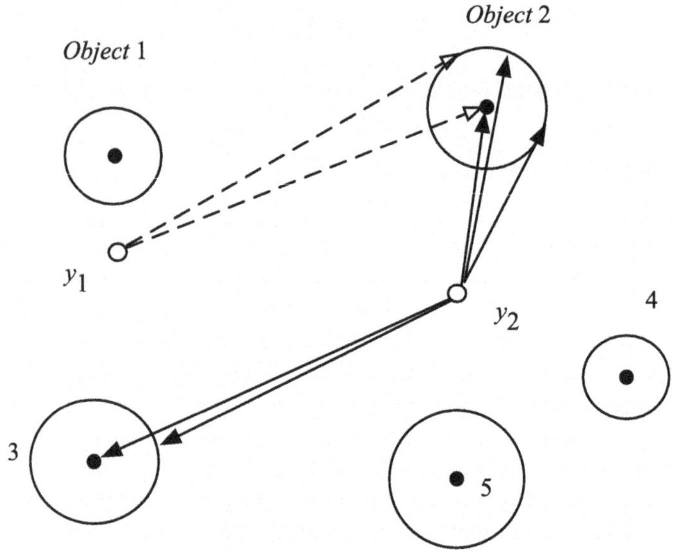

Fig. 2 Distances in the proposed model. The lines with an arrow show that distance will vary

so that:

$$\mathrm{E}\left(D_{ij}^2\right) = \mathrm{E}\left(\sum_{t=1}^{p}\left(y_{it} - \mu_{jt}\right)^2 + pe_j^2\right).$$

Zinnes and Mackay (1983) discussed this formation. We assume that the non-centrality parameter:

$$\lambda_{ij} = \frac{\sum_{t=1}^{p}\left(y_{it} - \mu_{jt}\right)^2}{\sigma_j^2}$$

is large for all (i, j), so that the error variance, σ_j^2, for $j = 1, 2, \ldots, n$, is small compared with the true distance. By treating the third term on the right-hand side of the Equation (2) as being 0, we define the quantity between subject i and object j:

$$d_{ij}^* = \sqrt{\sum_{t=1}^{p}\left(y_{it} - \mu_{jt}\right)^2 + p\sigma_j^2}. \tag{3}$$

When a subject is rated his/her preference with uncertainty, this quantity d_{ij}^* is more appropriate than d_{ij} So, we estimate the set of values $\{d_{ij}^*\}$ from the observed preference data $\{z_{ij}\}$ for each subject by noting that:

$$z_{ij} \lesssim z_{ik} \text{ so that } d_{ij}^* \leq d_{ik}^*. \tag{4}$$

2.1 A Weighted Minimisation Framework

In general, the quantity $\{d^*_{ij}\}$ will not satisfy the condition (4). To estimate d^*_{ij}, for $i = 1, 2, \dots, N$ and $j = 1, 2, \dots, n$ which satisfy condition (4), we introduce $\hat{d}^*_{ij}$ such that:

$$\hat{d}^*_{ij} \leq \hat{d}^*_{ik} \text{ if } z_{ij} \lesssim z_{ik} \text{ for } j = 1, 2, \dots, n, \ k = 1, 2, \dots, n \tag{5}$$

for each subject, $i = 1, 2, \dots, N$. De Leeuw et al. (2010) proposed the pool adjacent violators algorithm framework for finding some quantities under ordered restriction. The disparities $\hat{d}^*_{ij}$ will be solved using this framework. Preference data are often treated being row-conditional data so that data are comparable within the same subject, and we calculate the disparities as they satisfy the condition (5) and minimise:

$$s^*_i = \sum_{j=1}^{n} \left(d^*_{ij} - \hat{d}^*_{ij} \right)^2 \text{ for } i = 1, 2, \dots, N,$$

for a given $\{d^*_{ij}, \ j = 1, 2, \dots, n\}$. As s^*_i is not invariant under the normalisation of configuration. So, some normalisation factor is needed. One will be:

$$t^*_{1i} = \sum_{j=1}^{n} \left(d^*_{ij} \right)^2.$$

2.2 Normalising Factor of s^*_i

Busing (2006) and van Deun et al. (2005) point out that the degenerated solution in the multidimensional unfolding model has often occurred, and so some appropriate normalising factor of s^*_i is needed. We chose:

$$t^*_{2i} = \sum_{j=1}^{n} \left(d^*_{ij} - \overline{d}^*_i \right)^2$$

as a normalising factor. So, a loss function for each subject will be defined by:

$$S^*_{ir} = \sqrt{\frac{s^*_i}{t^*_{2i}}}, \text{ for } i = 1, 2, \dots, N,$$

and an overall loss function may be defined by:

$$\sqrt{\sum_{i=1}^{N} S_{ir}^{*}}.$$

However, the value of this overall loss function increases as the number of subjects increases, without any upper limit. We therefore define the overall loss function S_{2r} by:

$$S_{2r} = \sqrt{\frac{1}{N}\sum_{i=1}^{N} S_{2ir}^{*}} = \sqrt{\frac{1}{N}\sum_{i=1}^{N} \frac{s_i}{t_{2i}^{*}}}, \tag{6}$$

where $0 \leq S_{2r} \leq 1$. The loss function S_{2r} is a variant of stress formula two (Kruskal, 1964) and is discussed by Roskam (1968).

3 Algorithm

For the pre-specified dimensionality, p, the model parameters $\mathbf{X}_p$, $\mathbf{Y}_p$, and $\sigma^2 = \left[\sigma_1^2, \sigma_2^2, \ldots, \sigma_n^2\right]$ will be estimated. An initial object configuration $n \times p$ matrix $\mathbf{X}_p^{(0)}$ and subject configuration $N \times p$ matrix $\mathbf{Y}_p^{(0)}$ are derived as follows:

1. Calculate the initial configurations, $\mathbf{X}_p^{(0)}$ and $\mathbf{Y}_p^{(0)}$. This calculation is done by executing the double-centring transformation for the observed preference data matrix $\mathbf{Z}$, having double-centred matrix $\mathbf{Z}^+ = \left[z_{ij}^+\right]$:

$$z_{ij}^+ = z_{ij} - \frac{1}{n}\sum_{j=1}^{n} z_{ij} - \frac{1}{N}\sum_{i=1}^{N} z_{ij} + \frac{1}{Nn}\sum_{i=1}^{N}\sum_{j=1}^{n} z_{ij},$$

and applying the singular value decomposition to this transformed matrix $\mathbf{Z}^+$:

$$\mathbf{Z}^+ = \mathbf{U}\mathbf{D}\mathbf{V}^t,$$

and $\mathbf{U}^t\mathbf{U} = \mathbf{I}$, $\mathbf{V}^t\mathbf{V} = \mathbf{I}$ and $\mathbf{D} = \mathrm{diag}\left(d_1, d_2, \ldots, d_{\min(N-1, n-1)}\right)$. We then calculate $\mathbf{X}^{(0)}$ and $\mathbf{Y}^{(0)}$ so that:

$$\mathbf{X}^{(0)} = \mathbf{V}\mathbf{D}^{1/2}, \ \mathbf{Y}^{(0)} = \mathbf{U}\mathbf{D}^{1/2},$$

of the dimensionality $\min(N-1, n-1)$. We then adopted a lower-dimensional approximation using the first p components:

$$\hat{\mathbf{X}}_p^{(0)} = \begin{bmatrix} x_{11}^{(0)} & \cdots & x_{1p}^{(0)} \\ \vdots & \ddots & \vdots \\ x_{n1}^{(0)} & \cdots & x_{np}^{(0)} \end{bmatrix}, \hat{\mathbf{Y}}_p^{(0)} = \begin{bmatrix} y_{11}^{(0)} & \cdots & y_{1p}^{(0)} \\ \vdots & \ddots & \vdots \\ y_{N1}^{(0)} & \cdots & y_{Np}^{(0)} \end{bmatrix}$$

are adopted them as an initial configuration.

2. Calculate an initial estimate of $\widehat{\sigma}_j^2$, denoted by $\widehat{\sigma}_j^{2(0)}$, as follows:

$$\widehat{\sigma}_j^{2(0)} = \frac{1}{N-1} \sum_{j=1}^{N} \left(z_{ij}^* - \sum_{t=1}^{p} \hat{y}_{it}^{(0)} \hat{x}_{jt}^{(0)} \right)^2,$$

and define $\hat{\sigma}^{2(0)} = \left[\hat{\sigma}_1^{2(0)}, \hat{\sigma}_2^{2(0)}, \ldots, \hat{\sigma}_n^{2(0)} \right]$.

3. Iterative optimisation: Before starting the iteration optimisation process, three convergence criteria are pre-specified. The iterative process will be terminated when one of the following criteria is satisfied:

- minimum value of $S_{2r}^{(l)}$ is less than 0.01, where l denotes the iteration number,
- the absolute difference $\mid S_{2r}^{(l)} - S_{2r}^{(l-1)} \mid$ is smaller than 0.00001,
- the iteration number l is exceeded the number of maximum iterations 100.

a. Calculate the disparities, $\{d_{ij}^{*(l)}\}$.
b. Check whether the pre-specified convergence criteria are satisfied or not. If one of the convergence criteria is satisfied, exit the iterative loop
c. Calculate the gradient vectors:
$\mathbf{G}_{\hat{\mathbf{X}}}^{(l)}$ for $\hat{\mathbf{X}}_p^{(l)}$, $\mathbf{G}_{\hat{\mathbf{Y}}}^{(l)}$ for $\hat{\mathbf{Y}}_p^{(l)}$, and $G_{\hat{\sigma}^{2(l)}}^{(l)}$ for $\hat{\sigma}^{2(l)}$.
d. Update the configurations and variance, σ^2, where:

$$\hat{\mathbf{X}}_p^{(l+1)} = \hat{\mathbf{X}}_p^{(l)} - \text{step}_{\mathbf{X}}^{(l)} \times \mathbf{G}_{\mathbf{X}}^{(l)},$$

$$\hat{\mathbf{Y}}_p^{(l+1)} = \hat{\mathbf{Y}}_p^{(l)} - \text{step}_{\mathbf{Y}}^{(l)} \times \mathbf{G}_{\mathbf{Y}}^{(l)},$$

$$\widehat{\sigma}^{2(l+1)} = \widehat{\sigma}^{2(l)} - \text{step}_{\sigma^2}^{(l)} \times \mathbf{G}_{\sigma^2}^{(l)},$$

with $\widehat{\sigma}^{2(l+1)} > 0$ where $\text{step}_{\mathbf{X}}^{(l)}$, $\text{step}_{\mathbf{Y}}^{(l)}$, and $\text{step}_{\sigma^2}^{(l)}$ are the step-size for each of $\mathbf{X}_p^{(l)}$, $\mathbf{Y}_p^{(l)}$, and $\sigma^{2(l)}$, calculated using the quadratic search method.
e. Repeat steps a. to d.

4 A Simulation Study

We checked the validity of the present procedure through a simulation study. The data were generated as follows:

1. Generate true configurations for the dimensionality $p = 2$:
   The true two-dimensional configurations **X** and **Y** were generated using a uniform distribution on [0,1]. They were normalised so that the mean of each dimension is 0, and their total sum-of-squares is $n + N$. That is:

$$\sum_{i=1}^{n} \left(x_{jt} + y_{it}\right) = 0, \quad t = 1, 2, \ldots, p,$$

$$\sum_{t=1}^{p} \left(\sum_{j=1}^{n} x_{jt}^2 + \sum_{i=1}^{N} y_{it}^2 \right) = n + N.$$

2. Add error terms:
   An error is added to the above object configuration under a normal distribution by multiplying the mean of squared distances with a random number that is generated under the Normal distribution with mean zero and variance equal to the $(erl)^2 \times \overline{d}_j^2$:

$$X_{jt} \sim N\left(x_{jt}, \sigma_j^2\right),$$

$$\sigma_j^2 = (erl)^2 \times \overline{d}_j^2$$

   where

$$\overline{d}_j^2 = \frac{1}{N} \sum_{i=1}^{N} d_{ij}^2,$$

   and erl is one of $\{0.3, 0.5\}$.
3. Execute analysis:
   These distances with errors were used to derive the preference ranks which served as preference data.

4.1 Simulation Design and Goodness-of-Fit

We combine an error level $\{0.3, 0.5\}$ with the number of objects $n = \{10, 20\}$, the number of subjects $N = \{30, 60\}$, and the solution dimensionality of $p = 2$. Hence, under these $2 \times 2 \times 2 = 8$ conditions, we generated 50 preference data matrices for each combination of conditions.

For a dissimilarity matrix with n objects:
Spence and Ogilvie (1973) performed a Monte Carlo study on the stress formula one:

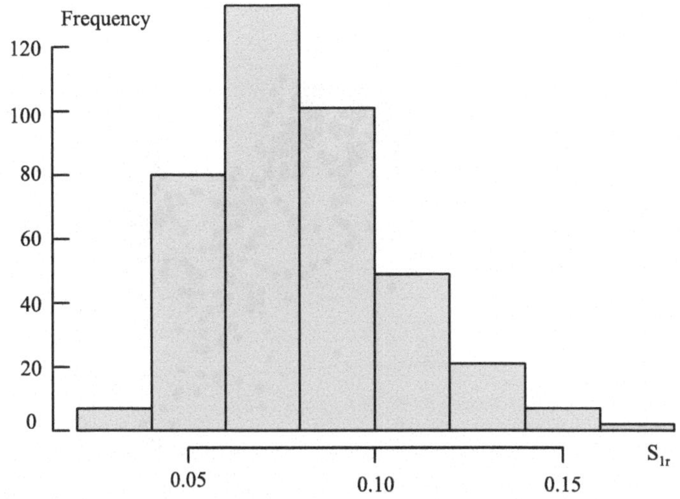

Fig. 3 Histogram of S_{1r}

$$\text{Stress}_1 = \sqrt{\sum_{j=2}^{n}\sum_{k<j}^{n-1}\left(d_{jk} - \hat{d}_{jk}^*\right)^2 / \sum_{j=2}^{n}\sum_{k<j}^{n-1}\left(d_{jk}\right)^2},$$

and concluded that $\text{Stress}_1 > 0.21$ is expected under the null hypothesis of no structure in the data. Figure 3 gives the histogram of the S_{1r} values for our simulation study. The maximum value of $S_{1r} = \sqrt{\frac{1}{N}\sum_{i=1}^{N} s_i^*/t_{1r}^*}$ was 0.1776, and the values of S_{ir} are less than 0.21.

A pseudo-correlation coefficient, $\bar{r}$, between preference data and the recovered preference data is calculated as follows:

1. Calculate the correlation coefficient between preference data and the recovered preference data of subject i, r_i.
2. Transform to Fisher's z:
   r_i is transformed to Fisher's z_i so that:

$$z_i = \frac{1}{2}\ln\left(\frac{1+r_i}{1-r_i}\right), \quad \text{for } i = 1, 2, \ldots, N.$$

3. Calculate the mean of the z_i values:

$$\bar{z} = \frac{1}{N}\sum_{i=1}^{N} z_i .$$

4. Transform $\bar{z}$ inversely:

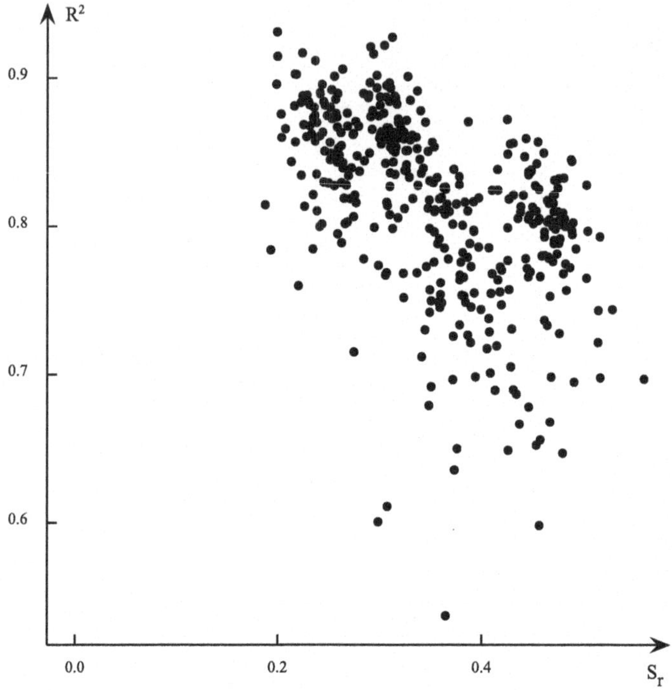

Fig. 4 Scatter plot between S_r and R^2

This mean $\bar{z}$ is inversely transformed by:

$$\bar{r} = \frac{e^{2\bar{z}} - 1}{e^{2\bar{z}} + 1}.$$

Therefore, we have 400 S_{2r} and 400 $\bar{r}$ values. The scatter diagram between S_{2r} and the squares of $\bar{r}$, $R^2 = r^2$, is shown in Fig. 4; the mean of R^2 was 0.8155 and its first quantile was 0.7813.

The histogram, scatter diagram, and the mean of R^2 show that our procedure recovered the proper configuration.

To find out which factors contribute to the results, we analysed the variance for 400$\bar{z}$, of which the mean was 1.8313, and the standard deviation was 0.3405. However, we need to consider the degrees of freedom of the recovered preference data, which is $\hat{n^*} = Nn - 2(n + N - 1) - \frac{2(2+1)}{2} - 3$. We then adjust the standard error of z by $\hat{n^*}$ instead of $\sqrt{Nn - 1}$ where:

$$z = \left(\frac{1}{2} \ln\left(\frac{1+r}{1-r}\right)\right) / \sqrt{\hat{n^*}}.$$

Table 2 Table of analysis of variance results

Factor	Sum of squares	df	Mean square	F	p
n	23569.56	1	23569.56	3328.148	<0.001
N	19293.58	1	19293.58	2724.357	<0.001
error level	2668.35	1	2668.35	376.785	<0.001
$n \cdot N$	476.00	1	476.00	67.214	<0.001
$n\cdot$ error level	1.18	1	1.18	0.166	0.684
$N\cdot$ error level	6.65	1	6.65	0.939	0.333
$n \cdot N\cdot$ error level	15.84	1	15.84	2.237	0.136
Residuals	2776.10	392	7.08		
Total	48807.26	399			

Moreover, a summary of results from this ANOVA of the eight simulated conditions is shown in Table 2.

Table 2 shows that the number of objects and subjects are critical factors of the recovery compared with the error level.

5 Application

We applied the present model to a real data set, sushia.5000.order data set, which was collected by Kamishima (2003), and analysed by using a clustering method that Kamishima and Akaha (2009) proposed. He collected these data as follows:

1. Selected sushi brands:
   Ten sushi brands were selected. They were shrimp, sea eel, tuna, squid, sea urchin, salmon roe, egg, fatty tuna, tuna roll, and cucumber roll.
   In sushi shops in Japan, sea urchin, salmon roe, and fatty tuna are generally expensive, while cucumber roll, egg, and squid are inexpensive.
   The tuna sushi, tuna, fatty tuna, and tuna roll are preferred to other sushi brands, in general.
2. Number of subjects:
   Kamishima and Akaha (2009) randomly selected five thousand subjects to take part in his study.
3. Data collection:
   Subjects were asked to the rank sushi according to their preference for them.

We tried to reduce the number of rows of the data matrix so that the number of subjects wasn't too large. We aggregated the 5000×10 data matrix by applying the Ward method based on the squares of the Euclidean distances. The dendrogram obtained is shown in Fig. 5, and the five thousand subjects are aggregated into 50

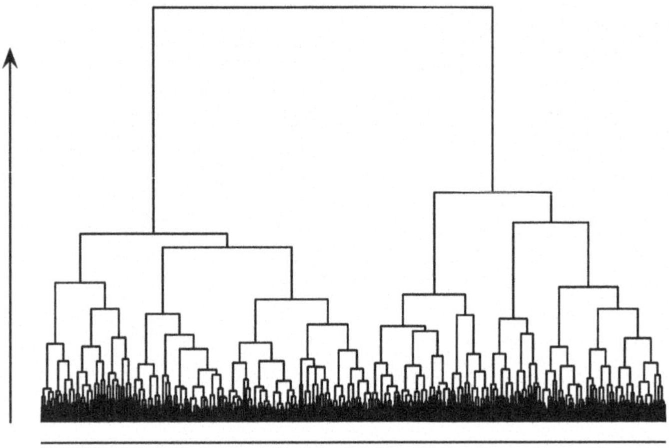

Fig. 5 Dendrogram of 5000 subjects by clustered by the ward method in R using the option ward.D2

clusters, and the means of z_{ij} were calculated for each cluster, so that, for the gth cluster:

$$\bar{z}_{gj} = \frac{1}{N_g} \sum_{i \in M(g)} z_{ij}\,;\ j = 1, 2, \ldots, n,\ g = 1, 2, \ldots, 50,$$

where N_g is the number of subjects in cluster g and $M(g)$ is the set of member's in cluster g. The aggregated data matrix $\bar{\mathbf{Z}} = \left[\bar{z}_{gj}\right]$ was analysed using the present model. We analysed this sushi preference data for each of $p = 1, 2, \ldots, 5$ so that the number of sushi brands is 10. The values of S_{2r} from the 5-dimensional to 1-dimensional solution were 0.3811, 0.4607, 0.4371, 0.4382, and 0.4427, respectively. The following values of stress formula one were also calculated by:

$$S_{1r} = \sqrt{\frac{1}{N} \sum_{i=1}^{N} s_i^* / \sum_{j=1}^{n} (d^*)_{ij}^2}$$

and these values were 0.0444, 0.0872, 0.05433, 0.0602, and 0.1132, respectively. We selected a 2-dimensional solution since Kruskal (1964) showed that a solution with a stress formula one that is less than 0.10 is fair. Each object is represented as a point, and the estimated standard deviation of the object is represented as a circle with its radius being equal to the estimated standard deviation of the object in Fig. 6.

The radius of the dashed circle's around each point is one standard deviation of the corresponding object. While an orthogonal rotation of this object configuration is possible, we can easily interpret this object configuration without any rotation. As sea urchin and salmon roe are positioned on the right-hand side which aligns with those brands with a high price, cucumber roll and squad are positioned on the left hand side, where brands with a low price lie. Therefore, Dimension 1 is interpreted

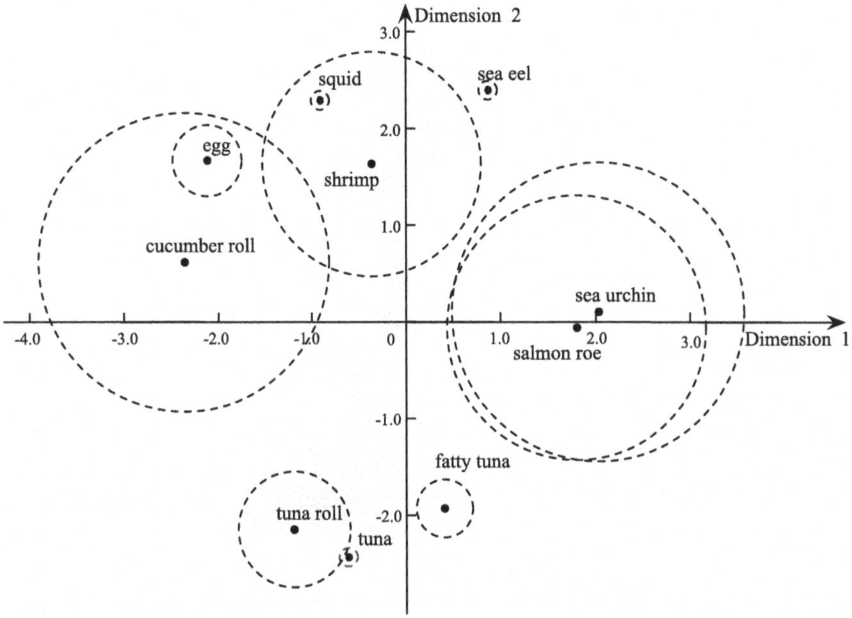

Fig. 6 Object configuration with standard deviation

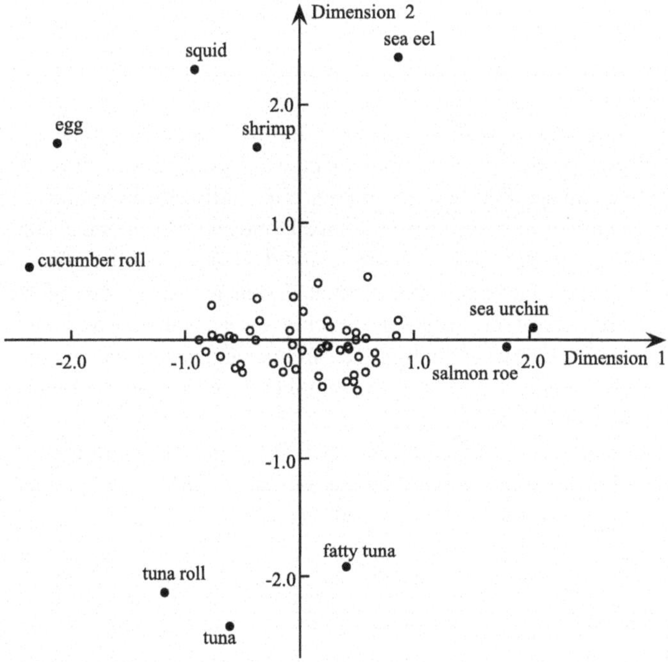

Fig. 7 Joint configuration

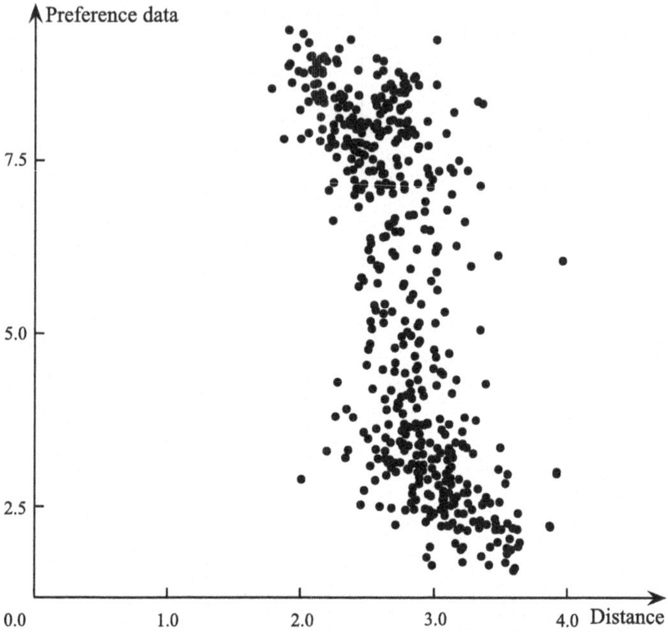

Fig. 8 Scatter diagram between distance and the aggregated preference data

as the expensive sushi vs inexpensive sushi dimension. Dimension 2 is interpreted as the tuna sushi vs other fish (including sushi, shrimp, squid, and sea ell) dimension.

The joint configuration is shown in Fig. 7. In Fig. 7, the open circle represents the cluster of subjects; hence, we find all subjects near the centre of the configuration. From Fig. 6, we can see that there are two groups. All subjects prefer sushi on the dimension 2, a kind of tuna, shrimp, and sea ell, one member of one group prefer the expensive sushi sea urchin and salmon roe, and the member of the other group does not like to this expensive sushi. The estimated standard deviations of object show that (1) there are consistent ratings in preference for cucumber and squid, salmon roe, and shrimp, (2) but ratings for tuna, sea urchin, egg, and sea ell are more differnt, likely due to preference. The scatter diagram between distance and the aggregated preference data is in Fig. 8

The scatter diagram of Fig. 8 shows that the obtained solution of the dimensionality $p = 2$ is not a degenerated configuration, and an exponential function can approximate the relation between data and distance.

6 Conclusion

We proposed an unfolding distance model in which preference for the object is affected by the variability of that object. The obtained solution reveals the preference structure of ten sushi well. It is easy to modify and apply the present model to the compensatory model.

Final Comments

Dear Nishisato-sennsei, Happy birthday for your 88th, "Beijyu" in Japanese.

I first met Dr. Nishisato, Nishisato-sennsei, about 35 years ago, at an academic lecture on data analysis at the Institute of Statistical Mathematics when I was still a young researcher. At that time, I was very impressed with the elegant Dual Scaling methodology by Nishisato-sensei. In addition to this, Dr. Nishisato's personality was charming and kind, and he was willing to answer various questions. He was always willing to talk with me when I met him at international conferences and other occasions.

References

Borg, I., Groenen, J.F.P.: Modern Multidimensional Scaling, 2nd edn. Springer, New York (2005)

Busing, F.M.T.A.: Avoiding degeneracy in metric unfolding by penalizing the intercept. Br. J. Math. Stat. Psychol. **59**, 419–427 (2006)

Carroll, J. D.: Individual differences and multidimensional Scaling. In: Shepard, R.N., Romney, A.K., Nerlove, S. (eds.) Multidimensional Scaling: Theory and Applications in the Behavior Sciences, Vol. I: Theory, pp. 105–155. Seminar Press, New York (1972)

Carroll, J. D.: Models and methods for multidimensional analysis of preferential choice (or other dominance) data. In: Lantermann, E.D., Feger, H. (eds.) Proceedings of Aachen Symposia on Decision Making and Multidimensional Scaling. Springer Verlag, Berlin (1980)

Coombs, C.H.: A Theory of Data. Wiley (1964)

Coombs, C.H.: Psychological scaling without a unit of measurement. Psychol. Rev. **57**(3), 145–158 (1950)

De Leeuw, J., Hornik, K., Mair, P.: Isotone optimization in R: pool-adjacent-violators algorithm (PAVA) and active set methods. J. Stat. Softw. **32**(5), 1–24 (2010)

De Soete, G., Carroll, J.D.: A maximum likelihood method for fitting the wandering vector model. Psychometrika **48**, 553–566 (1983)

DeSarbo, W.S., De Soete, G., Jedidi, K.: Probabilistic multidimensional scaling models for analyzing consumer choice behavior. Commun. Cogn. **20**, 93–116 (1987)

Kamishima, T.: Nantonac collaborative filtering: Recommendation based on order responses. In: KDD2003: Proceedings of the Ninth ACM SIGKDD International Conference on Knowledge Discovery and Data Mining, pp. 583–588. Association for Computing Machinery, New York (2003)

Kamishima, T., Akaha, S.: Efficient clustering for orders. In: Zighed, D.A., Tsumoto, S., Ras, Z.W., Hacid, H. (eds.) Mining Complex Data, pp. 261–280. Springer, Berlin (2009)
Kruskal, J.B.: Multidimensional scaling by optimizing goodness of fit to a nonmetric hypothesis. Psychometrika **29**, 1–27 (1964)
Mackay, D.: Internal multidimensional unfolding about a singleidealA probabilistic solution. J. Math. Psychol. **51**, 305–318 (2007)
Roskam, E.E.C.I.: Metric Analysis of Ordinal Data in Psychology. VAM, Voorschoten (1968)
Schönemann, P.H., Wang, M.M.: An individual difference model for the multidimensional analysis of preference data. Psychometrika **37**, 275–309 (1972)
Spence, I., Ogilvie, J.C.: A table of expected stress values for random rankings in nonmetric multidimensional scaling. Multivar. Behav. Res. **8**, 511–517 (1973)
Van Deun, K., Groenen, P.J.F., Heiser, W.J., Busing, F.M.T.A., Delbeke, L.: Interpreting degenerate solutions in unfolding by use of the vector model and the compensatory distance model. Psychometrika **70**, 23–47 (2005)
Zinnes, J.L., Mackay, D.B.: Probabilistic multidimensional scaling: complete and incomplete data. Psychometrika **48**, 27–48 (1983)

Analysis of Contingency Table by Two-Mode Two-Way Multidimensional Scaling with Bayesian Estimation

Jun Tsuchida and Hiroshi Yadohisa

1 Introduction

The analysis of contingency tables plays an important role in many research fields. These methods are particularly useful for analysing data on human behaviour, which often involves many qualitative variables. Correspondence analysis (Beh and Lombardo 2014) and dual scaling (Nishisato 2022) are representative methods for visualising the relationship between variables using a contingency table. These methods analyse the residuals when fitting a model that assumes independence of observations. The residuals are visualised by expressing them as the inner product of coordinate vectors. A model that expresses deviations from a symmetric model, such as the inner product of coordinate vectors, rather than deviations from a model that assumes independence, has also been proposed (Beh and Lombardo 2022).

Log-linear models that express interaction terms as distances have also been proposed. For example, a model that expresses the frequency of contingency tables as a multiplication of the distance and a constant term has been proposed; see Takane (1987). In addition, when expressing the interaction terms of log-linear models as a function of distance, multidimensional scaling (MDS) methods (Borg and Groenen 2005) that estimate the coordinate vector that recovers the distances have also been proposed; see, for example, De Rooij and Heiser (2001, 2003, 2005). These models represent the distances between categories of row and column variables in the contingency table. In contrast, these models do not represent the distances between categories for the same variables. Hence, interpreting the distance between estimated coordinate vectors of categories for the same variable is difficult because the estimated coordinate vectors do not consider the distances between categories for such variables. Interpreting the distance between the categories of the same variable is possible, but how the contingency table depends on the distance remains unclear.

J. Tsuchida (✉)
Department of Data Science, Kyoto Women's University, Kyoto, Japan
e-mail: tsuchidj@kyoto-wu.ac.jp

H. Yadohisa
Department of Culture and Information Science, Doshisha University, Kyoto, Japan
e-mail: hyadohis@mail.doshisha.ac.jp

E. J. Beh et al. (eds.), *Analysis of Categorical Data from Historical Perspectives*,
Behaviormetrics: Quantitative Approaches to Human Behavior 17,
https://doi.org/10.1007/978-981-99-5329-5_16

In this paper, we consider Bayesian estimation of the coordinate vector of MDS. Using Bayesian estimation, prior knowledge of the distance between categories for the same variable can be incorporated. When the distance between categories for the same variable is regarded as a missing value, the computation of the posterior distribution and missing value imputation can be performed simultaneously in the framework of Bayesian estimation. The Bayesian estimation method for MDS has been proposed by Oh and Raftery (2001). Our estimation method is derived based on Oh and Raftery's method.

The paper is organised as follows. Section 2 describes the model equation and the estimation method of the proposed method, and describes several extensions of the proposed method. Section 3 describes the numerical experiments conducted to evaluate the performance of the proposed method. In Sect. 4, we present results obtained by applying the proposed method to real data and their interpretations. Section 5 summarises the paper and discusses future work.

2 Model and Estimation

In this section, we describe the model formula of the proposed method and the parameter estimation method, specifically a Markov Chain Monte Carlo (MCMC) method for parameter estimation. Moreover, some extensions of the proposed method are also described.

2.1 Model Formula

Let $\boldsymbol{D} = \left(\delta_{ij}\right)$, for $i = 1, 2, \ldots, R$ and $j = 1, 2, \ldots, C$ be a contingency table. We assume that δ_{ij} is mutually independent and follows a Poisson distribution with parameter μ_{ij}, so that $\delta_{ij} \sim \text{Po}\left(\mu_{ij}\right)$. Therefore, the log-linear model we consider is:

$$\log \mu_{ij} = \lambda + \lambda_i^{(\mathrm{R})} + \lambda_j^{(\mathrm{C})} - \|\boldsymbol{x}_i - \boldsymbol{y}_j\|^2, \tag{1}$$

where $\lambda \in \mathbb{R}$ is the intercept, $\lambda_i^{(\mathrm{R})}$ is the main effect of the ith row category, and $\lambda_j^{(\mathrm{C})}$ is the main effect of the jth column category. The terms $\boldsymbol{x}_i \in \mathbb{R}^p$ and $\boldsymbol{y}_j \in \mathbb{R}^p$ are the coordinate vectors of the ith row category and jth column category, respectively. The term p is the number of dimensions of coordinate vectors, which is chosen by the analyst before applying the proposed method to the contingency table.

The model formula of the proposed method (1) corresponds to the log-linear model for contingency tables, and is the same as the model described by De Rooij and Heiser (2003). The difference between the proposed model and the log-linear model is the interaction term. The interaction term of the proposed method is a squared Euclidean distance. The similarity between the ith row and jth column category f_{ij} is defined as $f_{ij} = \log \mu_{ij} - \left(\lambda + \lambda_i^{(\mathrm{R})} + \lambda_j^{(\mathrm{C})}\right)$. The equation $f_{ij} = -\|\boldsymbol{x}_i - \boldsymbol{y}_j\|^2$ holds. This equation shows that the model formula of the proposed method corresponds to classical multidimensional scaling for two-mode, two-way data.

2.2 Estimation Algorithm

We assume that the prior distributions of parameters are mutually independent. The prior of $\lambda, \lambda_i^{(\mathrm{R})}, \lambda_j^{(\mathrm{C})}$ is assumed as follows:

$$\begin{aligned}
\lambda &\sim \mathrm{N}\left(0,\ \sigma_{(\mathrm{I})}^2\right), \\
\lambda_i^{(\mathrm{R})} &\sim \mathrm{N}\left(0,\ \sigma_{(\mathrm{MR})}^2\right), \quad (i = 1,\ 2,\ \ldots,\ R) \\
\lambda_j^{(\mathrm{C})} &\sim \mathrm{N}\left(0,\ \sigma_{(\mathrm{MC})}^2\right), \quad (j = 1,\ 2,\ \ldots,\ C)
\end{aligned}$$

where $\mathrm{N}\left(\mu,\ \sigma^2\right)$ is a normal distribution with mean parameter μ and variance parameter σ^2. The terms $\sigma_{(\mathrm{I})}^2$, $\sigma_{(\mathrm{MR})}^2$ and $\sigma_{(\mathrm{MC})}^2$ are hyper-parameters of the variance of parameters, which are defined before they are applied to the contingency table. We assume that each prior of $\boldsymbol{x}_i$ and $\boldsymbol{y}$ follows a multivariate normal distribution $\mathrm{N}\left(\boldsymbol{\mu},\ \boldsymbol{\Sigma}\right)$, where $\boldsymbol{\mu} \in \mathbb{R}^p$ and $\boldsymbol{\Sigma} \in \mathbb{R}^{p \times p}$ are a mean vector and covariance matrix, respectively. Specifically, no information about the coordinate vector is available, and so we set the prior distribution of $\boldsymbol{x}_i$ and $\boldsymbol{y}_j$ as follows:

$$\begin{aligned}
\boldsymbol{x}_i &\sim \mathrm{N}\left(\boldsymbol{0},\ \boldsymbol{\Sigma}_R\right),\ \boldsymbol{\Sigma} = \mathrm{diag}\left(\eta_{(\mathrm{R})}^2\right), \quad (i = 1,\ 2,\ \ldots,\ R) \\
\boldsymbol{y}_j &\sim \mathrm{N}\left(\boldsymbol{0},\ \boldsymbol{\Sigma}_C\right),\ \boldsymbol{\Sigma} = \mathrm{diag}\left(\eta_{(\mathrm{C})}^2\right), \quad (j = 1,\ 2,\ \ldots,\ C).
\end{aligned}$$

This is similar to the Bayesian MDS of Oh and Raftery (2001). We use the Metropolis-Hastings (MH) algorithm to calculate the posterior distribution. We also use a (multivariate) normal distribution as the proposal distribution of the MH algorithm follows.

One advantage of using Bayesian estimation pertains to the imputation of missing data. From the contingency table, we interpret the relationship between the row and column categorical variables. In contrast, we do not interpret the relationship between the categories of the same categorical variable from the contingency table. We regard contingency tables of row categorical variables and column categorical variables as missing data; then, we impute these missing data. Figure 1 illustrates missing and observed data. Now, we term this contingency table an extended contingency table, $\boldsymbol{D}^*$. We assume δ_{ij}^* represents missing data in the contingency table $\boldsymbol{D}^*$. The estimation procedure is the same as the procedure of De Tibeiro and Murdoch (2010):

- Step 1: The imputation of missing values—impute the missing $(i,\ j)$th cell of $\boldsymbol{D}^*$ by sampling $\mathrm{Po}\left(\mu_{ij}\right)$.

- Step 2: The sampling parameter step—sampling parameters given $\boldsymbol{D}^*$ that have no missing value.

We calculate the posterior distribution of the parameters by repeating Steps 1 and 2.

Let $\boldsymbol{\theta} = \left(\lambda, \lambda_1^{(R)}, \lambda_2^{(R)}, \ldots, \lambda_R^{(R)}, \lambda_1^{(C)}, \lambda_2^{(C)}, \ldots, \lambda_C^{(C)}\right), \boldsymbol{X} = (\boldsymbol{x}_1, \boldsymbol{x}_2, \ldots, \boldsymbol{x}_R)'$, and $\boldsymbol{Y} = \left(\boldsymbol{y}_1, \boldsymbol{y}_2, \ldots, \boldsymbol{y}_C\right)'$. Since $\boldsymbol{\theta}$ is similar to the log-linear model, there is indeterminacy, but the effect is not strong, depending on the prior distribution settings. Even if $\boldsymbol{X}$ and $\boldsymbol{Y}$ are multiplied by the same rotation matrix, the value of $\|\boldsymbol{x}_i - \boldsymbol{y}_j\|^2$ does not change. Furthermore, there is no problem if the dimensions are swapped, which leads to a dimension switching problem. To address this problem, we use the method of Okada and Mayekawa (2011). To use this method, $\boldsymbol{Z}$ is defined as follows using $\boldsymbol{X}$ and $\boldsymbol{Y}$:

$$\boldsymbol{Z} = \begin{pmatrix} \boldsymbol{X} \\ \boldsymbol{Y} \end{pmatrix}.$$

Let $\boldsymbol{Z}^{(t)}$ be the tth sample of $\boldsymbol{Z}$. Then, we find the rotation matrix $\boldsymbol{H}^{(t)}$ that minimises the following objective function:

$$g(\boldsymbol{H}^{(t)} \mid \bar{\boldsymbol{Z}}, \boldsymbol{Z}^{(t)}) = \|\bar{\boldsymbol{Z}} - \boldsymbol{Z}^{(t)}\boldsymbol{H}^{(t)}\|_F^2, \tag{2}$$

where, $\bar{\boldsymbol{Z}}$ is the mean of the sample, and $\|\cdot\|_F$ denotes the Frobenius norm. Then, the samples are rotated a posteriori.

	Row variable	Column variable
Row variable	**Missing**	**Observed**
Column variable	**Observed**	**Missing**

Fig. 1 Image of extended contingency table $\boldsymbol{D}^*$. The missing values are imputed by Bayesian estimation

2.3 Some Extensions

Here, we describe some extensions of the proposed method based on the model formula. The model represents the second-order interaction terms between the variables to be attached to the distances. Hence, it can be easily extended to a multi-way contingency table. Indeed, De Rooij and Heiser (2001) attempted an extension to a ternary contingency table. When a ternary contingency table is one-mode, three-way, defining triadic distance is necessary. We simply use the triadic distance (Nakayama 2005) and the n-way metric (Warrens 2010).

In the two-mode case, several patterns are possible. We term the variable that is used to estimate the coordinate vector as a pointing variable, and the other variables as condition variables. If the contingency table is obtained by pointing variable $\times$ conditional variable $\times$ conditional variable, appending pointing variables is difficult. This is because the distance between pointing variables must be defined by the main effect. However, defining the distance is difficult, because distance is defined for pair of objects. Next, in the case of pointing variable $\times$ pointing variable $\times$ conditional variable, the interaction of pointing variable $\times$ pointing variable is considered as the distance. Other combinations can be estimated as in the usual log-linear model.

Two patterns are possible in the three-mode case: in the case of pointing variable $\times$ pointing variable $\times$ condition variable, the distance between categories of the pointing variable are used as the interaction terms. In the case of pointing variable 1 $\times$ pointing variable 2 $\times$ condition variable, the estimation is the same as in the present method and in the two-mode case. In the case of pointing variable 1 $\times$ pointing variable 2 $\times$ pointing variable 3, three pointing variables can be appended by using the second-order interaction as distance. If we consider the triadic distance as the third-order interaction, we need to define this distance, as in the one-mode case.

It is also possible to model asymmetric MDS by considering the difference in the main effects of the ith row and ith column categories as a measure of asymmetry. when the coordinate matrix $\boldsymbol{X} = \boldsymbol{Y}$ holds, the second-order interaction shows symmetry. Hence, $\log\left(\mu_{ij}/\mu_{ji}\right) = \left(\lambda_i^{(R)} - \lambda_i^{(C)}\right) + \left(\lambda_j^{(R)} - \lambda_j^{(C)}\right)$, and the asymmetry can be expressed by the difference of the main effects.

Tsuchida and Yadohisa (2016) present a symmetric MDS for general n-way tables. If the distances defined in that study are used, simple Bayesian estimation is possible. However, it should be noted that the number of parameters increases when the number of mode and way also increase. Hence, whether to include higher-order interaction terms is debatable. In addition, for higher-order contingency tables, the number of zero cells often increases. These can be treated as missing data, and complementation may be considered. As an alternative, we use the zero-inflated model.

3 Numerical Example

We conducted a numerical experiment to investigate performance of the proposed method. The methods compared are the proposed method, the method of De Rooij and

Table 1 True distance between variables

	X1	X2	X3	X4	Y1	Y2	Y3	Y4
X1								
X2	2							
X3	1	1						
X4	1	1	2					
Y1	1	5	4	2				
Y2	2	4	5	1	1			
Y3	5	1	4	2	8	5		
Y4	5	5	2	8	10	13	10	

Heiser (2003), and unfolding (e.g. Borg and Groenen 2005) with Bayesian estimation based on Oh and Raftery (2001). We selected these three methods for comparison because they enabled us to investigate model performance and estimation methods. The model formula of De Rooij and Heiser (2003) is the same as the proposed method. The estimation method of unfolding with Bayesian estimation is similar to the proposed method. Hence, we investigate the utility of the proposed method by comparing it with these extant methods. The evaluation index is the correlation coefficient between the distance based on the estimated coordinate vector and the true distance. The correlation coefficient was used because the scale of each distance is different. We calculated the correlations with all distances and those for which data were available.

The data were generated using the following procedure. First, the elements in Table 1 were set as the true distance d_{ij} values. Then, $\mu_{ij} = \exp\left(\max\{6 - d_{ij},\ 1\}\right)$ was defined. Next, δ_{ij} was generated as random numbers from a Poisson distribution with parameter μ_{ij}. Only data corresponding to the rows Y and the columns X in Table 1 were used.

The number of samples selected was set to 20,000. The burn-in time was 10,000. The initial value of each parameter was generated from the uniform distribution from -1 to 1. The variance of the prior distribution of each parameter was set to 100. The posterior mean was used as the estimate.

The first row of Table 2 is the mean of the correlation coefficients between the true distance and the estimated distance for the observed data used. The second row of Table 2 is the correlation coefficient between all distances, including the missing values. The proposed method has the best result for both correlation coefficients.

Table 2 Mean of the correlation between true and estimated distance for each iteration

	Proposed method	Bayesian unfolding	De Rooij and Heiser
Distance of observed data	0.918	0.747	0.759
All distances including missing values	0.674	0.577	0.500

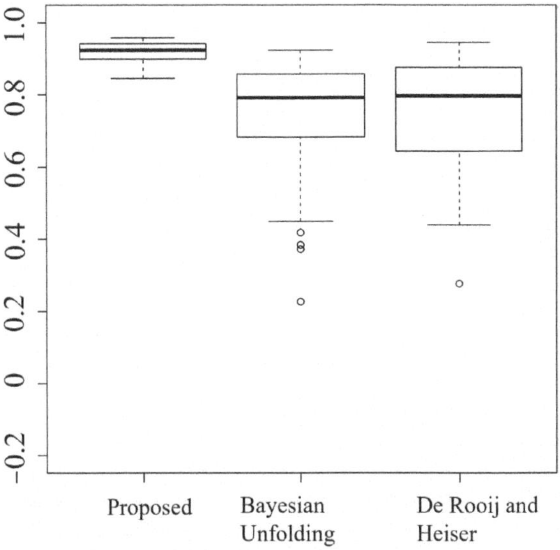

Fig. 2 Boxplot of correlation between true and estimated distance for each iteration, for observed data

Comparing the method of De Rooij and Heiser and unfolding with Bayesian estimation, the correlation coefficient for the observed data is better for the De Rooij and Heiser model, while Bayesian unfolding is better for all distances. The proposed method could reflect the advantages of both since the model is the same as that of De Rooij and Heiser and the estimation method is similar to unfolding with Bayesian estimation.

Figures 2 and 3 show the boxplots of the correlation between the true and the estimated distance for each iteration. From Figs. 2 and 3, observe that the proposed method has a smaller interquartile range than the other methods and is more stable.

4 Real Data Example

We used the proposed method to analyse survey data of green tea purchases in Japan. The survey period was from 1 January 2013 to 31 December 2013, and the target population was 6000 people aged 20–60 years in the Tokyo metropolitan area. To investigate the relationships among green tea purchases, age and sex, we defined the age-sex categories as follows: M1, M2, and M3 as males aged 20–34 years, 35–49 years, and 50–60 years, respectively; and F1, F2, and F3 as females aged 20–34 years, 35–49 years, and 50–60 years, respectively.

Table 3 shows the frequency of purchasing green tea brands for each age-sex category. The frequency of purchase was taken as the total frequency of purchase for

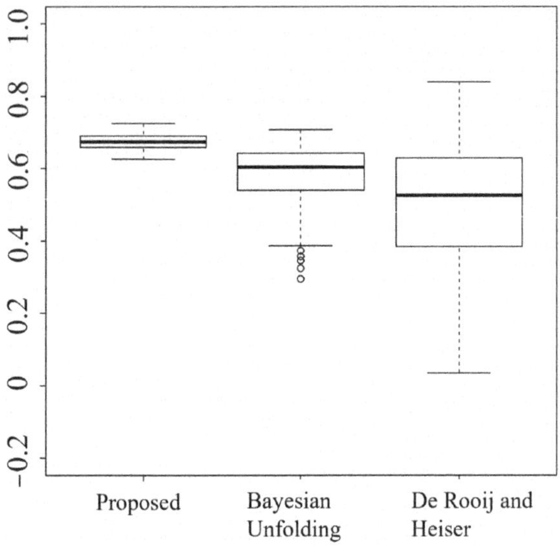

Fig. 3 Boxplot of correlation between true and estimated distance for each iteration for all distances, including missing values

Table 3 Frequency of purchases of green tea brands in each demographic category

	M1	M2	M3	F1	F2	F3
A	148	411	278	46	28	27
B	22	144	150	6	3	3
C	833	1080	1643	279	394	619
D	1321	2075	2739	492	961	1256
E	73	66	162	50	37	55
F	1151	1533	2912	527	602	1345
G	19	85	335	11	15	60
H	46	96	110	37	69	71
I	63	77	91	39	12	38
J	15	57	235	10	2	22
K	1346	1647	3708	567	632	1584
L	187	287	811	81	49	82

those who belonged to each age-sex category. The initial values of the coordinate vectors were estimated by ordinary unfolding, and the initial values of the intercept and main effect were estimated by a log-linear model. The variance parameter was set as 100. The burn-in count was set to 30,000 and the total number of samples to 50,000.

The results are shown in Fig. 4. Several obervations may be made. Firstly, M2 and M3 are very close to the private label and the inexpensive label of the convenience

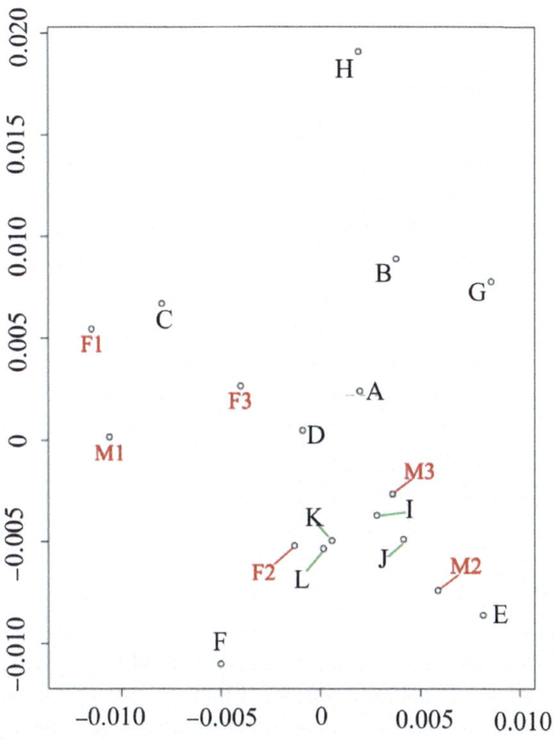

Fig. 4 Scatter plot of posterior mean of coordinate vector

store. This result is likely because some M2 and M3 work on weekdays and purchase a lunch box with green tea at a convenience store. F2 is associated with popular brands and Food for Specified Health Uses (FOSHU). Green tea with FOSHU is currently very popular in Japan. However, in this survey period, it was not as popular. The younger age groups of M1 and F1 are relatively close to brand C. Brand C has a long history in plastic bottle packaging and an association with popular characters. In 2013, the inclusion of famous characters on the bottle was controversial, which may have promoted purchasing among the younger generation.

The brands B, H, and G are located far from the age-sex category, and are far from the main products of the green tea brand. This indicates that the number of purchases is low. This may be because these brands are not sold in convenience stores and vending machines, which are the usual channels for green tea purchases (Table 4).

Table 4 Characteristics of each brand

Brand	Characteristic
A	Private label of major supermarket
B	Low-priced carton green tea
C	Major brand made by popular beverage corporation
D	Major brand made by popular beverage corporation
E	Low-priced major brand
F	Major brand made by popular beverage corporation
G	Low-priced major brand
H	Private label of major supermarket
I	Private label of major convenience store
J	Private label of major convenience store
K	One of the most popular brands of green tee
L	FOSHU and high-priced

5 Discussion

In this paper, we propose a Bayesian estimation method for applying MDS to contingency tables. The numerical example shows that the proposed method reproduces distances in the sense of correlation coefficients better than existing methods do. This may be because the distance between the categories in the rows and columns is reproduced appropriately, which may be attributed to the combination of Bayesian estimation and imputation.

Using a Bayesian prior distribution including for the estimated results based on the contingency table is also possible. Although not performed in this study, calculating Bayesian credit intervals for the coordinate vectors is also possible. This may provide useful information, including the stability of the estimation. As mentioned in Sect. 3, Bayesian estimation can provide useful information when applying asymmetric MDS or other methods in contingency table analysis.

Three issues are needed to be addressed in the future. The first is the selection of the number of dimensions, which is required in MDS. One direction is the Oh and Raftery (2001) method. Whether this method is also applicable to contingency table analysis must be checked.

The second issue is to clarify the relationship with existing methods. By considering the inner product model, it would be possible to consider the correspondence with the applicable extant analysis methods. Since the model of the proposed method uses the square of the Euclidean distance between objects, it would be possible to naturally rewrite the method as an inner product model by using double centralisation. A further advantage of using an inner product is that negative interactions can be represented. Furthermore, by incorporating this into the double-centred modelling, it could be written within the constraints of the log-linear model. However, the interpretation of imputation would then become difficult. When considered as a distance model, the relationship with the method of Takane (1987) is also an issue to be addressed in the future.

The third issue is the relationship between modelling for higher-order contingency tables and multiple correspondence analysis. In the model of the proposed method, the main effects can be thought of as a correction term that transforms similarity into dissimilarity to form MDS. Since correspondence analysis measures the deviation from independence, it is related to the residuals when the main effects are subtracted in the independent model. However, whether such a relationship applies for higher-order contingency tables must be checked.

References

Beh, E.J., Lombardo, R.: Correspondence Analysis: Theory, Practice and New Strategies. Wiley, Chichester (2014)

Beh, E.J., Lombardo, R.: Visualising departures from symmetry and Bowker's X^2 statistic. Symmetry **14**, 1103, 25 pages (2022)

Borg, I., Groenen, P.J.F.: Modern Multidimensional Scaling: Theory and Applications, 2nd edn. Springer, New York (2005)

De Rooij, M.: Distance association models for the analysis of repeated transition frequency tables. Statistica Neerlandica **55**, 157–181 (2001)

De Rooij, M., Gower, J.C.: The geometry of triadic distances. J. Classifi. **20**, 181–220 (2003)

De Rooij, M., Heiser, W.J.: A distance representation of the quasi-symmetry model and related distance models. In: Yanai, H., Okada, A., Shigemasu, K., Kano, Y., Meulman, J. (eds.) New Developments in Psychometrics: Proceedings of the International Meeting of the Psychometric Society (IMPS2001), pp. 487–494. Springer (2003)

De Rooij, M., Heiser, W.J.: Graphical representations and odds ratios in a distance-association model for the analysis of cross-classified data. Psychometrika **70**, 99–122 (2005)

De Tibeiro, J.J., Murdoch, D.J.: Correspondence analysis with incomplete paired data using Bayesian imputation. Bayesian Anal. **5**, 519–532 (2010)

Kroonenberg, P.M.: Applied Multiway Data Analysis. Wiley Hoboken, NJ (2008)

Le Roux, B., Rouanet, H.: Multiple Correspondence Analysis. Sage (2010)

Nakayama, A.: A multidimensional scaling model for three-way data analysis. Behaviormetrika **32**, 95–110 (2005)

Nishisato, S.: Optimal Quantification and Symmetry. Springer, Singapore (2022)

Oh, M.-S., Raftery, A.E.: Bayesian multidimensional scaling and choice of dimension. J. Am. Stat. Assoc. **96**, 1031–1041 (2001)

Okada, K., Mayekawa, S.: Bayesian nonmetric successive categories multidimensional scaling. Behaviormetrika **38**, 17–31 (2011)

Takane, Y.: Analysis of contingency tables by ideal point discriminant analysis. Psychometrika **52**, 493–513 (1987)

Tsuchida, J., Yadohisa, H.: Asymmetric multidimensional scaling of n-mode m-way categorical data using a log-linear model. Behaviormetrika **43**, 103–138 (2016)

Warrens, M.J.: n-way metrics. J. Classifi. **27**, 173–190 (2010)

On Correspondence Analysis and Related Methods

What's in a Name? Correspondence Analysis … Dual Scaling … Quantification Method III … Homogeneity Analysis …

Michael Greenacre

1 Introduction

For Nishi's *Festschrift*, celebrating his 88th birthday, I thought I would avoid a technical paper and rather reflect about how we create terms and names for new methodologies and how these names impact their development and dissemination. There is no doubt that a name has the characteristics of a brand label for a research product, and that the"packaging" of the product does influence how it is perceived and used.

I first give a simple example of how names can affect and clarify subsequent usage. When I wrote my first book, *Theory and Applications of Correspondence Analysis* (Greenacre 1984), in order to avoid the repetition of referring all the time to "coordinates that have (weighted) sum of squares equal to 1", I decided to call them "standard coordinates". This echoed the idea of standardisation, since the coordinates that were standardised in the usual sense of "average sum of squares [of deviations from the mean] equal to 1" were exactly what was meant by the term standard coordinates. Averaging was equivalent to assigning weights all equal to $1/n$, as in regular principal component analysis (PCA), for example. Or it could just as easily involve differential weighting with weights summing to 1, as in correspondence analysis. Similarly, instead of "coordinates that have (weighted) sum of squares equal to lambda, the eigenvalue that is the variance on a principal axis", I called these simply "principal coordinates", the same as the coordinates in Gower's principal coordinate analysis. This might sound like fairly trivial inventions, but the names have stuck and are now quite generally used in the context of solutions from correspondence analysis and related methods. Terms with a clear "branding" are recognisable and assist in the communication process between researchers.

M. Greenacre (✉)
Department of Economics and Business, Universitat Pompeu Fabra, Barcelona, Spain

Barcelona School of Management, Barcelona, Spain
e-mail: michael.greenacre@upf.edu

E. J. Beh et al. (eds.), *Analysis of Categorical Data from Historical Perspectives*, Behaviormetrics: Quantitative Approaches to Human Behavior 17,
https://doi.org/10.1007/978-981-99-5329-5_17

Talking about branding and correspondence analysis, I find in the marketing literature that "the key to success will be picking a memorable word with sounds that convey the right emotions or ideas to your audience". So I ask: "What's in a name?" when it comes to the main subject of this volume.

2 One Method, Many Names

We all know that Benzécri's correspondence analysis, Nishisato's dual scaling, Hayashi's quantification method III and de Leeuw's homogeneity analysis, are all equivalent methods (although homogeneity analysis is usually the alternative name for multiple correspondence analysis, which has correspondence analysis as a special case). So why then, at the time of writing, has "correspondence analysis" got 212,000 results on Google Scholar (as of 23 October 2022), "dual scaling" 5190, "quantification method III" 262, and "homogeneity analysis" 10,300? For the record, "multiple correspondence analysis" has 23,900, and "canonical correspondence analysis" has 48,100, which would all no doubt be included among those for "correspondence analysis"—see Fig. 1A.

The name correspondence analysis, usually abbreviated as CA, comes from the French *analyse des correspondances* (Benzécri 1973), or more fully *analyse factorielle des correspondances*, that is factorial analysis of correspondences—notice the plural. For Benzécri, a correspondence table, *un tableau de correspondance*, was the matrix of nonnegative data, or what is equivalent as far as CA is concerned, that matrix divided by its grand total. Such a correspondence table, or simply a "correspondence", quantified the association between two categorical variables, the categories of which defined the rows and columns of the table. Thus, *analyse des correspondances*, was the analysis of correspondences (plural), and this term had been in use at least since 1961 when Benzécri worked in Rennes.

In the earliest English publication of Benzécri that I am aware of (in 1969), "Statistical analysis as a tool to make patterns emerge from data" (Benzécri 1969), Sect. 3 was indeed titled with the plural, "Analysis of Correspondences (Principles)", whereas Sect. 4 used the singular, "Analysis of Correspondence (Examples)". The statistical ecologist Mark Hill, in his paper "Correspondence analysis: a neglected multivariate method" (Hill 1974), says in the abstract that:

> R.A. Fisher's canonical analysis of contingency tables ... is designated by another author's name "correspondence analysis".

Thus, it seems that Hill converted "analysis of correspondence" by "another author" (Benzécri) to "correspondence analysis", and that name has stuck to this day.

Nishisato's earliest work on the topic seems to be about 1975–1976 with his work on optimal scaling, with the terms dual scaling emerging in 1978. In fact, in the recent chapter Personal Reflections, published in the book *Modern Quantification Theory* Nishisato et al. (2021) and with this chapter signed by the book's four editors, there is a Sect. 1.4 Names for Quantification Theory where "quantification theory"

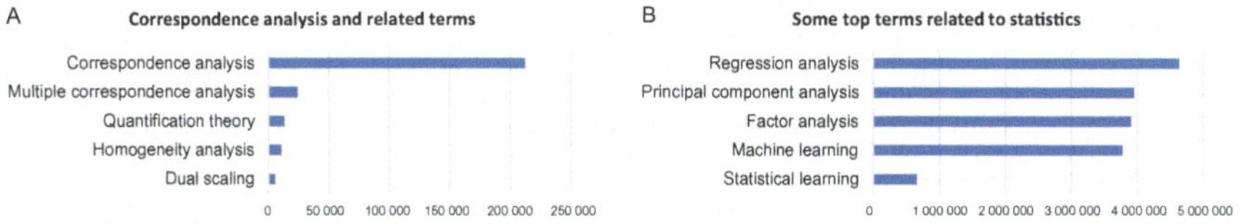

Fig. 1 A Citations in Google Scholar (as of 22 December 2022) of terms related to this essay. B Citations of more popular statistical terms

is used as a generic term for all the methods mentioned before and even more. Here is a passage:

> Because quantification theory has appealed to a large number of researchers in different disciplines and countries, it has acquired many aliases such as gradient method, reciprocal averaging, simultaneous linear regression, Guttman scaling, Hayashi's theory of quantification, optimal scaling, principal component analysis of categorical data, correspondence analysis, homogeneity analysis, dual scaling and nonlinear multidimensional descriptive analysis.

The term "quantification theory" has 13,300 mentions in Google Scholar (see Fig. 1A), but the problem is that the term means something completely different. The most cited article, "A computing procedure for quantification theory" (4195 citations, almost a third of the total) explains it as a central problem of mathematical logic, and the *Encyclopaedia Britannica* defines quantification in logic as the attachment of signs of quantity to a proposition, such as universal quantification (using the sign ∀, "for all") and existential quantification (using the sign ∃, "there exists."). Clearly, "quantification theory" means something different to a mathematician, and the term could cause confusion as an example of brand name similarity, and—if we operated in a business world—trademark infringement! (Howard et al. 2000)

Figure 1B shows the number of citations for some of the top terms in statistics, all an order of magnitude higher, in the millions. The term "principal component analysis" includes the use of "principal components analysis" in the plural, which is the less preferred term, being used about a quarter of the time. In the preface of his book, Joliffe (2002) explains why he chose the term with "component" in the singular (for example, "factors analysis" is never used), also pointing out that the growth of usage of the singular form has been much stronger than the plural form. While on the topic of singular and plural, in the recent 50th anniversary jubilee edition of the Journal of Multivariate Analysis, Farebrother (1922) uses "principal components analysis" and, strangely, "canonical correlations analysis" (the latter being used 0.7% of the time, compared to 99.3% for the familiar singular form, amongst the 923,000 citations). Finally, another interesting difference in Fig. 1B is the occurrence of the two versions of the buzzword "learning". The widely used "machine learning" is set to outstrip "regression analysis" in the near future. The alternative term "statistical learning", introduced by Trevor Hastie and co-authors (Hastie et al. 2009; James et al. 2013), and quite rightly so, will probably not do as well, with only about one-sixth of the citations of "machine learning" at present. This is in spite of their books being the best references in the field, and emphasising statistical criteria more than in typical machine learning applications. Could this be a case of data scientists preferring the engineering resonance of "*machine* learning"?

3 The Branding of a Method

When it comes to explain branding in business, it is clear that trying to justify the name as the correct one for substantive reasons can sometimes be challenging: take "Apple" as a brand name, for example, where the product has no relation to fruit, except perhaps appealing to a healthy Californian lifestyle in Silicon Valley ("An apple a day keeps the doctor away"). In the same chapter mentioned above, the authors try very hard to justify their preference for the term "dual scaling":

> As of today, many researchers appear to prefer the name correspondence analysis to dual scaling. However, in the current book, we will see that dual scaling may be a more appropriate name than correspondence analysis for two reasons: (1) the optimal weights for rows and columns are symmetrically scaled (note: symmetrical=dual) and (2) the object of quantification is to find multidimensional coordinates for rows and columns in dual space (to be defined later). We will find it later that the solution to the perennial problem of joint graphical display can be found in this dual space, not in space for the contingency table, as used in correspondence analysis.

It is not the purpose of this chapter to defend or justify any name as being better than any other, only to point out the present reality that correspondence analysis, originating in Benzécri's French term, has clearly become the accepted name for the method. It has been adopted, arguably correctly or incorrectly, in the terms "detrended correspondence analysis", "canonical correspondence analysis", "nonsymmetric correspondence analysis", "taxicab correspondence analysis", as well as "multiple correspondence analysis" rather than homogeneity analysis. For example, a name such as "nonsymmetric dual scaling" would represent a contradiction of the symmetry of the optimal weights given as a justification of the term "dual scaling". Hayashi's series of quantification methods, might need something like "Quantification method V" to describe this variation of CA. Also, one wonders if practitioners know (or need to know) anything at all about "dual spaces". The biplot as a least-squares approximation of a matrix, sometimes weighted least-squares, is by far the most digestible way of explaining and interpreting these methods, with all the nice scaling properties of the biplot coordinates falling out naturally as a consequence, where everything is in one space.

As for the possible confusion generated by the name "correspondence analysis", clearly the method has nothing to do with a linguistic analysis of people's letters sent by post or by email or social media! However, CA can actually be used in the textual analysis of such communications sent by "correspondence"—here lies an unintended ambiguity that reminds one of the origins of *analyse des correspondances* by the linguist Benzécri, who originally used the method to analyse textual data.

4 Scaling the Results

Reflecting on the "dual space" aspect and the "symmetrical scaling" referred to in the above quote, my personal experience with correspondence analysis is far from the simple contingency table applications, rather being totally dominated by results that are not symmetrically scaled. Most nonnegative data matrices being analysed by correspondence analysis are naturally asymmetric, with rows being sampling units and columns being variables, hence playing different roles. Here there is the endless and tiresome confusion around the joint display and the "scalings" used to define the row and column coordinates. This variety of choices is divided into several camps:

- The purists who insist on asymmetric scaling where the singular values of the method are assigned either to the rows or to the columns, to obtain a true biplot—after many attempts I finally called this plot, with one set (usually the samples, or rows) in principal coordinates and the other in standard coordinates, an "asymmetric biplot".
- A small camp who prefer to assign the square roots of the singular values to rows and columns, which I call the "symmetric biplot".
- The pragmatists, mostly French and Spanish, who have no problem in assigning the singular values to both rows and columns, that is resulting in a joint representation of two sets of points in principal coordinates. This is not a true biplot and so I have called this the "symmetric map", since each configuration is an approximate distance representation, with the same weighted variance (i.e. inertia). A true biplot has a scalar product interpretation between the jointly represented row and column configurations. The practise of interpreting a symmetric map as an approximate biplot has been partially justified by Ruben Gabriel (2002).

All of the above terms, asymmetric biplot, symmetric biplot, and symmetric map, as well as a version of the asymmetric biplot I have called the "contribution biplot" (Greenacre 2013) (which I mostly prefer, being very useful when there are large numbers of variables and one wants to downplay those with low contributions), are applicable to all other methods that use the singular-value decomposition: principal component analysis (PCA), canonical variate analysis (CVA), redundancy analysis (RDA), canonical correspondence analysis (CCA), logratio analysis (LRA, both unweighted and weighted), and so on. This has not stopped the proliferation of other terms, for example in PCA asymmetric biplots are often called either "form biplots" or "covariance biplots", depending on whether the rows or columns are scaled by the singular values. Gabriel originated terms for the same pair of biplots, respectively called "row-metric-preserving biplots" and "column-metric preserving biplots", although these are hardly used today.

5 More Names that Mean the Same Thing

There is a possible further confusion of methods when it comes to the topic of "forced classification" in the dual scaling literature. Forced classification in dual scaling is achieved by increasing the weight of certain rows of the data so that they are "forced" into lying on the major principal axes and thus dominating the solution. In correspondence analysis, on the other hand, this is achieved using the concepts of "active" and "passive" (also called "supplementary") points (Greenacre 2016), which have been an integral part of correspondence analysis since its earliest definition (Benzécri 1973). Passive points have zero weight, so play no role in establishing the principal axes of the solution, which is determined by the active points. In other words, they make no contribution to the solution, but they still have a position with respect to the axes and can be visualised. Not being aware of this fact would possibly make users think forced classification is a completely different method. As a further comment, if the idea is to perform a type of discriminant analysis between group means, given a classification of the rows, this can be achieved either by adding the means as additional rows to the data and declaring them active (with weights proportional to the sample sizes of the groups) and all the individual rows as passive (with zero weight), or equivalently by performing a canonical correspondence analysis where the grouping factor is specified as an explanatory (i.e. constraining) categorical variable.

Coming to multiple correspondence analysis, or MCA, this term has been more widely accepted than homogeneity analysis (see Fig. 1A), although they are equivalent. MCA's popularity has probably rubbed off from that of correspondence analysis (CA), an example of brand association. The technical term "homogeneity" is perhaps not well understood as a concept in the context of categorical data analysis. It is not clear how dual scaling, with the properties listed above, especially that of "symmetrical scaling", can be extended to more than two categorical variables. In fact, it is still not crystal clear how CA itself can be extended to more than two variables, since in its most utilised form, MCA does not have CA as an exact special case. Here too there has been some confusion. MCA is usually defined as the CA of the samples-by-categories indicator matrix, which is a matrix of zeros and ones with dummy variables as columns. Having all the categories as columns led to the heated debate between Carroll, Green and Shaffer on the one hand (Carroll et al. 1986), and myself on the other (Greenacre 1989), in the *Journal of Marketing Research*, where the former authors proposed a distance interpretation between all the column category points. I pointed out the fallacy of this type of multidimensional scaling interpretation by actually computing the between-category distances in the "full space" of the columns, where these distances had little interest, and the percentages of variance displayed were necessarily very low. As a way of clarifying the interpretation, I subsequently defined two alternative ways of computing the MCA solution, first by adjusting the scales of the axes so that the multivariable MCA had bivariable CA as a special case, and secondly by proposing "joint correspondence analysis" (JCA) (Greenacre 1987) as the true generalisation of CA, where all two-way cross-tables of the categorical variables were jointly optimised in the solution. To cut a long story short, one should think of MCA (or JCA) as the average analysis of all two-way CAs, in much the same way as PCA of a covariance matrix is the analysis of all bivariate covariances. In fact, each two-way cross-table is the categorical analogy of a covariance.

6 Superfluous Names, Meaningful Names

A correspondence analysis term that I have tended to avoid recently in my classes to marine biologists and other practitioners, is "inertia". In French "inertie" was used by Benzécri to signify the weighted sum of squared distances of the row points, or the column points, to their respective centroids, where distance was the chi-square distance. As is well-known, if one computes the Pearson chi-square statistic on a table of nonnegative numbers, even if the table is not a contingency table, the inertia is the value of the chi-square divided by the grand total of the table. It measures the amount of variance in the correspondence table. This term undoubtedly originates in physics, where the "moment of inertia" of a solid object is defined as the integral of mass times squared distance to the centroid. Maybe this is a useful concept for the variation of the point masses (row points or column points) in their respective spaces, but I have found that students accept the simple term "variance", or "weighted variance" without needing a completely new name for the weighted case. Percentages of explained variance are familiar, whereas percentages of explained inertia are rather abstract.

The chi-square distance which underlies the geometric spaces in CA is a useful term, due to Louis Guttman (1941), since it involves standardisation by dividing by the square root of the mean (i.e. square root of the expected value, as in the classic chi-square statistic) for ratio-scale data, as opposed to the square root of the variance (i.e. the standard deviation) for interval-scale data. This distance function has beautiful properties. It ensures the principle of distributional equivalence: for example, two rows with the same relative values (i.e. equivalent distributions) can be simply amalgamated, by summing them, without the chi-square distances between the columns being affected, and vice versa. Also, as I have shown in a series of papers, the chi-square distance on power-transformed data, specifically the Box-Cox transformation, converges to the logratio distance used in compositional data analysis as the power tends to 0. This surprising result means that the chi-square standardisation offers an alternative to the logratio transformation, and needs no replacement of zeros, which are the bane of the logratio approach.

Mentioning the Box-Cox transformation brings me to a final comment about naming terms and methods after people. Box-Cox is an exception, in my opinion, because of the rhyming and the combination of two great statisticians' names. It is clearly recognisable, especially if one qualifies it as the Box-Cox power transformation. Otherwise, I am not personally in favour of using people's names, rather preferring some technical meaning embedded in the name. I would be unhappy if the chi-square distance had been called the Guttman distance or the Benzécri distance. Benzécri called the horseshoe, or arch, effect in CA, "l'effet Guttman", which luckily has not survived in English as the Guttman effect. Working in the field of compositional data analysis, a field mostly attributed to John Aitchison, I find researchers inventing terms such as the "Aitchison distance", "Aitchison space" and "Aitchison geometry". Unless you are in the inner circle of this field, these terms convey nothing to you, and it is not even clear that Aitchison himself favoured his name being used in this way, since he did not use these terms himself, at least not in his own single-author publications. Hence, in my writings in this area I prefer the terms "logratio distance" and "logratio space", which do convey meaning, since they are the distance and space defined on logratio-transformed compositional data.

7 Conclusion

In conclusion, names and terms convey meaning and familiarity with them is important for researchers in their understanding and communication with others. Careful thought is needed when inventing names and terms and new "brands" should not be advertised to "consumers" unless they serve a useful purpose. Confusion is best avoided to keep concepts and methods clear and unambiguous.

Acknowledgements I would like to express my thanks to the editors of this volume of papers for their invitation to contribute, for the smooth publication process, and especially to Nishi for his long and fruitful dedication to multivariate data analysis.

References

Benzécri, J.-P.: Statistical analysis as a tool to make patterns emerge from data. In: Watanabe, S. (ed.) Methodologies of Pattern Recognition, pp. 35–74. Hawaii University, Honololu, HI (1969)

Benzécri, J.-P.: L'Analyse des Données. L'Analyse des Correspondances. Dunod, Paris, Tôme II (1973)

Carroll, J.D., Green, P.E., Schaffer, C.M.: Interpoint distance comparisons in correspondence analysis. J. Mark. Res. **23**, 271–280 (1986)

Davis, M., Putnam, H.: A computing procedure for quantification theory. J. ACM **7**, 201–215 (1960)

Encyclopedia Britannica: Quantification. Available online at https://www.britannica.com/topic/quantification. Last accessed 22 Dec. 2022

Farebrother, R.W.: Notes on the prehistory of principal components analysis. J. Multivariate Anal. **188**, 104814, 9 pages (2022)

Gabriel, K.R.: Goodness-of-fits biplots and correspondence analysis. Biometrika **89**, 423–426 (2002)

Greenacre, M.J.: Theory and Applications of Correspondence Analysis. Academic Press, London (1984)

Greenacre, M.J.: Correspondence analysis of multivariate categorical data by weighted least-squares. Biometrika **75**, 457–467 (1987)

Greenacre, M.J.: The Carroll-Green-Schaffer in correspondence analysis: a theoretical and empirical appraisal. J. Mark. Res. **26**, 358–365 (1989)
Greenacre, M.: Contribution biplots. J. Comput. Graph. Stat. **22**, 107–122 (2013)
Greenacre, M.: Correspondence Analysis in Practice, 3rd edn. Chapman & Hall/CRC Press, Boca Raton, FL (2016)
Greenacre, M.: Compositional Data Analysis in Practice. Chapman & Hall/CRC Press, Boca Raton, FL (2018)
Greenacre, M.: Compositional data analysis. Ann. Rev. Stat. Appl. **8**, 271–299 (2021)
Guttman, L.: The quantification of a class of attributes?: a theory and method of scale construction. In: Horst, P., Wallin, P., Guttman, L. (eds.) The Prediction of Personal Adjustment, pp. 321–348. Social Science Research Council, New York (1941)
Hastie, T., Tibshirani, R., Friedman, J.H.: The Elements of Statistical Learning, 2nd edn. Springer, New York (2009)
Hill, M.O.: Correspondence analysis—a neglected multivariate method. J. Royal Stat. Soc. Ser. C (Applied Statistics) **23**, 340–354 (1974)
Howard, D.J., Kerrin, R.A., Gengkler, C.: The effects of brand name similarity on brand source confusion: implications for trademark infringement. J. Public Policy Mark. **19**, 250–264 (2000)
James, G., Witten, D., Hastie, T., Tibshirani, R.: An Introduction to Statistical Learning: With Applications in R. Springer, New York (2013)
Joliffe, I.T.: Principal Component Analysis. Springer, New York (2002)
Nishisato, S., Beh, E.J., Lombardo, R., Clavel, J.G.: Modern Quantification Theory: Joint Graphical Display, Biplots, and Alternatives. Springer, Singapore (2021)

History of Homogeneity Analysis Based on Co-Citations

Jan L. A. van Rijckevorsel

1 Introduction

The prehistory of correspondence analysis (CA) is a confusing period. Different varieties with mutually different criteria, names, and objectives to be optimised arose more or less independently of each other. The aliases *Homogeneity analysis (HOMALS)* and *(multiple) Correspondence analysis (MCA)* occur therefore also interchangeably in this contribution. The development of ideas is obscured due to selective reporting and the technical possibilities of the time. Before 1970, statisticians had little access to other people's work compared to now resulting in sparse citations that were limited to one's own small circle. It was simply impossible to cite parallel developments if one was not aware of their existence. In addition, it was customary to honour the early developers of the technique you were working on as Nishisato also pointed out in his "personal reflections" on page 8 of Nishisato et al. (2021). The idea that people stood on each other's shoulders was part of a tradition and that this had to be accounted for also played a role. Therefore, the analysis of "who" was cited and which citations these forefathers had in common in said period are relevant and interesting. It shows us how knowledge and which knowledge was spread at the time and how this statistical technique has developed over time.

It is also interesting that the co-citations are analysed by using CA itself. By rearranging rows and columns in such a manner that as much marginal weight as possible is placed on or around the diagonal of the matrix, groups of authors can be identified whose members have the same co-citation pattern. The technique that optimises such a diagonalisation happens to be CA. This seems appropriate for a *Festschrift* in honour of Shizuhiko Nishisato, one who has a soft spot for historiography himself given his "personal reflections".

J. L. A. van Rijckevorsel (✉)
Amsterdam School of Economics, University of Amsterdam, Amsterdam, The Netherlands
e-mail: rijckejl@planet.nl

E. J. Beh et al. (eds.), *Analysis of Categorical Data from Historical Perspectives*, Behaviormetrics: Quantitative Approaches to Human Behavior 17,
https://doi.org/10.1007/978-981-99-5329-5_18

After 35 years, I have not gotten the impression that the following historical analysis based on co-citations has any reputation. The document has never been available digitally to my knowledge, so relatively few researchers had access to the content. This does not exclude that the analysis and its results may not be worthwhile and therefore not quoted. Read and judge for yourself.

2 Other Reviews

The most important reviews on the history of homogeneity analysis in 1987 were De Leeuw (1973, 1983), Nishisato (1980), Gifi (1981), Benzécri (1977), which later was included in an edition published by Bordas in 1982, and Tenenhaus and Young (1985). De Leeuw (1973) observes two historically different developments: the principal components analysis of categorical data and the bivariate analysis of a contingency table. Nishisato (1980) does not divide the historical development into mainstreams, instead he gives an interesting account on the history of dual scaling instead which covers the development of homogeneity analysis as well. Benzécri (1977) and De Leeuw (1983) discuss the prehistory of homogeneity analysis. In particular, Benzécri (1977) contains a useful report on the state of the art (in 1977 however!) of French correspondence analysis and its generalisations. De Leeuw (1983) shows that in 1905, Pearson almost discovered correspondence analysis and, if so, he (Pearson) probably would have been a fervent user. Gifi (1981) gives a comprehensive review on nonlinear multivariate analysis with many details but somewhat unevenly scattered over 300 densely written pages.

Tenenhaus and Young (1985) discuss four different methods all leading to the same basic characteristic equations: reciprocal averaging, the ANOVA type approach of maximising variance, principal components analysis of categorical data and generalised canonical correlation analysis. We discuss each of these in turn. First, reciprocal averaging is a natural criterion for deriving scores conceived by Richardson and Kuder in 1933 that stayed largely unnoticed and thus historically not very discriminating. Brigitte Escoffier, also known by her maiden name Cordier (1963, 1965, 1969) and Benzécri (1964, 1969, 1973), made it a cornerstone of their technique and a guideline in the intuitively attractive interpretation of geometrical representations in homogeneity analysis, calling it *le principe barycentrique*; see also Hill (1973, 1974) for this interpretation. The algorithmic aspects of reciprocal averaging already observed by Fisher (1940) and programmed by Mosier (1946) and by Baker (1960) came into full use with the availability of the 1970 generation of computers that could process large arrays effectively in an iterative way; see Hill (1973), De Leeuw (1976), Lebart et al. (1977) and van Rijckevorsel et al. (1978). Second, manipulating the data in such a way that a convenient between sum-of-squares is maximised, while the total sum-of-squares is kept constant, is a general and thus not discriminating way of presenting least squares problems. Guttman (1941) and Fisher (1940) were the first to use this formulation in homogeneity analysis. Third, the term "generalised canonical analysis" is proposed by McKeon (1966). One can interpret the first eigenvalue of homogeneity analysis of m variables as the average of the squared canonical correlations between *m* sets, each set containing just one variable. Both Fisher (1940) and Guttman (1941) were familiar with this interpretation, which is originally mostly used in the bivariate approach. Gittins (1985) reviews this interpretation of homogeneity analysis in a general framework of all kinds of canonical analysis. In France, generalised canonical analysis developed into the nonlinear canonical analysis of Masson (1974, 1980), Dauxois and Pousse (1976), Saporta (1980), and Leclerc (1980).

3 Selecting Papers for a Co-citation Analysis

Instead of reiterating the available material into another historical review, we will analyse the mutual citations of some selected publications by means of correspondence analysis. The first problem encountered is: Which are the most representative papers? The basic list of publications is provided by the references of other (historical) reviews on homogeneity analysis. Because history of referencing and not today's state of the art is important here, we start with the publication in 1933 by Richardson and Kuder and stop in 1975. We have the impression that after, say 1975, things were not what they used to be in homogeneity analysis. The different approaches were quickly integrating into a single more general framework during the early 1970s and the same bundle of papers is cited too often to be of any significance in citation analysis. Some publications within this interval are not used because of their relative lack of new relevant material, which is not already covered by other authors or because of their obscure status (such as being programme descriptions, internal reports), or last, but not least, because they are not cited by other authors. The latter does of course also apply to the pioneering papers on homogeneity analysis which are included in our analysis where possible. The papers that are excluded on the foregoing grounds are the following: Mosier (1946), Mosteller (1949), Johnson (1950), Lubin (1950), Slater (1960), Baker (1960), Kendall and Stuart (1973), Shiba (1965), Nishisato (1972, 1973), and Saito (1973).

Two other well-known papers are also excluded from our analysis because they are not directly related to homogeneity analysis. The publications in question are the several editions (1938, 1948, 1955) of Fisher's book "Statistical Methods for Research Workers", and the paper by Maung (1941b). Fisher (1938, pp. 285–289) is the first one to use the words *appropriate scoring* when finding numerical scores for categorical variables that maximise additivity in analysis of variance and not particularly in the context of maximising the correlation between scores for rows and columns of a contingency table (i.e. homogeneity analysis). Fisher introduced the latter idea in his 1940 paper in the *Annals of Eugenics.* The confusion arises because optimal scaling or *appropriate scoring* is not necessarily identical to homogeneity analysis; see Nishisato (1980). Bock (1960), for example, refers to Fisher (1938), Johnson (1950), Lancaster (1957) and Williams (1952) in the same breath. The first two authors however were interested in scoring for maximal additivity, while Lancaster and Williams discuss scoring to maximise correlation. Even Guttman

(1959) included Fisher (1955) in his list of references, while his text mentions neither Fisher nor maximising additivity. A similar situation exists for two closely dated papers by Khint Maung (1941a, 1941b) that both appeared in the *Annals of Eugenics*. The first paper deals with discriminant analysis applied on Tocher's data on the hair and eye colour of Scottish children (pp. 64–76). As such, it fits into the Fisher (1938) and Johnson (1950) work on discriminant analysis. In Maung's second paper that year, he explicitly applies Fisher's scoring technique for maximising correlation on Tocher's data (pp. 191–223). Both papers by Maung are sometimes referenced indiscriminately. Fisher (1938, 1948, 1955) and Maung (1941b) are not included in the citation analysis.

The first French publications on homogeneity analysis by Escofier (1969), Benzécri (1973), and Naouri (1970) are a problem for a citation analyst because of their non-standard way of citing. Benzécri refers sparsely to non-French work on homogeneity analysis in these publications, and he does not include Guttman (1941) in his list of references. Because it sounds not likely that Benzécri would not have known Guttman (1941) we searched Benzécri's (1973) text and on page 25 it reads:

> ..., Guttman avait d'abord envisage de faire notre analyse (dont l 'analyse des scalogrammes est un cas particulier) mais n'en vint jamais aux calculs probablement pour la seule raison qu'en 1941 (sic) les ordinateurs n'existaient pas

which, when translated to English, reads:

> ..., Guttman had first considered doing our analysis (of which the analysis of scalograms is a special case) but never came to the calculations probably for the sole reason that in 1941 (sic) computers did not exist.

This is the most explicit reference to Guttman (1941) made by Benzécri. The remark itself is not true by the way because the same supplement (no B3) containing Guttman's (1941) paper includes a contribution of Ledyard R. Tucker (1941) titled: *A note on a machine method for the quantification of attributes.* It is also highly probable that the term *Chi Square Metric* used by Benzécri (1973) originates with Guttman (1941, p. 330). Whether there exists more of such references we do not know. Benzécri (1973) promises on page 27 to give a complete bibliography but, apart from scattered references throughout the text, he does not get any further than an author's index in Vol. II and we had to wait until 1977 for his more extended bibliography. The papers by Escofier (1965, 1969) and Naouri (1970) are French doctoral theses that hardly refer to other work. Occasionally, several papers by one author are indiscriminately cited as if they carried the same message. Therefore, we sometimes better speak of an oeuvre, a body of work, that includes several separate publications. We consider the following publications as oeuvres: in the citation matrix, Lancaster refers to Lancaster (1957) and Lancaster (1958); Benzécri refers to Benzécri (1964), Benzécri (1969), and Benzécri (1973); Hayashi refers to Hayashi (1950), Hayashi (1952), and Hayashi (1954); De Leeuw refers to De Leeuw (1968) and De Leeuw (1973); Hill refers to Hill (1973) and Hill (1974); and Lingoes refers to Lingoes (1963), Lingoes (1964), and Lingoes (1968). These considerations lead to the following list of publications to be used in a citation analysis in order of time

Table 1 Transaction matrix between cited (= columns) and citing (= rows) papers on homogeneity analysis

	Ho	Ri	Hi	Ho	Fi	Gu	Ma	Bu	Gu	Ha	Wi	Bu	Gu	La	Lo	To	Gu	Bo	Li	Mc	Es	Na	
Guttman 41	1	1		1																			3
Maung 41					1																		1
Guttman 50	1					1																	2
Hayashi						1			1														2
Williams 52					1		1																2
Burt 53								1	1			1	1										4
Guttman 53						1		1	1			1	1										5
Lancaster			1		1		1				1												4
Lord 58				1		1			1														3
Torgerson 58				1		1			1														3
Guttman 59			1			1	1	1	1		1		1										7
Bock 60						1			1		1			1									4
McKeon 66				1	1	1					1			1	1	1							7
Lingoes						1			1							1							4
McDonald 68				1	1	1								1	1								5
Benzécri						1			1	1						1					1	1	6
De Leeuw	1		1	1	1	1	1	1	1	1	1	1	1	1	1	1	1	1	1	1	1		20
Hill			1		1	1	1				1			1		1					1	1	9
Total	3	1	4	6	7	13	5	4	10	2	6	3	5	5	3	5	1	1	1	1	3	2	

of publication: Richardson and Kuder (1933), Hotelling (1933), Hirschfeld (1935), Horst (1936), Fisher (1940), Guttman (1941), Maung (1941b), Burt (1950), Guttman (1950), Hayashi, Williams (1952), Burt (1953), Guttman (1953), Lancaster, Lord, (1958), Torgerson (1958), Guttman (1959), Bock (1960), McKeon (1966), Lingoes (1968), McDonald (1968), Escofier (1969), Naouri (1970), and the work of Benzécri, De Leeuw, and Hill. Their mutual citations are collected in a binary transaction matrix of citing (rows) versus cited (columns) publications; see Table 1.

4 The Historical Development of Homogeneity Analysis

4.1 An Overview

The visual inspection of the transaction matrix, combined with the interpretation of the content of the papers, leads to a reconstruction of the historical development of homogeneity analysis. Figure 1 sketches the developments over the period 1933 to

1975 that have contributed to homogeneity analysis. It is important to notice that the independent progress in some quarters remains unobserved in other quarters during a considerable interval of time. Hirschfeld's paper on maximising correlation and linearising regressions from 1935, for instance, is first mentioned by Lancaster (1958), and the work of Guttman in 1941 is greatly ignored by the bivariate school.

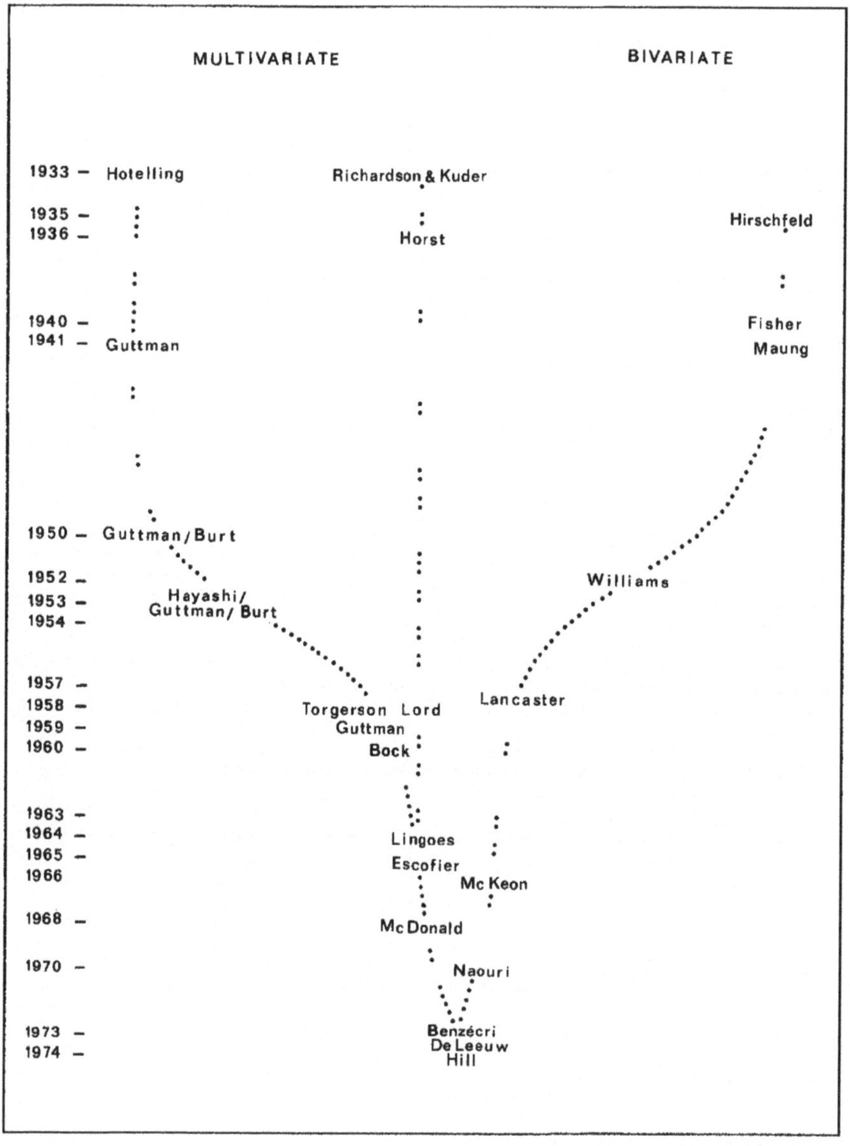

Fig. 1 Sketch of the historical development of homogeneity analysis

The historical mainstreams are, in our opinion, the bivariate approach that lies within the English statistical tradition and the multivariate approach that is greatly oriented to the USA (including Israel). The latter development can be divided in a one-dimensional variety (i.e. differential weighting) and in a multi-dimensional variety (i.e. principal components analysis of categorical data); see De Leeuw (1973).

The first integration of the different approaches begins with Guttman (1959). Next is McKeon (1966). At the same time, Escofier (1963, 1965) writes her thesis wherein the French geometric style is introduced and l'analyse des correspondances (bivariate) and l'analyse des correspondances multiples (multivariate) are combined into a single framework. Three years later, McDonald (1968) produces his definitive review on differential weighting including the work of Lancaster (1957) on the bivariate distribution. The first authors to cover the whole field are De Leeuw (1973) and Benzécri (1973). The last integrator of the interval that we study is Hill (1973, 1974) who, as a representative of the English bivariate school, integrates the French approach with some American results. Cyril Burt (1950, 1953), the only Anglo-Saxon (a Scotsman actually), did not fit in to the bivariate tradition at the time. He discusses factor analysis of categorical data in the psychometric tradition.

We will give a brief characteristic of every member of each school and start with the bivariate approach.

4.2 The Bivariate Approach

Hirschfeld (1935), better known as H. O. Hartley, discovers that the scores assigned to the rows and columns of a contingency table to maximise their correlation, also, linearise their bivariate regressions, which holds for categorical variables, continuous variables, and for bivariate normals as well. Although he is a founding father of the bivariate approach, he is not recognised as such until 1957 by Lancaster. Hirschfeld's very first result is also due to the second founding father: Louis Guttman (1941). A third founding father of the bivariate approach is Fisher (1940), who derives homogeneity analysis as a special case of multiple regression or, discriminant analysis with just two variables. Implicitly, Fisher proposes the system of reciprocal averaging equations and realises that there are min $(k_j, k_z) - 1$ orthogonal solutions to the bivariate homogeneity analysis. Fisher does not cite Hirschfeld's paper, and so his (Fisher's) work can be regarded as another independent discovery of homogeneity analysis. Maung (1941b) is the first to apply Fisher's technique. Williams (1952) refers to Fisher and Maung and elaborates on the relationship with the chi-square statistic. Lancaster (1957, 1958) continues the development that started with Hirschfeld and Fisher and shows the relationship between the chi-square statistic, continuous variables, and bivariate densities. In this way, Lancaster precedes Naouri and Lancaster's results relate directly to recent work on order dependence and oscillating eigen-vectors in homogeneity analysis by van Rijckevorsel (1987). McKeon (1966) is the first to include multivariate results. He also introduces the interpretation of homogeneity analysis as a form of generalised canonical analysis. He is the predecessor to Gittins (1985) and anticipates the French development marked by a.o. Masson (1974, 1980) and Dauxois and Pousse (1976). Naouri (1970) is the dense, mathematical French double of Lancaster, includes multivariate results as well, and refers not to other publications. Hill (1973, 1974) is the first Anglo-Saxon author of the bivariate school to integrate bivariate and multivariate results with French geometry publishing in an accessible international journal *Journal of the Royal Statistical Society, Series C (Applied Statistics)*. He reintroduces reciprocal averaging, and he made reciprocal averaging the cornerstone of his approach, just as Benzécri and Escofier did before him.

4.3 The Multivariate School

The multivariate school can be divided into the one-dimensional and the multi-dimensional approach. This distinction is historically not well established, however. The use of homogeneity analysis as a nonlinear method to obtain optimal weights that maximise the variance is called *differential weighting*. The interpretation of the extraneous solutions of homogeneity analysis as factors or principal components like in linear principal components analysis is known as the factor analysis or principal components analysis of categorical data. The multi-dimensional interpretation into axes or factors is by far the most popular and most flexible interpretation. Notwithstanding its intuitive appeal as a simple way to represent complex relationships advocated as such by Burt (1950, 1953), Bock (1960), Lingoes (1963, 1964, 1968), and Benzécri (1969, 1973), the analytical significance and meaning of these extraneous axes have troubled many, an author including Guttman (1941, 1950, 1959), De Leeuw (1982), Bekker (1983) and Schriever (1985). The problem does not occur in the bivariate approach where Williams (1952) introduces the interpretation of further axes in terms of chi-square decomposition and in which Lancaster (1958) formulates the canonical theory to (continuous) marginal distributions.

The multivariate school in MCA started with Hotelling's historic paper on the statistical use of principal components as a data reduction technique in 1933. Only Guttman (1941, 1950) refers to this paper. Guttman (1941, p. 346) was also aware of differential weighting because he referred to Wilks (1938) and Edgerton and Kolbe (1936) who showed that homogeneity and discrimination in numerical data are optimised by their principal components. He also refers to Richardson and Kuder (1933) so he was familiar with reciprocal averaging. In the same publication, Guttman warns of the undefined analytical meaning of subsequent axes. All in all, Guttman's (1941) paper is a comprehensive algebraic treatment of homogeneity analysis that had all the qualities to be the alpha and omega of homogeneity analysis for the next thirty years. Ironically, many of Guttman's results were later to be rediscovered independently. Burt (1950) derives homogeneity analysis explicitly as the factorial analysis of categorical data. His approach is multivariate and multi-dimensional from the start. The paper is conceived within the framework and tradition of factor analysis in psychometry, and Burt does not refer to other work on homogeneity analysis. In his

1950 paper, Guttman gives the complete analytical, theoretical, and practical discussion of the multivariate and multi-dimensional homogeneity analysis of binary data. Hayashi published several papers on homogeneity analysis in the early 1950s; see Hayashi (1950, 1952, 1954). Hayashi mentions Guttman but does not refer to him in the context of scale analysis, intensity analysis, and paired comparisons. We may safely conclude that Hayashi knew the publications of Guttman on paired comparisons in 1946 and on scale analysis in 1950. Only in his 1952 paper did Hayashi refer to other non-Japanese work on homogeneity analysis. Despite the title *Multi-dimensional quantification with the application to analysis of social phenomena,* Hayashi (1954) does not seem to consider further axes in homogeneity analysis. The multi-dimensionality that Hayashi refers to does relate to the use of several variables in partitioning the one-dimensional homogeneity analysis solution.

In his reaction to Burt's (1950), paper Guttman (1953) explains the dangers involved when using other than the first axis in homogeneity analysis. The problem here is the presence of nonlinear regressions on the principal components of categorical data. The classical interpretation of principal components of categorical data as independent latent variables can be misleading because of these nonlinearities. Later work by Lancaster (1958), De Leeuw (1984), and Schriever (1983, 1985) confirm this argument; see also van Rijckevorsel (1987). Burt replies in 1953 on Guttman (1953) and he (Burt) defended his use of further axes in homogeneity analysis in the style of factor analysis with "*Why Not?*". However, Guttman had already answered the "*Why Not?*" question; see above. This discussion between Guttman and Burt is important because it links principal components analysis (sometimes incorrectly called factor analysis) with the scaling of categorical data, and it shows the first signs of the existence of the horseshoe problem. Torgerson (1958) in his widely distributed handbook describes homogeneity analysis as a deterministic (as opposed to probabilistic) way to scale categorical data. His analytical derivation is based on Mosteller's (1949) matrix notation of Guttman (1941). Guttman (1959) is the first member of the American school to mention the Anglo-Saxon bivariate approach. He refers to Hirschfeld (1935) and to Fisher (1955). It can be regarded as Guttman's ultimate report on nonlinear principal components analysis. Together with his 1941 and 1950 papers, Guttman (1959) covers all of the important developments up to 1959. Bock (1960) introduces the word optimal scaling with respect to homogeneity analysis. Lingoes proves throughout his publications between 1963 and 1968 to be the perfect Guttman scholar, who programmed all Guttman's ideas into a computer (IBM 7090). Although Lingoes discusses the linearisation of the bivariate regressions and is the first one to use the characteristic regression plots of homogeneity scores with both regression lines, he does not refer to the bivariate approach explicitly. The original (unpublished) thesis by Escofier (then known under her maiden name Cordier) in 1965 is based on material developed by Benzécri (1964) and a mimeo by Cordier (1963) for a course in linguistics in Rennes in 1964 and 1965. Escofier was Benzécri's programmer, and she is the first to introduce the word "*Analyse Factorielle des Correspondances*" in a publication. Her treatment is in the French geometric style and relates directly to both principal components analysis and to the bivariate approach. Her thesis was officially published in 1969. Benzécri

(1973) compiled an enormous erudite compendium of theoretical results and applications in two lengthy volumes. He refers indirectly to Guttman (1941) and directly to Hayashi (1952), Guttman (1950), Torgerson (1958), Escofier (1969), and Naouri (1970). Benzécri rediscovers and reformulates many at that time known facts into the French framework of for example inertia, clouds of points, metrics, etc. De Leeuw (1973) covers the whole field in an unpublished doctoral thesis, officially published in 1984. He coins the term "indicator matrix" and proposes others forms of nonlinear multivariate analysis in relation to homogeneity analysis and is the direct precursor to Gifi (1980, 1981). It is the first of a series of textbooks in the English language on homogeneity analysis during the early 1980s which include Nishisato (1980), Gifi (1981), Greenacre (1984), and Lebart et al. (1984).

We can be relatively brief on one-dimensional homogeneity analysis called differential weighting. Richardson and Kuder (1933) wrote down an intuitive technique to give one-dimensional weights to categories to obtain a scale that measures, which they called reciprocal averaging. Only Horst (1935) and, through him, Guttman (1941) initially noticed their paper. Fisher (1940) independently mentioned the same idea. Horst (1936) gives the basics of differential weighting but does not relate to homogeneity analysis. The latter is done for the first time by Guttman (1941). Lord (1958) elaborates on Horst (1936) and Guttman (1941). He (Lord) shows that the first eigenvalue in homogeneity analysis is directly related to Cronbach's alpha, the all-time workhorse in the construction and analysis of questionnaires even today. McDonald (1968) reviews all the work on differential weighting and links to the bivariate school.

4.4 The Analysis of Co-citations

In correspondence analysis, we interpret the co-citations as the joint frequencies of references common to two papers. The data matrix is complete symmetric, and the numbers of relevant references per paper are on the diagonal; see Table 2. The vectors with row- and column sums are equal. Because there exists no prevalence of rows over columns or vice versa, we use the symmetric scaling.

The first axis of the correspondence analysis of the co-citation matrix (the first scaled eigenvalue equals 0.48) is plotted against the timescale in Fig. 2, and the resulting graph resembles the earlier historical sketch (Fig. 1).

Because some papers have no common references with other work, we end up with fewer points than in the historical sketch in Fig. 1. The difference between the multivariate and the bivariate approach is however easily recognised, while the difference between the use of one and more dimensions in the multivariate case is lost on the first axis. The corresponding reordering of rows and columns by correspondence analysis runs from multivariate to bivariate; see Table 3.

The second axis (second scaled eigenvalue equals 0.37, not given) is dominated by the ubiquitous early papers of Guttman (1941, 1950). Apart from this effect, the papers on differential weighting can be discerned as a slightly separated group

Table 2 Symmetric co-citation matrix of papers on homogeneity analysis

	Gu	Ma	Gu	Ha	Wi	Bu	Gu	La	Lo	To	Gu	Bo	Mc	Li	Mc	Be	De	Hi
Guttman 41	3		1						1	1			1		1		2	
Maung 41		1			1			1					1		1		1	1
Guttman 50	1		2	1			1		1	1	1	1	1	1	1	1	2	1
Hayashi			1	2		1	2		2	2	2	2	1	2	1	2	2	1
Williams 52		1			2			2			1		1		1		2	2
Burt 53				1		4	4		1	1	3	1		2		1	4	
Guttman 53			1	2		4	5		2	2	4	2	1	3	1	2	5	1
Lancaster		1			2			4			3	1	2		1		4	4
Lord 58	1		1	2		1	2		3	3	2	2	2	2	2	2	3	1
Torgerson 58	1		1	2		1	2		3	3	2	2	2	2	2	2	3	1
Guttman 59			1	2	1	3	4	3	2	2	7	3	2	3	1	2	7	4
Bock 60			1	2		1	2	1	2	2	3	4	3	2	2	2	4	3
McKeon 66	1	1	1	1	1		1	2	2	2	2	3	7	2	5	2	7	5
Lingoes			1	2		2	3		2	2	3	2	2	4	1	3	4	2
McDonald 68	1	1	1	1	1		1	1	2	2	1	2	5	1	5	1	5	3
Benzécri			1	2		1	2		2	2	2	2	2	3	1	6	5	4
De Leeuw	2	1	2	2	2	4	5	4	3	3	7	4	7	4	5	5	20	8
Hill		1	1	1	2		1	4	1	1	4	3	5	2	3	4	8	9

with exclusively positive scores. The latter effect is evoked by the commonly unique reference to Horst's paper from 1936 by Guttman (1941), Lord (1958), Torgerson (1958), McKeon (1966) and McDonald (1968). The distinction between the use of the first axis/dimension only and the use of multiple axes in the PCA tradition, known from the discussion of Guttman versus Burt from 1953, is not reflected by the citation behaviour. This is reflected in both the transaction matrix and in the plot of the first axis against time. It seemed better not to include De Leeuw because his referencing is too widespread to be called discriminating.

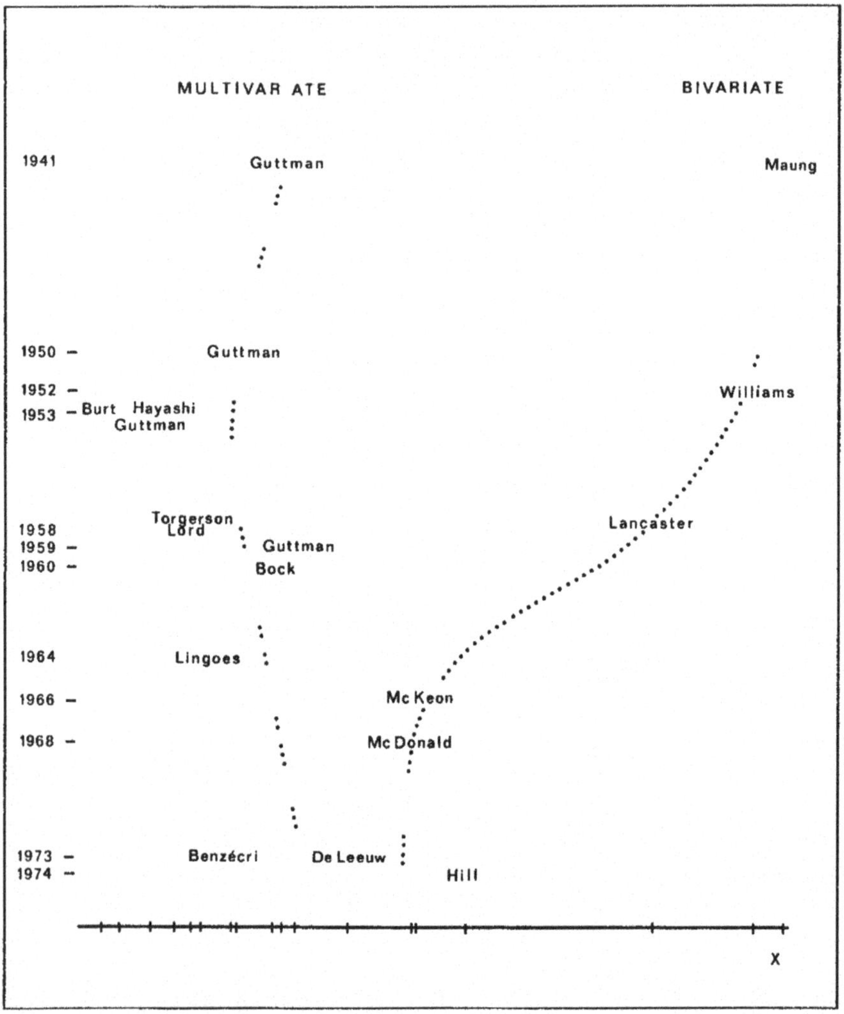

Fig. 2 First homogeneity axis of co-citations plotted versus the timescale

5 Discussion

One may wonder "What are the benefits of a co-citation analysis?". The answer is that citations give the "hard" explicit indication that researchers know and use each other's work and when they do not recognise the work of others. One need not to speculate to infer the convergence of ideas.

But the idiosyncrasy of co-citation analysis is also that key researchers who do not cite or are not cited are missed. The cells of the co-citation matrix contain the number of citations two authors have in common. Authors who were not able to

Table 3 Re-ordered co-citations

	Bu	Gu	Ha	Li	To	Lo	Gu	Be	Gu	Bo	Gu	De	Mc	Mc	Hi	La	Wi	Ma
Burt 53	4	4	1	2	1	1	0	1	0	1	3	4	0	0	0	0	0	0
Guttman 53	4	5	2	3	2	2	1	2	0	2	4	5	1	1	1	0	0	0
Hayashi	1	2	2	2	2	2	1	2	0	2	2	2	1	1	1	0	0	0
Lingoes	2	3	2	4	2	2	1	3	0	2	3	4	1	2	2	0	0	0
Torgerson 58	1	2	2	2	3	3	1	2	1	2	2	3	2	2	1	0	0	0
Lord 58	1	2	2	2	3	3	1	2	1	2	2	3	2	2	1	0	0	0
Guttman 50	0	1	1	1	1	1	1	2	1	1	1	2	1	1	1	0	0	0
Benzécri	1	2	2	3	2	2	1	6	0	2	2	5	1	2	4	0	0	0
Guttman 41	0	0	0	0	1	1	1	0	3	0	0	2	1	1	0	0	0	0
Bock 60	1	2	2	2	2	2	1	2	0	4	3	4	2	3	3	1	0	0
Guttman 59	3	4	2	3	2	2	1	2	0	3	7	7	1	2	4	3	1	0
De Leeuw	4	5	2	4	3	3	2	5	2	4	7	20	5	7	8	4	2	1
McDonald 68	0	1	1	1	2	2	1	1	1	2	1	5	5	5	3	1	1	1
McKeon 66	0	1	1	2	2	2	1	2	1	3	2	7	5	7	5	2	1	1
Hill 74	0	1	1	2	1	1	1	4	0	3	4	8	3	5	9	4	2	1
Lancaster	0	1	0	0	0	0	0	0	0	1	3	4	1	2	4	4	2	1
Williams 52	0	0	0	0	0	0	0	0	0	0	1	2	1	1	2	2	2	1
Maung 41	0	0	0	0	0	0	0	0	0	0	0	1	1	1	1	1	1	1

quote anyone from the selection in the transaction matrix, including the forerunners Hirschfeld, Hotelling, Richardson and Kuder and others or who are not quoted by anyone during the early this period of the development of homogeneity analysis, are excluded from further analysis. This exclusion is made regardless of their position from a historical perspective. If Nishisato (1980) had been included in this citation analysis, which was not possible due to the time window studied, he probably would have met the same fate as De Leeuw. Their citations are so widely spread that they no longer discriminate in the analysis. His location in the plot would be very close to the origin, just like De Leeuw's.

Substantive reasons why researchers cite selectively are the (school of the) country in which they work and the native language they speak. Traditions in a field of application such as psychometry, biometry, ecology, lexicology, and different technical starting points, such as geometry, probability, matrix algebra, or combinations of these, have also influenced citation behaviour.

On the other hand, analytical differences between statistical methods are not (or not necessarily) fully reflected in the citation behaviour of their authors. Different approaches or aliases of CA are not mutually exclusive and can be united in one person.

Similar studies with "prehistory of CA" in the title differ from this study in the sense that they are not based on mutual citations, but on the analytical differences between techniques and thus convergence of ideas. In addition, they are limited to a specific period—see De Leeuw (1983)—or school—see Benzécri (1977). In a review of Benzécri (1977) called "*Histoire et Préhistoire de l'Analyse des données par J.P. Benzecri: un cas de généalogie retrospective*" (History and Prehistory of Data Analysis by J. P. Benzécri: a case of retrospective genealogy), Armatte (2008) shows the idiosyncrasy of language and culture bound ideas in MCA. It is a textbook example of historiography in statistics. Interestingly, he (Armatte) believes that the history of statistics should preferably not be studied by statisticians saying (p. 21):

> L'histoire des sciences est une discipline bien trop importante pour la laisser aux scientifiques…que l'on veut étudier!

which, in English, says:

> The history of science is far too important a discipline to leave it to scientists…that we want to study!

This quote means there is also a potential flaw of our study. Recent historical overviews of CA covering a longer period are expansive bibliographies that show the convergence of ideas to a lesser extent. It may look like the world of CA has become too large for that. The most recent and comprehensive study on the history of (variants of) CA by Beh and Lombardo (2019) confirms just how big CA has become. They come to more than 30 new variants of CA since 1975 and half of them developed after the turn of this century. In addition, they also formulate new concepts to structure this proliferation in their commendable effort.

The historical development based on the visual inspection of the transaction matrix as illustrated in the sketch in Fig. 1 agrees with the common view of the history of CA at the time (i.e. in 1987). The sketched development is partly recovered by the first gradient of the correspondence analysis of the co-citations plotted against time in Fig. 2 and accounts for nearly half (45.7%) of the inertia. The difference between the bivariate and the multivariate approach in CA is striking on the first CA axis. It reflects the development of the psychometric test and questionnaire tradition on the one hand and the statistical tradition of the analysis of the contingency table on the other. The initial emergence of differential weighting and reciprocal averages in psychometrics was a part of that. Starting with Horst in 1936, and a common denominator for works by Guttman, Lord, Torgerson, McKeon, and McDonald, differential weighting and reciprocal averaging are weaker constructs on the first gradient of the citation matrix. Note that Guttman (1953, 1959) and Burt (1953) have so much in common that their discussion in 1953 does not show up. The position of De Leeuw near the origin is caused not only by his extensive citation of others but also by the high degree of self-citation compared to others. He often cites informal reports where he recorded his first ideas.

Understandably, the need to test the table of co-citations for perfect symmetry was less felt. The analysis of co-citations in this study is limited to just one axis because only one gradient, namely time, was sought.

Judging by the co-citations, the early days of CA seem to be determined by multivariate psychometricians and bivariate statisticians. This conclusion fits in with the existing narrative about the origin of CA discussed by Beh and Lombardo (2012) and Nishisato et al. (2021), but it is more parsimonious. However, given the selective limitations of co-citations, it would go too far to follow Ockham's razor in all its consequences here.

References

Armatte, M.: Histoire en Préhistoire de l'Analyse des donnnées par J.P. Benzécri: un cas de généalogie retrospective. J. Electronique d'Histoire des Probabilités et de la Statistique **4**, 1–22 (2008)

Baker, F.B.: Univac scientific computer program for scaling of psychological inventories by the method of reciprocal averages. CPA 22. Behav. Sci. **5**, 268–9 (1960)

Bekker, P.: Relations Between Various Forms of Non-linear Principal Components Analysis. Unpublished Masters thesis, Department of Data Theory, Leiden University, Leiden, The Netherlands (1983)

Beh, E.J., Lombardo, R.: A genealogy of correspondence analysis. Aust. N. Z. J. Stat. **54**(2), 137–168 (2012)

Beh, E.J., Lombardo, R.: A genealogy of correspondence analysis: Part 2, The variants. Electron. J. Appl. Stat. Anal. **12**(2), 552–603 (2019)

Benzécri, J.P.: Analyse factorielle des proximités. Publications de l'Institut de Statistique de l'Université de Paris **13**, 235–282 (1964)

Benzécri, J.P.: Histoire et Préhistoire de l'Analyse des Correspondances. Bordas, Paris (1982)

Benzécri, J.P., et al.: L'Analyse des Données: II. L'Analyse des Correspondances. Dunod, Paris (1973)

Benzécri, J.P.: Statistical analysis as a tool to make patterns emerge from data. In: Watanabe, S. (eds.) Methodologies of Patterns Recognition. Academic Press, New York (1969)

Benzécri, J.P.: Histoire et préhistoire de l'analyse des correspondances. Cahiers de l'Analyse des Données **2**, 9–40 (1977)

Bock, R.D.: Methods and Applications of Optimal Scaling. The University of North Carolina Psychometric Laboratory Research Memorandum, no: 25 (1960)

Burt, C.: The factorial analysis of qualitative data. Brit. J. Psychol. (Statistical Section) **3**, 166–185 (1950)

Burt, C.: Scale analysis and factor analysis. Comments on Dr. Guttman's paper. Brit. J. Psychol. **6**, 5–23 (1953)

Cordier, B.: L'Analyse Factorielle des Correspondances. Faculté des sciences Université de Rennes, mimeographed paper (1963)

Cordier, B.: L'Analyse Factorielle des Correspondances. Thése de 3ième cycle, Université de Rennes (1965)

Cours de linguistique mathématique. Lecons sur l'analyse factorielle et la reconaissance des formes 1e, 2e, 3e, 4e, 5e leçons, Rennes (1964–1965)

Dauxois, J., Pousse, A.: Les analyses Factorielles en Calcul des Probabilités et en Statistique. Essai d'étude Synthetique. Thèse d'Etat. Université de Toulouse III (1976)
Edgerton, H.A., Kolbe, L.E.: The method of minimum variation for the composite criteria. Psychometrika **1**, 183–187 (1936)
Escofier-Cordier, B.: L'analyse Factorielles des Correspondances. Re-issued in 1969 by Cahiers du BURO, vol. 13, pp. 25–59 (1965)
Fisher, R.A.: Statistical Methods for Research Workers, 7th edn. Oliver and Boyd, London (1938)
Fisher, R.A.: The precision of discriminant functions. Ann. Eugen. **10**, 422–429 (1940)
Fisher, R.A.: Statistical Methods for Research Workers. Hafner, New York (1948)
Fisher, R.A.: Statistical Methods for Research Workers, 12th edn. Oliver and Boyd, London (1955)
Gifi, A.: Niet Lineaire Multivariate Analyse. Department of Data Theory, Leiden University, Leiden, The Netherlands (1980)
Gifi, A.: Nonlinear Multivariate Analysis. Department of Data Theory, Leiden University, Leiden, The Netherlands. Republished in 1990 by Wiley, New York (1981)
Gittins, R.: Canonical Analysis. Springer, New York (1985)
Greenacre, M.J.: Theory and Applications of Correspondence Analysis. Academic Press, London (1984)
Guttman, L.: The quantification of a class of attributes: a theory and method of scale construction. In: Horst, P., Wallin, P., Guttman, L. (eds.) The Prediction of Personal Adjustment, pp. 319–348. Social Science Research Council, New York (1941)
Guttman, L.: An approach for quantifying paired comparisons and rank order. Ann. Math. Stat. **17**, 144–163 (1946)
Guttman, L.: The principal components of scale analysis. In: Stouffer, S.A., Guttman, L., Suchman, E.A., Lazarsfeld, P.F., Star, S., Clausen, J.A. (eds.) Measurement and Prediction, pp. 312–361. Princeton University Press, Princeton, NJ (1950)
Guttman, L.: A note on Sir Cyril Burt's factorial analysis of qualitative data. Br. J. Psychol. **6**, 1–4 (1953)
Guttman, L.: Metricizing rank-ordered and ordered data for a linear factor analysis. Sankhyã **21**, 257–268 (1959)
Hayashi, C.: On the quantification of qualitative data from the mathematico-statistical point of view. Ann. Inst. Stat. Math. **2**, 35–47 (1950)
Hayashi, C.: On the prediction of phenomena from qualitative data and the quantification of qualitative data from the mathematico-statistical point of view. Ann. Inst. Stat. Math. **3**, 69–98 (1952)
Hayashi, C.: Multidimensional quantification with the applications to analysis of social phenomena. Ann. Inst. Stat. Math. **5**, 121–143 (1954)
Hill, M.O.: Reciprocal averaging: an eigenvector method of ordination. J. Ecol. **61**, 237–249 (1973)
Hill, M.O.: Correspondence analysis: a neglected multivariate method. J. Royal Stat. Soc. C (Applied Statistics) **23**, 340–354 (1974)
Hirschfeld, H.O.: A connection between correlation and contingency. Cambridge Philos. Soc. Proc. **31**, 520–524 (1935)
Horst, P.: Measuring complex attitudes. J. Soc. Psychol. **6**, 369–374 (1935)
Horst, P.: Obtaining a composite measure from a number of different measures of the same attribute. Psychometrika **1**, 53–60 (1936)
Hotelling, H.: Analysis of complex of statistical variables into principal components. J. Educ. Psychol. **24**, 417–441 and 498–520 (1933)
Johnson, P.O.: The quantification of qualitative data in discriminant analysis. J. Am. Stat. Assoc. **45**, 65–76 (1950)
Kendall, M.G., Stuart, A.: The Advanced Theory of Statistics, vol. 3, 3rd edn. Griffin, London (1973)
Lancaster, H.O.: Some properties of the bivariate normal distribution considered in the form of a contingency table. Biometrika **44**, 289–292 (1957)
Lancaster, H.O.: The structure of bivariate contributions. Ann. Math. Stat. **29**, 719–736 (1958)

Lebart, L., Morineau, A., Tabard, N.: Techniques de la Description Statistique: Méthodes et Logiciels pour l'Analyse des Grands Tableaux. Dunod, Paris (1977)
Lebart, L., Morineau, A., Warwick, K.M.: Multivariate Descriptive Statistical Analysis. Wiley, New York (1984)
Leclerc, A.: Quelques proprietés optimales en analyse de données en terme de correlation entre variables. Mathematiques et Sciences Humaines **18**, 51–67 (1980)
De Leeuw, J.: Canonical analysis of relational data. Department of Data Theory, RN 007-68, Leiden University, Leiden, The Netherlands (1968)
De Leeuw, J.: Canonical analysis of categorical data. Unpublished Doctoral Dissertation, Leiden University, Leiden, The Netherlands (1973)
De Leeuw, J.: HOMALS. Paper Presented at the Symposium on Optimal Scaling, 1976 Spring Meeting of the Psychometric Society, Bell Laboratories, Murray Hill, NJ (1976)
De Leeuw, J.: Nonlinear principal component analysis. In: Caussinus, H., Ettinger, P., Tomassone, R. (eds.) COMPSTAT1982 Proceedings in Computational Statistics, pp. 77–86. Physica Verlag, Vienna (1982)
De Leeuw, J.: On the prehistory of correspondence analysis. Stat. Neerl. **37**, 161–164 (1983)
De Leeuw, J.: Canonical analysis of categorical data. DSWO Press, Leiden University, Leiden, The Netherlands (1984)
Lingoes, J.C.: Multivariate analysis of contingencies. Comp. Rep. Univ. Michigan **2**, 1–24 (1963)
Lingoes, J.C.: Simultaneous linear regressions: a 7090 Program for analyzing metric/nonmetric or linear/nonlinear data. Behav. Sci. **9**, 87–88 (1964)
Lingoes, J.C.: The multivariate analysis of qualitative data. Multivar. Behav. Res. **3**, 61–94 (1968)
Lord, F.M.: Some relations between Guttman's principal components of scale analysis and other psychometric theory. Psychometrika **23**, 291–296 (1958)
Lubin, A.: Linear and nonlinear discriminant functions. Br. J. Psychol. **3**, 90–104 (1950)
Masson, M.: Méthodologies générales de traitement statistique de l'information de masse. Cedic-Fernand Nathan, Paris (1980)
Masson, M.: Processus linéaires et analyse des données non linéares. Thèse de 3ème cycle. Université Pierre et Marie Curie, Paris (1974)
Maung, K.: Discriminant analysis of Tocher's eye colour data. Ann. Eugen. **11**, 64–76 (1941a)
Maung, K.: Measurement of association in contingency tables with special reference to the pigmentation of hair and eye colours of Scottish children. Ann. Eugen. **11**, 189–223 (1941b)
McDonald, R.P.: A unified treatment of the weighting problem. Psychometrika **33**, 351–381 (1968)
McKeon, J.J.: Canonical analysis: some relations between canonical correlation, factor analysis, discriminant function analysis and scaling theory. In: Psychometric Monograph, No. 13 (1966)
Mosier, C.I.: Machine methods in scaling by reciprocal averages. In: Endicott, N.Y. (ed.) Proceedings, Research Forum, pp. 35–39. International Business Corporation (1946)
Mosteller, F.: A theory of scalogram analysis, using noncumulative types of items; a new approach to Thurstone's method of scaling attitudes. Harvard University, Laboratory of Social Relations, Report No 9 (1949)
Naouri, J.C.: Analyse factorielle des correspondances continues. Publications de l'Institutde Statistique de l'Université de Paris **19**, 1–100 (1970)
Nishisato, S.: Analysis of Categorical Data: Dual Scaling and Its Applications. The University of Toronto Press, Toronto (1980)
Nishisato, S., Beh, E.J., Lombardo, R., Clavel, J.G.: Modern Quantification Theory. Springer, Singapore (2021)
Nishisato, S.: Optimal Scaling and Its Generalizations. I. Methods. Department of Measurement and Evaluation, OISE, University of Toronto, Toronto (1972)
Nishisato, S.: Optimal Scaling and Its Generalizations. II. Applications. Department of Measurement and Evaluation, OISE, University of Toronto, Toronto (1973)
Richardson, M., Kuder, G.F.: Making a rating scale that measures. Pers. J. **12**, 36–40 (1933)
van Rijckevorsel, J., de Leeuw, J.: An outline to HOMALS. Department of Data Theory, RN 002-78, Leiden University, Leiden, The Netherlands (1978)

van Rijckevorsel, J.: The Applications of Fuzzy Coding and Horseshoes in Multiple Correspondence Analysis. DSWO Press, Leiden University, Leiden, The Netherlands (1987)

Saito, T.: Quantification of Categorical Data by Using the Generalised Variance, pp. 61–80. Nippon UNIVAC Sogo Kenkyu-sho Inc., Soken Kiyo (1973)

Saporta, G.: About some remarkable properties of generalised canonical analysis. Paper Presented at the Second Meeting of the Psychometric Society. Groningen, The Netherlands (1980)

Schriever, B.F.: Scaling of order dependent categorical variables with correspondence analysis. Int. Stat. Rev. **51**, 225–238 (1983)

Schriever, B.F.: Order Dependence. Unpublished Doctoral Thesis, Free University of Amsterdam, The Netherlands (1985)

Shiba, S.: A method for scoring multicategory items. Jpn. Psychol. Res. **7**, 75–79 (1965)

Slater, P.: The analysis of personnel preferences. Br. J. Psychol. **3**, 119–135 (1960)

Tenenhaus, M., Young, F.W.: An analysis and synthesis of multiple correspondence analysis, optimal scaling, dual scaling, homogeneity analysis and other methods for quantifying categorical data. Psychometrika **50**, 91–119 (1985)

Torgerson, W.S.: Theory and Methods of Scaling. Wiley, New York (1958)

Tucker, L.R.: A note on the machine method for the quantification of attributes (Supplementary study B3). In Horst, P., Wallin, P., Guttman, L. (eds.) The Prediction of Personal Adjustment. Social Science Research Council, New York (1941)

Wilks, S.S.: Weighting system for linear function of correlated variables when there is no dependent variable. Psychometrika **3**, 23–40 (1938)

Williams, E.J.: Use of scores for the analysis of association in contingency tables. Biometrika **39**, 274–289 (1952)

Low Lexical Frequencies in Textual Data Analysis

Ludovic Lebart

1 Introduction

The description of tables cross-tabulating vocabulary and texts is commonly performed through correspondence analysis (CA), well adapted to frequency profiles and lexical tables, thanks to the distributional equivalence property of the chi-squared (χ^2) distance. CA is then complemented by clustering, often using additive trees (AT). It is difficult to trace CA's history accurately, due to its various variants and names; see, for example, Hayashi (1950), Benzécri (1973), and more recently, the two genealogy papers of Beh and Lombardo (2012, 2019). Nishisato (1980) has been a milestone in the history of CA as the first textbook in English, but also because of its original point of view.

Many francophone linguists use the distances of Evrard (1966) that are based on the presence or absence of words (or lemmas) and directly derived from the Phi (ϕ) coefficient of Yule-Pearson (Yule 1912). Such distances may provide more meaningful representations for discriminating between texts or for authorship attributions. Brunet et al. (2021) have shown from a large corpus of 50 novels written by 25 authors of the twentieth century (two novels per author) that a flawless pairing of novels by author could be obtained from the Evrard distance matrix. This matrix is easily deduced from the correlation matrix of binary variables (presence-absence). The corresponding binary data table can be described through principal component analysis (PCA) (Hotelling 1933). Section 2 is both a reminder and a review of the measurement of association between binary variables in exploratory analyses of text, whereas Sect. 3 deals with some solutions proposed in practise (information retrieval, open-ended questions in sample surveys, ...). Section 4 shows, with a full sized example, how PCA can provide a complementary point of view to that of CA,

L. Lebart (✉)
Centre National de la Recherche Scientifique (CNRS), Paris, France
e-mail: ludovic@lebart.org

E. J. Beh et al. (eds.), *Analysis of Categorical Data from Historical Perspectives*, Behaviormetrics: Quantitative Approaches to Human Behavior 17,
https://doi.org/10.1007/978-981-99-5329-5_19

emphasising the role played by the presence or absence of words. Finally, we illustrate how to modulate the distances according to the dimension of the principal space (number of kept axes) and thus providing an enrichment of the usual approaches.

2 Pearson-Yule ϕ and Pearson r

In statistics, the phi coefficient (ϕ) is a measure of association between two binary variables. Based on the correlation coefficient r of Pearson (1900), this measure has been proposed by Yule (1912) who had previously published a similar measure of association (Yule 1900). This coefficient is closely related to the chi-square (χ^2) calculated on the same contingency table to test the independence between rows and columns. It coincides with the Pearson correlation coefficient r between two binary variables.

Two binary variables x and y are considered positively associated if the data concentrates on the diagonal cells and considered negatively associated if they concentrate outside the diagonal. If we have a 2×2 table for two texts and adopt the notations outlined in Table 1, the coefficient ϕ which describes the association of x and y is given by the following formula, with the notations of Table 1:

$$\phi\,(1,\ 2) = \frac{n_{11}n_{00} - n_{10}n_{01}}{\sqrt{n_{1\bullet}n_{0\bullet}n_{\bullet 0}n_{\bullet 1}}}\,. \tag{1}$$

Note that as early as 1900, Yule proposed a similar formula:

$$Q_{\text{Yule}} = \frac{n_{11}n_{00} - n_{10}n_{01}}{n_{11}n_{00} + n_{10}n_{01}}\,. \tag{2}$$

Cohen (1960) proposed replacing the geometric mean of the denominator of (1) by an arithmetic mean:

$$s\,(1,\ 2) = \frac{2(n_{11}n_{00} - n_{10}n_{01})}{n_{1\bullet}n_{0\bullet} + n_{\bullet 0}n_{\bullet 1}}\,. \tag{3}$$

Table 1 2×2 contingency table confronting two texts

Words	Present in Text2	Absent in Text2	Total
Present in Text1	n_{11}	n_{10}	$n_{1\bullet}$
Absent in Text1	n_{01}	n_{00}	$n_{0\bullet}$
Total	$n_{\bullet 1}$	$n_{\bullet 0}$	n

Table 2 Incidence table, **X**, with general term x_{ij} (x_{ij} = 1 if word i is present in text j)

Words	Text1	Text2
Word 1	1	0
Word 2	1	1
Word 3	0	0
Word 4	1	0
⋮	⋮	⋮
Word n	0	1
Total	$n_{1\bullet}$	$n_{\bullet 1}$

The reader can consult Warren (2008) and Baulieu (1989) for an overview of the set of coefficients of association for 2 × 2 tables proposed over the years across many varying disciplines.

2.1 Link Between ϕ and χ^2

The square of the coefficient r is linked to Pearson's χ^2 statistic for the same 2 × 2 contingency table by the classical relationship (where n is the total number of observations: here number of distinct words):

$$\phi^2 = \frac{\chi^2}{n} \tag{4}$$

since we have:

$$n\phi^2 = \frac{n\left(n_{11}n_{00} - n_{10}n_{01}\right)^2}{n_{1\bullet}n_{0\bullet}n_{\bullet 0}n_{\bullet 1}} \tag{5}$$

being the classical formula of χ^2 for a 2 × 2 table, with 1 degree of freedom.

2.2 Equivalence of ϕ with Pearson's r

The classic Pearson correlation coefficient r calculated on the binary data of the incidence table, **X**, (Table 2) (a licit calculation in the case of two variables with two categories) coincides with the χ^2 coefficient:

$$r_{12} = \frac{1}{n}\sum_{i=1}^{n}\frac{\left(x_{i1} - \bar{x}_1\right)\left(x_{i2} - \bar{x}_2\right)}{s_1 s_2}, \tag{6}$$

with

$$\bar{x}_1 = \frac{1}{n}\sum_{i=1}^{n} x_{i1} = \frac{n_1}{n}, \qquad \bar{x}_2 = \frac{1}{n}\sum_{i=1}^{n} x_{i2} = \frac{n_2}{n}$$

and, for instance, for text 1:

$$s_1^2 = \frac{1}{n}\sum_{i=1}^{n} (x_{i1} - \bar{x}_1)^2 .$$

From (6) and Table 2, we find:

$$r_{12} = \frac{n_{11}n - n_{1\bullet}n_{\bullet 1}}{\sqrt{n_{1\bullet}n_{0\bullet}n_{\bullet 0}n_{\bullet 1}}}$$

and we get (1) by replacing n, $n_{1\bullet}$ and $n_{\bullet 1}$ with their values as functions of n_{11}, n_{01}, n_{10} and n_{00}. This equivalence with the classical χ^2 test of independence together with the identity of $\phi\,(1,\ 2)$ with the linear correlation coefficient r_{12} give the coefficient ϕ a special position among association measures.

2.3 The Chi-square Distance (χ^2)

The chi-square distance (χ^2 distance) used in CA is an approximation of a measure of mutual information (derived from the theory of Shannon (1948)) evaluating the information provided by an empirical contingency table with respect to the hypothesis of independence of rows and columns; see, for instance, Benzécri (1973). This distance shares with a few others the property of *distributional equivalence* which ensures stability of results by aggregating rows or columns with the same profiles. CA has become one of the basic tools for describing lexical tables:

$$d^2\,(j,\ j') = \sum_{i=1}^{n} \frac{n}{n_{i\bullet}} \left(\frac{n_{ij}}{n_{\bullet j}} - \frac{n_{ij'}}{n_{\bullet j'}}\right)^2 . \tag{7}$$

As shown in (7) the χ^2 distance involves inverses of frequencies which can be problematic[1] in the case of very low frequencies (in CA, the criterion of fit which gives each point a weight equal to its frequency partially compensates for this weakness).

[1] This can be addressed using the Freeman-Tukey distance measure as Cuadras and Cuadras (2006, 2015), Cuadras et al. (2006), Beh et al. (2018) and Beh and Lombardo (2024) do.

3 Low Frequencies and Frequency Discrepancies

In this section, we briefly review two approaches which aim to deal with strong frequency disparities or to involve binary coding of words.

3.1 Logarithmic Analysis

Logarithmic analysis (LA) also complies with the distributional equivalence property of CA for any arrays of positive numbers. Kazmierczak (1985) based the LA on the principle of Yule (1912) according to which one does not change the distance between two rows or between two columns of a table by replacing the rows and columns of this table by other proportional rows and columns (generalisation of distributional equivalence). In fact, this method dates back to Aitchison (1983) in a different setting. Note that similar, but not identical, variant had been proposed initially under the name of Spectral Analysis by Lewi (1976), then by Greenacre and Lewi (2009). LA consists in taking the logarithms of the data, after the possible addition of a constant (the smallest number to ensure values ≥ 1) in the case of negative or zero value. After having centred the data both in rows and in columns, LA submits the obtained table to an unstandardised principal components analysis (PCA), which coincides in such a case with a mere singular value decomposition (SVD). If $\mathbf{X}$ is a (n, m) data matrix, and if $\mathbf{A}$ and $\mathbf{B}$ are two diagonal matrices respectively of dimensions (n, n) and (m, m) with positive diagonal elements, the logarithmic analysis of the new array $\mathbf{AXB}$ coincides with that of $\mathbf{X}$. This property of strong invariance, together with the shrinking effect of the logarithm function, makes this technique robust, well suited to applications to massive data, for which the frequency disparities (from 1 to 1000 for example) constitute a technical obstacle. Beh and Lombardo (2024) also show that the total inertia of LA is based on the modified log-likelihood ratio statistic.

3.2 TF-IDF Coefficients and LSA

The elements of the (words $\times$ texts) lexical table can be replaced by the coefficient, Term Frequency $\times$ Inverse of Document Frequency, or TF-IDF (Salton and McGill 1983). Popular in Text Mining applications, the coefficient TF-IDF is the product of the frequency of a term (TF) by the logarithm of the quotient giving the "total number of documents/number of documents in which the term is present".[2] This quotient (IDF) therefore involves the inverse of the proportion of documents in which the term appears. The logarithm, as with the LA method mentioned above, helps to cushion extreme situations, such as when the term is only present in one document

[2] Note that in the context of information retrieval, "term" is often used instead of "word", and "document" instead of "text".

out of thousands. In other words, the TF-IDF coefficient combines an indicator of the dominance of the term (TF component) with an indicator of its specialisation in the corpus (IDF component), the latter indicator varying from 0 (the term is in all the documents) to a maximum (when the term is in a single document) which depends on the size of the set of documents. In information retrieval, the aim is to find one or more documents in a database (where documents are both short and numerous) using a few terms. One must then penalise the documents which do not contain these terms (element TF in the formula). If we denote by d the number of documents, $d(i)$ the number of documents which contain the term i, by f_{ij} the frequency of the term i in the document j, $f_{i\bullet}$ the total frequency of term i, and $f_{\bullet j}$ the total frequency of document j, we have:

- frequency of term i in document j:

$$\mathrm{TF}\,(i,\ j) = \frac{f_{ij}}{f_{\bullet j}},$$

- logarithm of the inverse of the frequency of documents containing the term i:

$$\mathrm{IDF}\,(i,\ j) = \log\left(\frac{d}{d(i)}\right).$$

Like CA and LA, Latent Semantic Analysis (LSA) [or Latent Semantic Indexing (LSI)] (Deerwester et al. 1990) is a singular values decomposition (SVD) of a transformed lexical table. Here, SVD applies to the matrix **T**, the general term of which is the TF-IDF coefficient:

$$t\,(i,\ j) = \frac{f_{ij}}{f_{\bullet j}} \log\left(\frac{d}{d\,(i)}\right). \tag{8}$$

We also know that CA can be deduced from the SVD of the matrix **W** with the general term:

$$w\,(i,\ j) = \frac{f_{ij}}{\sqrt{f_{i\bullet} f_{\bullet j}}} \tag{9}$$

Note that it can also be performed using:

$$\alpha\,(i,\ j) = \frac{f_{ij}}{f_{i\bullet} f_{\bullet j}} \tag{10}$$

which Goodman (1996), Beh (2004), and Beh and Lombardo (2014, 2021) call a "Pearson ratio", whereas Greenacre (2009) calls it a "contingency ratio". The term $w\,(i,\ j)$ can also be written:

$$w(i,\ j) = \frac{f_{ij}}{f_{\bullet j}} \left(\sqrt{\frac{f_{\bullet j}}{f_{i\bullet}}} \right). \tag{11}$$

Equations (8) and (11) differ by the factors represented by their right parentheses which both penalise the words (index i) that are frequent in the corpus: by the number of documents $d(i)$ which contain them for $t(i,\ j)$ in (8), by their overall frequency $f_{i\bullet}$ for $w(i,\ j)$ in (11). The concepts of number d of documents and of number $d(i)$ of documents containing a word i are especially operative for numerous and short documents.

General remark:

We have seen that low frequencies occur naturally in short texts whether they are documents or abstracts in a database, fragments or units of context, pages of novels, or even answers to open questions. Presence-absence coding is then an acceptable and empirically proven option. It can also be modulated by thresholding ("present" if more than s occurrences, for example). On the other hand, for applications to large corpus of texts, coding the presence or absence of a word could be a deliberate option which provides a specific point of view on the texts of a corpus, complementary to the global processing of original frequencies.

4 Illustrative Example

To show the relevance of presence-absence coding in textual data analysis, and the interest of PCA in this case, we will use the classical STATE OF THE UNION corpus which brings together the speeches on the State of the Union delivered by the American presidents in office before Congress, from George Washington (1789) to Barack Obama (2009) [42 speeches].

The corpus used here comprises 1,746,702 occurrences and 25,246 distinct words.

For this methodological example, we will work on the original words (or: word-forms) of the plain text (without lemmatisation). We are talking here about illustration rather than application because this corpus is meant as a benchmark allowing comparisons and is not an object of study in itself. Its strong chronological structure means that other methodologies can be applied with profit, and the problematic authorship of certain speeches would require interpretative precautions which go beyond the present example. The process of global description of the corpus after coding in the form of presence-absence of words will be schematised by five graphical displays involving repectively PCA (Figs. 1 and 2), CA (Fig. 3) and AT (Figs. 4 and 5).[3]

[3] We specify that the significance of the coordinates of the points, the stability of the observed patterns in the scattering diagrams of Figs. 1, 2 and 3 have been confirmed through extensive Bootstrap validations.

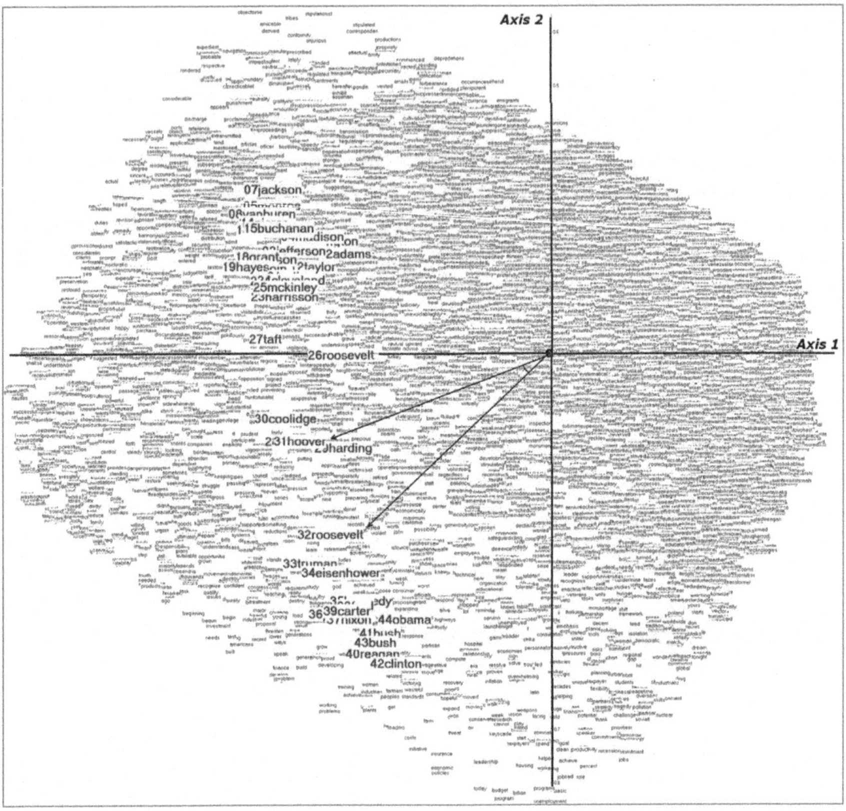

Fig. 1 Sketch of the plane (1, 2) of the PCA of the binary table: superimposition of 10,030 Words and 42 Presidents (after rescaling the cloud of words). The 10,030 words are not easily readable at this scale. Only the shape of their scattering diagram is clearly visible. As an example, the cosine of the angle between the 2 vectors joining the origin to both Presidents 23 and 32 (Hoover and Roosevelt) is an approximation of their correlation coefficient (related to a χ^2 with one degree of freedom, according to Sects. 2.1 and 2.2)

4.1 Principal Component Analysis of the Binary Table

The lexical table is similar to Table 2 (Sect. 1) but has 42 columns (Presidents) and 10,030 rows (distinct words). The number of rows is smaller than the 26,246 distinct words of the original corpus because the rows must have at least two 1s (presence) (such constraint eliminates the hapaxes[4] and at least two 0s (absence) (this second

[4] For many texts, the order of magnitude of the number of hapaxes (words with a single occurrence) is about half of the number of distinct words. More generally, the frequency of words in some corpus of natural language follows an empirical distribution known in lexicometrics under the name of Zipf's law. It states that the frequency of any word is inversely proportional to its rank in the frequency table. It accounts for the drastic reduction of the number of words when discarding

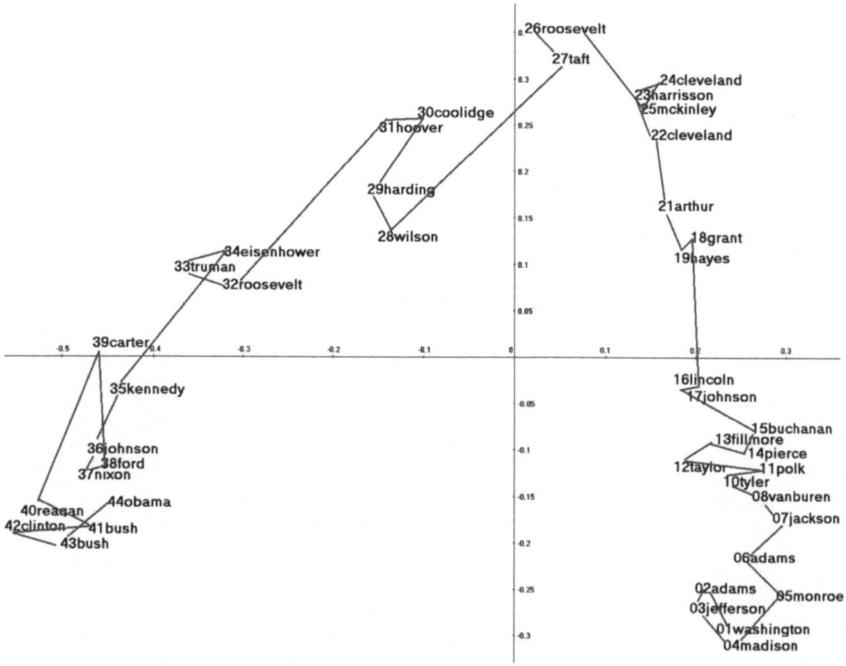

Fig. 2 Sketch of the plane (2, 3) of the PCA of the binary table (10,030 × 42) Words × Presidents

constraint eliminates the terms present in all the texts or absent only in a single text). Such trimming has the effect of reducing the size of the table by removing many usual tool-words (or function words) and auxiliaries, as well as a lot of common terms. The loss of raw information can seem considerable, but the only option that interests us at this point is to deal with meaningful distances.

4.2 *The "Size Factor", Axis 1 of the PCA*

In PCA, the origin of the principal axes is the mean-point of the coordinates of individuals (here, words) in one space, but it is not the mean-point of the variables in the other space (one of the important differences between PCA and CA). When there is a positive correlation between all the pairs of variables (here, Presidents) we obtain a "size factor". This is the famous "general aptitude factor" (supposed to measure intelligence) already described by Spearman (1904): some students have good marks in all subjects, and the first dimension puts them against those who have bad marks in all subjects (schematic situation largely discussed since). Here, some words are common among all Presidents.

the low frequencies; see, for instance, Lebart et al. (1988) for these statistical properties of textual data.

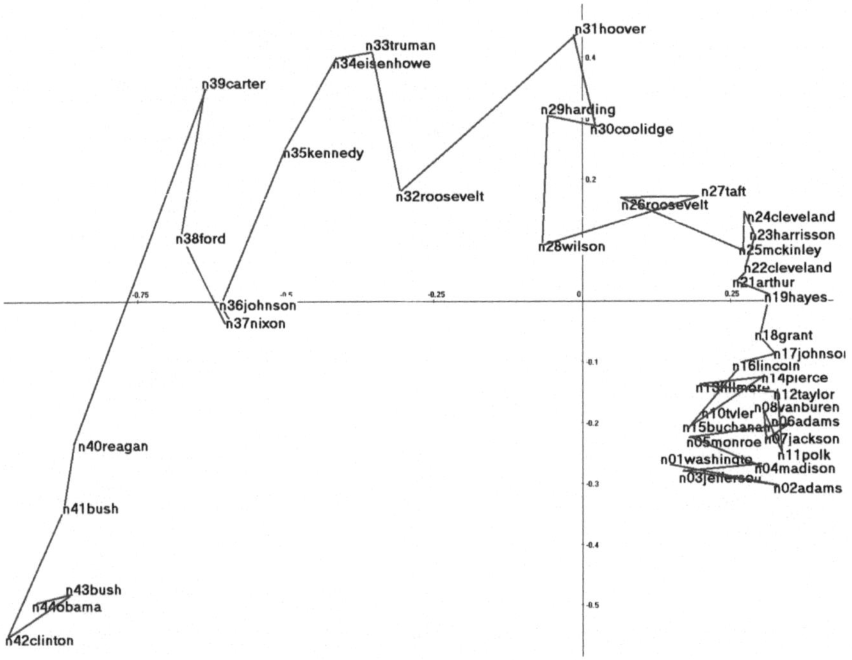

Fig. 3 Sketch of the plane (1, 2) from the CA of the lexical table (10,682 × 42) Words × Presidents

Figure 1 shows that the first axis is a consensus axis (axis absent from a CA which is based on profiles that are conditional frequencies). This horizontal axis roughly tells us that the Presidents all speak the same language (share most of the words, quite simply because these are frequent in the language), while the second vertical axis tells us that they do not all say the same thing (hence, the chronological pattern). Out of more that 10,000 words, it is anecdotal to select a few characteristic words. Let us mention for example , among the words common to all Presidents (left hand side of axis 1) [*treaty, effect, subject, constitution, enterprise, labour, influence, ...*]. Such selection is even more anecdotal on the right hand side which deals with the numerous unfrequent words [*prescription, backgrounds, terrorist, sanctuary, Kuwait, ...*] . In fact, the role of words describing a historical context is exacerbated on the vertical axis which is strongly chronological. We observe a mixture of style and historical events. On the top [*hereafter, amicable, tribes, Spain, stipulation, expedient, injurious, liberty*] and in the lower part [*America, budget, programme, unemployment, jobs, rural, freedom, Soviet, ...*]. Let us remind that these few words are quoted simply to illustrate the statistical calculations. They should be complemented by characteristic lines of texts, or phrases, or repeated segments to take into account their contexts; see Lebart et al. (1988, 2019). A complete interpretation, with all the methodological precautions that it implies, would require the volume of a book or of a dissertaion.

Although all the Presidents occupy the negative half of axis 1, we can detect however a slight but significant trend towards the more recent Presidents. The previous approximation: "they speak the same language" needs to be revised: they do not speak quite the same language, due to an evolution in the vocabulary. This is evident given the historical length of the period.

4.3 The Plane (2, 3) of the PCA

Figure 2 presents the plane spanned by axes 2 and 3 to ensure a comparison with the plane (1, 2) of the CA which follows. The parabolic shape of the sequence of Presidents in Figs. 2 and 3 is not just a pure Guttman effect (or horseshoe effect). The cloud of 10,030 words (not represented here) has a different shape and the central area contains many words common to extreme periods. The chronological pattern is obvious, with some noteworthy irregularities (such as President Carter, on the left hand side, recognised as a President out of the ordinary). We will not dwell on the interpretation of other interesting details in the methodological framework of this contribution. The reader can find more analyses involving the same corpus in Lebart et al. (2019).

4.4 Comparison with the CA of the Entire Lexical Table

Figure 3 shows the principal plane (1, 2) of a CA of the original lexical table (discarding the hapaxes is also mandatory with CA). The sequence of the first twenty Presidents (right side of the figure) is less clearly represented in this space. The high disparities between frequencies are not favourable to the distances involved in CA; see Sect. 2.3. In fact, in the present application, the plane (1, 2) of CA is not exactly comparable to the plane (2, 3) of PCA. The axis 1 of PCA, dealt with in the previous subsection, contains pieces of information that cannot be detected from the conditional frequencies involved in CA.

4.5 Modulations of Additive Trees According to Dimensions

The principles of Additive Tree (AT) date back to Buneman (1971). A useful algorithm has been provided by Sattath and Tversky (1977), under the name of *Neighbour Joining Algorithm*, and a free software implementation by Huson and Bryant (2006). The fundamental property of AT is the following: The original distance between two objects (represented by vertices of the additive tree) is, as much as possible, the length of shortest path in the tree between these two vertices.

The modulations of distances according to the number of principal axes kept which is described below concern all the principal axes methods mentioned (PCA, CA, logarithmic analysis, LSA). They contribute here to the clarity of interpretation of distances on binarised data as complementary to frequency data. The synergy of additive trees with principal axes techniques allows the researcher to go further on in the exploration of multidimensional spaces.

Figure 4 presents an additive tree (AT) constructed by taking all the main axes of PCA on presence-absence data.[5] These data reconstruct the correlation matrix built from binary data. The proximities (on the graph) are therefore interpreted in terms of coefficients ϕ, given by (1) of Sect. 2, or in terms of coefficient r, given by (6). ϕ and r are easier to conceive, conceptualise and interpret than a χ^2 distance. Figure 5 produces a similar tree, but the reconstitution of the correlation matrix is limited to the first 4 axes, making it possible to highlight a specific branch of the tree (lower left side of the tree) corresponding to a particular period (in this case, period known as *Gilded Age*: reconstruction period after the end of the civil war, industrial development, massive immigration, ...). This period corresponds to Presidents 18 (Grant, 1869) to 27 (Taft, 1913). Among the most characteristic words of this period, we find: *silver, gold, department, tariff, channel, Cuba, Venezuela, Chinese, Indian.* We will now focus on some other noticeable parts of the tree.

The small cluster at the bottom right of the same additive tree comprises the first four Presidents, the so-called *founding fathers*, from the first President (Washington, 1790) to the fourth (Madison, 1797). We note an exceptional use of function words such as *the, of, which*, phrases such as *Gentlemen of the house of representatives* (which also contain *the* and *of*), with also the words *tribes, British, Spain, barbary, execution, squadron, manufactures, fortifications*. The pronoun *we* is almost absent. It will become the most characteristic word of the corpus after the Second World War.

For the small branch of the tree going from Presidents 28 (Wilson, 1913) to 31 (Hoover, 1929), the most characteristics words are: *prohibition, agriculture, depression, veterans, railways*.

For the upper part of the tree, from Presidents 41 (G. H. Bush, 1989) to 44 (Obama, 2009), the most characteristic words are: *we, America, tonight, children, parents, medicare, health, terrorists, Iraq, ...*

The deformations of the additive tree (AT) do not exclude the consultation of principal planes, but the AT has a considerable advantage over them: it summarises subspaces having more than two dimensions, as in the space of Fig. 5 generated by the 4 first principal axes.

[5] All the analyses presented here (building of lexical tables from raw texts, PCA, CA, Bootstrap validations, Additive Tree) have been performed with the free software DtmVic that can be downloaded from: https://www.dtmvic.com. An updated version of the Additive Tree software Splitstree by Huson and Bryant (2006) can also be directly downloaded from https://software-ab.cs.uni-tuebingen.de/download/splitstree4/welcome.html.

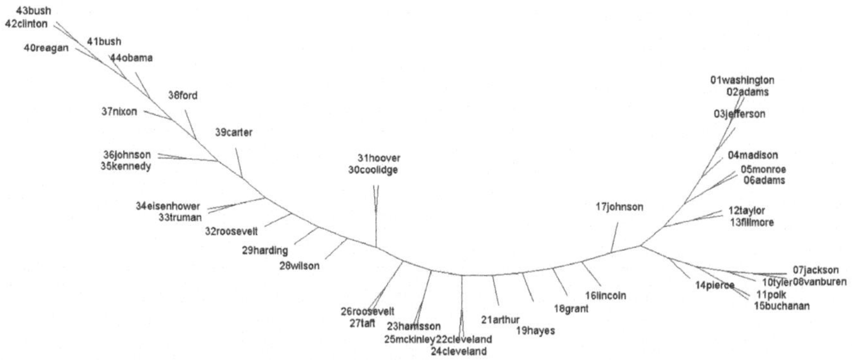

Fig. 4 Additive Tree computed from all axes from the CA of the lexical table (10,030 × 42) Words × Presidents

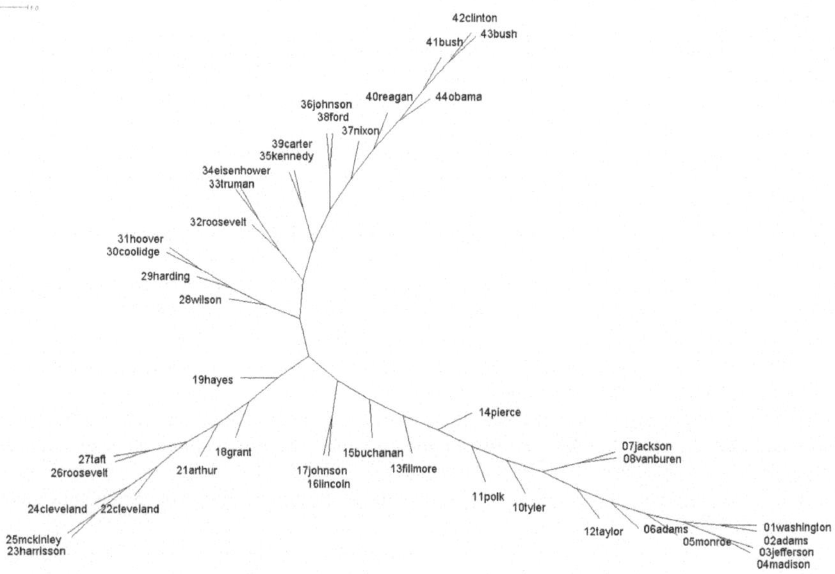

Fig. 5 Additive tree calculated on the first 4 axes of the same CA, highlighting the specificity of the period 1870–1910 (bottom left of the figure) (modulations of distances according to the number of kept axes)

5 Conclusion

The use of coefficients ϕ and r, like that of χ^2 (for 2 × 2 tables) makes it possible to work on the distances derived from the binary coding of the lexical tables, option imposed by the nature of the texts or deliberately selected to provide a specific point of view on these texts. At the confluence of several statistical approaches, naturally linked to PCA, these distances have a descriptive and discriminating power attested by numerous applications. The explicit formulation of the coefficients ensures transparency and quality of communication of the results.

References

Aitchison, J.: Principal component analysis of compositional data. Biometrika **70**(1), 57–65 (1983)

Baulieu, F.: A classification of presence/absence based dissimilarity coefficients. J. Classif. **6**(1), 233–246 (1989)

Beh, E.J.: Simple correspondence analysis: a bibliographic review. Int. Stat. Rev. **72**, 257–284 (2004)

Beh, E.J., Lombardo, R.: A genealogy of correspondence analysis. Aust. New Zealand J. Stat. **54**, 137–168 (2012)

Beh, E.J., Lombardo, R.: Correspondence Analysis: Theory, Practice and New Strategies. Wiley, Chichester (2014)

Beh, E.J., Lombardo, R., Alberti, G.: Correspondence analysis and the Freeman-Tukey statistic: a study of archaeological data. Comput. Stat. Data Anal. **128**, 73–86 (2018)

Beh, E.J., Lombardo, R.: A genealogy of correspondence analysis: Part 2—the variants. Electron. J. Appl. Stat. Anal. **12**, 552–603 (2019)

Beh, E.J., Lombardo, R.: An Introduction to Correspondence Analysis. Wiley, Chichester (2021)

Beh, E.J., Lombardo, R.: Correspondence analysis and the Cressie-Read family of divergence statistics. Int. Stat. Rev. (in press) (2024)

Benzécri, J.P.: L'Analyse des Données. Tôme 2: L'Analyse des Correspondances. Dunod, Paris (1973)

Brunet, E., Lebart, L., Vanni, L.: Littérature et intelligence artificielle. In: Mayaffre, D., Vanni, L. (eds.) L'Intelligence Artificielle des Textes, pp. 73–130. Champion, Paris (2021)

Buneman, P.: The recovery of trees from measurements of dissimilarity. In: Hodson, F.R.D., Kendall, G., Tautu, P. (eds.) Mathematics in the Archeological and Historical Sciences, pp. 387–395. Edinburgh University Press, Edinburgh (1971)

Cohen, J.: A coefficient of agreement for nominal scales. Educ. Psychol. Meas. **20**, 37–46 (1960)

Cuadras, C.M., Cuadras, D.: A parametric approach to correspondence analysis. Linear Algebra Appl. **417**, 64–74 (2006)

Cuadras, C.M., Cuadras, D.: A unified approach for the multivariate analysis of contingency tables. Open J. Stat. **5**, 223–232 (2015)

Cuadras, C.M., Cuadras, D., Greenacre, M.: A comparison of different methods for representing categorical data. Commun. Stat. Simul. Comput. **35**, 447–459 (2006)

Deerwester, S., Dumais, S.T., Furnas, G.W., Landauer, T.K., Harshman, R.: Indexing by latent semantic analysis. J. Am. Soc. Inform. Sci. **41**(6), 391–407 (1990)

Evrard, E.: Etude statistique sur les affinités de cinquant-huit dialectes bantous. In: de Strasbourg, Colloque (ed.) Statistique et Analyse Linguistique, pp. 85–94. Presses Universitaires de France, Paris (1966)

Goodman, L.A.: A single general method for the analysis of cross-classified data: reconciliation and synthesis of some methods of Pearson, Yule, and Fisher, and also some methods of correspondence analysis and association analysis. J. Am Stat. Assoc. **91**, 408–428 (1996)

Greenacre, M.: Power transformations in correspondence analysis. Comput. Stat. Data Anal. **53**(8), 3107–3116 (2009)

Greenacre, M., Lewi, P.: Distributional equivalence and subcompositional coherence in the analysis of compositional data, contingency tables and ratio-scale measurements. J. Classif. **26**(1), 29–54 (2009)

Hayashi, C.: On the quantification of qualitative data from the mathematico-statistical point of view. Ann. Inst. Stat. Math. **2**, 35–47 (1950)

Hotelling, H.: Analysis of a complex of statistical variables into principal components. J. Educ. Psychol. **24**, 498–520 (1933)

Huson, D.H., Bryant, D.: Application of phylogenetic networks in evolutionary studies. Molecular Biol. Evol. **23**(2), 254–267 (2006)

Kazmierczak, J.B.: Analyse logarithmique: deux exemples d'application. Revue de Statistique Appliquée **33**(1), 13–24 (1985)

Lebart, L., Pincemin, B., Poudat, C.: Analyse des Données Textuelles. Presses de l'Université du Québec, Québec, Canada (2019)

Lebart, L., Salem, A., Berry, E.: Exploring Textual Data. Kluwer Ac. Publisher, Dordrecht, The Netherlands (1998)

Lewi, P.J.: Spectral mapping, a technique for classifying biological activity profiles of chemical compounds. Arzneimittelforschung (Drug Research) **26**, 1295–1300 (1976)

Nishisato, S.: Analysis of Categorical Data: Dual Scaling and its Applications. University of Toronto Press, Toronto, Canada (1980)

Pearson, K.: Mathematical contributions to the theory of evolution. VII. On the correlation of characters not quantitatively measurable. Philos. Trans. Royal Soc. London Ser. A **195**, 1–47 (1900)

Salton, G., McGill, M.J.: Introduction to Modern Information Retrieval, International Student Edition. McGraw Hill, New York (1983)

Sattath, S., Tversky, A.: Additive similarity trees. Psychometrika **42**(3), 319–345 (1977)

Shannon, C.E.: A mathematical theory of communication. Bell Syst. Tech. J. **27**, 379–423 (1948)

Spearman, C.: General intelligence, objectively determined and measured. Am. J. Psychol. **15**, 201–293 (1904)

Warren, M.J.: On association coefficients for 2 $\times$ 2 tables and properties that does not depend on the marginal distributions. Psychometrika **73**, 777–789 (2008)

Yule, G.U.: On the association of attributes in statistics. Philos. Trans. Royal Soc. A **75**, 257–319 (1900)

Yule, G.U.: On the methods of measuring the association between two attributes. J. Royal Stat. Soc. **75**, 579–642 (1912)

Correspondence Analysis with Pre-Specified Marginals and Goodman's Marginal-Free Correspondence Analysis

Vartan Choulakian and Smail Mahdi

1 Introduction

We dedicate this contribution to Prof. Nishisato, a pioneer in dual scaling also known as correspondence analysis. This paper describes a particular development in the history of correspondence analysis for the analysis of two-way contingency tables.

Correspondence analysis (CA) and logratio analysis (LRA) are two popular methods for the analysis and visualisation of a contingency table (two-way frequency counts data having I rows and J columns). Benzécri (1973) is the reference book on CA and Nishisato (1980) on dual scaling. Beh and Lombardo (2014) present a panoramic review of CA and its variants.

LRA includes two independent and well-developed methods:

(a) RC association models for the analysis of contingency tables discussed by Goodman (1979, 1981a, b, 1991, 1996), and
(b) a set of compositional vectors (CoDA); see Aitchison (1986).

CA and LRA are based on three different principles:

(a) Benzécri's *distributional equivalence principle* in CA, termed by Nishisato (1984) as the *principle of equivalent partitioning*,
(b) RC association models on Yule's *scale invariance principle*, and
(c) Aitchison's *subcompositional coherence principle* in CoDA.

A recent discussion of these three principles can be found in Choulakian et al. (2023).

V. Choulakian (✉)
Département de Mathématiques et de Statistique, Université de Moncton, Moncton, NB, Canada
e-mail: vartan.choulakian@umoncton.ca

S. Mahdi
Department of Computer Science, Mathematics and Physics, Faculty of Science and Technology, University of the West Indies, CaveHill Campus, Barbados
e-mail: smail.mahdi@cavehill.uwi.edu

E. J. Beh et al. (eds.), *Analysis of Categorical Data from Historical Perspectives*, Behaviormetrics: Quantitative Approaches to Human Behavior 17,
https://doi.org/10.1007/978-981-99-5329-5_20

Goodman (1996) referred to his equation (46) as "marginal-free correspondence analysis" (mfCA) where his principal aim was to reconcile Pearson's correlation measure with Yule's association measure for the analysis of contingency tables. Underlying mfCA is the assumption that the two categorical variables consist of two equi-probable categories. In this paper, we show that mfCA is a particular case of CA with pre-specified marginals, which has been studied since the early 1980s under the direction of Benzécri. In Benzécri's edited journal *Les Cahiers de l'Analyse des Données*, the following papers appeared; Madre (1980), Choulakian (1980), Choulakian (1984), Benzécri (1983a, b), Benzécri et al. (1980) and Moussaoui (1987). Furthermore, we show that mfCA is also a particular first-order approximation of LRA analysis with uniform weights.

This paper is organised as follows. Section 2 presents three different basic ways of representing the concept of association in a contingency table. Section 3 discusses the important consequences of Yule's scale invariant association index. Section 4 presents the main result, and Sect. 5 discusses an example. Finally we make some concluding remarks in Sect. 6. The R code used to perform the computations is displayed in the Appendix.

2 Preliminaries on the Analysis of Contingency Tables

We consider a two-way contingency table $\mathbf{N} = (n_{ij})$ for $i = 1, \ldots, I, j = 1, \ldots J$, and define $\mathbf{P} = \mathbf{N}/n = (p_{ij})$ that is of size $I \times J$ to be the associated correspondence matrix (probability table) of the contingency table $\mathbf{N}$. We define as usual $p_{i+} = \sum_{j=1}^{J} p_{ij}$, $p_{+j} = \sum_{i=1}^{I} p_{ij}$, the vector $\mathbf{r} = (p_{i+}) \in \mathbb{R}^I$, the vector $\mathbf{c} = (p_{+j}) \in \mathbb{R}^J$, and $\mathbf{M}_I = \text{Diag}(\mathbf{r})$ to be the diagonal matrix having elements p_{i+}, similarly, $\mathbf{M}_J = \text{Diag}(\mathbf{c})$. We suppose that $\mathbf{M}_I$ and $\mathbf{M}_J$ are positive definite metric matrices of size $I \times I$ and $J \times J$, respectively; this means that the diagonal elements of $\mathbf{M}_I$ and $\mathbf{M}_J$ are strictly positive.

2.1 Independence of the Row and Column Categories

(a) If the I row categories and the J column categories are mutually independent, then:

$$\sigma_{ij} = p_{ij} - p_{i+}p_{+j} = 0 \tag{1}$$

for $i = 1, \ldots, I, j = 1, \ldots, J$ and where (σ_{ij}) is the residual matrix of (p_{ij}) with respect to the independence model $(p_{i+}p_{+j})$.

Remark 1 The contingency table $\mathbf{N} = (n_{ij})$ can also be represented (coded) as an indicator matrix $\mathbf{Z} = [\mathbf{Z}_I \ \mathbf{Z}_J] = [(\mathbf{z}_{\alpha i}) \ (\mathbf{z}_{\alpha j}))]$ of size $n \times (I + J)$ where $z_{\alpha i} = 0$ if individual α does not have level i of the row variable, $z_{\alpha i} = 1$ if individual α

has level i of the row variable; $z_{\alpha j} = 0$ if individual α does not have level j of the column variable, $z_{\alpha j} = 1$ if individual α has level j of the column variable. Note that $\mathbf{N} = \mathbf{Z}_I'\mathbf{Z}_J$ and $\sigma_{ij} = p_{ij} - p_{i+}p_{+j}$ is the covariance between the i-th column of $\mathbf{Z}_I$ and the j-th column of $\mathbf{Z}_J$.

(b) The independence assumption, $\sigma_{ij} = 0$, can also be interpreted in another way as:

$$\Delta_{ij} = \frac{1}{p_{+j}}\left(\frac{p_{ij}}{p_{i+}} - p_{+j}\right) = 0 \tag{2}$$

and is the column and row homogeneity model. Benzécri (1973, p. 31) named the conditional probability vector (p_{ij}/p_{+j} for $i = 1, \ldots, I$ and $j = 1, 2, \ldots, J$ fixed) the profile of the j-th column. He also referred to the element $p_{ij}/\left(p_{i+}p_{+j}\right)$ as the density function of the probability measure (p_{ij}) with respect to the product measure $\left(p_{i+}p_{+j}\right)$. The element $p_{ij}/\left(p_{i+}p_{+j}\right)$ is referred to as a Pearson ratio in Goodman (1996) and Beh and Lombardo (2014, p. 123).

(c) A third way to represent the independence assumption, $\sigma_{ij} = 0$, and the row and column homogeneity models, $\Delta_{ij} = 0$, is via the following weighted loglinear formulation for $\left(w_i^R, w_j^C\right)$ and assuming $p_{ij} > 0$ and defining $G_{ij} = \log(p_{ij})$:

$$\lambda_{ij} = G_{ij} - G_{i+} - G_{+j} + G_{++} = 0\,. \tag{3}$$

Here $G_{i+} = \sum_{j=1}^{J} G_{ij} w_j^C$, $G_{+j} = \sum_{i=1}^{I} G_{ij} w_i^R$ and $G_{++} = \sum_{i=1}^{I}\sum_{j=1}^{J} G_{ij} w_j^C w_i^R$; $w_j^C > 0$ and $w_i^R > 0$, satisfying $\sum_{j=1}^{J} w_j^C = \sum_{i=1}^{I} w_i^R = 1$ are a priori fixed or data dependent probability weights. Two popular weights are the marginal weights $\left(w_i^R = p_{i+}, w_j^C = p_{+j}\right)$ and the uniform weights $\left(w_i^R = 1/I, w_j^C = 1/J\right)$. This is implicit in Goodman (1996, eq. (7)) and Goodman (1991, eq. (2.2.6)) and is made explicit in Egozcue et al. (2015). Equation (3) is equivalent to the logratios:

$$\log\left(\frac{p_{ij}p_{i_1 j_1}}{p_{i j_1}p_{i_1 j}}\right) = 0 \text{ for } i \neq i_1 \text{ and } j \neq j_1,$$

which Goodman (1979, eq. (2.2)) refers to as the "null association model". Equation (3) is also equivalent to:

$$p_{ij} = \frac{\exp(G_{i+})\exp(G_{+j})}{\exp(G_{++})},$$

from which we deduce that under the independence assumption the marginal row probability vector (p_{i+}) is proportional to the vector of weighted geometric means $(\exp((G_{i+}))$. A similar property is true also for the columns; see, for example, Egozcue et al. (2015).

2.2 Interaction Factorisation

Suppose the independence-homogeneity-null association models are not true. Then each of the three equivalent model formulations (1), (2) and (3) can be generalised to explain the nonindependence-nonhomogeneity-association, named interaction, among the I rows and the J columns by adding k bilinear terms, where $k = \text{rank}(\mathbf{N}) - 1$. We designate any one of the interaction indices (1), (2) and (3) by τ_{ij}.

Benzécri (1973, Vol. 2, pp. 31 – 32) emphasised the importance of row and column weights or metrics in multidimensional data analysis; this is the reason why, in the French data analysis circles, any study starts with a triplet ($\mathbf{X}$, $\mathbf{M}_I$, $\mathbf{M}_J$), where $\mathbf{X}$ represents the data set, $\mathbf{M}_I = \text{Diag}\left(m_i^r\right)$ is the metric matrix defined for the rows and $\mathbf{M}_J = \text{Diag}\left(m_j^c\right)$ is the metric matrix defined for the columns. We follow the same procedure here but $\mathbf{X}$ is the pre-processed data, where:

(a) In covariance analysis, $\mathbf{X} = \left(\tau_{ij}\right) = \left(\sigma_{ij}\right)$ and:

$$(\mathbf{M}_I, \mathbf{M}_J) = (\text{Diag}(1/I), \text{Diag}(1/J))$$

(b) In CA, $\mathbf{X} = \left(\tau_{ij}\right) = \left(\Delta_{ij}\right)$ and:

$$(\mathbf{M}_I, \mathbf{M}_J) = \left(\text{Diag}(p_{i+}), \text{Diag}\left(p_{+j}\right)\right)$$

(c) In LRA, $\mathbf{X} = \left(\tau_{ij}\right) = \left(\lambda_{ij}\right)$ and:

$$(\mathbf{M}_I, \mathbf{M}_J) = \left(\text{Diag}\left(w_i^R\right), \text{Diag}\left(w_j^C\right)\right),$$

with $\sum_{i=1}^{I} w_i^R = \sum_{j=1}^{J} w_j^C = 1$.

We factorise the interactions in (1), (2) and (3) by singular value decomposition (SVD) or taxicab SVD (TSVD) as:

$$\tau_{ij} = \sum_{\alpha=1}^{k} \frac{f_\alpha(i)\, g_\alpha(j)}{\delta_\alpha}, \tag{4}$$

where $f_\alpha(i)$ is the i-th row principal coordinate $g_\alpha(j)$ is the j-th column principal coordinate along the α-th principal direction. Also δ_α is the dispersion measure of the α-th principal axis.

Remark 2 (a) In the SVD case the parameters $(f_\alpha(i), g_\alpha(j), \delta_\alpha)$ satisfy the conditions: for $\alpha, \beta = 1, \ldots, k$:

$$\delta_\alpha^2 = \sum_{i=1}^{I} f_\alpha^2(i)\, m_i^r = \sum_{j=1}^{J} g_\alpha^2(j)\, m_j^c,$$

where

$$\sum_{i=1}^{I} f_\alpha(i)\, m_i^r = \sum_{j=1}^{J} g_\alpha(j)\, m_j^c = 0$$

and

$$\sum_{i=1}^{I} f_\alpha(i)\, f_\beta(i)\, m_i^r = \sum_{j=1}^{J} g_\alpha(j)\, g_\beta(j)\, m_j^c = 0$$

for $\alpha \neq \beta$.

(b) In the TSVD case the parameters $(f_\alpha(i),\ g_\alpha(j),\ \delta_\alpha)$ satisfy the conditions: for $\alpha, \beta = 1, \ldots, k$:

$$\delta_\alpha = \sum_{i=1}^{I} \mid f_\alpha(i) \mid m_i^r = \sum_{j=1}^{J} \mid g_\alpha(j) \mid m_j^c$$

where

$$\sum_{i=1}^{I} f_\alpha(i)\, m_i^r = \sum_{j=1}^{J} g_\alpha(j)\, m_j^c = 0$$

and

$$\sum_{i=1}^{I} f_\alpha(i)\, \mathrm{sign}\left(f_\beta(i)\right) m_i^r = \sum_{j=1}^{J} g_\alpha(j)\, \mathrm{sign}\left(g_\beta(j)\right) m_j^c = 0$$

for $\alpha > \beta$.

A description of TSVD can be found in Choulakian (2006, 2016).

Remark 3 (a) In the case where $(\tau_{ij}) = (\sigma_{ij})$, the bilinear decomposition (4) is also named "interbattery analysis" and was first proposed by Tucker (1958). Later, Tenenhaus and Augendre (1996) reintroduced it within the context of correspondence analysis, where they showed that the Tucker approach (SVD of the covariance matrix (σ_{ij})) of some correspondence tables produced more interesting (interpretable) structure than CA.

(b) In the case where $(\tau_{ij}) = (\Delta_{ij})$, the CA decomposition has many interpretations. Essentially, for data analysis purposes, Benzécri (1973) interpreted it as weighted principal components analysis of the row and column profiles. Another useful interpretation of CA, comparable to Tucker's "interbattery analysis", is its links with canonical correlation analysis (Hotelling 1936); see Lancaster (1958) and Goodman (1991, 1996).

(c) In the case where $(\tau_{ij}) = (\lambda_{ij})$ and $(\mathbf{M}_I, \mathbf{M}_J) = \left(\mathrm{Diag}\left(w_i^R\right), \mathrm{Diag}\left(w_j^C\right)\right)$, where $\left(w_i^R, w_j^C\right)$ are pre-specified; we note this case leads to TLRA or LRA.

For the important case where $(\mathbf{M}_I, \mathbf{M}_J) = (\text{Diag}(1/I), \text{Diag}(1/J))$, we get uniformly weighted (or taxicab) logratio analysis, or uwLRA (or uwTLRA). For an example of general pre-specified weights see Egozcue and Pawlowsky-Glahn (2016).

(d) In the case where $\left(\tau_{ij}\right) = \left(\lambda_{ij}\right)$ and $(\mathbf{M}_I, \mathbf{M}_J) = \left(\text{Diag}(p_{i+}), \text{Diag}\left(p_{+j}\right)\right)$, where $\left(w_i^R, w_j^C\right) = \left(p_{i+}, p_{+j}\right)$ are data dependent, we get marginally weighted (or taxicab) logratio analysis, or mwLRA (or mwTLRA).

3 Yule's Principle of Scale Invariance

To really understand Yule's principle of scale invariance, we start by quoting Goodman (1996, Section 10):

> Pearson's approach to the analysis of cross-classified data was based primarily on the bivariate normal. He assumed that the row and column classifications arise from underlying continuous random variables having a bivariate normal distribution, so that the sample contingency table comes from a discretised bivariate normal; and he then was concerned with the estimation of the correlation coefficient for the underlying bivariate normal. On the other hand, Yule felt that, for many kinds of contingency tables, it was not desirable in scientific work to introduce assumptions about an underlying bivariate normal in the analysis of these tables; and for such tables, he used, to a great extent, coefficients based on the odds-ratios (for example, Yule's Q and Y), coefficients that did not require any assumptions about underlying distributions. The Pearson approach and the Yule approach appear to be wholly different, but a kind of reconciliation of the two perspectives was obtained in Goodman (1981a).

An elementary exposition of these ideas with examples can also be found in Mosteller (1968).

In the notation of our paper, Goodman's reconciliation is based on defining the a priori weights in the association index (3), $\lambda_{ij} = \lambda\left(p_{ij}, w_j^C, w_i^R\right)$, where by its decomposition into bilinear terms, mwLRA corresponds to Pearson's approach, while uwLRA corresponds to Yule's approach, since:

$$\log\left(\frac{p_{ij}\,p_{i_1 j_1}}{p_{ij_1}\,p_{i_1 j}}\right) = \lambda_{ij} + \lambda_{i_1 j_1} - \lambda_{i_1 j} - \lambda_{ij_1}$$
$$= \sum_{\alpha=1}^{k} \frac{\left(f_\alpha(i) - f_\alpha(i_1)\right)\left(g_\alpha(j) - g_\alpha(j_1)\right)}{\delta_\alpha}, \tag{5}$$

where the principal factor scores satisfy the marginally or uniformly weighted relations, see Remark's 2(a) and 2(b).

We also note that Kazmierczak (1985, 1987) also tried to reconcile CA and uwLRA by proposing the "generalised principle of distributional equivalence" by encompassing the principles of Benzécri and Yule; for further details, see Choulakian et al. (2023).

To have a clear picture of LRA with general a priori pre-specified weights $\left(w_j^C, w_i^R\right)$, we first study the properties of the association index λ_{ij} that distinguishes it from the interaction indices (1) and (2).

3.1 Scale Invariance of an Interaction Index

We are concerned with the property of scale dependence or independence of the three interaction indices (1), (2) and (3). We note that in (1), (2) and (3), p_{ij} depends on n_{ij} since $p_{ij} = n_{ij}/\sum_{i,j} n_{ij}$. To emphasise this dependence, we express an interaction index by $\tau_{ij}\left(n_{ij}\right) = \tau\left(p_{ij}, m_i^R, m_j^C\right)$ where, in the case of the association index, $\tau_{ij}\left(n_{ij}\right) = \lambda_{ij}$ is defined by (3). In the case of the nonhomogeneity index $\tau_{ij}\left(n_{ij}\right) = \Delta_{ij}$ is defined by (2), and in the case of the nonindependence index $\tau_{ij}\left(n_{ij}\right) = \sigma_{ij}$ is defined by (1). Following Yule (1912), we state the following:

Definition 1 An interaction index $\tau_{ij}\left(n_{ij}\right)$ is scale invariant if $\tau_{ij}\left(n_{ij}\right) = \tau_{ij}\left(a_i n_{ij} b_j\right)$ for arbitrary scales $a_i > 0$ and $b_j > 0$.

It is important to note that Yule's principle of scale invariance concerns a function of four interaction terms—see (5)—while in Definition 1 the invariance concerns each interaction term.

It is evident that neither the interaction indices (1) and (2), nor (3), with data dependent marginal weights $\left(p_{i+}, p_{+j}\right)$ are scale invariant.

Concerning the association index (3) we have the following two lemmas:

Lemma 1 *The association index (3) with pre-specified weights* $\left(w_i^R, w_j^C\right)$ *is scale invariant.*

Proof Let

$$n^* = \sum_{i=1}^{I}\sum_{j=1}^{J} a_i n_{ij} b_j$$

and $w_j^C > 0$ and $w_i^R > 0$ satisfying:

$$\sum_{j=1}^{J} w_j^C = \sum_{i=1}^{I} w_i^R = 1\,.$$

Then

$$
\begin{aligned}
\tau_{ij}\left(a_i n_{ij} b_j\right) &= \lambda\left(\frac{a_i n_{ij} b_j}{n^*},\ w_i^R,\ w_j^C\right) \\
&= \log\left(\frac{a_i n_{ij} b_j}{n^*}\right) - \sum_{j=1}^{J} w_j^C \log\left(\frac{a_i n_{ij} b_j}{n^*}\right) \\
&\quad - \sum_{i=1}^{I} w_i^R \log\left(\frac{a_i n_{ij} b_j}{n^*}\right) + \sum_{j=1}^{J}\sum_{i=1}^{I} w_j^C w_i^R \log\left(\frac{a_i n_{ij} b_j}{n^*}\right) \\
&= \lambda\left(n_{ij},\ w_i^R,\ w_j^C\right) \\
&= \lambda\left(p_{ij},\ w_i^R,\ w_j^C\right) \\
&= \tau_{ij}\left(n_{ij}\right) \\
&= \lambda\left(a_i p_{ij} b_j,\ w_i^R,\ w_j^C\right).
\end{aligned}
\tag{6}
$$

Lemma 2 *To a first-order approximation*:

$$
\lambda_{ij} \approx \frac{p_{ij}}{w_j^C w_i^R} - \frac{p_{i+}}{w_i^R} - \frac{p_{+j}}{w_j^C} + 1 .
$$

Proof The average value of the density function $p_{ij}/\left(w_j^C w_i^R\right)$ with respect to the product measure $w_j^C w_i^R$ is 1. So the IJ values of $p_{ij}/\left(w_j^C w_i^R\right)$ are distributed around 1. By Taylor series expansion of $\log(x)$ in the neighbourhood of $x = 1$, we have the first-order approximation $\log(x) \approx x - 1$. Therefore, substituting $a_i = 1/w_i^R$ and $b_j = 1/w_j^C$ into (6), and by using:

$$
\log\left(\frac{p_{ij}}{w_j^C w_i^R}\right) \approx \frac{p_{ij}}{w_j^C w_i^R} - 1
$$

we get:

$$
\begin{aligned}
\lambda\left(p_{ij},\ w_j^C,\ w_i^R\right) &= \lambda\left(\frac{p_{ij}}{w_j^C w_i^R},\ w_j^C,\ w_i^R\right) \\
&= \log\left(\frac{p_{ij}}{w_j^C w_i^R}\right) - \sum_{j=1}^{J} w_j^C \log\left(\frac{p_{ij}}{w_j^C w_i^R}\right) \\
&\quad - \sum_{i=1}^{I} w_i^R \log\left(\frac{p_{ij}}{w_j^C w_i^R}\right) + \sum_{j=1}^{J}\sum_{i=1}^{I} w_j^C w_i^R \log\left(\frac{p_{ij}}{w_j^C w_i^R}\right) \\
&\approx \frac{p_{ij}}{w_j^C w_i^R} - 1 - \left(\frac{p_{i+}}{w_i^R} - 1\right) - \left(\frac{p_{+j}}{w_j^C} - 1\right) + 0,
\end{aligned}
$$

which is the required result.

Remark 4 Lemma 2 provides a first-order approximation to mwTLRA and uwTLRA, where we see that both first-order approximations are marginal-dependent but in different ways.

(a) In the case $(a_i, b_j) = (1/p_{i+}, 1/p_{+j})$ and $\left(w_j^C, w_i^R\right) = (p_{+j}, p_{i+})$ in Lemma 2:

$$\lambda_{ij} = \lambda\left(p_{ij}, p_{+j}, p_{i+}\right) \approx \frac{p_{ij}}{p_{+j}p_{i+}} - 1$$

which implies that CA (or TCA) is a first-order approximation of LRA (or TLRA) with pre-specified weights (p_{i+}, p_{+j}) of a data set $(a_i n_{ij} b_j)$ where (p_{i+}, p_{+j}) are the marginals of (n_{ij}), a result stated in Cuadras et al. (2006). Or, it can be deduced explicitly for data dependent mwLRA by the fact that:

$$\lambda\left(p_{ij}, p_{+j}, p_{i+}\right) = \lambda\left(\frac{p_{ij}}{p_{+j}p_{i+}}, p_{+j}, p_{i+}\right)$$

which can be found in Goodman (1996).

(b) In the case where $(a_i, b_j) = (I, J)$ and $\left(w_j^C, w_i^R\right) = (1/I, 1/J)$ in Lemma 2:

$$\lambda_{ij} = \lambda\left(p_{ij}, \frac{1}{J}, \frac{1}{I}\right) \approx IJp_{ij} - Ip_{i+} - Jp_{+j} + 1$$

which implies that the bilinear expansion of the right hand side of TSVD (or SVD) is a first-order approximation of uwTLRA (or uwLRA).

In this subsection, we discussed the approximation of LRA (or TLRA) to CA (or TCA) related methods. Greenacre (2009) posed the reciprocal question: "when do CA related methods converge to LRA?". To answer this question he stated two results which we discuss in the following subsection.

3.2 *Box-Cox Transformation*

Theoretically CA and LRA have been presented in a unified mathematical framework via the Box-Cox transformation by Goodman (1996), where the bilinear terms have been estimated by SVD. Goodman's framework was further considered, among others, by Cuadras et al. (2006), Cuadras and Cuadras (2015), Greenacre (2009, 2010) and Beh and Lombardo (2024).

Consider the triplet $(\mathbf{X}, \mathbf{M}_I, \mathbf{M}_J)$, where $\mathbf{X} = (x_{ij})$ represents the data set with $x_{ij} > 0$, and $(\mathbf{M}_I, \mathbf{M}_J) = \left(\text{Diag}\left(w_i^R\right), \text{Diag}\left(w_j^C\right)\right)$ with $\sum_{j=1}^J w_j^C = \sum_{i=1}^I w_i^R = 1$. Let α be a nonnegative real number. Following Goodman (1996, eqs. (3), (4) and (5)), we define the interaction index:

$$\begin{aligned}\operatorname{Int}\left(\frac{x_{ij}^{\alpha}}{\alpha}, w_j^C, w_i^R\right) &= \frac{x_{ij}^{\alpha}}{\alpha} - \sum_{j=1}^{J} w_j^C \frac{x_{ij}^{\alpha}}{\alpha} - \sum_{i=1}^{I} w_i^R \frac{x_{ij}^{\alpha}}{\alpha} + \sum_{j=1}^{J}\sum_{i=1}^{I} w_j^C w_i^R \frac{x_{ij}^{\alpha}}{\alpha} \\ &= \left(\frac{x_{ij}^{\alpha}-1}{\alpha}\right) - \sum_{j=1}^{J} w_j^C \left(\frac{x_{ij}^{\alpha}-1}{\alpha}\right) - \sum_{i=1}^{I} w_i^R \left(\frac{x_{ij}^{\alpha}-1}{\alpha}\right) \\ &\quad + \sum_{j=1}^{J}\sum_{i=1}^{I} w_j^C w_i^R \left(\frac{x_{ij}^{\alpha}-1}{\alpha}\right). \end{aligned} \tag{7}$$

Using the well-known result based on L'Hôpital's rule (or the Box-Cox transformation):

$$\lim_{\alpha \to 0} \left(\frac{x_{ij}^{\alpha}-1}{\alpha}\right) = \log(x_{ij})$$

then (7) converges to:

$$\begin{aligned}\lambda\left(x_{ij}, w_j^C, w_i^R\right) &= \log\left(x_{ij}\right) - \sum_{j=1}^{J} w_j^C \log\left(x_{ij}\right) - \sum_{i=1}^{I} w_i^R \log\left(x_{ij}\right) \\ &\quad + \sum_{j=1}^{J}\sum_{i=1}^{I} w_j^C w_i^R \log\left(x_{ij}\right). \end{aligned} \tag{8}$$

We consider two cases of (7) and (8):

(a)

$$\lambda\left(x_{ij}, w_j^C, w_i^R\right) = \lambda\left(p_{ij}, p_{+j}, p_{i+}\right) = \lambda\left(\frac{p_{ij}}{p_{+j}p_{i+}}, p_{+j}, p_{i+}\right)$$

is the interaction term of mwLRA (or mwTLRA), and equivalent to Result 2 in Greenacre (2010).

(b)

$$\begin{aligned}\lambda\left(x_{ij}, w_j^C, w_i^R\right) &= \lambda\left(p_{ij}, \frac{1}{J}, \frac{1}{I}\right) = \lambda\left(IJp_{ij}, \frac{1}{J}, \frac{1}{I}\right) &(9)\\ &= \lambda\left(\frac{p_{ij}}{p_{+j}p_{i+}}, \frac{1}{J}, \frac{1}{I}\right) &(10)\end{aligned}$$

is the interaction term of uwLRA (or uwTLRA); this is similar to Result 1 in Greenacre (2010).

Equation (7) can also be applied in a general way. To show this, consider the following argument:

In (7) we replace w_j^C and w_i^R with:

$$w_j^C(\alpha) = \frac{m_j^c \sum_{i=1}^{I} m_i^r x_{ij}^\alpha}{\sum_{i=1}^{I} \sum_{j=1}^{J} m_i^r m_j^c x_{ij}^\alpha}$$

and

$$w_i^R(\alpha) = \frac{m_i^r \sum_{j=1}^{J} m_j^c x_{ij}^\alpha}{\sum_{i=1}^{I} \sum_{j=1}^{J} m_i^r m_j^c x_{ij}^\alpha},$$

respectively, where $m_i^r > 0$, $m_j^c > 0$ and $\sum_{j=1}^{J} m_j^c = \sum_{i=1}^{I} m_i^r = 1$. We see that:

$$\lim_{\alpha \to 0} w_j^C(\alpha) = m_j^c.$$

Similarly

$$\lim_{\alpha \to 0} w_i^R(\alpha) = m_i^r.$$

Therefore, we get:

$$\lim_{\alpha \to 0} \text{Int}\left(x_{ij}^\alpha, w_j^C(\alpha), w_i^R(\alpha)\right) = \lambda\left(x_{ij}, m_j^c, m_i^r\right),$$

which is the interaction term of LRA with a priori weight $\left(m_i^r, m_j^c\right)$.

Two special cases of this argument are as follows:

Case a) Setting $\left(m_i^r, m_j^c\right) = (1/I, 1/J)$ we get:

$$\begin{aligned} \lim_{\alpha \to 0} \text{Int}\left(x_{ij}^\alpha, w_j^C(\alpha), w_i^R(\alpha)\right) &= \lambda\left(p_{ij}, \frac{1}{J}, \frac{1}{I}\right) \\ &= \lambda\left(IJp_{ij}, \frac{1}{J}, \frac{1}{I}\right) \\ &= \lambda\left(\frac{p_{ij}}{p_{+j}p_{i+}}, \frac{1}{J}, \frac{1}{I}\right), \end{aligned}$$

which is the interaction term of uwLRA (or uwTLRA).

Case b) Setting $\left(m_i^r, m_j^c\right) = \left(p_{i+}, p_{+j}\right)$ we get:

$$\begin{aligned} \lim_{\alpha \to 0} \text{Int}\left(x_{ij}^\alpha, w_j^C(\alpha), w_i^R(\alpha)\right) &= \lambda\left(p_{ij}, p_{+j}, p_{i+}\right) \\ &= \lambda\left(\frac{p_{ij}}{p_{+j}p_{i+}}, p_{+j}, p_{i+}\right), \end{aligned}$$

which is the interaction term of mwLRA (or mwTLRA).

4 CA with Pre-Specified Marginals and Goodman's mfCA

Suppose we observe a probability table $\mathbf{P} = (p_{ij})$ of size $I \times J$. Let $\mathbf{Q}$, of size $I \times J$, be an *unknown* probability table with *known* marginals q_{i+} and q_{+j}. The CA of $\mathbf{P}$ with the pre-specified marginals of $\mathbf{Q}$ is done in the following two steps:

Step 1: We construct $\mathbf{Q}$ which is in a sense "nearest to $\mathbf{P}$". Two general criteria are:
(1)

$$\mathrm{Int}\left(q_{ij},\ q_{+j},\ q_{i+}\right) = \lambda\left(q_{ij} = a_i p_{ij} b_j,\ q_{+j}, q_{i+}\right)$$

based on (3), or
(2)

$$\min_{q_{ij}} \sum_{j=2}^{J} \sum_{i=1}^{I} q_{i+}q_{+j} \left(\frac{q_{ij}}{q_{i+}q_{+j}} - \frac{p_{ij}}{p_{i+}p_{+j}} \right)^2$$

based on (2).

Step 2: We apply CA to the constructed probability $\mathbf{Q}$ such that:

$$\frac{q_{ij} - q_{i+}q_{+j}}{q_{i+}q_{+j}} = \sum_{\alpha=1}^{k} \frac{f_\alpha(i) g_\alpha(j)}{\delta_\alpha},$$

which represents the CA of $\mathbf{P}$ with pre-specified marginals $\left(q_{i+},\ q_{+j}\right)$. Choulakian (1980) presents an example where both criteria described in Step 1 have been applied and similar results have been obtained.

In the case where $\mathrm{Int}\left(q_{ij},\ 1/J,\ 1/I\right) = \lambda\left(q_{ij} = a_i p_{ij} b_j,\ 1/J,\ 1/I\right)$, we get Goodman's mfCA; see Goodman (1996, eq. (46)). $\mathbf{Q} = \left(q_{ij}\right)$ is related to $\mathbf{P} = \left(p_{ij}\right)$ via the strictly positive scales $\left(a_i,\ b_j\right)$ that keeps Yule's association between the i-th row and the j-th column unchanged. The iterative proportional fitting algorithm (IPFA) is used to construct $\mathbf{Q}$. That is, the constructed probability table $\left(q_{ij}\right)$ has uniform marginals $q_{+j} = 1/J$ and $q_{i+} = 1/I$. So in Step 2, the CA representation is:

$$IJq_{ij} - 1 = \sum_{\alpha=1}^{k} \frac{f_\alpha(i) g_\alpha(j)}{\delta_\alpha},$$

which represents a first-order approximation to both uwLRA and mwLRA by Remark 3. Furthermore, by Remark 2 we see that mfCA can be interpreted both the decompositions of Tucker and Hotelling.

5 Example

We consider the *rodent* data set of size 28 $\times$ 9 found in the R package `TaxicabCA`. Its source is in Quinn and Keough (2002). This is an abundance data set of 9 species of rats in 28 cities in California. Choulakian (2017) analysed it by comparing the CA and TCA maps; furthermore Choulakian (2021) showed that it has a quasi-2-block diagonal structure. Here we compare the dispersion results along the first two principal dimensions for the 4 methods: CA, TCA, mfCA and mfTCA:

In CA: $\delta_1 = \text{corr}\,(f_1\,(i)),\ g_1\,(j)) = 0.864$ and $\delta_2 = \text{corr}\,(f_2\,(i)\,,\ g_2\,(j)) = 0.678$.
In mfCA: $\delta_1 = \text{corr}\,(f_1\,(i)\,,\ g_1\,(j)) = 0.827$ and $\delta_2 = \text{corr}\,(f_2\,(i)\,,\ g_2\,(j)) = 0.679$.
In TCA: $\delta_1 = 0.478$ and $\delta_2 = 0.422$.
In mfTCA: $\delta_1 = 0.743$ and $\delta_2 = 0.541$.

The R code in the Appendix produced the following four maps: CA, mfCA, TCA and mfTCA. To interpret these maps, essentially, the distances between the profiles (conditional probabilities) of two columns, or two rows, are assessed.

In CA the chi-squared distance between columns j and j_1 is:

$$\sum_{i=1}^{I} \frac{1}{p_{i+}} \left(\frac{p_{ij}}{p_{+j}} - \frac{p_{ij_1}}{p_{+j_1}} \right)^2 = \sum_{\alpha=1}^{k} (g_\alpha\,(j) - g_\alpha\,(j_1))^2\,. \tag{11}$$

By letting $p_{i+} = 1/I$ and $p_{+j} = 1/J$, and replacing p_{ij} with q_{ij} in (11) we get the chi-squared distance between columns j and j_1 in mfCA:

$$I \sum_{i=1}^{I} \left(\frac{p_{ij}}{1/J} - \frac{p_{ij_1}}{1/J} \right)^2 = \sum_{\alpha=1}^{k} (g_\alpha\,(j) - g_\alpha\,(j_1))^2\,. \tag{12}$$

In TCA the taxicab distance between columns j and j_1 is:

$$\sum_{i=1}^{I} \left| \frac{p_{ij}}{p_{+j}} - \frac{p_{ij_1}}{p_{+j_1}} \right| \leq \sum_{\alpha=1}^{k} \left| g_\alpha\,(j) - g_\alpha\,(j_1) \right|. \tag{13}$$

The taxicab distance between columns j and j_1 in mfCA is:

$$\sum_{i=1}^{I} \left| \frac{p_{ij}}{1/J} - \frac{p_{ij_1}}{1/J} \right| \leq \sum_{\alpha=1}^{k} \left| g_\alpha\,(j) - g_\alpha\,(j_1) \right|. \tag{14}$$

Comparing these four equations, the last three are similar; (11) is dissimilar when the weights $1/p_{i+}$ are substantially different from $1/I$. This fact seems visually apparent in the four figures (Figs. 1, 2, 3 and 4).

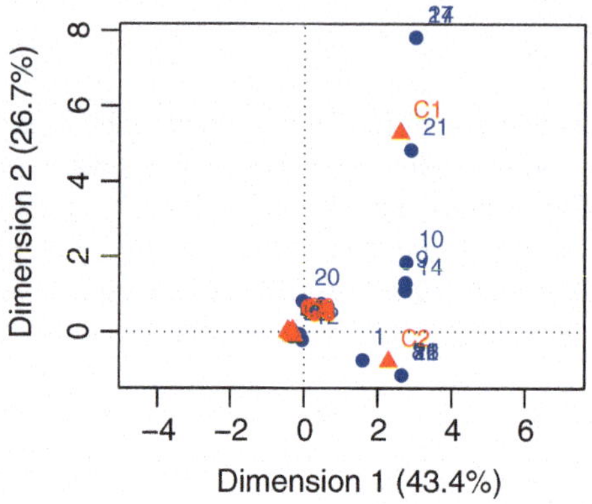

Fig. 1 CA map of rodent data

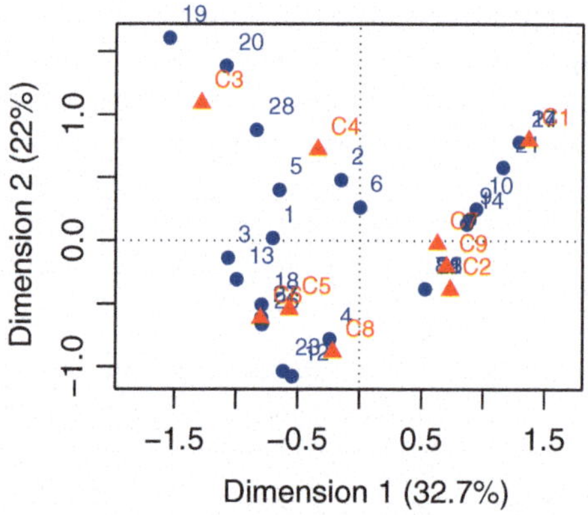

Fig. 2 mfCA map of rodent data

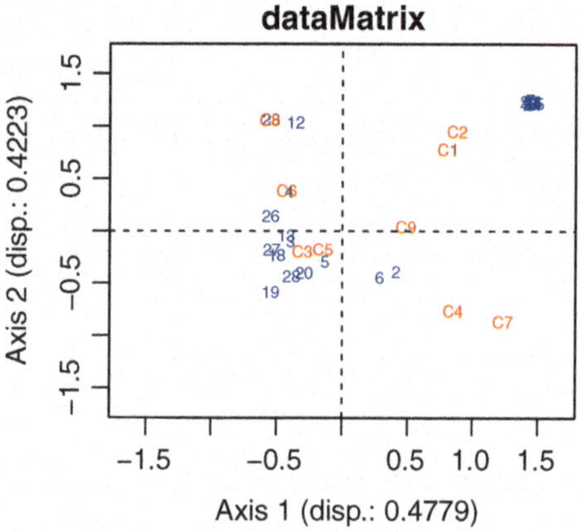

Fig. 3 TCA map of rodent data

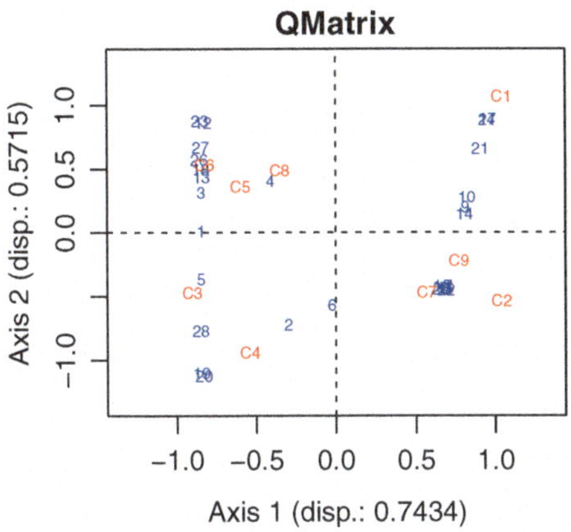

Fig. 4 mfTCA map of rodent data

6 Conclusion

Goodman (1996) introduced marginal-free correspondence analysis where his principal aim was to reconcile Pearson's correlation measure with Yule's association measure for the analysis of contingency tables. We showed that marginal-free correspondence analysis is a particular case of correspondence analysis with pre-specified weights studied in the beginning of the 1980s by Benzécri and his students. mfCA seems to be more robust than the ordinary CA; further applications are needed to see its practical usefulness.

Appendix

The execution of the following R code will produce the four maps displayed in Figs. 1, 2, 3 and 4 for the rodent data set. The code uses the following three packages: the `ipfr` package of Ward and Macfarlane (2020), the `ca` package of Greenacre et al. (2022) and the `TaxicabCA` of Allard and Choulakian (2019). The `ipfr` package applies the iterative proportional fitting algorithm to produce the table (q_{ij}) which has uniform row and column marginals. The `ca` package performs the CA and produces the CA map and the `TaxicabCA` package produces the TCA map.

```
# install packages
install.packages(c("ipfr", "ca", "TaxicabCA"))

library(TaxicabCA)
dataMatrix = as.matrix(rodent)
nRow <- nrow(dataMatrix)
nCol <- ncol(dataMatrix)
ssize <- sum(dataMatrix)
```

```
# Computation of Q matrix of rodent
library(ipfr)
mtx <- dataMatrix
row_targets <- rep(ssize/nRow, nRow)
column_targets <- rep(ssize/nCol, nCol)
QMatrix <- ipu_matrix(mtx, row_targets, column_targets)

rownames(dataMatrix) <- rownames(QMatrix)
                                  <- paste("", 1:nRow, sep = "")
colnames(dataMatrix) <- colnames(QMatrix)
                                 <- paste("C", 1:nCol, sep = "")

# CA map of rodent dataset
library(ca)
plot(ca(dataMatrix))

# mfCA map of rodent
plot(ca(QMatrix))

# TCA map of rodent
tca.Data <- tca(dataMatrix, nAxes = 2, algorithm = "exhaustive")
plot(
  tca.Data,
  axes = c(1, 2),
  labels.rc = c(1, 1),
  col.rc = c("blue", "red"),
  pch.rc = c(10, 10, 0.1, 0.1),
  mass.rc = c(F, F),
  cex.rc = c(0.6, 0.6),
  jitter = c(F, T)
)

# mfTCA map of rodent
tca.DataQ <- tca(QMatrix, nAxes = 2, algorithm = "exhaustive")
plot(
  tca.DataQ,
  axes = c(1, 2),
  labels.rc = c(1, 1),
  col.rc = c("blue", "red"),
  pch.rc = c(10, 10, 0.1, 0.1),
  mass.rc = c(F, F),
  cex.rc = c(0.6, 0.6),
  jitter = c(T, F)
)
```

Acknowledgements Choulakian's research has been supported by NSERC of Canada. The authors thank the editors E.J. Beh, R. Lombardo and J.G. Clavel for their critical reading of the paper and providing constructive criticisms.

References

Allard, J., Choulakian, V.: Package `TaxicabCA` analysis in R. Available online on the CRAN at https://CRAN.R-project.org/package=TaxicabCA. Last accessed 1 May 2023 (2019)

Aitchison, J.: The Statistical Analysis of Compositional Data. Chapman and Hall, London (1986)

Beh, E.J., Lombardo, R.: Correspondence Analysis: Theory. Practice and New Strategies. Wiley, Chichester (2014)

Beh, E.J., Lombardo, R.: Correspondence analysis and the Cressie-Read family of divergence Statistics. Int Stat Rev (in press) (2024)

Benzécri, J.P.: L'Analyse des Données: Vol. 2: L'Analyse des Correspondances. Dunod, Paris (1973)

Benzécri, J.P.: Ajustement d'un tableau à des marges sous l'hypothèse d'absence d'interaction ternaire. Les Cahiers de l'Analyse des Données **8**(2), 227–233 (1983)

Benzécri, J.P.: Sur une généralisation du problème de l'ajustement d'une mesure à des marges. Les Cahiers de l'Analyse des Données **8**(3), 359–370 (1983)

Benzécri, J.P., Bourgarit, C., Madre, J.L.: Problème: ajustement d'un tableau à ses marges d'après la formule de reconstitution. Les Cahiers de l'Analyse des Données **5(l)**, 163 – 172 (1980)

Choulakian, V.: Un exemple d'application de diverses méthodes d'ajustement d'un tableau à des marges imposées. Les Cahiers de l'Analyse des Données **5**(2), 173–176 (1980)

Choulakian, V.: Méthodes et critères pour l'ajustement d'un tableau à des marges imposées. Les Cahiers de l'Analyse des Données **9(l)**, 113 – 117 (1984)

Choulakian, V.: Taxicab correspondence analysis. Psychometrika **71**, 333–345 (2006)

Choulakian, V.: Matrix factorizations based on induced norms. Stat. Optim. Inform. Comput. **4**, 1–14 (2016)

Choulakian, V.: Taxicab correspondence analysis of sparse contingency tables. Italian J. Appl. Stat. **29**(2–3), 153–179 (2017)

Choulakian, V.: Quantification of intrinsic quality of a principal dimension in correspondence analysis and taxicab correspondence analysis. Available online at https://arxiv.org/abs/2108.10685. Last accessed 1 May 2023 (2021)

Choulakian, V., Allard, J., Mahdi, S.: Taxicab correspondence analysis and taxicab logratio analysis: a comparison on contingency tables and compositional data. Aust. J. Stat. **52**, 39–70 (2023)

Cuadras, C.M., Cuadras, D., Greenacre, M.: A comparison of different methods for representing categorical data. Commun. Stat. Simul. Comput. **35**(2), 447–459 (2006)

Cuadras, C.M., Cuadras, D.: A unified approach for the multivariate analysis of contingency tables. Open J. Stat. **5**, 223–232 (2015)

Egozcue, J.J., Pawlowsky-Glahn, V.: Changing the reference measure in the simplex and its weighting effects. Aust. J. Stat. **45**(4), 25–44 (2016)

Egozcue, J.J., Pawlowsky-Glahn, V., Templ, M., Hron, K.: Independence in contingency tables using simplicial geometry. Commun. Stat. Theor. Methods **44**, 3978–3996 (2015)

Goodman, L.A.: Simple models for the analysis of association in cross-classifications having ordered categories. J. Am. Stat. Assoc. **74**, 537–552 (1979)

Goodman, L.A.: Association models and the bivariate normal for contingency tables with ordered categories. Biometrika **68**, 347–355 (1981)

Goodman, L.A.: Association models and canonical correlation in the analysis of cross-classifications having ordered categories. J. Am. Stat. Assoc. **76**, 320–334 (1981)

Goodman, L.A.: Measures, models, and graphical displays in the analysis of cross-classified data. J. Am. Stat. Assoc. **86**, 1085–1111 (1991)

Goodman, L.A.: A single general method for the analysis of cross-classified data: reconciliation and synthesis of some methods of Pearson, Yule, and Fisher, and also some methods of correspondence analysis and association analysis. J. Am. Stat. Assoc. **91**, 408–428 (1996)

Greenacre, M.: Power transformations in correspondence analysis. Comput. Stat. Data Anal. **53**, 3107–3116 (2009)

Greenacre, M.: Log-ratio analysis is a limiting case of correspondence analysis. Math. Geosci. **42**, 129–134 (2010)

Greenacre, M., Nenadic, O., Friendly, M.: Package `ca` in R. Available online on the CRAN at https://CRAN.R-project.org/package=ca. Last accessed 1 May 2023 (2022)

Hotelling, H.: Relations between two sets of variables. Biometrika **28**, 321–377 (1936)

Kazmierczak, J.B.: Analyse logarithmique?: deux exemples d'application. Revue de Statistique Appliquée **33**, 13–24 (1985)

Kazmierczak, J.B.: Sur l'usage d'un principe d'invariance pour aider au choix d'une métrique. Statistique et Analyse des Données **12**(3), 37–57 (1987)

Lancaster, H.O.: The structure of bivariate distributions. Ann. Math. Stat. **29**, 719–736 (1958)

Madre, J.L.: Méthodes d'ajustement d'un tableau à des marges. Les Cahiers de l'Analyse des Données **5**(1), 87–99 (1980)

Mosteller, F.: Association and estimation in contingency tables. J. Am. Stat. Assoc. **63**, 1–28 (1968)
Moussaoui, A.E.: Sur la reconstruction approchée d'un tableau de correspondance à partir du tableau cumulé par blocs suivant deux partitions des ensembles I et. J. Les Cahiers de l'Analyse des Données **12**(3), 365–370 (1987)
Nishisato, S.: Analysis of Categorical Data: Dual Scaling and its Applications. University of Toronto Press, Toronto (1980)
Nishisato, S.: Forced classification: a simple application of a quantification method. Psychometrika **49**, 25–36 (1984)
Quinn, G., Keough, M.: Experimental Design and Data Analysis for Biologists. Cambridge University Press, Cambridge, UK (2002)
Tenenhaus, M., Augendre, H.: Analyse factorielle inter-batteries de Tucker et analyse canonique aux moindres carres partiels. In: Recueil des Résumés des Communications des 28'eme Journées de Statistique, 693–697 (1996)
Tucker, L.R.: An inter-battery method of factor analysis. Psychometrika **23**, 111–136 (1958)
Ward, K., Macfarlane, G.: Package `ipfr` in R. Available online on the CRAN at https://CRAN.R-project.org/package=ipfr. Last accessed 1 May 2023 (2020)
Yule, G.U.: On the methods of measuring association between two attributes. J. Royal Stat. Soc. **75**, 579–642 (1912)

Group and Time Differences in Repeatedly Measured Binary Symptom Indicators: Matched Correspondence Analysis

Se-Kang Kim

1 Introduction

Typically, when a researcher, such as an applied psychologist or statistician, compares the mean score differences of an outcome in repeated measurements, the researcher compares the mean score differences of an outcome. Assuming the measurement scale is interval or ratio, the researcher uses paired t-test or repeated-measures analysis of variance (ANOVA) to test the mean differences between or among groups. When measurements are not continuous but dichotomously scored (for instance, "0" = symptom absent and "1" = symptom present), it is not possible to use a paired t-test or a repeated-measures ANOVA. The chi-squared test can be repeated multiple times to test the group difference for each binary variable (e.g. Binomial test). However, if the binary variables are significantly correlated, the chi-squared test results for each binary variable will be biased if they are reported separately.

Furthermore, the Pearson product moment correlation cannot be used to estimate the correlation between binary variables. Rather, tetrachoric correlation is frequently estimated using binary variables, with each binary variable's underlying structure assumed to be normal (Vaswani 1950). Pearson's correlation is based on the assumption of bivariate normality, which can be easily tested through the creation of a histogram or Q-Q plot for each variable. However, no research has investigated how to test the normality assumption for binary variables' underlying structure (Demirtas 2016). Consequently, it is impossible to determine whether the bivariate normality assumption for the tetrachoric correlation is met, and the results with unwarranted normality may be biased. In addition, estimating relationships between interrelated

S.-K. Kim (✉)
Psychology Division, Department of Pediatrics, Baylor College of Medicine, Texas Children's Hospital, Houston, TX, USA
e-mail: Se-Kang.Kim@bcm.edu

E. J. Beh et al. (eds.), *Analysis of Categorical Data from Historical Perspectives*, Behaviormetrics: Quantitative Approaches to Human Behavior 17,
https://doi.org/10.1007/978-981-99-5329-5_21

binary variables becomes more challenging when these binary variables are repeatedly measured at two different time points (such as admission and discharge in this study), given that these binary variables are interrelated at each time point and over time.

In this paper, a novel application of a variant of the conventional correspondence analysis technique, known as correspondence analysis (CA) of matched matrices (henceforth referred to as matched CA), is presented by taking into account the statistical complications that arise when examining the group and time differences and their relationships in repeatedly measured binary variables. The matched matrices are constructed from two independent groups that share the same row and column properties in their two-way contingency tables. In contrast to other CA variants, the matched CA identifies two types of dimensions: sum dimensions and difference dimensions. The sum dimensions represent aggregated effects of row/column properties included in matched groups, while the difference dimensions represent differences in row/column properties in matched groups, as introduced by Greenacre (2003) for analysis of independent group differences (e.g. gender).

This study, unlike Greenacre's original matched CA paradigm, incorporates both between-and within-group designs, enabling researchers to examine not only potential differences between two independent groups, but also changes in the members of each group over time. The utility of matched CA is demonstrated using patients with anorexia and bulimia who respond to binary indicators of psychiatric symptoms at admission and discharge. A group of anorexia patients and a group of bulimia patients are between-group factors, whereas a pre-test condition (admission) and a post-test condition are within-group factors (discharge). This mixed design aims to examine not only the impact of different eating disorders (anorexia vs. bulimia) on psychiatric symptom indicators, but also the impact of time (admission vs. discharge) on the same symptom indicators. In fact, the time difference effect describes the efficacy of treatment upon discharge. This study employs a quasi-experimental between-group design because patients are not randomly assigned to either type of eating disorder.

When constructing a matched contingency table in a mixed design, the ages are included as a row categorical variable (e.g. ages 12–14, 15,..., 26–39). In this study, age is used as a covariate for symptom indicators because it is known to be a significant covariate for clinical diagnosis; see, for example, Kim (2020) and Kim et al. (2021b). Thus, an age groups $\times$ symptom indicators contingency table is created for each group (anorexia or bulimia) and then used to generate a matched table. A table like this is created by horizontally stacking anorexia next to bulimia, followed by horizontally stacking bulimia next to anorexia. From the concatenated table, the group- and time-difference dimensions are identified, and then their statistical stability is evaluated. Since group differences (anorexia versus bulimia) and time differences (discharge versus admission) are scaled to estimate the dimensions, their coordinates can be interpreted as group and time differences in the binary symptom indicators if the dimensions are statistically stable; see Kim (2020) for details.

In addition, a biplot with only the statistically stable dimensions is constructed to visually inspect the association between categories (Gower and Hand 1996; Gower et al. 2011), and the visually inspected association is estimated with correlation coefficients to enhance the interpretation. When conducting differential diagnosis for patients, the magnitudes of correlation coefficients, such as between their age groups and specific symptom indicators, would be clinically useful; see, for example, Kim and Annunziato (2020). In cross-sectional studies, the algorithm for estimating Pearson's correlation has been used to quantify intra-and inter-categorical variable association; see, for example, Kim et al. (2022), Kim et al. (2020, 2021a), Kim and Grochowalski (2019), and Kim and Frisby (2019). This study applies this correlation estimation algorithm to the analysis of repeated binary indicator measurements.

2 Method

For analysis of interrelated binary indicators using CA, multiple correspondence analysis (MCA) may be considered. MCA is analogous to principal component analysis (PCA) of indicator variables, which is designed to identify latent dimensions of binary data (Beh and Lombardo 2014, 2021; Le Roux and Rouanet 2010; Lebart et al. 1984), but not to estimate dimensions of either between-group or within-group differences. As a result, MCA is incompatible with the objectives of this study. Among the numerous CA variants (Beh and Lombardo 2021), matched CA is one that can estimate the between-group and within-group differences investigated in this study.

2.1 Example Data: Binary Psychiatric Symptom Indicators

Sample. This study included female patients with anorexia ($n = 1177$) and bulimia ($n = 752$) who met Diagnostic and Statistical Manual of Mental Disorders (DSM-IV text revision; APA, 2000) criteria for a primary eating disorder diagnosis at the Remuda Ranch Programmes for Eating Disorders in Wickenburg, Arizona. Anorexia and bulimia are both eating disorders with symptoms similar to distorted body image. They are distinguished, however, by distinct food-related behaviours. People suffering from anorexia, for example, severely restrict their food intake in order to lose weight. Whereas people who have bulimia eat an excessive amount of food in a short period of time, then purge or use other methods to prevent weight gain. The age range was 12–39 years for both anorexia ($M = 19.57$ years, SD $= 5.76$) and bulimia ($M = 22.26$ years, SD $= 6.07$). The sample was 93.6% Caucasian, 2.7% Mixed/Unknown, 2.1% Hispanic, 0.9% Asian, 0.7% African American, and 0.2% Native American.

Six binary symptom indicators. All anorexia and bulimia patients were assessed according to the following six psychiatric symptoms: (1) Major depressive disorder (hereafter denoted as Ma), (2) Depression not otherwise specified (De), (3) Obsessive compulsive disorder (Ob), (4) Generalised anxiety disorder (Ge), (5) Anxiety disorder not otherwise specified (An), and (6) Social phobia (So) are originally recorded with (0, 1) indicators, where "0" represents symptom absence and "1" represents symptom presence. However, to avoid any redundancy, only the symptom presence indicators are analysed. In correspondence analysis, an absence indicator for a given symptom represents the opposite of that symptom's presence indicator.

Discretisation of age. Age is discretised into eight groups in order to include it as a covariate in a contingency table with the symptom indicators. Anorexia patients were classified into the following age groups: 12–14 ($n = 156$), 15 ($n = 147$), 16 ($n = 143$), 17 ($n = 127$), 18–19 ($n = 173$), 20–21 ($n = 127$), 22–25 ($n = 129$), and 26–39 ($n = 175$). Bulimia patients were divided into the following age groups: 12–14 ($n = 25$), 15 ($n = 41$), 16 ($n = 51$), 17 ($n = 57$), 18–19 ($n = 139$), 20–21 ($n = 108$), 23–25 ($n = 124$), and 26–39 ($n = 207$). Thus, at admission and discharge, each group (anorexia or bulimia) would have an 8 (age groups) 6 (symptom presence indicators) contingency table.

2.2 Matched CA of Two Block Circulant Matrices

Generating concatenated tables for group and time differences. For each time point, an 8×6 table is generated for anorexia patients and an 8×6 table is generated for bulimia patients. $\mathbf{A}_{ad}$ and $\mathbf{B}_{ad}$ shall represent the admission tables for patients with anorexia and bulimia, respectively. For matched CA, $\mathbf{A}_{ad}$ is stacked on top of $\mathbf{B}_{ad}$ and then concatenated to create a 16×12 concatenated matrix, $\mathbf{C}_{ad}$, for the admission group, where "*ad*" refers to admission:

$$\mathbf{C}_{ad} = \begin{bmatrix} \mathbf{A}_{ad} & \mathbf{B}_{ad} \\ \mathbf{B}_{ad} & \mathbf{A}_{ad} \end{bmatrix}. \tag{1a}$$

Similarly, a second 16×12 concatenated matrix, $\mathbf{C}_{di}$, is created, where "*di*" refers to discharge:

$$\mathbf{C}_{di} = \begin{bmatrix} \mathbf{A}_{di} & \mathbf{B}_{di} \\ \mathbf{B}_{di} & \mathbf{A}_{di} \end{bmatrix}. \tag{1b}$$

Combining these two concatenated matrices yields a 32×24 super concatenated matrix:

$$\mathbf{C} = \begin{bmatrix} \mathbf{C}_{di} & \mathbf{C}_{ad} \\ \mathbf{C}_{ad} & \mathbf{C}_{di} \end{bmatrix} = \begin{bmatrix} \mathbf{A}_{di} & \mathbf{B}_{di} & \mathbf{A}_{ad} & \mathbf{B}_{ad} \\ \mathbf{B}_{di} & \mathbf{A}_{di} & \mathbf{B}_{ad} & \mathbf{A}_{ad} \\ \mathbf{A}_{ad} & \mathbf{B}_{ad} & \mathbf{A}_{di} & \mathbf{B}_{di} \\ \mathbf{B}_{ad} & \mathbf{A}_{ad} & \mathbf{B}_{di} & \mathbf{A}_{di} \end{bmatrix}. \tag{2}$$

This new 32×24 concatenated matrix $\mathbf{C}$ repeats the rows and columns twice to account for group- and time-difference effects, as well as their interaction. Using matched CA, the sum (**A** + **B**) and group-difference (**A** – **B**) dimensions will be identified from **C** for anorexia and bulimia patients. Furthermore, because **A** and **B** are nested within time (discharge and admission), the time-difference (discharge subtracted by admission) dimensions will also be identified alongside group × time interaction dimensions. Since the primary goal of this study is to examine the differences in symptoms between anorexia and bulimia patients, the sum (**A** + **B**) dimensions will not be investigated further.

Standardising C for singular value decomposition (SVD). **C** is a super 32 × 24 concatenated matrix composed of two sub-concatenated matrices: $\mathbf{C}_{di}$ (at discharge), an $\mathbf{A}_{di}\mathbf{B}_{di}\mathbf{B}_{di}\mathbf{A}_{di}$ circulant matrix and $\mathbf{C}_{ad}$ (at admission), an $\mathbf{A}_{ad}\mathbf{B}_{ad}\mathbf{B}_{ad}\mathbf{A}_{ad}$ circulant matrix. These two block circulant matrices, which are nested with two time points (discharge and admission), are designed to represent the group differences (anorexia vs. bulimia). When estimating the group differences, the time variants are aggregated. When estimating the time differences, the group differences are aggregated. When applying an SVD to **C**, it has to be standardised because the rows and columns must be weighted differently; the different weights are relative proportions of their respective margins or marginal proportions.

The standardisation procedure. Several steps are involved in the standardisation process. The matrix **C** is first converted to the correspondence matrix, $\mathbf{P} = (1/n)\mathbf{C}$, where $n = \mathbf{1}^{\mathrm{T}}\mathbf{C}\mathbf{1}$ is a grand total of **C**. The ith row and jth column marginal proportions are defined by $r_i = \sum_{j=1}^{J} p_{ij}$ and $c_j = \sum_{i=1}^{I} p_{ij}$, respectively, so that $\mathbf{r} = \mathbf{P}\mathbf{1}$ and $\mathbf{c} = \mathbf{P}^{\mathrm{T}}\mathbf{1}$. Since **C** is a circulant matrix, the vector of the eight-age group marginal proportions (as row weights) and the vector of the six psychological symptom indicator marginal proportions (as column weights) are assigned to each of four blocks, $\begin{bmatrix} \mathbf{C}_{di} & \mathbf{C}_{ad} \\ \mathbf{C}_{ad} & \mathbf{C}_{di} \end{bmatrix}$ in Equation (2). The diagonal matrices of the row and column marginal proportions are also defined as $\mathbf{D}_r = \mathrm{diag}(\mathbf{r})$ and $\mathbf{D}_c = \mathrm{diag}(\mathbf{c})$, respectively. The subsequent definitions and results are given in terms of these relative quantities $\mathbf{P} = \{p_{ij}\}$, $\mathbf{r} = \{r_i\}$, and $\mathbf{c} = \{c_j\}$, whose elements add up to 1 in each case. The standardised matrix **C**, $\mathbf{Z}_{\mathrm{C}}$, is as follows:

$$\mathbf{Z}_{\mathrm{C}} = \mathbf{D}_r^{-1/2}(\mathbf{P} - \mathbf{r}\mathbf{c}^{\mathrm{T}})\mathbf{D}_c^{-1/2}. \tag{3}$$

When the difference of these matrices, $(\mathbf{P} - \mathbf{r}\mathbf{c}^{\mathrm{T}})$, is close to zero, there is little association between rows and columns.

Standard and principal coordinates. The row and column standard coordinates are defined as $\mathbf{\Phi} = \{\phi_{ij}\} = \mathbf{D}_r^{-1/2}\mathbf{U}$ and $\mathbf{\Gamma} = \{\gamma_{ij}\} = \mathbf{D}_c^{-1/2}\mathbf{V}$, respectively, from the SVD of the $\mathbf{Z}_{\mathrm{C}}$ matrix: $\mathbf{Z}_{\mathrm{C}} = \mathbf{U}\Sigma\mathbf{V}^{\mathrm{T}}$. **U** and **V** (where $\mathbf{U}^{\mathrm{T}}\mathbf{U} = \mathbf{I}$ and $\mathbf{V}^{\mathrm{T}}\mathbf{V} = \mathbf{I}$) are the left and right singular vector matrices of (**A** – **B**) included in the $\mathbf{Z}_{\mathrm{C}}$. The principal coordinates are numerically equal to the standard coordinates multiplied by the singular values from the diagonal matrix of the singular values arranged in descending order. The row categories in a row isometric biplot (Beh and Lombardo 2014; Greenacre 2010) are represented by principal coordinates defined by $\mathbf{F} = \Phi\Sigma = \{f_{ij}\}$, where $\mathbf{\Sigma} = \mathrm{diag}(\sigma)$ which is the diagonal matrix of the singular values of $\mathbf{Z}_C$. The column categories are represented by standard coordinates, $\boldsymbol{\Gamma}$, which are scaled to have a weighed mean of 0 and variance of 1 in each dimension so that $\Gamma \mathbf{D}_c = 0$ and $\Gamma \mathbf{D}_c \boldsymbol{\Gamma}^{\mathrm{T}} = \mathbf{I}$, and references can be made to each row that lies at the centre of gravity of the associated column vectors, which are represented by the standard coordinates.

Dimensions of within-group differences. Matched CA is developed to optimise the scaling of any between-group (e.g. gender) differences in the categorical variables (e.g. job involvement status); see, for example, Greenacre (2017). However, in this study, time is included as a within-group factor in addition to a between-group factor due the repeated measurement of binary response outcomes at admission and discharge. Thus, within-group dimensions are also identified. Moreover, because both between- and within-group factors are involved, their interaction dimensions can be determined.

2.2.1 Test Statistical Stability of Dimensions

Generating random tables. To test the statistical stability of (group or time) difference dimensions and interaction dimensions, 10,000 random contingency tables of the same size as the original contingency table **C** are generated using a series of permutations of **C**. A contingency table is a cross-tabulation matrix comprising rows and columns of categorical variables. Follow these steps to generate random contingency tables: (a) generate new tables containing row and column categorical variables from the original contingency table; (b) independently permute either variable of the original contingency table; (c) cross-tabulate the permuted data to generate a random contingency table; (d) repeat steps (a) to (c) 10,000 times to create 10,000 random contingency tables; (e) perform a matched CA of each random contingency table to estimate dimensional eigenvalues (also known as principal inertia); and (f) compute a Monte-Carlo *p*-value for a given dimension by counting the random eigenvalues greater than the observed eigenvalue. For example, if 490 randomly simulated eigenvalues (out of 10,000) are larger than an observed eigenvalue (for each dimension) estimated from the original contingency table, the empirical *p*-value for the eigenvalue of the corresponding dimension would be 0.049 which is less than a predetermined significance level alpha at $\alpha = 0.05$, and the dimension is statistically stable at $\alpha = 0.05$; however, if there are more than 500 random eigenvalues greater than an observed eigenvalue for any dimension, the dimension is considered statistically unstable and its interpretation is excluded.

Interpreting coordinates as differences in binary indicators. Because group and time differences of the binary indicators of psychiatric symptoms are scaled with matched CA, their dimensional coordinates (e.g. standard coordinates) can be interpreted as group and time differences in binary indicators of psychiatric symptoms. However, if dimensions are not statistically stable, interpreting their coordinates as group or time differences becomes problematic. Therefore, only statistically stable dimensions' coordinates are interpreted as group and time differences.

Row isometric biplots with statistically stable dimensions. It is possible to create a (row isometric) biplot with any pair of dimensions (e.g. Kim and Annunziato 2020), but its interpretation would be suspect if the dimensions were not statistically stable (Kim and Grochowalski 2019). Therefore, only statistically stable dimensions will be utilised when creating a biplot. The optimally scaled coordinates of a biplot maximise the association between rows and columns (Gabriel 1971; Gabriel and Odoroff 1990; Gower and Hand 1996; Gower et al. 2011; Greenacre 2010; Nishisato 1994, 2007).

To improve the legibility of a row isometric biplot. The conventional row isometric biplot is formed by jointly displaying the row principal coordinates and the column standard coordinates. However, the row principal coordinates are weighted with singular values, and their values are much too small in comparison to the column standard coordinates for visual evaluation in the same biplot. Thus, the row isometric biplot employed in this study is generated using the row age groups' principal coordinates of, but the column clinical symptoms' standard coordinates $\{\gamma_{jk}\}$ are multiplied by $\left\{c_j^{1/2}\right\}$ to bring the column coordinate scale closer to the row coordinate scale (Greenacre 2017, pp. 101–102). The term "improved biplot" will henceforth be used to describe this particular type of row isometric biplot. This is done without affecting the statistical properties of the biplot in order to make the row principal coordinates more readable visually (relative to the column standard coordinates); for a more detailed explanation, see the following section.

Re-expressing Pearson ratios for the improved row isometric biplot. The Pearson ratios (Beh 2004) for a row isometric biplot with the first two dimensions can be expressed without an error term such as: $p_{ij}/(r_j c_j) - 1 = \sum_{k=1}^{2} f_{ik}\gamma_{jk}$. This equation can be rewritten in terms of the jth element of the row profile as: $\left(p_{ij}/r_i - c_j\right)/c_j = \sum_{k=1}^{2} f_{ik}\gamma_{jk}$. In fact, $\sum_{j=1}^{J}\left(p_{ij}/r_i - c_j\right)$ represents how much row ith. Profile, which is $\mathbf{p}_i = (p_{i1}, \ldots, p_{iJ})$, deviates from the average row profile which is the vector of column marginal proportions, $\mathbf{c}$. In the biplot, each column vertex point defines the line onto which row profile is projected, and $\left(p_{ij}/r_i - c_j\right)/c_j = \left(p_{ij}/r_i - c_j\right)/\left(c_j^{1/2}c_j^{1/2}\right) = \sum_{k=1}^{2} f_{ik}\gamma_{jk}$ can be re-expressed as: $\left(p_{ij}/r_i - c_j\right)/c_j^{1/2} = \sum_{k=1}^{2} f_{ik}\left(c_j^{1/2}\gamma_{jk}\right)$, where $\left(c_j^{1/2}\gamma_{jk}\right)$ is used as a column coordinate for the improved biplot but f_{ik} remains the i's row principal coordinate along dimension k (Greenacre 2017, p. 102).

Visual inspection of category relationships in an improved biplot. To visually inspect relationships between row and column categories in an improved (row isometric) biplot, the column vertex points are typically expressed with projections from the origin (0, 0) and pass through their standard coordinates, which are depicted as lines, but the row profiles are depicted as dot points at the locations specified by their principal coordinates. If imaginary lines from the origin are drawn to the rows' principal coordinates, the magnitude and direction of the association can be approximated by examining the angle between row and column coordinates; see, for example, Beh and Lombardo (2014). Trigonometrically:

- if the angle between f_{ik} and $c_j^{1/2}\gamma_{jk}$ is close to zero, the correlation is close to + 1;
- if the angle is close to 90°, the correlation is close to zero;
- if the angle is close to 180°, the correlation is close to −1;
- if the angle between 0° and 90°, the correlation is between 0 and + 1;
- if the angle is between 90° and 180°, the correlation is between 0 and −1.

The angles between row coordinates and the angles between column coordinates can be examined as well.

Estimate the category relationship using correlation. The row and column categories are projected onto this biplot using the formulas $\mathbf{f}_i = (f_{i1} f_{i2})$ and $\mathbf{f}_{i'} = (f_{i'1} f_{i'2})$ for the principal coordinates of row category i and i', respectively. In Euclidean geometry, the scalar product of two vectors $\mathbf{f}_i$ and $\mathbf{f}_{i'}$ is denoted by $\mathbf{f}_i^{\mathrm{T}}\mathbf{f}_{i'}$, which is equal to the product of the lengths of the two vectors multiplied by the cosine of the angle between them, such as $\mathbf{f}_i^{\mathrm{T}}\mathbf{f}_{i'} = \|\mathbf{f}_i\| \cdot \|\mathbf{f}_{i'}\| \cdot \cos\theta_{ii'}$, where $\|\mathbf{f}_i\|$ denotes the length of the vector $\mathbf{f}_i$ and accordingly, $\cos\theta_{ii'} = \mathbf{f}_i^{\mathrm{T}}\mathbf{f}_{i'}/\|\mathbf{f}_i\| \cdot \|\mathbf{f}_{i'}\|$, where $\cos\theta_{ii'}$ is the correlation estimate between rows i and i'. Similarly, the correlation between columns j and j' is estimated in the given plane using the improved standard coordinates of columns j and j', $\tilde{\gamma}_j = c_j^{1/2}(\gamma_{j1}\gamma_{j2})$ and $\tilde{\gamma}_{j'} = c_{j'}^{1/2}(\gamma_{j'1}\gamma_{j'2})$ so that $\cos\theta_{jj'} = \tilde{\gamma}_j^{\mathrm{T}}\tilde{\gamma}_{j'}/\|\tilde{\gamma}_j\| \cdot \|\tilde{\gamma}_{j'}\|$. Likewise, the correlation between row i and column j can be estimated. However, because a biplot in this case is row isometric, the correlation between the row i principal coordinate and the column j improved (standard) coordinates is calculated using the formula $\cos\theta_{ij} = \mathbf{f}_i^{\mathrm{T}}\tilde{\gamma}_j/\|\mathbf{f}_i\| \cdot \|\tilde{\gamma}_j\|$ (Kim and Grochowalski 2019).

Dimensionality. With 32 rows and 24 columns in the super concatenated matrix, **C**, theoretically, $\min(32-1, 24-1) = 23$ dimensions provide an optimal display of the association. The 23 dimensions can be divided into four distinct sets: the first is for the group-difference dimensions, the second for the time-difference dimensions, the third for the group × time interaction dimensions, and the fourth for the sum (or average) dimensions.

How to identify group, time, interaction, and sum dimensions. Matched CA partitions the total inertia of the 32 × 24 matrix into four different types of eigenvalues: group, time, group × time interaction, and sum (or average) dimensions. To identify each type of dimension, the following sign patterns must be applied to the pattern of dimensional coordinates.

To help understand these signs in Table 1, the four columns (1, 2, 3, 4), are labelled as follows: "1" refers to the fact that if the sign of each set of coordinates alternates (i.e. + – + –) along a dimension, this dimension is defined as a group-difference dimension; "2" refers to the fact that if the signs of the first two sets of coordinates are the same, but the signs of the other two sets are opposite to the first two sets (i.e. + + – –), this dimension is defined as a time-difference dimension; "3" refers to that if the sign of the first set of coordinates is the same as the sign of the last set of coordinates, but the signs of the two sets are opposite to those of the first and the last sets of the coordinates (i.e. + – – +) along a dimension, then this dimension is defined as a group × time interaction dimension; and "4" refers to the fact that

Table 1 Dimensional coordinate patterns

		Dimension			
		1	2	3	4
Time 1	A	+	+	+	+
	B	–	+	–	+
Time 2	A	+	–	–	+
	B	–	–	+	+
		Group (G)	Time (T)	T × G	Sum

if the signs of the four sets of the coordinates are the same (i.e. + + + +) along a given dimension, then this dimension is defined as a sum (or average) dimension. As previously stated, the sum (or average) dimensions that aggregate group and time differences will not be investigated further because they are irrelevant to the purpose of this study.

3 Results

3.1 Initial Matched CA Results

Identifying statistically stable dimensions. Since the maximum dimensionality is 23, their observed eigenvalues are subjected to a permutation test to determine the statistical stability of their respective dimensions. The empirical *p*-values for the eigenvalues of the first, second, third, fourth, fifth, and up to the twenty-third dimensions are as follows: $p < 0.0001$, $p < 0.0001$, $p < 0.0001$, $p < 0.0011$, $p = 0.9067$, ..., $p = 0.9772$. The permutation test results show that the first four eigenvalues of the dimensions are statistically significant at $\alpha = 0.01$, and they will be further investigated.

Identifying different types of dimensions. To identify the group-difference, time-difference, and its interaction dimensions, the improved standard coordinates, $\left\{c_j^{1/2}\gamma_{jk}\right\}$, of the first two dimensions are examined, as they account for largest amount of the total variance.

- Among the first four statistically significant dimensions, the first and third are identified as the group-difference (anorexia—bulimia) dimensions and account for 81.1% of the total variance. These two group-difference dimensions are used to construct a group-difference biplot.
- The fourth and ninth dimensions are time-difference (admission—discharge) dimensions that account for 4% of the total variance. The eigenvalue of the fourth dimension is statistically significant at $\alpha = 0.01$ ($p < 0.0001$), but the ninth dimension is not significant ($p = 0.9997$) and will not be considered further. A

time-difference biplot cannot be constructed due to the requirement for two stable dimensions.

- The sixth and twelfth dimensions are group × time interaction dimensions, but neither is statistically significant; therefore, they will not be considered further.

3.2 Evidence of Group and Time Differences

The coordinates of group-difference dimensions. Since both group-difference dimensions, dimensions 1 and 3, are statistically stable, their coordinates represent group differences (anorexia—bulimia) in the six psychiatric symptom indicators. The following are the group differences for dimension 1: 0.2890 for {De}, 0.2745 for {An}, 0.0929 for {So}, 0.1524 for {Ob}, 0.1409 for {Ge}, and 0.2004 for {Ma}. Positive values indicate that all six psychiatric symptoms are more severe in patients with anorexia than in patients with bulimia, while negative values indicate symptoms are less severe. This is because CA in this study only scales the psychiatric symptom present indicators. The group differences for dimension 3 are as follows: 0.2303 for {De}, 0.1763 for {An}, −0.0203 for {So}, −0.1148 for {Ob}, −0.1927 for {Ge}, and −0.3385 for {Ma}. The negative values for {So, Ob, Ge, Ma} means that these four symptoms are less severe in anorexic patients than in bulimic patients.

The coordinates of a time-difference dimension. Since only dimension 4 is statistically stable, its coordinates represent time differences (discharge—admission) in the six psychiatric symptom indicators. The following are the time differences: −0.2348 for {De}, 0.0335 for {An}, 0.2459 for {So}, 0.2744 for {Ob}, 0.2261 for {Ge}, and 0.0812 for {Ma}. Similar to the interpretation of the group-difference coordinates, the positive values represent symptom deterioration or the absence of a treatment effect, whereas the negative values indicate symptom improvement or treatment efficacy. The treatment efficacy is only observed for {De} because it is the only symptom with a negative time difference.

The group-difference biplot. In Fig. 1, a pair of statistically stable group-difference dimensions are used to construct an improved biplot, and the association between age groups and symptom indicators is visually examined. Understanding such a relationship can aid in diagnosing which age group is associated with which symptom indicator. Figure 1 shows the group-difference points with bulimia points anchored at the origin: Therefore, the greater the difference in symptoms between anorexia and bulimia patients, the longer the projection. Note that for Fig. 1: $a1$ = ages 12–14; $a2$ = age 15; and $a3$ = age 16; $a4$ = age 17; $a5$ = ages 18–19; $a6$ = ages 20–21; $a7$ = ages 22–25; and $a8$ = ages 26–39. De = Depression not otherwise specified; An = Anxiety disorder not otherwise specified; So = Social phobia; Ob = Obsessive compulsive disorder; Ge = Generalised anxiety disorder; and Ma = Major depressive disorder.

Since this is a group-difference biplot, the time variant (admission—discharge) is aggregated and therefore does not appear in the biplot. On the right side of Fig. 1 are the six column projections, {De, An, So, Ob, Ge, Ma}. This demonstrates the

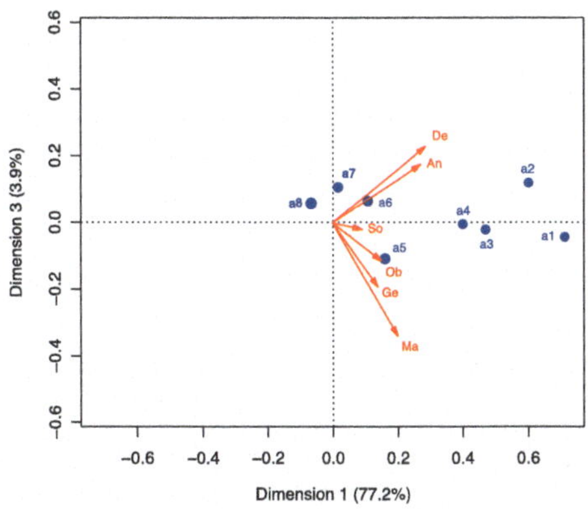

Fig. 1 The group-difference (Anorexia—Bulimia) biplot

apparent group differences between patients with anorexia and bulimia on dimensions 1 and 3. Refer to the section titled "The coordinates of group-difference dimensions"; these group differences account for 81% of the total variance. However, dimension 1 explains the majority of the group differences, as it accounts for 95% ($= 77.2/81.1 \times 100$) of the group-difference variance in Fig. 1.

Visual inspection of association in a group-difference biplot. Figure 1 illustrates the group differences in the CA-scaled indicators of symptom presence. The distances of the symptom points from the origin, as well as the symptom points anchored at the origin for bulimia patients, represent residuals between the responses of anorexia and bulimia patients (anorexia—bulimia). Although there are age-related group differences, they are irrelevant to the current investigation because age is considered a covariable with symptom indicators. The clinical utility of studying the association between age groups and symptom indicators is of interest. All category interactions in Fig. 1 should be interpreted in terms of patients with anorexia. For instance, in Fig. 1, the age group of 20–21 years ($a6$) is adjacent to the symptoms De and An, indicating that anorexia patients of this age are positively associated with the De and An symptoms. Also observed is a strong and positive association between anorexia patients aged 18–19 ($a5$) and the three symptoms, Ob, Ge, and Ma. Also, there is a strong positive association between anorexia patients aged 12–17 ($a1 - a4$) and the symptom (So), because the angles between these lines and the symptom indicators are close to zero if imaginary lines are drawn from the age groups to the origin. However, if this positive association is interpreted in terms of bulimia patients, it will be a negative association, which is consistent with the analysis of an **ABBA** circulant matrix (where **A** = anorexia patients and **B** = bulimia patients). Similarly, anorexia patients aged 22–25 ($a7$) and 26–39 ($a8$) are negatively associated with the symptom indicators, So, Ob, Ge, and Ma, due to the obtuse angles between these categories. For bulimia patients, however, these same age groups of $a7$ and $a8$ would be positively associated with So, Ob, Ge, and Ma. In the following section, correlation coefficients will be estimated to verify these visually determined results.

Correlation analysis to determine visually inspected association structure. The visual inspection of the category configurations in the group-difference biplot of Fig. 1 permits the assessment of the nature of association between age groups and symptom indicators. However, the visual inspection does not provide any numerical summaries of the association's structure. This section supplements the visually inspected association structure with correlation coefficients. Table 2 shows a summary of these coefficients, where the correlation coefficients equal to or larger than r = 0.71 were highlighted in bold and regarded as significant because they represented at least 50% ($0.71^2 = 0.50$) of the shared variance between categories.

Correlation between age groups and symptom indicators. Consistent with visual inspection, the angles between *a*6 and {An, De} in Fig. 1 are virtually zero, so their respective correlations are $r = 1.00$ and $r = 0.99$. Likewise, the angles between a5 and {Ob, Ge, So} are either zero or close to zero, and their respective correlations are $r = 1.00$, $r = 0.94$, and $r = 0.93$. In addition, the correlation between {So} and [*a*1, *a*2, *a*3, *a*4] ranges from $r = 0.93$ to $r = 0.99$, as shown in the "So" column of Table 2, indicating that anorexia patients aged 12–17 are strongly associated with social phobia. Remember that these correlations should be interpreted in terms of anorexia patients, but for bulimia patients, all positive correlations should be interpreted as negative correlations.

Table 2 The Pearson correlation coefficients between age groups and symptom indicators in the group difference biplot

		Symptom indicator				
	Ma	De	Ob	Ge	An	So
*a*1 (ages 12–14)	0.56	**0.74**	**0.83**	0.64	**0.81**	**0.99**
*a*2 (age 15)	0.33	**0.89**	0.66	0.42	**0.93**	**0.92**
*a*3 (age 16)	0.55	**0.75**	**0.82**	0.62	**0.82**	**0.99**
*a*4 (age 17)	0.52	**0.77**	**0.80**	0.60	**0.83**	**0.98**
*a*5 (ages 18–19)	0.90	0.30	**1.00**	**0.94**	0.40	**0.93**
*a*6 (ages 20–21)	0.00	**0.99**	0.37	0.09	**1.00**	**0.73**
*a*7 (ages 22–25)	−0.77	**0.74**	−0.47	−0.71	0.67	−0.06
*a*8 (ages 26–39)	−0.95	−0.17	−1.00	−0.98	−0.27	−0.87

Note that for Table 2: Ma = Major depressive disorder; De = Depression not otherwise specified; Ob = Obsessive compulsive disorder; Ge = Generalised anxiety disorder; An = Anxiety disorder not otherwise specified; and So = Social phobia. The correlation coefficients equal to or larger than the positive correlation $r = 0.71$ were bolded and interpreted because they represented at least 50% ($0.71^2 = 0.5$) of the shared variance between categories

4 Discussion

Clinical Meaningfulness in the Matched CA Results. The weighted Euclidean distances computed for the symptom indicators represent the differences computed from the anorexia patients' residuals of their responses on the binary symptom indicators, after subtracting the responses of the bulimia patients. Therefore, any projected lines of the symptom indicators from the origin (0, 0) in the group-difference biplot indicate that the group differences (in the symptom indicators) should be interpreted in terms of anorexia patients. Similarly, the correlation between the projected age lines and the symptom indicators must be interpreted in terms of anorexia patients.

Nevertheless, since the circulant matrix **C** is analysed for matched CA, if the correlations are to be interpreted in terms of bulimia patients, the signs of the correlation coefficients in Table 2 must be reversed. For instance, the correlations of the "$a6$" age group of anorexia patients with {De, An} were $r = 0.99$ and $r = 1.00$, but when interpreted for bulimia patients, they are $r = -0.99$ and $r = -1.00$. Note that correlation coefficients are maintained with two decimal places. When comparing the severity of symptoms between anorexia and bulimia patients, the majority of anorexia patient age groups have a strong positive correlation with the symptom indicators. Therefore, one may conclude that anorexia patients exhibit more severe psychiatric symptoms than bulimia patients. Nevertheless, the negative correlations of the age groups 22–25 years ($a7$) and 26–39 years ($a8$) indicate that anorexia patients in these age ranges display less severe psychiatric symptoms (see Ma, Ob, Ge, and So in Table 2) than bulimia patients in the same age ranges.

When a researcher conducts matched CA with cross-sectional or repeatedly measured data, the researcher must carefully determine the ordering of the circulant matrix. To examine treatment efficacy for psychiatric symptoms, the responses of anorexia and bulimia patients at discharge are placed intentionally before their responses at admission for the analysis of the circulant matrix in this study. Anorexia patients have more severe eating disorder symptoms, particularly weight loss, than bulimia patients. Therefore, when examining binary psychiatric symptom indicators, group differences should be interpreted in terms of anorexia patients; bulimia patients' response points are anchored at (0, 0), where the origin indicates no group differences in their responses. Since correlation coefficients are calculated using the coordinates of the group-difference biplot, they should also be interpreted in terms of the anorexia patients.

Additional comments on group- and time-difference coordinates. The group-difference coordinates signify the degree of deviation between the responses of anorexia patients to binary psychiatric symptom indicators and those of bulimia patients. However, the first group-difference dimension (dimension 1) accounted for 95.2% of the total group-difference variance, while the second group-difference dimension (dimension 3) did only 4.8% of the total group-difference variance. Therefore, the first dimensional coordinates should be primarily interpreted as group differences in symptom indicators: The severity of the six symptom indicators experienced by anorexia patients is found to be significantly greater than that of bulimia patients.

Only the first one (dimension 4) of the time-difference dimensions is statistically stable; its coordinates represent the time differences between the symptom indicators. If the coordinate values are negative, these time differences can be used to evaluate the efficacy of a treatment. Only the coordinate of De (Depression not otherwise specified) is negative to the findings of the present study. Thus, the psychiatric symptoms for both the anorexia and bulimia groups do not improve significantly following treatment, which is somewhat consistent with the findings of a recent study (Kim and Annunziato 2020).

When researchers apply a matched CA to repeated-measures data, they need to consider the following five steps: (1) Identify group- and time-difference, and interaction dimensions; (2) Test the statistical stability of the dimensions with a permutation test; (3) Interpret the coordinates of the statistically stable dimensions as group or time differences (in terms of categorical variables of interest); (4) Construct a biplot with a pair of statistically stable dimensions to visually inspect any association between the rows and columns; and (5) Complement visually inspected associations with correlation coefficients to improve interpretation.

Limitation. In this study, matched CA is applied to repeated measurements with two time points (pre- and post-treatment) and two independent groups (patients with anorexia and bulimia); however, matched CA can also be used to analyse data with more than two time points and groups. Nonetheless, a circulant matrix comprised of multiple time points and groups will be quite large, making it considerably more difficult to interpret the results. In addition, continuous repeated-measures data must be properly discretised for matched CA. Discretisation of continuous data has been performed for correspondence analysis; see, for example, Kim and Frisby (2019), but when data are discretised, a researcher should consider theoretical justifications (e.g. Kim et al. 2022).

Acknowledgements It gives me great pleasure and honour to contribute to Professor Nishisato's *Festschrift*. Since the beginning of my first academic position in 1999, I have known Nishi. Multidimensional scaling was the subject of my dissertation, and I have always been interested in scaling techniques in general. I came across Nishi's *Element of Dual Scaling: An Introduction to Practical Data Analysis* (1994) while looking for scaling methods and was so enthralled by it that I read it cover to cover in a few days, yellow highlighting the majority of the sentences. I then absorbed the same enthusiasm into Nishi's *Multidimensional Nonlinear Descriptive Statistics* (MUNDA 2006). I still remember that instead of going to the conference, I spent most of the time in my hotel room reading Nishi's MUNDA, and its review was published in *Applied Psychological Measurement* (2011). I recently reviewed Nishi's *Optimal Quantification and Symmetry* (2022), and with the same excitement and joy, I'm looking forward to doing the same for *Measurement, Mathematics, and New Quantification Theory* (2023). Reading and learning about Nishi's dual scaling inspired me to engage in his quantification or correspondence analysis domains. Sadly, I'm probably one of the few people in the United States who has published an application of quantification theory (or correspondence analysis) to applied psychology, despite my belief that these domains are gold mines for applied psychology and other social science research. I sincerely wish that more American researchers incorporate quantification theory into their own research.

References

Beh, E.J.: Simple correspondence analysis: a bibliographic review. Int. Stat. Rev. **72**, 257–284 (2004)
Beh, E.J., Lombardo, R.: Correspondence Analysis: Theory, Practice and New Strategies. Wiley, Chichester (2014)
Beh, E.J., Lombardo, R.: An Introduction to Correspondence Analysis. Wiley, Chichester (2021)
Demirtas, H.: A note on the relationship between the phi coefficient and the tetrachoric correlation under nonnormal underlying distributions. Am. Stat. **70**(2), 143–148 (2016)
Gabriel, K.R., Odoroff, C.I.: Biplots in biomedical research. Stat. Med. **9**, 469–485 (1990)
Gabriel, K.R.: The biplot graphic display of matrices with application to principal component analysis. Biometrika **58** (3), 453–467 (1971)
Gower, J.C., Hand, D.J.: Biplots. Chapman & Hall, London (1996)
Gower, J., Lubbe, S., le Roux, N.: Understanding Biplots. Wiley, Chichester (2011)
Greenacre, M.J.: Singular value decomposition of matched matrices. J. Appl. Stat. **30**, 1101–1113 (2003)
Greenacre, M.J.: Biplots in Practice. Fundación BBVA, Madrid (2010)
Greenacre, M.: Correspondence Analysis in Practice, 3rd edn. Chapman & Hall/CRC, Boca Raton, FL (2017)
Kim, S.-K.: Test treatment effect differences in repeatedly measured symptoms with binary values: the matched correspondence analysis approach. Behav. Res. Meth. **52**, 1480–1490 (2020)
Kim, S.-K., Annunziato, R.A.: Can eating disorder treatment also alleviate psychiatric comorbidity: matched correspondence analysis? Psychother. Psychosom. **89**(2), 125–127 (2020)
Kim, S.-K., Frisby, C.L.: Gaining from discretization of continuous data: the correspondence analysis biplot approach. Behav. Res. Meth. **51**(2), 589–601 (2019)
Kim, S.-K., Grochowalski, J.H.: Exploratory visual inspection of category associations and correlation estimation in multidimensional subspaces. J. Classif. **36**(2), 177–199 (2019)
Kim, S.-K., McKay, D., Storch, E., Bussing, R., Goodman, W.K., Small, B., McNamara, J., Murphy, T.: Anxiety and symptom impact mediate outcome in cognitive-behavior therapy and pharmacotherapy for childhood obsessive-compulsive disorder: the correspondence analysis approach. J. Psychopathol. Behav. Assess. **42**, 739–750 (2020)
Kim, S.-K., McKay, D., Tolin, D.F.: Examining the generality and specificity of gender moderation in obsessive compulsive beliefs: stacked prediction by correspondence analysis. Br. J. Clin. Psychol. **61**(3), 613–628 (2021a)
Kim, S.-K., McKay, D., Murphy, T.K., Bussing, R., McNamara, J.P., Goodman, W.K., Storch, E.C.: Age moderated–anxiety mediation for multimodal treatment outcome among children with obsessive-compulsive disorder: an evaluation with correspondence analysis. J. Affect. Disord. **282**, 766–775 (2021b)
Kim, S.-K., McKay, D., Cepeda, S.L., Schneider, S.C., Wood, J., Storch, E.C.: Assessment of improvement in symptom severity for children with autism spectrum disorders: the matched correspondence analysis approach. J. Psychiatr. Res. **145**, 175–181 (2022)
Le Roux, B., Rouanet, H.: Multiple Correspondence Analysis. Sage, Thousand Oaks, CA (2010)
Lebart, L., Morineau, A., Warwick, K.M.: Multivariate Descriptive Statistical Analysis. Wiley, New York (1984)
Nishisato, S.: Elements of Dual Scaling: An Introduction to Practical Data analysis. Erlbaum, Hillsdale, NJ (1994)
Nishisato, S.: Multidimensional Nonlinear Descriptive Analysis. Chapman & Hall/CRC, Boca Raton, FL (2007)
Vaswani, S.: Assumptions underlying the use of the tetrachoric correlation coefficient. Sankhyā: Indian J. Stat. **10(3)**, 269–276 (1950)

Trust of Nations

Represented by Hayashi's Quantification Method III

Ryozo Yoshino

1 Introduction: Birth of "Statistical Mathematics"—Philosophy of Statistics

The late Chikio Hayashi (1918–2002), as a key member of the Institute of Statistical Mathematics (ISM), initiated and developed a longitudinal and cross-national comparative survey that has lasted for more than six decades from 1953 to the present: "Nihonjin no Kokuminsei Chosa [the Japanese National Character Survey (JNCS)]" and "Ishiki no Kokusai-Hikaku [the Comparative Survey of People's Attitudes and Awareness]". (See https://www.ism.ac.jp/ism_info_e/kokuminsei_e.html and Note 2). The present author has been a member of the survey team for more than three decades.

The JNCS was closely linked to the re-organisation of official statistics and the establishment of statistical public opinion polls to develop Japan's post-war democracy after the WWII. It also symbolises the development of Japanese statistical philosophy such as "Statistical Mathematics" that began in the field of Japanese statistics in the early 1940s. Traditional "mathematical statistics" rely on the mathematical assumptions of probability distribution theory which are almost impossible to verify directly. Several groups of statisticians criticised this and aimed to develop new statistical approaches to solving social problems in a practical way (Midzuno 2003). One group led to the establishment of the Institute of Statistical Mathematics. Since then, this idea has been successively developed as "Hayashi's quantification theory" in the 1950s and 1960s (Hayashi 1950), "Multidimensional Data Analysis", "Behaviormetrics", and "Science of Survey" in the 1970s and the 1980s, and "Science of Data" since the late 1980s (Hayashi, 1984, 1998a, b, 2001; Yoshino 2001, 2021; Yoshino and Hayashi 2002, 2010).

R. Yoshino (✉)
The Institute of Statistical Mathematics, Tokyo, Japan
e-mail: ryoshino@mail.doshisha.ac.jp

E. J. Beh et al. (eds.), *Analysis of Categorical Data from Historical Perspectives*, Behaviormetrics: Quantitative Approaches to Human Behavior 17,
https://doi.org/10.1007/978-981-99-5329-5_22

Some explanation is needed for "Science of Data [Deita no Kagaku in Japanese]" in this context. This term was coined by Hayashi in the 1980s. At a keynote speech by the International Federation of Classification Society (IFCS) held in Kobe in 1996, Hayashi explained that conventional hypothesis testing, numerical models, and statistical models were not suitable to the study of complex and ambiguous phenomena such as in human sciences and social sciences (Hayashi 1998b). He proposed to construct a "Science of Data" based on a data-driven, exploratory and holistic approaches that deals with such complex and ambiguous phenomena in human and social sciences (Hayashi 2001; Osumi 2003). His idea is closely linked to Tukey's "Exploratory Data Analysis" or Benzécri's approach. In exemplifying his idea, Hayashi often made use of "Hayashi's quantification method III (QMIII)" that he invented in 1950s with Hiroshi Midzuno (Note 3). The method is mathematically the same as Benzécri's correspondence analysis, Bock's optimal scaling, or Nishisato's dual scaling, although these have all been independently developed in different fields; see Nishisato (2007), Matsumoto (2022) and Osumi (2003) for a more detailed explanation while Nishisato (2023) also explains this issue from his point of view.

In the early 1970s, Hayashi began conducting overseas survey under the paradigm of *"Cultural Link Analysis"* in order to study the Japanese national character at a higher level. It was later developed by the present author as "*Cultural Manifold Analysis*" *(CULMAN)* that has been used for the longitudinal and cross-national comparative survey under the statistical philosophy of "Science of Data" (Yoshino 2021).

Over the past 65 years, our research team has collected statistical random sampling data of people's attitudes, opinions and values for the development of our paradigm to justify longitudinal and cross-national research. The purpose of our research is to promote global mutual understanding of people's attitudes, behavioural manners, religion, values, etc. Mutual understanding is the key to avoiding unnecessary conflicts between countries and developing a peaceful and prosperous world; see Fig. 3 for a global manifold of local communities.

This paper, as part of our recent research, presents various applications of Hayashi's QMIII on interpersonal trust. Due to page restrictions, I will leave the details of our paradigm and data analysis to the book (Yoshino 2021), which is a summary of our long-term research on various topics and discussions. The detailed discussion on people and trust is therefore not repeated here; see instead Yoshino (2019, 2021). Nevertheless, I still believe that some of the results presented here on QMIII provide basic information for readers to develop more sophisticated studies of trust based on each country's culture, economy, and political system.

2 Hayashi's Quantification Method III and Cross-National Comparison

Since 1971, the JNCS has been expanded to include cross-national surveys. The survey items were selected to compare people's social values, ways of thinking, emotions, religious attitudes, etc. These aspects may provide information on the psychological distance between countries or races.

Cross-national surveys need to overcome the multifaceted methodological problems of cross-national comparability. These include problems concerning:

1. translation (survey questions need to be created to maintain the same meaning in different languages),
2. the comparison of datasets collected using different sampling procedures in different countries, and
3. descriptions and characteristics of the compared countries in terms of a common logic (or framework of thinking).

Comparing people from different cultures makes these problems even more problematic.

Multidimensional analysis techniques, such as QMIII, may give a specific solution to these problems. For example, Yoshino (2021) describes the robustness of multidimensional analysis for slightly different wordings of questions. Furthermore, Yoshino and Hayashi (2002) show that one can disregard differences in sampling methods in a total configuration obtained by QMIII when comparing data from many countries with respect to a group of items in contrast to an examination of only a single item. In addition, Yoshino (1992a, b, c) showed that one could even detect falsified data by applying multidimensional scaling, called the "super-culture model". The model is closely related to the Cultural Consensus Theory (CCT) by Batchelder and Romney (1988); a method that the present author contributed to in the early stages of its development as a research assistant.

QMIII also works to solve the weight adjustment problem for the sampling probabilities of respondents. As an example, take our 1991 Japanese–Brazilian Survey. For the Brazil data, there was a great deal of bias among Japanese–Brazilian's due to the complex ethnicities that made it difficult to find the right weight adjustment. The weighting coefficients could be large enough to make the data less reliable; i.e. the variance could become too large (Yamamoto et al. 1993). However, in the QMIII output, the difference between the dataset with weighting adjustment and the dataset without weighting adjustment is so small that the effect of weighting can be ignored in response pattern analysis when comparing data for multiple items in many countries. Thus, multidimensional analysis may offset differences in item wordings, sampling methods, and weight adjustments, to provide a stable macro pattern. Multidimensional data analysis provides a consistent overall analysis, losing some of the details of individual item data. In a sense, QMIII is a reasonably sensitive and reasonably insensitive scaling for survey data analysis. This is one of the "principle of complementarity" in our "Science of Data" (Yoshino 2021).

In a cross-national comparative survey, comparing completely different countries from the beginning is not the best way to make a meaningful comparison in our type of questionnaire survey. By comparing pairs of countries (or groups) with some similarities and differences, such as language and ethnicity, and identifying similarities and differences in their response patterns, you can reveal more meaningful statistical comparisons. Gradually connecting these comparison links (country pairs) will expand the chain of links and ultimately allow for global comparisons. This idea was developed as a research paradigm called "Cultural Link Analysis" (CLA) and eventually integrated three types of linkages:

1. longitudinal (temporal) linkage,
2. cross-national (spatial) linkage, and
3. thematic linkage (item-structure linkage).

Furthermore, we have developed a paradigm called "Cultural Manifold Analysis" (CULMAN) to introduce a hierarchical structure into the three types of linkages. This study confirms that when comparing response patterns across countries (or across multiple groups), applying multidimensional analysis to response data for a number of items from people in multiple countries is effective in obtaining stable results, even if the response data sets are collected with different sampling methods and in different languages; see Yoshino (2021, Sect. 3). Such stability couldn't be obtained with pairwise comparisons.

Under these research paradigms, our early research in cross-national survey revealed attitudes and social values particular to the Japanese people, such as interpersonal relationships and religion. In addition, some survey results have been reported on the general response tendency and the degree of self-disclosure particular to each country. For example, the Japanese people tend to avoid extreme answers and choose a category near the middle of the options or say "undecided" or "don't know". French people tend to give negative or critical answers to any question while Indians in general tend to give positive or optimistic answers. However, it should be noted that the same Japanese person, for example, may have different response patterns when in Japan and when abroad. The idea of CULMAN is expected to serve as a framework for developing empirical social sciences to understand the rise and fall of a civilisation and promote mutual understanding between different cultures.

3 Trust and People

During the Cold War between the US and Soviet Union (1947–1991), Rotter (1971, p. 443) stated:

> It seems clear that disarmament will not proceed without an increase in trust on one or both sides of the iron curtain.

The iron curtain was torn down in 1993, but new local conflicts have been occurring incessantly all over the world. Mutual understanding and mutual respect among countries are important for the peaceful development and economic prosperity of the world, and "trust" is the key to mutual understanding and respect.

In the 1990s the Japanese witnessed significant social change under the rapid reform of the economic and political system called "globalisation". This change brought us the collapse of even basic interpersonal relationships at home, at school, and at work. This seems to be one of the main reasons why Japan was unable to recover in almost three decades after the "burst of the bubble economy" from 1991 to 1993. Its aftereffects continue to this day.

Furthermore, our studies of interpersonal trust, institutional trust, and other social values depicted features of people of several countries and Japanese immigrants overseas. Some universal social values on human bonds such as the importance of family was recognised, although the styles of family may be different across countries or time. On the other hand, our studies have shown that trust scales, and perhaps subjective scales in general, are not simple across cultures, and that scales and scale objects are complementary or interactive in their measurement operations (Yoshino 2001; Yoshino and Hayashi 2002). As Dogan (2000, p. 258) wrote:

> decline in trust in authority can sometimes be a sign of political maturity that the critical spirit of most citizens has improved.

Perhaps trust and distrust may not be diametrically opposed on a one-dimensional scale. They are closely related to a kind of multidimensional mind structure in each culture and each social condition.

4 People's Sense of Trust and CULMAN

Some researchers say that "trust" cannot be measured directly (Fukuyama 1995). There may probably be no universal measure of trust across cultures and times. Even if there is one, it may not necessarily be linear with respect to various factors such as income and social class. In other words, with social factors, as with medicines, the right amount, neither too much nor too little, is important for a good effect. However, by properly analysing the longitudinal patterns and cross-national patterns of questionnaire survey responses, it is possible to identify certain important aspects of people's sense of trust. Here it is important to consider the data, taking into account each country's social situation and the general response tendency (Yoshino 2021, Sect. 3.4), rather than comparing the data superficially.

The past decades have developed psychological studies of measures of interpersonal trust. Among others, a set of three question items from the General Social Survey (GSS) by National Opinion Research Centre at the University of Chicago has been often used to measure people's sense of trust; for some history on their use of the three items, see Yoshino (2001, 2019, 2021). Although the GSS started as an American version of JNCS, we have adopted the three items from the GSS for our survey since 1978. The three question items are stated as follows (for the Japanese phrasing of the three items, see https://www.ism.ac.jp/kokuminsei/index.html):

Q36. Would you say that, most of the time, people try to be helpful, or that they are mostly just looking out for themselves? (1) Try to be helpful, (2) Look out for themselves,
Q37. Do you think that most people would try to take advantage of you if they got the chance, or would they try to be fair? (1) Take advantage, (2) Try to be fair,
Q38. Generally speaking, would you say that most people can be trusted or that you can't be too careful in dealing with people? (1) Can be trusted, (2) Can't be too careful.

These items are often used as measures of reciprocity norms, fairness norms, and generalised interpersonal trust, respectively. However, as noted previously, the author is sceptical of using these items as such superficial measures and often pays attention to multidimensional analysis of response patterns. As for the data of these three items, Yoshino (2021) considered a multidimensional structure of local charts (a cluster of countries) that appear according to the degree of similarity between the countries. The main purpose of applying QMIII here is to analyse the mutual relations (similarity/dissimilarity) between countries in the multidimensional structure, rather than just to discuss the level of "trust" on the unidimensional distribution of responses for each of the three question items (Q36, Q37, and Q38). In the multidimensional study, each local chart is created according to the similarity between countries, and all charts form a manifold (hierarchical structure of local charts). The manifold depends on the range of question items and the range of countries to be compared. Yoshino (2016, 2019, 2021) reported some results on the following survey data:

a. The Seven Country Survey (Japan, USA, UK, France, Italy, West Germany, and The Netherlands [1987–1993]).
b. The East Asian Values Survey (EAVS) (Japan, China [Beijing, Shanghai, Hong Kong], Taiwan, South Korea, and Singapore [2002–2005]).
c. The Pacific-Rim Values Survey (PRVS) (Japan, China [Beijing, Shanghai, Hong Kong], Taiwan, South Korea, Singapore, USA, Australia, and India [2004–2009]).
d. The Asia-Pacific Values Survey (APVS) (Japan, China [Beijing, Shanghai, Hong Kong], Taiwan, South Korea, Singapore, USA, Australia, India, and Vietnam [2010–2014]).

For more information on these surveys, simple tabulations of the response data for each country in each survey, and cross-tabulations by gender, age, etc., one may visit see the ISM URL: https://www.ism.ac.jp/~yoshino/index_e.html.

I have already described the results from the applications of QMIII to the data obtained from surveys in some detail; see Yoshino (2019, 2021). Figure 1 shows an example on the APVS data.

In order to examine the similarity/dissimilarity of people's sense of trust in these countries, a group of local charts was created, taking into account the approximate mutual distance between countries on the QMIII output (circles surrounding countries in each figure), but the structure of the local charts can be considered from various perspectives, so the representation here should be considered tentative.

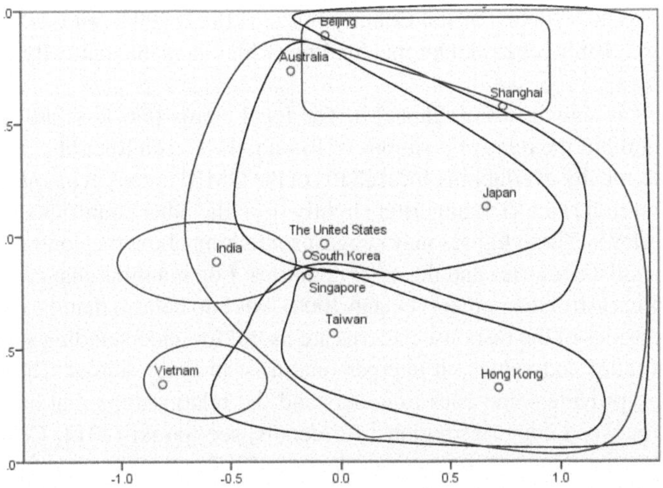

Fig. 1 Quantification Method III (QMIII) on GSS 3 trust Items in the APVS (2010–2014): Japan, South Korea, Beijing, Shanghai, Hong Kong, Taiwan, Singapore, USA, Australia, India, and Vietnam. The clustering in this figure is drawn as an example of a local chart of a manifold, with the relationships among the countries clustered on a trial basis. The way a manifold looks depends on the perspective from which the system is viewed, such as cultural, economic, or political. Eigenvalues of the first and the second dimension are, respectively, 1.64 and 1.57

QMIII simultaneously analyses the structure of the data obtained from the three questions. Many of these countries/regions have been repeatedly surveyed in our studies (see the list of countries surveyed in a, b, c, and d above), allowing long-term comparisons. However, selections of slightly different countries/regions can result in sometimes fairly consistent output patterns of QMIII, and sometimes different patterns. For the details, see Yoshino (2021).

5 Cross-National Analysis of Interpersonal Trust in Europe, Russia, and Asia

The survey team directed by M. Sasaki, A. Ishikawa, V. Davydenko, A.B. Kpreychenko, N. Dryakhlov, Zh.T. Toshchenko, and V.D. Shadrikov (Sasaki 2014) carried out a trust survey project targeting Russia, the Czech Republic, Finland, Germany, the United States, Japan, Taiwan, and Turkey. Sasaki (2014, Chaps. 5 and 6) report on the research and analysis. As a member of the team, I had access to the data. Not only are some countries geographically distant from each other, but also their cultural backgrounds and political systems are so diverse that a direct comparison in terms of CULMAN may not make much sense. Therefore, in order to get an idea of the overall composition in line with CULMAN's concept, I applied QMIII to the combined data of the Sasaki's survey, Seven Country Survey (1987–1993), and APVS (2010–2014). (I used SPSS ver.24.0, Optimal Scaling, which is mathematically equivalent to QMIII).

The result is shown in Fig. 2a and b. The local charts (clusters) are tentatively drawn to highlight the relative positions of Russia, the Czech Republic, and Turkey on the hierarchically overlapping local charts of the QMIII output. It has already been observed that India and Vietnam differ slightly from the other countries surveyed by PRVS and APVS. These figures may suggest something about the contrast between previous socialist countries and the other countries. I do not think that certain countries are more trustful than others (Dogan 2000), but I do believe that the similarities and dissimilarities of the response patterns are useful for understanding some differences of attitudes and values on interpersonal trust in many senses. These configurations may provide some cues to understand the relationships that have existed or do exist between countries/regions; for details, see Sasaki (2014, Chap. 9). As repeatedly observed in our past surveys, clusters of UK and USA, and of France and Italy are found in these configurations too. Figure 2a and b also suggest that Japan and Germany are similar in some aspects and different in others. For example, in a survey of seven countries, it was observed that in interpersonal relations at home and at work the two are similar, but in attitudes towards science to overcome psychological and social problems, Germany is positive and Japan is negative (Hayashi et al. 1998).

Note that Sasaki's survey data is several years old, and that the three GSS items are not directly linked to geography or international politics. However, given the current international situation in Russia and Ukraine, the position of France, Italy, Russia, and Turkey in these configurations may appear to be closely related to complex international political relations, including economic sanctions against Russia.

For the eight countries in Sasaki's project, he developed a data analysis that included three items of trust as well as many items of trust and distrust, concluding his findings with the following four observations (Sasaki 2014, p. 262):

1. For all the eight countries compared, the parental socialisation of trust/distrust is the most important factor for the children to develop their sense of trust in later lives.
2. For all the eight countries, various degrees of parents keeping promises during childhood are related with parental socialisation of trust/distrust in childhood.
3. For five of the countries (USA, Finland, Germany, the Czech Republic, and Taiwan), experience of being betrayed is linked with later development of sense of trust distrust. But the combined data of eight countries, the effect is not significant.
4. In each of the six countries, with the exception of the USA and Germany, there was no significant relationship between the extent to which parents kept their promises to their children during childhood and the subsequent development of children's sense of trust. However, a modest significant correlation was found in the data for the eight countries combined.

Overall, we confirmed several preceding studies that children's experience of observance with parents leads to their sense of trust or distrust.

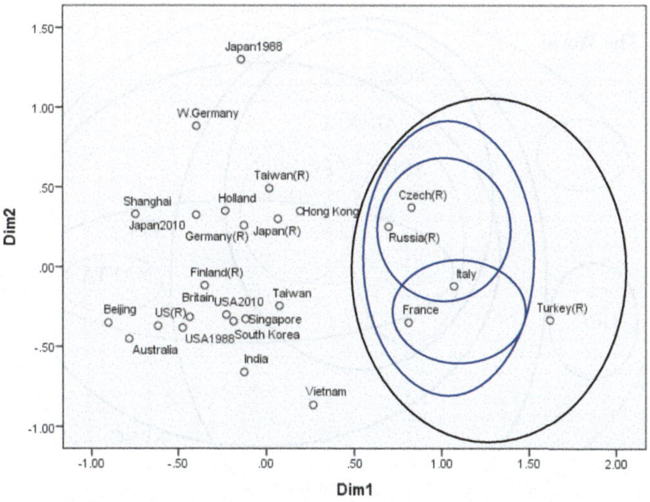

a

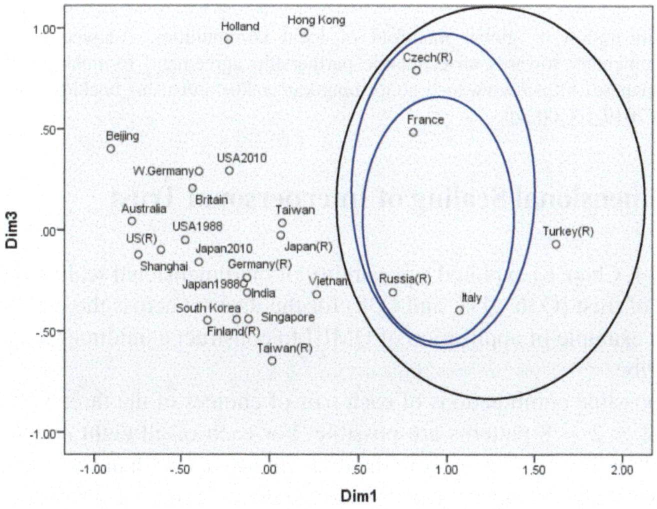

b

Fig. 2 **a** Quantification Method III (MMIII) on GSS 3 trust items for the combined data of the seven country survey (1987–1993), the APVS (2010–2014) and the Russian Survey Project (2008–2012) (Russia, the Czech Republic, Finland, US, Japan, Taiwan, and Turkey Survey): the first and third dimension. Note: (R) indicates part of a Russian research project. The clustering in this figure is drawn as an example of a local chart of a manifold by trial clustering the relationships among countries. The way a manifold looks depends on the perspective of the system, such as the cultural, economic or political system. Eigenvalues of the first dimension and the second are, respectively, 1.80 and 1.66. **b** Quantification Method III (QMIII) on GSS 3 Trust Items for the combined data of the Seven Country Survey (1987–1993), the APVS (2010–2014) and the Russian Survey Project (2008–2012) (Russia, the Czech Republic, Finland, US, Japan, Taiwan, and Turkey Survey): the 1st and the 3rd dimension. See this figure with Fig. 2a. Note: (R) indicates part of a Russian research project. Eigenvalues of the first dimension and the third are, respectively, 1.80 and 1.46

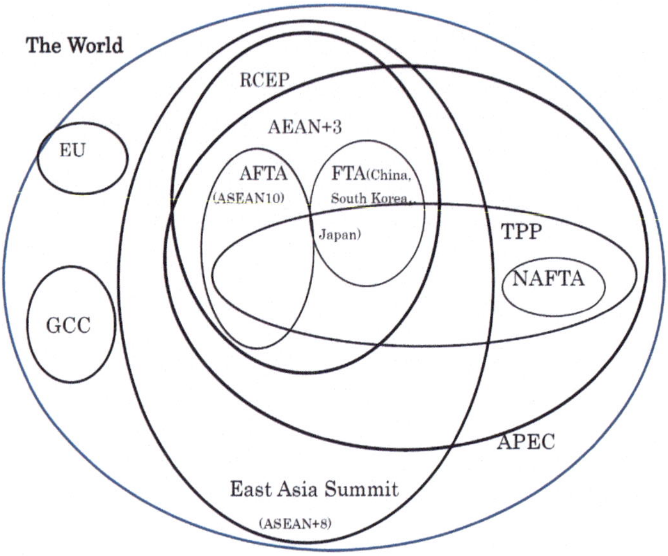

Fig. 3 An illustration of global manifold of local communities. Adapted from p. 17 of "General commentary towards an economic partnership agreement: formulation of the discipline" (in Japanese) https://www.meti.go.jp/shingikai/sankoshin/tsusho_boeki/fukosei_boeki/report_2019/pdf/2019_03_00.pdf

6 Unidimensional Scaling of Interpersonal Trust

Sasaki (2014, Chap. 8) explored a possibility of unidimensional scaling on the three GSS items of trust (Q36, Q37, and Q38) for the data set across the eight countries. It shows an example of application of QMIII to construct a unidimensional scale as I now describe.

For all possible combinations of each pair of choices of the three GSS items on trust, $2 \times 2 \times 2 = 8$ patterns are possible. For each of all eight countries, those who gave all optimistic responses to the three items are less than some 5%, whereas those who gave all pessimistic responses to the three items vary across the countries such as 4.7% of respondents in Turk, 6.1% in USA, 11.5% in Germany, 12.1% in the Czech Republic, 12.6% in Russia, 19.2% in Finland, 20.9% in Taiwan, and 29.4% in Japan. Note that "1", "2", and "1" to Q36, Q37, and Q38, respectively, are optimistic responses whereas "2", "1", and "2" are pessimistic responses. Here, for convenience to explore possible Likert scales let's change the coding so that "1" is an optimistic response and "2" is the pessimistic response. Sasaki examined the validity of constructing a Likert scale such that Type A includes "111", "112", "212", and "222" (in that order) and Type B ("111", "211", "212", and "222") with help by Fumi Hayashi who is another member of our survey project. As a QMIII technical specialist, Fumi Hayashi was Chikio Hayashi's assistant for decades before his death. They share the same surname but have no kinship.

Sasaki applied QMIII to the data of the eight response patterns across the eight countries, and the result yielded three clusters: Germany and USA; the Czech Republic, Turkey, and Russia; Japan, Finland, and Taiwan. He then applied QMIII to the raw data of the three items for the eight countries and found that Finland and Taiwan are located far away from the other six countries. After some trials and errors, he concluded that Likert scale can be constructed as a unidimensional scale of the three items of trust (A or B type) in each of the six countries except Finland and Taiwan. Although the three items of the GSS are often used in many countries, Sasaki's study cautions us to use them as a trust scale across all countries. This is consistent with our past studies of comparative surveys that covered Asia–Pacific countries, European countries, India, and USA as well as Japanese immigrants in Hawaii, on the West Coast of USA, and Brazil (Yoshino 2021).

7 For a Peaceful and Prosperous World

Finally, Fig. 3 shows an illustration of the Global Manifold of local communities as of 2021. On this manifold, each local chart (local community) represents a dynamic change in international relations. New local charts emerge and others disappear; two local charts may merge into a larger local chart. Several pairs of local charts overlap and as a whole constitute a global manifold. In order to maintain stable, peaceful, and prosperous development, a set of "soft" rules linking pairs of local charts, rather than a single restrictive global regulation, is necessary. The leaders of the modern world need to be mediators, resolving regional conflicts not through military force but through global cooperation and harmony.

I sincerely hope that our survey research can contribute to our mutual understanding for the development of world peace and prosperity.

Appendix

Note (1) This paper is a revision of some parts of Yoshino (2019, 2021), with the addition of a new data analysis.

Note (2): For details of our past surveys, please refer to the series of ISM Survey Reports and the following related websites:

- http://www.ism.ac.jp/editsec/kenripo/index.html (ISM Survey Research Reports in Japanese).
- http://www.ism.ac.jp/editsec/kenripo/index_e.html (ISM Survey Research Reports in English)
- http://www.ism.ac.jp/ism_info_j/kokuminsei.html (JNCS in Japanese)
- http://www.ism.ac.jp/ism_info_e/kokuminsei_e.html (JNCS in English)
- http://www.ism.ac.jp/~yoshino/ (ISM Cross-National Studies in Japanese)

- http://www.ism.ac.jp/~yoshino/index_e.html (ISM Cross-National Studies in English)
  For a detailed list of related publications https://www.ism.ac.jp/~yoshino/references_e.html

Note (3): As mention in the text, in Japan, a series of statistical philosophies, starting with "Statistical Mathematics", followed by "quantification theory", "Behaviormetrics", "Survey Science", and "Science of Data". Hayashi's quantification theory was originally developed to solve important social problems related to economic reconstruction and post-war democratisation after the defeat in the WWII. His theory produced several techniques called Hayashi's Quantification Method I, II, III, IV, etc. that deal with categorical data for solving problems concerning reform of the Japanese language, personnel assessment in state-owned enterprises, market surveys, etc. (Komazawa et al. 1998). The 1970s and 1980s saw numerous applications of the quantification methods by many Japanese researchers. This proceeded in parallel with the development of the Behaviormetric Society of Japan, which was established in 1972. During this period, some researchers attempted to publish papers in foreign journals, but unfortunately, these were rarely accepted as there was little understanding of Hayashi's quantification theory among foreign researchers.

Hayashi's quantification theory was eventually sublimated into Hayashi's "Science of Data", and from around the late 1980s, discussions with researchers from various countries began to develop at International Federation of Classification Society (IFCS) (Hayashi 1998b).

For Hayashi's research team, the "science of data" means gathering comprehensive information on research topics that should lead to solutions to important social issues, in a series of processes from survey design, sampling design, preliminary survey, main survey, data cleaning, data analysis, and research reports for policy making. Our research is refined by repeating this series of processes and getting feedback at each stage. We believe that approaches such as hypothesis testing, computer simulations with virtual data and simple theory building are too naïve to deal with complex phenomena in the human and social sciences. Therefore, we emphasise statistical survey data-driven research (not "big data", which cannot be assessed for statistical accuracy) and holistic approaches. In this respect, QMIII is used diversely and effectively (Yoshino 2021).

References

Batchelder, W.H., Romney, A.K.: Test theory without an answer key. Psychometrika **53**(1), 71–92 (1988)

Dogan, M.: Deficit of confidence within European democracies. In: Halller, M. (ed.) The Making of the European Union, pp. 243–261. Springer, Paris (2000)

Fukuyama, F.: Trust. Free Press, New York (1995)

Hayashi, C.: On the quantification of qualitative data from the mathematico-statistical point of view (an approach for applying this method to the parole prediction). Ann. Inst. Stat. Math. **2**, 35–47 (1950)
Hayashi, C.: Chosa no Kagaku (Science of Survey). Kodan-sha, Toyko (1984) (Reprint by Chikuma Academic Bunko in 2001)
Hayashi, C.: The quantitative study of national character: interchronological and international perspectives. In: Sasaki, M. (ed.) Values and Attitudes Across Nations and Time, pp. 91–114. Brill, Boston (1998a)
Hayashi, C.: What is data science? Fundamental concepts and a heuristics example. In: Hayashi, C., Yajima, K., Bock, H. H., Ohsumi, N., Tanaka, Y., Baba, Y. (eds.) Data Science, Classification, and Related Methods: Proceedings of the Fifth Conference of the International Federation of Classification Societies (IFCS-96), Kobe, Japan, 27–30 March 1996, pp. 40–51. Tokyo, Springer (1998b)
Hayashi, C.: Deta no Kagaku (Science of Data). Asakura-syoten, Tokyo (2001)
Hayashi, C., Yoshino, R., Suzuki, T., Hayashi, F., Kamano, S., Miyake, I., Murakami, M., Sasaki, M.: Kokuminsei nanaka-koku hikaku (Cross-National Comparison of Seven Nations), Idemitsu-syoten, Tokyo (1998)
Komazwa, T., Hashiguchi, K., Ishizaki, R.: Paso-kon suryouka bunseki (Hayashi's Quantification Methods using Personal Computers). Asakura-syoten, Tokyo (1998)
Matsumoto, W.: Deta-Saiensu no wasuremono (Lost property of Data Science). Jyohou-Kenkyu (J. Inf.) **54**, 81–93 (2022). (In-house publication of Kansai University)
Midzuno, H.: What is "statistical mathematics"? Jpn. J. Behavior. **30**, 191–192 (2003). (In Japanese)
Nishisato, S.: Multidimensional Nonlinear Descriptive Analysis. Chapman-Hall/CRC, London (2007)
Nishisato, S.: Propositions for quantification theory. In: Okada, A., Shigemasu, K., Yoshino, R., Yokoyama, A. (eds.) Facets of Behaviormetrics: The 50th Anniversary of the Behaviormetric Society, pp. 173–191. Springer, Singapore (2023)
Osumi, N.: In memory of Dr. Chikio Hayashi—from classification to data science. Japan Classification Society News, 26. Available online at http://bunrui.jp/pdf/kaiho26.pdf (2003). Last accessed 14 May 2023 (in Japanese)
Rotter, J.B.: Generalized expectations for interpersonal trust. Am. Psychol. **26**, 443–452 (1971)
Sasaki, M. (ed.): Cross-National Studies of Trust. Social Science Research Institute. Chuo University Press, Tokyo (2014)
Yoshino, R.: Superculture as a frame of reference for cross-national comparison of national characters. Behaviormetrika **19**, 23–41 (1992a)
Yoshino, R.: Extension of the "test theory without answer key by Batchelder and Romney" and its application to an analysis of data of national consciousness. Proc. Inst. Stat. Math. **37**, 171–188 (1992b)
Yoshino, R.: The unbiased BIGHT model and its application to the distinction of responses to a free answer question in a social survey. Behaviormetrika **19**, 83–96 (1992c)
Yoshino, R.: Kokoro wo hakaru (Measuring the Mind). Asakura-syoten, Tokyo (2001)
Yoshino, R.: Trust of nations: looking for more universal vales for interpersonal and international relationships. Behaviormetrika **42**, 131–166 (2016)
Yoshino, R.: People and trust. In: Imaizumi, T., Nakayama, A., Yokoyama, S. (eds.) Advanced Studies in Behaviormetrics and Data Science – Essays in Honor of Akinori Okada, pp. 453–472. Springer, Singapore (2019)
Yoshino, R.: Cultural Manifold Analysis on National Character: Methodology of Cross-National and Longitudinal Survey. Springer, Singapore (2021)
Yoshino, R., Hayashi, C.: An overview of cultural link analysis of national character. Behaviormetrika **29**, 125–142 (2002)

Yamamoto, K., Kawai, T., Wakisaka, K., Miyao, S., Mori, K., Hayashi, C., Midzuno, H., Suzuki, T., Hayashi, F., Yoshino, R.: Research on national character of Japanese Brazilian—1991–1992. ISM Research Report, No. 74. Available online at https://www.ism.ac.jp/editsec/kenripo/pdf/kenripo074.pdf (1993). Last accessed 13 May 2023 (in Japanese)

Yoshino, R., Hayashi, F., Yamaoka, K.: Kokusai- hikaku deta no kaiseki (Analysis of Comparative Survey Data). Asakura-shoten, Tokyo (2010)

Deconstructing Multiple Correspondence Analysis

Jan de Leeuw

1 Notation

Let us start by defining some of the notation used in this paper. We have $i = 1, \ldots, n$ observations on each of $j = 1, \ldots, m$ categorical variables, where variable j has k_j categories. We use $k_\star$ for the sum of the k_j, while the maximum number of categories over all variables is $k_+ = \max(k_1, \ldots, k_m)$. We also define m_s, with $s = 1, \ldots, k_+$, where m_s is the number of variables with $k_j \geq s$. Thus both m_1 and m_2 are always equal to m. Also $\sum_{s=1}^{k_+} m_s = k_\star$. The fact that variables can have a different number of categories is a major notational nuisance. If they all have the same number of categories k then $k_+ = k$, $k_\star = mk$, and all m_s are equal to m.

The data are coded as m indicator matrices $\mathbf{G}_j$, with $\{\mathbf{G}_j\}_{ik} = 1$ if and only if object i is in category k of variable j and $\{\mathbf{G}_j\}_{ik} = 0$ otherwise. The $\mathbf{G}_j$ are $n \times k_j$, zero-one, and column-wise orthogonal (because the categories are mutually exclusive). If we concatenate the $\mathbf{G}_j$ horizontally we have the $n \times k_\star$ matrix $\mathbf{G}$, which we also call the indicator matrix; in French data analysis it is the "tableau disjonctif complet", in Nishisato (1980) it is the "response-pattern table". The Burt table ("tableau de Burt"), is the $k_\star \times k_\star$ cross product matrix $\mathbf{C} = n^{-1}\mathbf{G}'\mathbf{G}$. The univariate marginals are in the diagonal matrix $\mathbf{D} = \text{diag}(\mathbf{C})$. The normalised Burt table is the matrix $\mathbf{E} = m^{\frac{1}{2}}\mathbf{D}^{-\frac{1}{2}}\mathbf{C}\mathbf{D}^{-\frac{1}{2}}$.

Although we introduced $\mathbf{G}$, $\mathbf{C}$, $\mathbf{D}$ and $\mathbf{E}$ as partitioned matrices of real numbers, it is also useful to think of them as matrices with matrices as elements. Thus $\mathbf{C}$, for example, is an $m \times m$ matrix with as elements the matrices $\mathbf{C}_{j\ell} = \mathbf{G}_j'\mathbf{G}_\ell$, and $\mathbf{G}$ is an $1 \times m$ matrix with as its m elements $\mathbf{G}_j$. Note that because we have divided the cross product by n, all $\mathbf{C}_{j\ell}$, and thus all $\mathbf{D}_j = \mathbf{C}_{jj}$, add up to one.

J. de Leeuw (✉)
Department of Statistics, University of California Los Angeles (UCLA), Los Angeles, CA, USA
e-mail: deleeuw@stat.ucla.edu

E. J. Beh et al. (eds.), *Analysis of Categorical Data from Historical Perspectives*,
Behaviormetrics: Quantitative Approaches to Human Behavior 17,
https://doi.org/10.1007/978-981-99-5329-5_23

In the paper we often use the direct sum of matrices. If $\mathbf{A}$ and $\mathbf{B}$ are matrices, then their direct sum is:

$$\mathbf{A} \bigoplus \mathbf{B} = \begin{bmatrix} \mathbf{A} & \mathbf{0} \\ \mathbf{0} & \mathbf{B} \end{bmatrix}, \tag{1}$$

and if $\mathbf{A}_r$ are s matrices, then $\bigoplus_{r=1}^{s} \mathbf{A}_r$ is block-diagonal with the $\mathbf{A}_r$ as diagonal submatrices.

2 Introduction

Multiple Correspondence Analysis (MCA) can be introduced in many different ways.

Mathematically: MCA is the Singular Value Decomposition (SVD) of $m^{-\frac{1}{2}}\mathbf{G}\mathbf{y} = \sqrt{\lambda}\mathbf{x}$ and $m^{-\frac{1}{2}}\mathbf{G}'\mathbf{x} = \sqrt{\lambda}\mathbf{D}\mathbf{y}$, the Eigen Value Decomposition (EVD) $\mathbf{E}\mathbf{y} = \lambda^2\mathbf{y}$ for the normalised Burt table, and the EVD of $m^{-1}\mathbf{G}'\mathbf{D}^{-1}\mathbf{G}\mathbf{x} = \lambda^2\mathbf{x}$, the average projector. Using m in the equations seems superfluous, but it guarantees that $0 \leq \lambda \leq 1$.

Statistically: MCA is a scoring method that minimises the within-individual and maximises the between-individuals variance, it is a graphical biplot method that minimises the distances between individuals and the categories of the variables they score in, it is an optimal scaling method that maximises the largest eigenvalue of the correlation matrix of the transformed variables, and that linearises the average regression of one variable with all the others. It can also be presented as a special case of Homogeneity Analysis, Correspondence Analysis, and Generalised Canonical Correlation analysis; see, for example, the review article by Tenenhaus and Young (1985).

It is of some interest to trace the origins of these various MCA formulations, and to relate them to an interesting exchange in the 1950s between two of the giants of psychometrics on whose proverbial shoulders we still stand. In 1950 Sir Cyril Burt published, in his very own British Journal of Statistical Psychology, a great article introducing MCA as a form of factor analysis of qualitative data (Burt 1950). There are no references in the paper to earlier occurances of MCA in the literature. This prompted Louis Guttman to point out in a subsequent issue of the same journal that the relevant equations were already presented in great detail in Guttman (1941). Guttman assumed Burt had not seen the monograph (Horst 1941) in which his chapter was published, because of the communication problems during the war, which caused "only a handful of copies to reach Europe" (Guttman 1953). Although the equations and computations given by both Burt and Guttman were identical, Guttman pointed out differences in interpretation between their two approaches. These differences will especially interest us in the present paper. They were also discussed in Burt's reaction to Guttman's note (Burt 1953). The three papers are still very readable and instructive, and in the first part of the present paper we'll put them in an historical context.

3 History

3.1 Prehistory

The history of MCA has been reviewed in De Leeuw (1973), Benzécri (1977b), Nishisato (1980, Sect. 1.2),Tenenhaus and Young (1985), Gower (1990), and Lebart and Saporta (2014), each from their own tradition and point of view. Although there is agreement on the most important stages in the development of the technique, there are some omissions and some ambiguities. Some of the MCA historians, in their eagerness to produce a long and impressive list of references, do not seem to distinguish multiple from ordinary Correspondence Analysis (CA), one-dimensional from multidimensional analysis, binary data from multicategory data, and data with or without a dependent variable.

What we call "prehistory" is MCA before Guttman (1941), and what we find in the prehistory is almost exclusively Reciprocal Averaging Analysis (RAA). We define RAA, in the present paper, starting from the indicator matrix G. Take any set of trial weights for the categories. Then compute the score for the individual by averaging the weights of the categories selected by that individual, and then compute a new set of weights for categories by averaging the scores of the individuals in the categories. These two reciprocal averaging steps are iterated until convergence is attained, that is when weights and scores do not change any more (up to a proportionality factor).

In various places it is stated, or at least suggested, that RAA (both the name and the technique) started with Richardson and Kuder (1933). This seems incorrect. That paper has no trace of RAA, although it does document a scale construction using Hollerith sorting and tabulation machines. What seems to be true, however, is that both the RAA name and the technique started at Proctor and Gamble in the early 1930s, in an interplay between Richardson and Horst, both Proctor and Gamble employees at the time. This relies on the testimony of Horst (1935), who does indeed attribute the name and basic idea of RAA to Richardson:

> The method which he suggested was based on the hypothesis that the scale value of each statement should be taken as a function of the average score of the men for whom the statement was checked and, further, that the score of each man should be taken as a function of the average scale value of the statements checked for the man.

The definition given by Horst is rather vague, because "a function of" is not very specific. It also does even mention the iteration of RAA to convergence (or, as Guttman would say, internal consistency). This iterative extension again seems to be due either to Horst or to Richardson. Horst was certainly involved at the time in the development of very similar techniques for quantitative data (Horst 1936; Edgerton and Kolbe 1936; Wilks 1938). For both quantitative and qualitative data these techniques are based on minimising within-person and maximising between-person variance, and they all result in computing the leading principal component of some data matrix. Horst (1935), starting from the idea to make linear combinations to maximise between individual variance, seems to have been the first one to realise that the equations defining RAA are the same as the equations describing Principal Component Analysis (PCA), and that consequently there are multiple RAA solutions for a given data matrix.

There are some additional hints about the history of RAA in the conference paper of Baker and Hoyt (1972). They also mostly credit Richardson, although they mention he never published a precise description of the technique, and it has been used "informally" without a precise justification ever since. They also mention that the first Hollerith type of computer implementation of RAA was by Mosier in 1942, the first UNIVAC programme was by Baker in 1962, and the first FORTRAN programme was by Baker and Martin in 1969.

We have not mentioned in our prehistory the work of Fisher (1938, 1940) and Maung (1941). These contributions, basically contemporaneous with Guttman (1941), clearly introduced the idea of optimal scaling for categorical data, of Correspondence Analysis of a two-way table, and even of non-linear transformation of the data to fit a linear (additive) model. They also came up with the first principal component of a Gramian matrix as a solution, realising there are multiple solutions to their equations. However, as pointed out by Gower (1990), they do not use MCA as it is currently defined. And, finally, although Hill (1973) seems to have independently come up with the RAA name and technique, its origins are definitely not in ecology.

3.2 Guttman 1941

RAA was used to construct a single one-dimensional scale, but Horst (1935) indicated already its extension to more than one dimension. The first publication of the actual formulas, using the luxuries of modern matrix algebra, was Guttman (1941), ironically in a chapter of a book edited by Horst. This is really where the history of MCA begins, although there are still some notable differences with later practise.

Guttman starts with the indicator matrix $\mathbf{G}$, and then generalises and systematises the analysis of variance approach to optimal scaling of indicator matrices. He introduces three criteria of internal consistency: one for the categories (columns), one for the objects (rows), and one for the entire table. All three criteria lead to the same optimal solution, which we now recognise as the first non-trivial dimension of MCA. We now also know, because we have been exposed to more matrix algebra than was common in the 1940s and 1950s, that this merely restates the fact that for any matrix $\mathbf{X}$ the non-zero eigenvalues of $\mathbf{X}'\mathbf{X}$ and $\mathbf{X}\mathbf{X}'$ are the same, and moreover they are equal to the squares of the singular values of $\mathbf{X}$. The left and right singular vectors of $\mathbf{X}$ are the eigenvectors of $\mathbf{X}'\mathbf{X}$ and $\mathbf{X}\mathbf{X}'$.

For our purposes in this paper the following quotation from Guttman's section five is important. When discussing the multiple solutions of the MCA stationary equations he says (pp. 330–331):

> There is an essential difference, however, between the present problem of quantifying a class of attributes and the problem of "factoring" a set of quantitative variates. The principal axis solution for a set of quantitative variates depends on the preliminary units of measurement

> of those variates. In the present problem, the question of preliminary units does not arise since we limit ourselves to considering the presence or absence of behaviour.

Thus Guttman, at least in 1941, shows a certain reluctance to consider the additional dimensions in MCA for data analysis purposes.

In addition to the stationary equations of MCA, Guttman also introduces the chi-square metric. He notes that the rank of **C**, and thus of **E**, is that of the indicator **G**, which is at most $1 + \sum_{j=1}^{m} \left(k_j - 1\right) = k_\star - (m - 1)$. Thus **C** has at least $m - 1$ zero eigenvalues, inherited from the linear dependencies in **G**. In addition **E** has a trivial eigen pair, independent of the data, with eigenvalue equal to 1. Suppose the vector **e** has all its $k_\star$ elements equal to $+1$. Then $\mathbf{Ce} = m\mathbf{De}$ and thus $\mathbf{Ey} = \mathbf{y}$, with $\mathbf{y} = \mathbf{D}^{\frac{1}{2}}\mathbf{e}$. If we deflate the eigenvalue problem by removing this trivial solution then the sum of squares of any off-diagonal submatrix of **C** is the chi-square for independence of that table.

Guttman also points out that the scores and weights linearise both regressions if we interpret the indicator matrix as a discrete bivariate distribution. This follows directly from the interpretation of MCA as a CA of the indicator matrix, because CA linearises regressions in a bivariate table. Of course interpreting the binary indicator matrix G as a bivariate distribution is quite a stretch. Both the chi-square metric and the linearised regressions were discussed earlier by Hirschfeld (1935) in the context of a single bivariate table. Neither Hirschfeld nor Fisher are mentioned in Guttman (1941).

There are no data and examples in Guttman's article. Benzécri (1977b) remarks:

> L. Guttman avait défini les facteurs mêmes calculés par l'analyse des correspondances. Il ne les avait toutefois pas calculés; pour la seule raison qu'en 1941 les moyens de calcul requis (ordinateurs) n'existaient pas.
>
> Translation: L. Guttman has defined the same factors as calculated by Correspondence Analysis. He did not calculate them, however, for the simple reason that in 1941 the necessary calculation tools (computers) did not exist.

That is not exactly true. In Horst (1941) the chapter by Guttman is followed by another chapter called "Two Empirical Studies of Weighting Techniques", which does have an empirical application in it. It is unclear who wrote that chapter, but the computations, which were carried out on a combination of tabulating and calculating machines, were programmed by nobody less than Ledyard R. Tucker.

3.3 Burt 1950

Guttman was reluctant to look at additional solutions of the stationary equations (additional "dimensions"), but Burt (1950) had no such qualms. After a discussion of the indicator matrix **G** and its corresponding cross product **C** (now known as the Burt table) Burt suggests a PCA of the normalised Burt table, i.e. solving the eigen problem $\mathbf{Ey} = m\lambda\mathbf{y}$. By the way, Burt discusses PCA as an alternative method of factor analysis, which is not in line with current usage clearly distinguishing PCA and FA.

Most of Burt's references are to previous PCA work with quantitative variables, and much of the paper tries to justify the application of PCA to qualitative data. No references to Guttman, Fisher, Horst, or Hirschfeld are given. The justifications that Burt presents are from the factor analysis perspective: **C** is a Gramian matrix, **E** is a correlation matrix, and the results of factoring **E** can lead to useful classifications of the individuals.

In the technical part of Burt's 1950 paper he discusses the rank, the trivial solutions, and the connection with the chi-squares of the bivariate subtables that we have already mentioned in our "Guttman (1941)" section.

3.4 Guttman 1953

As we saw in the introduction Guttman (1953) starts his paper with the observation that he already published the MCA equations in 1941. He gives this a positive spin, however, stating (p. 1):

> It is gratifying to see how Professor Burt has independently arrived at much the same formulation. This convergence of thinking lends credence to the suitability of the approach.

I will now insert a long quote from Guttman(1953, p. 2), because it emphasises the difference with Burt, and it is of major relevance to the present paper as well. Guttman really tells it like it is:

> My own article goes on to point out that, while the principal components here are formally similar to those for quantitative variables, nevertheless their interpretation may be quite different. The interrelations among qualitative items are not linear, nor even algebraic, in general. Similarly, the relation of a qualitative item to a quantitative variable is in general non-algebraic. Since the purpose of principal components–or any other method of factor analysis–is to help reproduce the original data, one must take into account this peculiar feature.
>
> The first principal component can possibly fully reproduce all the qualitative items entirely by itself: the items may be perfect, albeit non-algebraic, functions of this component. *Linear* prediction will not be perfect in this case, but this is not the best prediction technique possible for such data. Therefore, if the first principal component only accounts for a small proportion of the total variance of the data in the ordinary sense, it must be remembered that this ordinary sense implies linear prediction. If the correct, but *non-linear*, prediction technique is used, the whole variation can sometimes be accounted for by but the single component. In such a case, the existence of more than one principal component arises merely from the fact that a linear system is being used to approximate a non-linear one. (Each item is always a perfect linear function of *all* the principal components taken simultaneously).

This was written after the publication of Guttman (1950) in which the MCA of a perfect scale of binary items is discussed in impressive detail. The additional dimensions in such an analysis are curvilinear functions of the first, in regular cases in fact orthogonal polynomials of a single scale. Specifically, the second dimension is a quadratic, or quadratic-looking, function of the first, which creates the famous "horseshoe" or "arch" (in French: the "effect Guttman"). Since a horseshoe curves back in at its endpoints that name is often not appropriate, and we will call these non-linearities the Guttman effect. It seems that the second and higher curved dimensions are just mathematical artefacts, and much has been published since 1950 to explain them, interpret them, or to get rid of them (Hill and Gauch 1980).

In the rest of Guttman (1953) gives an overview of more of his subsequent work on scaling qualitative variables. This leads to material that goes beyond MCA (and thus beyond the scope of our present paper).

3.5 *Burt 1953*

Burt (1953, p. 5), in his reply to Guttman (1953), admits there are different objectives involved:

> If, as I gather, he cannot wholly accept my own interpretations, that perhaps is attributable to the fact that our starting-points were rather different. My aim was to factorise such data; his to construct a scale.

This does not answer the question, of course, if it is really advisable to apply PCA to the normalised Burt matrix. It also seems there also are some differences in national folklore, since Burt (1953, p. 6) goes on to say:

> In the chapters contributed to *Measurement and Prediction* both Dr. Guttman and Dr. Lazarsfeld draw a sharp distinction between the principles involved in these two cases. Factor analysis, they maintain, has been elaborated solely with reference to data which is quantitative *ab initio*; hence, they suppose, it cannot be suitably applied to qualitative data. On this side of the Atlantic, however, there has always been a tendency to treat the two cases together, and, with this double application in view, to define the relevant functions in such a way that they will (so far as possible) cover both simultaneously. British factorists, without specifying very precisely the assumptions involved, have used much the same procedures for either type of material. Nevertheless, there must of necessity be certain minor differences in the detailed treatment. These were briefly indicated in the paper Dr. Guttman has cited; but they evidently call for a closer examination. I think in the end it will be found that they are much slighter than might be supposed.

Burt then goes on to treat the case of a perfect scale of binary items, previously analysed by Guttman (1950). He points out that a PCA of a perfect scale gives (almost) the same results as those given by Guttman, and that consequently his approach of factoring a table works equally well as the approach that constructs a scale. Indeed, the differences between qualitative and quantitative factoring are "much slighter than might be supposed". Although Burt is correct, he does not discuss where the Guttman effect comes from, and whether it is desirable and/or useful.

3.6 Benzécri 1977

French data analysis ("Analyse des Données") views MCA as a special case of CA (Le Roux and Rouanet 2010). Benzécri (1977a) discusses the CA of the indicator matrix and gives a great deal of credit to Ludovic Lebart. Lebart (1975,1976) are usually mentioned as the first publications to actually use "analyse de correspondences multiples" and "tableau de Burt".

Benzécri also gives Lebart the credit for discovering that a CA of the indicator matrix **G** gives the same results as a CA of the Burt table **C**, which restates again our familiar matrix result that the singular value decomposition of a matrix gives the same results as the eigen decomposition of the two corresponding cross product matrices:

> L. Lebart en apporta la meilleure justification: les facteurs sur J issus de l'analyse d'un tel tableau I x J ne sont autres (à un coefficient constant près) que ceux issus de l'analyse du véritable tableau de contingence J x J suivant: k(j, j') = nombre des individus i ayant à la fois la modalité j et la modalité j'. Dès lors on rejoint le format original pour lequel a été conçue l'analyse des correspondances.
>
> Translation: L. Lebart has given the best justification: the factors on J from an analysis of an I x J table are the same as those from the analysis of the actual J x J contingency table with k(j, j') = the number of individuals i that are both in category j and j'. And thus we are back in the original format for which Correspondence Analysis was designed.

Benzécri also mentions the surprising generality and wide applicability of MCA:

> Le succès maintenant bien compris des analyses de tableaux en 0,1 mis sous forme disjonctive complète invite à rapprocher de cette forme, par un codage approprié, les données les plus diverses.
>
> Translation: The success, which we now understand well, of the analysis of (0,1) tables in disjunctive complete form invites us to apply this form, by suitable coding, to the most diverse forms of data.

This generality was later fully exploited in the book by Gifi (1990), which builds a whole system of descriptive multivariate techniques on top of MCA.

3.7 Gifi 1980

Gifi (1990) was mostly written in 1980–1981 from lecture notes for a graduate course in non-linear multivariate analysis, and builds on previous work in De Leeuw (1973). Throughout, the main engine of the Gifi approach to multivariate analysis minimises the meet-loss function:

$$\sigma\left(\mathbf{X};\ \mathbf{Y}_1,\ \ldots,\ \mathbf{Y}_m\right) = \sum_{j=1}^{m} \operatorname{tr}\ \left(\mathbf{X} - \mathbf{G}_j\mathbf{Y}_j\right)' \left(\mathbf{X} - \mathbf{G}_j\mathbf{Y}_j\right), \tag{2}$$

over the $n \times p$ matrices of scores $\mathbf{X}$ with $\mathbf{X}'\mathbf{X} = nI$ and over the $k_j \times p$ matrices of loadings $\mathbf{Y}_j$ that may or may not satisfy some constraints. Gifi calls this general approach Homogeneity Analysis (HA). Loss function (2) was partly inspired by Carroll (1968) who used this least squares loss function in generalised canonical analysis of quantitative variables.

The different forms of multivariate analysis in the Gifi framework arise by imposing additivity, and/or rank, and/or ordinal constraints on the $\mathbf{Y}_j$. See De Leeuw and Mair (2009) for a user's guide to the R package `homals`, which implements minimisation of meet-loss under these various sets of constraints.

If there are no constraints on the Y_j then minimising (2) computes the p dominant dimensions of an MCA. What makes the loss function (2) interesting in our comparative review of MCA is the distance interpretation and the corresponding geometry of the joint biplot of objects and categories. Gifi minimises the sum of the squared distances between an object and the categories of the variables that the object scores are in. If we make a separate biplot for each variable j it has n objects points and k_j category points. The category points are in the centroid of the object points in that category, and if we connect all those objects with their category points we get k_j star graphs in what Gifi calls the star plot. Minimising (2) means making the joint plot in such a way that the stars are as small as possible.

The homals package of De Leeuw and Mair (2009) actually computes the proportion of individuals correctly classified if we assign each individual to the category it is closest to (in p dimensions). In this way we can indeed find, like Guttman, that a single component can account for all of the "variance".

There are indications, especially in Gifi (1990, Sect. 3.9), that they are somewhat uncomfortable with the multidimensional scale construction aspects of MCA. They argue that each MCA dimension gives a quantification or transformation of the variables, and thus each MCA dimension can be used to compute a different correlation matrix between the variables. These correlation matrices, of which there are $k_\star - m$, can then all be subjected to a PCA. So the single indicator matrix leads to $k_\star - m$ PCA's. Gifi calls this "data production", and obviously does not like the outcome. Thus, as an alternative to MCA, they suggest using only the first dimension and the corresponding correlation matrix, which is very close to RAA and to Guttman (1941).

In the Gifi system the data production dilemma is further addressed in two ways. In the geometric framework based on the loss function (2) a form of non-linear PCA is defined in which we restrict the $k_j \times p$ category quantifications of a variable to have rank one, i.e. the points representing the categories of a variable are on a line through the origin. Gifi shows that this leads to the usual non-linear PCA techniques (De Leeuw 2006; Young et al. 1978). The second development to get away from the "data production" in MCA is the "aspect" approach (De Leeuw 1988a, b, 2004; Mair and De Leeuw 2010; De Leeuw et al. 1999). There we look for a single quantification or transformation of the variables that optimises any real valued function (or aspect) of the resulting correlation matrix. Non-linear PCA is the special cases in which we maximise the sum of the first p eigenvalues of the correlation matrix, and MCA chooses the scale to maximise the dominant eigenvalue. Other aspects lead to regression, canonical analysis, and structural equation models. In this more recent methodology based on aspects Guttman's one-dimensional scale construction approach has won out over Burt's multidimensional factoring method.

4 Deconstructing MCA

4.1 Introduction

We are left with the following questions from our history section, and from the Burt-Guttman exchange:

1. What, if anything, is the use of additional dimensions in MCA?
2. Where does the Guttman effect come from?
3. Is MCA really just PCA?
4. How many dimensions of MCA should we keep?
5. Which "variance" is "explained" by MCA?
6. How do we handle the "data production" aspects of MCA?

In De Leeuw (1982) several results are discussed that are of importance in answering these questions, and more generally for the interpretation (and deconstruction) of MCA. Additional, and more extensive, discussion of these same results is in Bekker and De Leeuw (1988) and De Leeuw (1988a).

To compute the MCA eigen decomposition we could, for example, use the Jacobi method, which diagonalises $\mathbf{E}$ by using elementary plane rotations. It builds up $\mathbf{Y}$ by minimising the sum of squares of the off-diagonal elements. Thus $\mathbf{E}$ is updated by iteratively replacing it by $J_{st}\mathbf{E}J_{st}$, where J_{st} with $s < t$ is a Jacobi rotation, that is, a matrix that differs from the identity matrix of order $k_\star$ only in elements (s, s) and (t, t), which are equal to u, and in elements (s, t) and (t, s) which are $+v$ and $-v$, where u and v are real numbers with $u^2 + v^2 = 1$. We cycle through all upper-diagonal elements $s < t$ for a single iteration, and continue iterating until the $\mathbf{E}$ update is diagonal (within some ϵ).

We shall discuss a different three-step method of *approximately* diagonalising $\mathbf{E}$, which, for lack of a better term, we call Deconstructed Multiple Correspondence Analysis (DMCA). It also works by applying elementary plane rotations to $\mathbf{E}$, but it is different from the Jacobi method because it is not intended to exactly diagonalise any arbitrary real symmetric matrix, or any normalised Burt matrix for that matter. It uses its rotations to eliminate all off-diagonal elements of all m^2 submatrices $\mathbf{E}_{kl}$, where $k, l = 1, \ldots, m$. If it cannot do this perfectly, it will try to find the best approximate diagonalisation. If DMCA does exactly diagonalise all submatrices, then some rearranging and additional computation finds the eigenvalues and eigenvectors of $\mathbf{E}$, and thus the MCA. The eigenvectors are, however, ordered differently (not by decreasing eigenvalues), and provide more insight in the inner workings of MCA. If an exact diagonalisation is not possible, the approximate diagonalisation often still provides this insight.

We first discuss some theoretical cases in which DMCA leads to the MCA, and after that some empirical examples are described in which the diagonalisation is only approximate and DMCA and MCA differ. As you will hopefully see, both types of DMCA examples show us what MCA as a data analysis technique tries to do, and how the results help in answering the six questions given above, arising from the Burt-Guttman exchange.

4.2 Mathematical Examples

4.2.1 Binary Data

Let's start with the case of binary data, i.e. indicator matrices for which all k_j are equal to two. The normalised Burt table $\mathbf{E} = m^{-1}\mathbf{D}^{-\frac{1}{2}}\mathbf{C}\mathbf{D}^{-\frac{1}{2}}$ consists of $m \times m$ submatrices $\mathbf{E}_{j\ell}$ of dimension 2×2. Suppose the marginals of variable j are p_{j0} and p_{j1}. For each j make the 2×2 table:

$$\mathbf{K}_j = \begin{bmatrix} +\sqrt{p_{j0}} & +\sqrt{p_{j1}} \\ +\sqrt{p_{j1}} & -\sqrt{p_{j0}} \end{bmatrix}, \tag{3}$$

and suppose $\mathbf{K}$ is the direct sum of the $\mathbf{K}_j$, i.e. the block-diagonal matrix with the $\mathbf{K}_j$ on the diagonal. Then $\mathbf{F} = \mathbf{K}'\mathbf{E}\mathbf{K}$ again has $m \times m$ submatrices of order two. For each $j, l = 1, \ldots, m$ the matrix $\mathbf{F}_{j\ell} = \mathbf{K}_j'\mathbf{E}_{j\ell}\mathbf{K}_\ell$ is diagonal, with element $(1, 1)$ equal to $+1$ and element $(2, 2)$ equal to the point correlation (or phi-coefficient) between binary variables j and ℓ (and thus also equal to $+1$ if $j = l$).

This means we can permute rows and columns of $\mathbf{F}$ using a permutation matrix $\mathbf{P}$ such that $\mathbf{R} = \mathbf{P}'\mathbf{F}\mathbf{P}$ is the direct sum of two correlation matrices $\mathbf{R}_{11}$ and $\mathbf{R}_{22}$, both of order m. $\mathbf{R}_{11}$ has all elements equal to $+1$, $\mathbf{R}_{22}$ has its off-diagonal elements equal to the phi-coefficients. We collect the $(1, 1)$ elements of all $\mathbf{F}_{jl}$, which are all $+1$, in $\mathbf{R}_{11}$ and the $(2, 2)$ elements in $\mathbf{R}_{22}$. Suppose $\mathbf{L}_1$ and $\mathbf{L}_2$ are the normalised eigenvectors of $\mathbf{R}_{11}$ and $\mathbf{R}_{22}$, and $\mathbf{L}$ is their direct sum. Then $\mathbf{\Lambda} = m^{-1}\mathbf{L}\mathbf{R}\mathbf{L}$ is diagonal, with on the diagonal the eigenvalues of $\mathbf{E}$ and with $\mathbf{KPL}$ the normalised eigenvectors of $\mathbf{E}$. Thus the eigenvalues of $\mathbf{E}$ are those of $m^{-1}\mathbf{R}_{11}$, i.e. one 1 and $m - 1$ zeros, together with those of $m^{-1}\mathbf{R}_{22}$. This restates the well-known result, mentioned by both Guttman (1941) and Burt (1950), that an MCA of binary data reduces to a PCA of the phi-coefficients.

4.2.2 Correspondence Analysis

Now let us look at Correspondence Analysis, that is, MCA with $m = 2$. There is only one single off-diagonal $p \times q$ cross table $\mathbf{C}_{12}$ in the Burt matrix. Suppose without loss of generality that $p \geq q$. Define $\mathbf{K}$ as the direct sum of the left and right singular vectors of $\mathbf{E}_{12}$. Then:

$$\mathbf{F} = \mathbf{K}'\mathbf{E}\mathbf{K} = \begin{bmatrix} \mathbf{I} & \Psi \\ \Psi' & \mathbf{I} \end{bmatrix}, \tag{4}$$

where Ψ is the $p \times q$ diagonal matrix of singular values of $\mathbf{E}_{12}$, and:

$$\mathbf{R} = \mathbf{P}'\mathbf{F}\mathbf{P} = \left\{ \bigoplus_{s=1}^{q} \begin{bmatrix} 1 & \psi_s \\ \psi_s & 1 \end{bmatrix} \right\} \bigoplus \mathbf{I}, \tag{5}$$

where the identity matrix at the end of Eq. (5) is of order $p - q$.

Thus the eigenvalues of $\mathbf{E}$ are $\frac{1}{2}(1 + \psi_s)$ and $\frac{1}{2}(1 - \psi_s)$ for all s, and DMCA indeed diagonalises $\mathbf{E}$. The relation between the eigen decomposition of $\mathbf{E}$ and the singular value decomposition of $\mathbf{E}_{12}$ is a classical result in Correspondence Analysis Benzécri (1977a), and earlier already in canonical correlation analysis of two sets of variables (Hotelling 1936).

4.2.3 Multinormal Distribution

Suppose we want to apply MCA to an m-variate standard normal distribution with correlation matrix $\mathbf{R} = \{\rho_{k\ell}\}$. Not to a sample, mind you, but to the whole distribution. This means we have to think of the submatrices $\mathbf{C}_{j\ell}$ as bivariate standard normal densities, having an infinite number of categories, one for each real number. Just imagine it as a limit of the discrete case (Naouri 1970).

In this case the columns of the $\mathbf{K}_j$, of which there is a denumerably infinite number, are the Hermite-Chebyshev polynomials $h_0, h_1, \ldots$ on the real line. We know that for the standard bivariate normal $\mathrm{E}_{j\ell}(h_s, h_t) = 0$ if $s \neq t$ and $\mathrm{E}_{j\ell}(h_s, h_s) = \rho_{j\ell}^s$. Thus $\mathbf{F} = \mathbf{K}'\mathbf{E}\mathbf{K}$ is an $m \times m$ matrix of diagonal matrices, where each $\mathbf{F}_{kl}$ submatrix is of denumerably infinite order and has all the powers of $\rho_{k\ell}$ along the diagonal. Then $\mathbf{R} = \mathbf{P}'\mathbf{F}\mathbf{P}$ is the infinite direct sum of elementwise powers of the matrix of correlation coefficients, or:

$$\mathbf{R} = \mathbf{P}'\mathbf{F}\mathbf{P} = \bigoplus_{s=0}^{\infty} \mathbf{R}^{(s)}, \tag{6}$$

and $\mathbf{\Lambda} = \mathbf{L}'\mathbf{R}\mathbf{L}$ is diagonal, with the first m eigenvalues of $\mathbf{R}^{(0)} = \mathbf{e}\mathbf{e}'$, then the m eigenvalues of $\mathbf{R}^{(1)} = \mathbf{R}$, then the m eigenvalues of $\mathbf{R}^{(2)} = \{\rho_{jl}^2\}$, and so on to $\mathbf{R}^{(\infty)} = \mathbf{I}$. Each MCA solution is composed of Hermite-Chebyshev polynomials of the same degree. Again, this restates a known result, already given in De Leeuw (1973).

These results remain true for what Yule called "strained multinormals", i.e. multivariate distributions that can be obtained from the multivariate normal by separate and generally distinct smooth monotone transformations of each of the variables. It also applies to mixtures of multivariate standard normal distributions with different correlation matrices (Sarmanov and Bratoeva 1967), to Gaussian copulas, as well as to other multivariate distributions whose bivariate marginals have diagonal expansions in systems of orthonormal functions (the so-called Lancaster probabilities, after Lancaster (1958, 1969).

The multinormal is a perfect example of the Guttman effect, that is, the eigenvector corresponding with the second largest eigenvalue usually is a quadratic function of the first, the next eigenvector usually is a cubic, and so on. We say "usually", because Gifi (1990, pp. 382–384) gives a multinormal example in which the first two eigenvectors of an MCA are both linear transformations of the underlying scale (i.e. they both come from $\mathbf{R}_{22}$). However, the Guttman effect is observed approximately in many (if not most) empirical applications of MCA, especially if the categories of the variables have some natural order and if the number of individuals is large enough.

4.2.4 Common Mathematical Structure

What do our three previous examples have in common mathematically? In all three cases there exist orthonormal $\mathbf{K}_j$ and diagonal $\Phi_{j\ell}$ such that $\mathbf{E}_{j\ell} = \mathbf{K}_j\Phi_{jl}\mathbf{K}'_\ell$. Or, in words, the matrices $\mathbf{E}_{j\ell}$ in the same row-block of $\mathbf{E}$ have their left singular vectors $\mathbf{K}_j$ in common, and matrices $\mathbf{E}_{j\ell}$ in the same column-block of $\mathbf{E}$ have their right singular vectors $\mathbf{K}_\ell$ in common. Equivalently, this requires that for each j the m matrices $\mathbf{E}_{j\ell}\mathbf{E}_{\ell j}$ commute.

Another way of saying this is that there are vectors $\mathbf{y}_1, \ldots, \mathbf{y}_m$ so that $\mathbf{C}_{j\ell}\mathbf{y}_\ell = \rho_{j\ell}\mathbf{D}_j\mathbf{y}_j$, i.e. so that all bivariate regressions are linear (De Leeuw 1988a). Not only that, we assume that such a set of weights exist for every dimension s, as long as $k_j \geq s$. If $k_j = 2$ then trivially all regressions are linear, because you can always draw a straight line through two points. If $m = 2$ all Correspondence Analysis solutions linearise the regressions in a bivariate table. In the multinormal example the Hermite polynomials provide the linear regressions. Simultaneous linearisability of all bivariate regressions seems like a strong condition, which will never be satisfied for observed Burt matrices. But our empirical examples, analysed below, suggest it will be approximately satisfied in surprisingly many cases. At the very least, assuming simultaneous linearisibility is a far-reaching generalisation of assuming multivariate normality.

In all three mathematical examples we used the direct sum of the $\mathbf{K}_j$ to diagonalise the $\mathbf{E}_{j\ell}$, then use a permutation matrix $\mathbf{P}$ to transform $\mathbf{F} = \mathbf{K}'\mathbf{E}\mathbf{K}$ into the direct sum $\mathbf{R} = \mathbf{P}'\mathbf{F}\mathbf{P}$ of correlation matrices, and then use the direct sum $\mathbf{L}$ to diagonalise $\mathbf{R}$ to $\Lambda = \mathbf{L}'\mathbf{R}\mathbf{L}$. This means that $\mathbf{KPL}$ has the eigenvectors of $\mathbf{E}$, but not ordered by decreasing or increasing eigenvalues. It also means that the eigenvectors have a special structure.

First, $\mathbf{F}$ is an $m \times m$ matrix of matrices $\mathbf{F}_{j\ell}$, which are $k_j \times k_\ell$. If all k_j are equal to, say, k, then $\mathbf{R}$ is a $k \times k$ matrix of matrices $\mathbf{R}_{st}$, which are all of order m. If the variables have a different number of categories, then $\mathbf{R}$ is a $k_+ \times k_+$ matrix of correlation matrices, with $\mathbf{R}_{st}$ of order $m_s \times m_t$, where m_s is defined as before as the number of variables with $k_j \geq s$.

KP is an orthonormal $m \times k_+$ matrix of matrices, in which column-block s is the direct sum of the m_s column vectors $\mathbf{K}_j\mathbf{e}_s$, with $\mathbf{e}_s$ unit vector s (equal to zero, except for element s, which is one). In a formula $\{\mathbf{KP}\}_{js} = \mathbf{K}_j\mathbf{e}_s\mathbf{e}_s'$ and $\{\mathbf{KP}\}_{js}\mathbf{L}_s = \mathbf{K}_j\mathbf{e}_s\mathbf{e}_s'\mathbf{L}_s$. Matrix $\{\mathbf{KPL}\}_{js}$ is the $k_j \times m_s$ outer product of column s of $\mathbf{K}_j$ and row s of $\mathbf{L}_s$. Each $\mathbf{R}_{ss}$ is computed with a single quantification of the variables, and there are only $k_+ - 1$ different non-trivial quantifications, instead of the $k_\star - m$ ones from MCA.

That the matrix **KPL** is blockwise of rank one connects DMCA with non-linear PCA, which is MCA with rank one restrictions on the category quatifications. We see that imposing rank one restrictions on MCA forces non-linear PCA to choose its solutions from the same $\mathbf{R}_{ss}$, thus preventing "data production".

5 The Chi-Square Metric

In the Correspondence Analysis of a single table it has been known since Hirschfeld (1935) that the sum of squares of the non-trivial singular values is equal to the chi-square (the total inertia) of the table. Although both Burt and Guttman pay homage to chi-square in the context of MCA, they do not really work through the consequences. In this section we analyse the total chi-square (TCS), which is the sum of all $m\,(m-1)$ off-diagonal bivariate chi-squares.

De Leeuw (1973, p. 32), shows that the TCS is related to the MCA eigenvalues by the simple equation:

$$\sum\sum_{1 \leq j \neq \ell \leq m} \mathcal{X}_{j\ell}^2 = n \sum_s (m\lambda_s - 1)^2\,, \tag{7}$$

where the sum on the right is over all $k_\star - m$ non-trivial eigenvalues. The same formula was given by Benzécri (1979). Equation (7), the MCA decomposition of the TCS, gives us a way to quantify the contribution of each non-trivial eigenvalue.

We now outline the DMCA decomposition of the TCS. An identity similar to (7) is:

$$\sum\sum_{1 \leq j \neq \ell \leq m} \mathcal{X}_{j\ell}^2 = \text{tr}\,\mathbf{E}^2 - (K + m\,(m-1))\;. \tag{8}$$

Equation (8) does not look particular attractive, until one realises that the constant subtracted on the right is the number of trivial elements in $\mathbf{F} = \mathbf{K}'\mathbf{E}\mathbf{K}$ (and thus in $\mathbf{R} = \mathbf{P}'\mathbf{K}'\mathbf{E}\mathbf{P}\mathbf{K}$) equal to one. There are K elements on the main diagonal, and $m\,(m-1)$ elements from the off-diagonal elements of the trivial matrix $\mathbf{R}_{11}g$.

Thus the TCS can be partitioned using $\mathbf{R}$, which is a $k_+ \times k_+$ matrix of matrices into $(k_+ - 1)^2$ non-trivial components. The most interesting ones are the $k_+ - 1$ sums of squares of the off-diagonal elements of the diagonal submatrices $\mathbf{R}_{22}, \ldots, \mathbf{R}_{k_+k_+}$, which is actually the quantity maximised by DMCA. And then there are the $(k_+ - 1)\,(k_+ - 2)$ sums of squares of the off-diagonal submatrices of $\mathbf{R}$, which is actually what DMCA minimises. The sum of squares of each diagonal block separately is its contribution to the DMCA fit, and total contribution to chi-square over all diagonal blocks shows how close DMCA is to MCA, i.e. how well DMCA diagonalises $\mathbf{E}$. In the mathematical examples from Sect. 4.2 DMCA is just a rearranged MCA, and all of the TCS comes from the diagonal blocks.

6 Computation

So, computationally, DMCA works in three steps. All three steps preserve orthonormality, guaranteeing that if DMCA diagonalisation works we have actually found eigenvalues and eigenvectors of $\mathbf{E}$, i.e. the MCA solution.

In the first step we compute the $\mathbf{K}_j$ by approximately diagonalising all off-diagonal $\mathbf{E}_{j\ell}$. This is is done in the mathematical examples by using known analytical results, but in empirical examples by Jacobi rotations that minimise the sum of squares of all off-diagonal elements of the off-diagonal $\mathbf{K}'\mathbf{E}\mathbf{K}$ (or, equivalently, maximise the sum of squares of the diagonal elements).

Each $\mathbf{K}_j$ is $k_j \times k_j$ and square orthonormal. We always set the first column of $\mathbf{K}_j$ equal to $n^{-\frac{1}{2}}\sqrt{\mathbf{d}_j}$, with d_j the marginals of variable j, to make sure the first column captures the non-zero trivial solution. This is done by setting the initial $\mathbf{K}_j$ to the left singular vectors of row-block j of E and not rotating pairs of indices (s, t) when s or t is one. This usually turns out to be a very good initial solution.

In the second step we permute the rows and columns of $\mathbf{F} = \mathbf{K}'\mathbf{E}\mathbf{K}$ into direct sum form. The $(1, 1)$ matrix $\mathbf{R}_{11}$ in $\mathbf{R} = \mathbf{P}'\mathbf{K}'\mathbf{E}\mathbf{K}\mathbf{P}$ has the $(1, 1)$ elements of all $\mathbf{F}_{j\ell}$, the $(1, 2)$ matrix $\mathbf{R}_{12}$ has the $(1, 2)$ elements of all $\mathbf{F}_{j\ell}$, and so on. Thus, if the first step has diagonalised all off-diagonal $\mathbf{E}_{j\ell}$, then all off-diagonal matrices in $\mathbf{R}$ are zero. The square symmetric matrices along the diagonal, of which there are k_+, are of order m, or of order m_s if not all k_j are equal. The first two, $\mathbf{R}_{11}$ and $\mathbf{R}_{22}$, are always of order m. $\mathbf{R}_{11}$ takes care of all m trivial solutions and has all its elements equal to one.

Then, in the third step, we diagonalise the matrices along the diagonal of $\mathbf{R}$ by computing their eigenvalues and eigenvectors. This gives $\mathbf{\Lambda} = \mathbf{L}'\mathbf{R}\mathbf{L}$, which is diagonal if the first step succeeded in diagonalising all off-diagonal $\mathbf{E}_{j\ell}$. All the loss that can make DMCA an imperfect diagonalisation method is in the first step, computing both $\mathbf{P}$ and $\mathbf{L}$ does not introduce any additional loss. Note again that the direct sums of $\mathbf{K}$ and $\mathbf{L}$ and the permutation matrix $\mathbf{P}$ are all orthonormal, and thus so are $\mathbf{KP}$ and $\mathbf{KPL}$.

Finally we compute $\mathbf{Y}'\mathbf{KPL}$, with $\mathbf{Y}$ the MCA solution, to see how close $\mathbf{Y}$ and $\mathbf{KPL}$ are, and which $\mathbf{R}_{ss}$ the MCA solutions come from. Note that $\mathbf{Y}'\mathbf{KPL}$ is also square orthonormal, which implies sums of squares of rows and columns add up to one, and squared elements can be interpreted as proportions of "variance explained".

DMCA has an interesting relationship with the Ordered Multiple Correspondence Analysis (OMCA) of Lombardo and Meulman (2010). DMCA choose the $\mathbf{K}_j$ that make the $\mathbf{E}_{jl}$ as diagonal as possible, in order to concentrate as much of the TCS in the diagonal correlation matrices $\mathbf{R}_{ss}$. In OMCA the $\mathbf{K}_j$ are chosen as orthogonal polynomials for variable j of degrees $0, \ldots, k_j - 1$, with again $\mathbf{K}$ their direct sum. Then compute $\mathbf{F} = \mathbf{K}'\mathbf{E}\mathbf{K}$ and $\mathbf{R} = \mathbf{P}'\mathbf{F}\mathbf{P}$ and $\Lambda = \mathbf{L}'\mathbf{R}\mathbf{L}$ as in DMCA. This gives the same type of partitioning of the TCS, and the same blockwise rank one approximate eigenvectors **KPL**, but of course with less of the total TCS concentrated on the diagonal. In the case of binary data and a continuous multinormal OMCA and DMCA are the same. If there are only two variables they are different, and the OMCA results are a rearrangement of those in Beh (1997). Of course if the $\mathbf{K}_j$ computed by DMCA are not polynomials, for example if categories are unordered nominal, the two methods can give very different results. But a more detailed comparison on various real examples would be useful. The web directory https://jansweb.netlify.app/post/code/ also has R code for MCA and OMCA.

6.1 The Programme

For the empirical examples in the present paper we use the R function `DMCA`, a further elaboration of the R function `jMCA` from De Leeuw and Ferrari (2008). The programme, and all the empirical examples with the necessary data manipulations, can be downloaded from https://jansweb.netlify.app/post/code/. The programme maximises the percentage of the TCS in the diagonal blocks of the DMCA. It is called with arguments:

- `burt`, the Burt matrix,
- `k`, the number of categories of the variables,
- `eps`, iteration precision, defaults to 1e-8,
- `itmax`, maximum number of iterations, defaults to 500,
- `verbose`, prints DMCA fit for all iterations, defaults to TRUE,
- `vectors`, DMCA eigenvectors, if FALSE only DMCA eigenvalues, defaults to TRUE,

and it returns a list with

- `kek`, the matrix $K'EK$,
- `pkekp`, the matrix $P'K'EKP$,
- `lpkekpl`, the matrix $L'P'K'EKPL$,
- `k`, the block-diagonal matrix K,
- `p`, the permutation P,
- `l`, the block-diagobal matrix L,
- `kp`, the matrtix KP,
- `kpl`, the matrix KPL,
- `chisquares`, the $m\,(m-1)$ chi-squares
- `chipartition`, the DMCA chi-partition,
- `chipercentages` = `chipartition` / TCS,

- `itel`, the number of iterations,
- `func`, the optimum value of trace of chipercentages

7 Empirical Examples

We analysed DMCA in our previous examples by relying solely on specific mathematical properties. There are some empirical examples in the last section of De Leeuw (1982), but with very little detail, and computed with a now tragically defunct APL programme. Showing the matrices $\mathbf{K}$, $\mathbf{P}$, $\mathbf{L}$ as well as $\mathbf{F}$, $\mathbf{R}$ and $\mathbf{\Lambda}$ in this paper would take up too much space, so we concentrate on how well DMCA reproduces the MCA eigenvalues. We also discuss which of the correlation matrices in $\mathbf{R}$ the first and last MCA vectors of weights (eigenvectors) are associated with, and we give the partitionings of the TCS.

7.1 Burt Data

The data for the example in Burt (1950) were collected by him in Liverpool in or before 1912, and are described in an outrageously politically incorrect paper (Burt 1912). Burt used $m = 4$, with variables hair-colour (fair, red, dark), eye colour (light, mixed, brown), head (narrow, wide), and stature (tall, short) for 100 individuals selected from his sample. This is not very interesting as a DMCA or MCA example because the data are so close to binary and thus there is not much room for DMCA to work with. We include the Burt data, using the Burt table from Burt (1950), for historical reasons.

The Burt table is of order $k_\star = 10$, so there are $k_\star - m = 6$ non-trivial eigenvalues. DMCA takes one single iteration cycle to convergence to fit 0.9462 from the initial SVD solution. Figure 1 plots the sorted MCA and DMCA non-trivial eigenvalues. In these plots we always remove the trivial points (0, 0) and (1, 1) because they would anchor the plot and unduly emphasise the closeness of the two solutions.

The matrix $\mathbf{R}$ has two diagonal blocks, $\mathbf{R}_{11}$ and $\mathbf{R}_{22}$, of order four and one block, $\mathbf{R}_{33}$, of order two. Thus the m_s are (4, 4, 2). The first non-trivial MCA solution correlates 0.9997 with the first non-trivial DMCA solution, which corresponds with the dominant eigenvalue of $\mathbf{R}_{22}$. The second MCA solution correlates -0.7319 with the second DMCA solution from $\mathbf{R}_{22}$ and -0.3749 and -0.5675 with the two DMCA solutions from $\mathbf{R}_{33}$. The fifth and sixth MCA solutions (the ones with the smallest non-trivial eigenvalues) correlate 0.9824 and 0.9937 with the remaining two DMCA solutions from $\mathbf{R}_{22}$. Thus, almost all the variation comes from $\mathbf{R}_{22}$, because with the k_j as small as (3, 3, 2, 2) we are very close to the case where all variables only take two values and all the variation is in the phi-coefficients in $\mathbf{R}_{22}$.

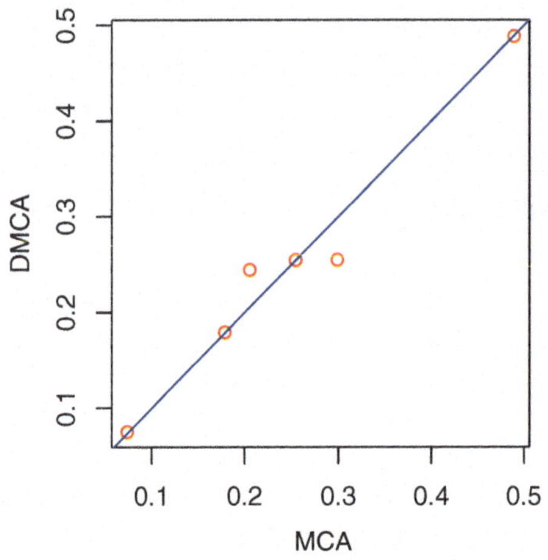

Fig. 1 Burt MCA/DMCA eigenvalues

We can further illustrate this with the chi-square partitioning. Of the TCS of 156.68 the diagonal blocks $\mathbf{R}_{22}$ and $\mathbf{R}_{33}$ contribute, respectively, 148.1664 (95%) and 0.08237 (0.05%), while the off-diagonal blocks contribute 8.4319 (5%).

7.2 GALO Data

The GALO data (Peschar 1975) are a mainstay Gifi example. The individuals are $n = 1290$ sixth grade school children in the city of Groningen, The Netherlands, about to go into secondary education. The $m = 4$ variables are gender (2 categories), IQ (9 categories), teachers advice (7 categories), and socio-economic status (6 categories). The Burt matrix is of order $k_\star = 24$, and thus there are $k_\star - m = 20$ non-trivial dimensions. Matrix $\mathbf{R} = \mathbf{P}'\mathbf{F}\mathbf{P}$ has 9 diagonal correlation blocks, with $\mathbf{R}_{11}$ and $\mathbf{R}_{22}$ of order four, $\mathbf{R}_{33}, \ldots, \mathbf{R}_{66}$ of order three, $\mathbf{R}_{77}$ of order two, and $\mathbf{R}_{88}$ and $\mathbf{R}_{99}$ of order one. DMCA takes 37 iteration cycles to a fit of 0.8689. The 20 sorted non-trivial MCA and DMCA eigenvalues are plotted in Fig. 2.

The strong Guttman effect in the GALO data is reflected in the close correspondence between the MCA and DMCA solutions. The first non-trivial MCA solution correlates 0.9967 with the dominant DMCA solution from $\mathbf{R}_{22}$, and the second MCA solution correlates 0.9915 with the dominant DMCA solution from $\mathbf{R}_{33}$. After that correlations become smaller, until we get to the smallest eigenvalues. The worst MCA solution correlates -0.9882 with the solution corresponding to smallest eigenvalue of $\mathbf{R}_{22}$, and the next worst correlates -0.9794 with the solution with the smallest eigenvalue of $\mathbf{R}_{33}$.

To illustrate graphically how close MCA and DMCA are we plot the 24 category quantifications on the first non-trivial dimension of the MCA solution (MCA dimen-

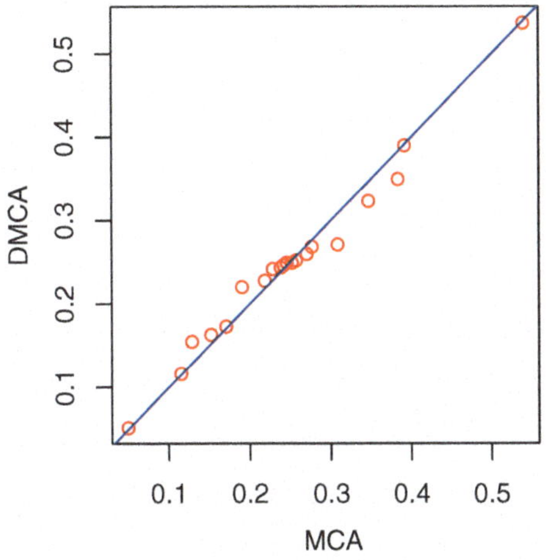

Fig. 2 GALO MCA/DMCA eigenvalues

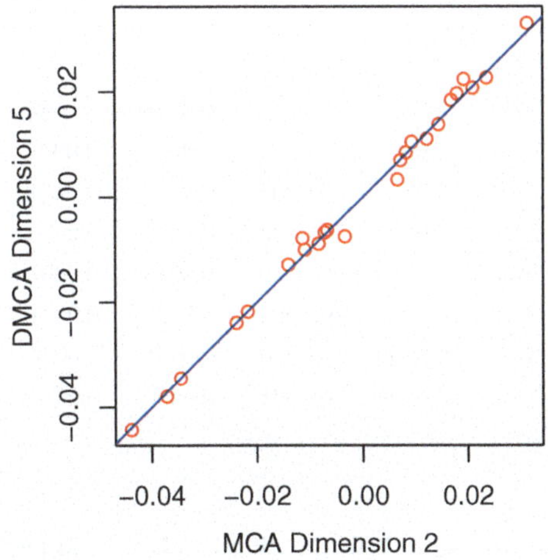

Fig. 3 GALO MCA/DMCA quantifications

sion two) and the first non-trivial dimension of DMCA (dimension five) in Fig. 3. Note the dominant MCA dimension is always the trivial one, so we need the second MCA dimension. For DMCA the first four dimensions correspond with the trivial $\mathbf{R}_{11}$, and thus the first interesting dimension is number $m + 1$, corresponding with the dominant eigenvalue of $\mathbf{R}_{22}$. In Fig. 4 we plot the corresponding MCA dimension three and DMCA dimension $2m + 1 = 9$, corresponding with the dominant eigenvalue of $\mathbf{R}_{33}$.

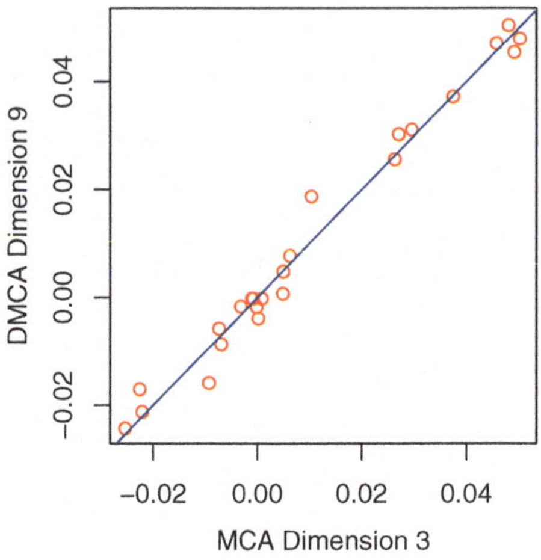

Fig. 4 GALO MCA/DMCA quantifications

Table 1 GALO TCS percentages

	DMCA2	DMCA3	DMCA4	DMCA5	DMCA6	DMCA7	DMCA8	DMCA9
DMCA2	0.5631	0.0020	0.0215	0.0172	0.0146	0.0024	7e-04	1e-04
DMCA3	0.0020	0.1638	0.0003	0.0001	0.0011	0.0001	0e+00	4e-04
DMCA4	0.0215	0.0003	0.0831	0.0003	0.0010	0.0001	0e+00	4e-04
DMCA5	0.0172	0.0001	0.0003	0.0492	0.0016	0.0008	0e+00	1e-04
DMCA6	0.0146	0.0011	0.0010	0.0016	0.0058	0.0001	5e-04	0e+00
DMCA7	0.0024	0.0001	0.0001	0.0008	0.0001	0.0041	0e+00	0e+00
DMCA8	0.0007	0.0000	0.0000	0.0000	0.0005	0.0000	0e+00	0e+00
DMCA9	0.0001	0.0004	0.0004	0.0001	0.0000	0.0000	0e+00	0e+00

The chi-square partitioning tells us the diagonal blocks of DMCA "explain" 87% of the TCS, with the blocks $\mathbf{R}_{22}, \ldots, \mathbf{R}_{77}$ contributing 56, 16, 8, 5, 0.5, and 0.4%. The complete partitioning is summarised in Table 1.

7.3 BFI Data

Our final example is larger, and somewhat closer to an actual application of MCA. The BFI data set is taken from the `psychTools` package (Revelle 2021). It has $n = 2800$ observations on $m = 25$ personality self report items. After removing persons with missing data there are $n = 2436$ observations left. Each item has $k = 6$ categories, and thus the Burt table is of order $m \times k = 150$. Matrix $\mathbf{R}$, excluding $\mathbf{R}_{11}$, has five diagonal blocks of order 25. DMCA takes 54 iterations for a DMCA

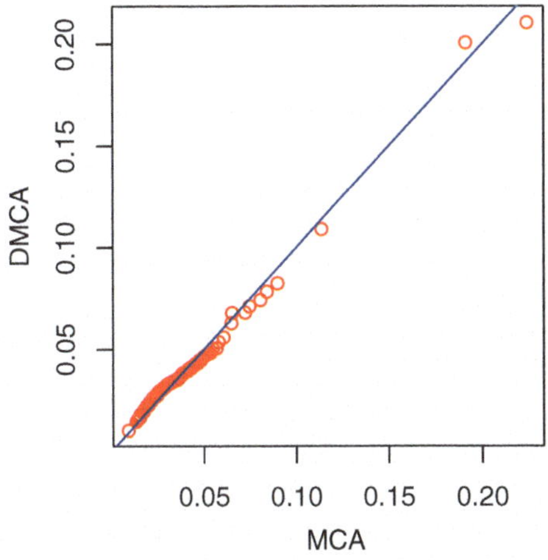

Fig. 5 BFI MCA/DMCA eigenvalues

Table 2 BFI TCS percentages

	DMCA2	DMCA3	DMCA4	DMCA5	DMCA6
DMCA2	0.4877	0.0153	0.0059	0.0055	0.0041
DMCA3	0.0153	0.3302	0.0053	0.0037	0.0035
DMCA4	0.0059	0.0053	0.0394	0.0049	0.0042
DMCA5	0.0055	0.0037	0.0049	0.0206	0.0046
DMCA6	0.0041	0.0035	0.0042	0.0046	0.0081

fit of 0.8860. The sorted non-trivial 125 MCA and DMCA eigenvalues are plotted in Fig. 5.

The percentages of the TCS from the non-trivial submatrices of **R** are summarised in Table 2.

8 Discussion

Our mathematical and empirical examples show that in a wide variety of circumstances MCA and DMCA eigenvalues and eigenvectors are very similar, although DMCA uses far fewer degrees of freedom for its diagonalisation. This indicates that DMCA can be thought of, at least in some circumstances, as a smooth version of MCA. The error is moved to the off-diagonal elements in the submatrices of $\mathbf{R} = \mathbf{P}'\mathbf{F}\mathbf{P}$ and the structure is concentrated in the diagonal correlation matrices.

We have also seen that DMCA is like MCA, in the sense that it gives very similar solutions, but it is also like non-linear PCA, because it imposes the rank one restrictions on the weights. Thus it is a bridge between the two techniques, and it clarifies their relationship.

DMCA also shows where the dominant MCA solutions originate, and indicates quite clearly where the Guttman effect comes from (if it is there). It suggest the Guttman effect, in a generalised sense, does not necessarily result in polynomials or arcs. As long as there is simultaneous linearisation of all bivariate regressions $\mathbf{E}$ is orthonormally similar to the direct sum of the $\mathbf{R}_{ss}$, and the principal components of the $\mathbf{R}_{ss}$ will give a generalised Guttman effect.

This allows us to suggest some answer for questions coming from the Burt-Guttman exchange. In many cases the principal components of MCA (beyond the first) come from the generalised Guttman effect, and should be interpreted as such. Thus the first principal component does have a special status, and thus justifies singling out RAA and Guttman scaling from the rest of MCA.

DMCA also reduces the amount of data production. Instead of $k_\star - m$ non-trivial correlations matrices of order m with their PCA's, we now have $k_+ - 1$ non-trivial correlation matrices of orders given by the m_s. That is still more than one single correlation matrix, as we have in non-linear PCA and the aspect approach, but the different correlation matrices may either be related by the Guttman effect or give non-trivial additional information.

We also mention some other attempts, besides (7) and DMCA, to deal with the influence of the diagonal blocks on the MCA solution. The first is Greenacre's Joint Correspondence Analysis or JMCA (Greenacre 1988), which minimises $\mathbf{E} - \mathbf{U}\mathbf{U}'$ not only over all $K \times p$ matrices $\mathbf{U}$ with $\mathbf{U}'\mathbf{U} = \mathbf{I}$, but in addition over the m diagonal blocks of $\mathbf{E}$. In JMCA the dominant trivial dimension is first removed. JMCA uses a variation of the Thomson's alternating least squares algorithm for least squares factor analysis, alternating the minimising over $\mathbf{U}$ for given $\mathbf{C}$ and the minimising over the diagonal blocks of $\mathbf{C}$ for given $\mathbf{U}$. The first minimisation is an MCA of the modified Burt matrix with the current diagonal blocks, the second minimisation replaces the diagonal blocks of $\mathbf{C}$ with the corresponding ones of $\mathbf{U}\mathbf{U}'$. As a result JMCA does optimise the fit to the TCS without the adjustments of (7). Nevertheless there are some problems with JMCA. It fixes the dimension p at a low value, and can compute separate un-nested solutions for each p. Thus it tends to "data production" in our sense, because we have to find a way to relate the solutions for different p. As in DMCA and MCA it would be advantageous to have a complete and simultaneous nested solution by always choosing $p = K - m$. The second problem with JMCA is that, when p becomes larger, Heywood cases may become more common, i.e. cases in which the reduced Burt matrix is no longer positive semi-definite. This potentially leads to complex numbers and negative variances.

The second way of dealing with the undesirable dimensionality and explained variances aspects of MCA is not to require $\mathbf{U}'\mathbf{U} = \mathbf{I}$ but $\mathbf{U}_j'\mathbf{U}_j = \mathbf{I}$ for all j. This is sometimes called strong orthogonality (Dauxois and Pousse 1976). We could call the resulting technique strong multiple correspondence analysis of SMCA. If $m = 2$ SMCA still gives MCA, and thus also CA and JMCA, but if $m > 2$ SMCA is only MCA or JMCA if we have simultaneous linearisability. SMCA tends to make all variables equally important; see the discussion in Nishisato and Sheu (1980). SMCA also has its problems. The constraint $\mathbf{U}_j'\mathbf{U}_j = \mathbf{I}$ limits the dimensionality of the non-trivial quantifications for variable j to $k_j - 1$, and it is unclear what to do with the higher dimensions in $\mathbf{E}$. In DMCA strong orthogonality constraints are imposed on the $\mathbf{K}_j$, but the columns of the $\mathbf{K}_j$ are distributed over different correlation matrices, and the resulting $\mathbf{U}_j$ are of rank one, but no longer orthonormal. The mathematical properties of both JMCA and SMCA deserve some further study.

This also seems the place to point out a neglected aspect of MCA. The smallest non-trivial solution gives a quantification or transformation of the data that maximises the singularity of the transformed data, i.e. the minimum eigenvalue of the corresponding correlation matrix. We have seen in our empirical examples that MCA and DMCA often agree closely in their smallest eigenvalue solutions, and that may indicate that it should be possible to give a scientific interpretation of these "bad" solutions. In fact, the smallest DMCA and MCA eigenvalues can be used in a regression interpretation in which we consider one or more of the variables as criteria and the others are predictors.

A complaint that many users of MCA have is that, say, the first two components "explain" such a small proportion of the "variance", by which they mean the trace of $\mathbf{E}$, which is $\mathbf{K}$, the total number of categories, and which, of course, has nothing to do with "variance". Equation (7) indicates how to quantify the contributions of the non-trivial eigenvalues. For the BFI data, for example, the first two non-trivial MCA eigenvalues "explain" 0.0832 percent of the "variance", but they "explain" 0.6305 percent of the TCS. Moreover DMCA shows us that we should really relate the eigenvalues to the $\mathbf{R}_{ss}$ they come from, and see how much they "explain" of their correlation matrices. It is even better to evaluate their contributions using the TCS and its partitioning described in Sect. 5 of this paper.

References

Baker, F.B., Hoyt, C.J.: The relation of the method of reciprocal averages to Guttman's internal consistency scaling method. Available online at https://eric.ed.gov/?id=ED062397 (1972). Last accessed 2 May 2023

Beh, E.J.: Simple correspondence analysis of ordinal cross-classifications using orthogonal polynomials. Biometrical J. **39**(5), 589–613 (1997)

Bekker, P., De Leeuw, J.: Relation between variants of nonlinear principal component analysis. In: Van Rijckevorsel, J.L.A., De Leeuw, J. (eds.) Component and Correspondence Analysis, pp. 1–31. Wiley, Chichester (1988)

Benzécri, J.P.: Histoire et préhistoire de l'analyse des données. Part V: l'Analyse des correspondances. Les Cahiers de l'Analyse Des Données **2(1)**, 9–40 (1977a)

Benzécri, J.P.: Sur l'analyse des tableaux binaires associés à une correspondance multiple. Les Cahiers de l'Analyse Des Données **2**(1), 55–71 (1977b)

Benzécri, J.P.: Sur le calcul des taux d'inertie dans l'analyse d'un questionaire. Les Cahiers de l'Analyse Des Données **4**(3), 377–378 (1979)

Burt, C.: The inheritance of mental characters. Eugenics Rev. **4**, 168–200 (1912)

Burt, C.: The factorial analysis of qualitative data. Br. J. Stat. Psychol. **3**, 166–85 (1950)

Burt, C.: Scale analysis and factor analysis. Br. J. Stat. Psychol. **6**, 5–23 (1953)

Carroll, J.D.: A generalization of canonical correlation analysis to three or more sets of variables. In: Proceedings of the 76th Annual Convention of the American Psychological Association, pp. 227–228. American Psychological Association, Washington, DC (1968)

Dauxois, J., Pousse, A.: Les Analyses Factorielles en Calcul des Probabilités et en Statistique: Essai d'Étude Synthétique. PhD thesis, Université Paul-Sabatier, Toulouse, France (1976)

De Leeuw, J.: Canonical Analysis of Categorical Data. PhD thesis, Leiden University, Leiden, The Netherlands (1973)

De Leeuw, J.: Nonlinear principal component analysis. In: Caussinus, H., Ettinger, P., Tomassone, R. (eds.) COMPSTAT 1982, pp. 77–86. Physika Verlag, Vienna, Austria (1982)

De Leeuw, J.: Multivariate analysis with linearizable regressions. Psychometrika **53**, 437–454 (1988a)

De Leeuw, J.: Multivariate analysis with optimal scaling. In: Das Gupta, S., Ghosh, J.K. (eds.) Proceedings of the International Conference on Advances in Multivariate Statistical Analysis, pp. 127–160. Indian Statistical Institute, Calcutta, India (1988b)

De Leeuw, J.: Least squares optimal scaling of partially observed linear systems. In: Van Montfort, K., Oud, J., Satorra, A. (eds.) Recent Developments in Structural Equation Models. Kluwer Academic Publishers, Dordrecht, The Netherlands (2004)

De Leeuw, J.: Nonlinear principal component analysis and related techniques. In: Greenacre, M., Blasius, J. (eds.) Multiple Correspondence Analysis and Related Methods, pp. 107–133. Chapman & Hall/CRC, Boca Raton, FL (2006)

De Leeuw, J., Ferrari, D.B.: Using Jacobi Plane Rotations in R. Preprint Series, vol. 556. UCLA Department of Statistics, Los Angeles, CA (2008)

De Leeuw, J., Mair, P.: Homogeneity analysis in R: the package `homals`. J. Stat. Software **31**(4), 1–21 (2009)

De Leeuw, J., Michailidis, G., Wang D.Y.: Correspondence analysis techniques. In: Ghosh, S. (ed.) Multivariate Analysis, Design of Experiments, and Survey Sampling, pp. 523–547. Marcel Dekker (1999)

Edgerton, H.A., Kolbe, L.E.: The method of minimum variation for the combination of criteria. Psychometrika **1**(3), 183–187 (1936)

Fisher, R.A.: Statistical Methods for Research Workers, 6th edn. Oliver & Boyd (1938)

Fisher, R.A.: The precision of discriminant functions. Ann. Eugenics **10**, 422–429 (1940)

Gifi, A.: Nonlinear Multivariate Analysis. Wiley, New York (1990)

Gower, J.C.: Fisher's optimal scores and multiple correspondence analysis. Biometrics **46**, 947–961 (1990)

Greenacre, M.J.: Correspondence analysis of multivariate categorical data by weighted least-squares. Biometrika **75**(3), 457–467 (1988)

Guttman, L.: The quantification of a class of attributes: a theory and method of scale construction. In: Horst, P., Wallin, P., Guttman, L. (eds.) The Prediction of Personal Adjustment, pp. 321–348. Social Science Research Council, New York (1941)

Guttman, L.: The principal components of scale analysis. In: Stouffer, S.A., Guttman, L., Suchman, E.A., Lazarsfeld, P.F., Star, S.A., Clausen, J.A. (eds.) Measurement and Prediction, pp. 312–361. Princeton University Press, Princeton, NJ (1950)

Guttman, L.: A note on Sir Cyril Burt's "factorial analysis of qualitative data.âŁž Br. J. Stat. Psychol. **6**, 1–4 (1953)

Hill, M.O.: Reciprocal averaging: an eigenvector method of ordination. J. Ecol. **61**(1), 237–249 (1973)

Hill, M.O., Gauch, G.: Detrended correspondence analysis: an improved ordination technique. Vegetatio **42**, 47–58 (1980)

Hirschfeld, H.O.: A connection between correlation and contingency. Proc. Camb. Philos. Soc. **31**, 520–524 (1935)

Horst, P.: Measuring complex attitudes. J. Soc. Psychol. **6**(3), 369–374 (1935)

Horst, P.: Obtaining a composite measure from a number of different measures of the same attribute. Psychometrika **1**(1), 54–60 (1936)

Horst, P.: The Prediction of Personal Adjustment. Social Science Research Council, New York (1941)
Hotelling, H.: Relations between two sets of variates. Biometrika **28**, 321–377 (1936)
Lancaster, H.O.: The structure of bivariate distributions. Ann. Math. Stat. **29**, 719–736 (1958)
Lancaster, H.O.: The Chi-Squared Distribution. Wiley (1969)
Le Roux, B., Rouanet, H.: Multiple Correspondence Analysis. Sage, Thousand Oaks, CA (2010)
Lebart, L.: L'orientation du dépouillement de certaines enquetes par l'analyse des correspondences multiples. Consommation **2**, 73–96 (1975)
Lebart, L.: Sur les calculs impliqués par la description de certains grands tableaux. Annales de l'Inséé **22**(23), 255–271 (1976)
Lebart, L., Saporta, G.: Historical elements of correspondence analysis and multiple correspondence analysis. In: Blasius, J., Greenacre, M. (eds.) Visualization and Verbalization of Data, pp. 31–44. CRC Press, Boca Raton, FL (2014)
Lombardo, R., Meulman, J.J.: Multiple correspondence analysis via polynomial transformations of ordered categorical variables. J. Classif. **27**, 191–210 (2010)
Mair, P., De Leeuw, J.: A general framework for multivariate analysis with optimal scaling: the R package `aspect`. J. Stat. Software **32**(9), 1–23 (2010)
Maung, K.: Discriminant analysis of Tocher's eye colour data for Scottish school children. Ann. Eugenics **11**, 64–76 (1941)
Naouri, J.C.: Analyse factorielle des correspondances continues. Publications de l'Institute de Statistique de l'Université de Paris **19**, 1–100 (1970)
Nishisato, S.: Analysis of Categorical Data: Dual Scaling and its Applications. University of Toronto Press, Toronto, Canada (1980)
Nishisato, S., Sheu, W.: Piecewise method of reciprocal averages for dual scaling of multiple-choice data. Psychometrika **45**, 467–478 (1980)
Peschar, J.L.: School, Milieu. Beroep. Tjeek Willink, Groningen, The Netherlands (1975)
Revelle, W.: `psychTools`: Tools to Accompany the 'psych' Package for Psychological Research. Available online on the CRAN at https://cran.r-project.org/web/packages/psychTools/ (2021). Last accessed 2 May 2023
Richardson, M.W., Kuder, G.F.: Making a rating scale that measures. Pers. J. **12**, 36–40 (1933)
Sarmanov, O.V., Bratoeva, Z.N.: Probabilistic properties of bilinear expansions of Hermite polynomials. Theo. Probabil. Appl. **12**(32), 470–481 (1967)
Tenenhaus, M., Young, F.W.: An analysis and synthesis of multiple correspondence analysis, optimal scaling, dual scaling, homogeneity analysis and other methods for quantifying categorical multivariate data. Psychometrika **50**, 91–119 (1985)
Wilks, S.S.: Weighting systems for linear functions of correlated variables when there is no dependent variable. Psychometrika **3**(1), 23–40 (1938)
Young, F.W., Takane, Y., De Leeuw, J.: The principal components of mixed measurement level multivariate data: an alternating least squares method with optimal scaling features. Psychometrika **45**, 279–281 (1978)

Generalised Canonical Correlation and Multiple Correspondence Analyses Reformulated as Matrix Factorisation

Kohei Adachi, Henk A. L. Kiers, Takashi Murakami, and Jos M. F. ten Berge

1 Introduction

Nishisato's (1980) pioneering book on dual scaling is included in the literature which interested the authors of this paper in multiple correspondence analysis (MCA) for multivariate categorical data. In this paper, we address its relationships to generalised canonical correlation analysis (GCCA) for multivariate numerical data rather than categorical data.

Let $\mathbf{X} = [\mathbf{X}_1, \ldots, \mathbf{X}_m]$ be an n-observations $\times$ K-variables block data matrix, whose j th block $\mathbf{X}_j$ is an $n \times K_j$ matrix for $j = 1, \ldots, m$ with $K = \sum_j K_j$ and $n > K_j$. GCCA refers to a multivariate analysis procedure for exploring the inter-relationships among the m sets of variables in $\mathbf{X} = [\mathbf{X}_1, \ldots, \mathbf{X}_m]$, when $\mathbf{X}$ contains numerical variables, particularly with $m \geq 3$. GCCA was originally proposed by Carroll (1968) and later formulated in some different manners as optimising functions of the correlations/variances defined for the linear composites of the variables, as reviewed by Kettenring (1971), van de Geer (1984), and Tenenhaus and Tenenhaus (2011). However, those authors have not treated least squares formulations of GCCA, which are considered in this paper.

K. Adachi (✉)
Graduate School of Human Sciences, Osaka University, Osaka, Japan
e-mail: adachi@hus.osaka-u.ac.jp

H. A. L. Kiers · J. M. F. ten Berge
Department of Psychology, University of Groningen, Groningen, The Netherlands
e-mail: h.a.l.kiers@rug.nl

J. M. F. ten Berge
e-mail: j.m.f.ten.berge@rug.nl

T. Murakami
Institute of Cultural Science, Chukyo University, Nagoya, Japan
e-mail: tandem06@sass.chukyo-u.ac.jp

E. J. Beh et al. (eds.), *Analysis of Categorical Data from Historical Perspectives*, Behaviormetrics: Quantitative Approaches to Human Behavior 17,
https://doi.org/10.1007/978-981-99-5329-5_24

The least squares (LS) formulation of GCCA is restricted to:

$$\min f_{\mathrm{H}}(\mathbf{F}, \mathbf{W}) = \sum_{j=1}^{m} \left\| \mathbf{X}_j \mathbf{W}_j - \mathbf{F} \right\|^2 \quad \text{s.t. } \frac{1}{n}\mathbf{F}'\mathbf{F} = \mathbf{I}_p \tag{1}$$

that is, minimising the LS loss function $f_{\mathrm{H}}(\mathbf{F}, \mathbf{W})$ over a $K \times p$ matrix $\mathbf{W} = \left[\mathbf{W}'_1, \ldots, \mathbf{W}'_m\right]'$ and an $n \times p$ matrix $\mathbf{F}$ subject to its column-wise orthonormality (Gifi 1990). Here, "s.t." is an abbreviation for "subject to", p is the specified dimensionality with $p \leq \min(n, K)$, $\mathbf{W}_j$ is a $K_j \times p$ matrix, and $\mathbf{I}_p$ denotes the $p \times p$ identity matrix. Problem (1) can be called a *homogeneity* (HMG) problem, as $\mathbf{X}_j\mathbf{W}_j$ is matched to a matrix $\mathbf{F}$ in (1) so that $\mathbf{X}_j\mathbf{W}_j$ are homogeneous across $j = 1, \ldots, m$. The formulation of GCCA with (1) is also found in Dahl and Næs (2006), Takane, Hwang, and Abdi (2008), van der Burg, de Leeuw, and Dijksterhuis (1994), Van de Velden and Bijmolt (2006), and Van de Velden and Takane (2012).

As illustrated on the left of Fig. 1, a purpose of this paper is to show that GCCA can be reformulated as two LS *matrix factorisation* problems. The first is:

$$\min f_{\mathrm{F}}(\mathbf{F}, \mathbf{W}) = \left\| \mathbf{X}\mathbf{C}^{-1/2} - \mathbf{F}\mathbf{W}'\mathbf{C}^{1/2} \right\|^2 \text{s.t. } \frac{1}{n}\mathbf{F}'\mathbf{F} = \mathbf{I}_p. \tag{2}$$

Here

$$\mathbf{C} = \begin{bmatrix} \mathbf{C}_1 & & \\ & \ddots & \\ & & \mathbf{C}_m \end{bmatrix} = \frac{1}{n}\begin{bmatrix} \mathbf{X}'_1\mathbf{X}_1 & & \\ & \ddots & \\ & & \mathbf{X}'_m\mathbf{X}_m \end{bmatrix}$$

is the $K \times K$ block diagonal matrix whose jth block $\mathbf{C}_j = n^{-1}\mathbf{X}'_j\mathbf{X}_j$ of dimension $K_j \times K_j$ is supposed to be positive definite. We call (2) a *full matrix* factorisation (FMF) problem to distinguish it from the second LS problem we now describe.

The second LS problem is formulated as follows:

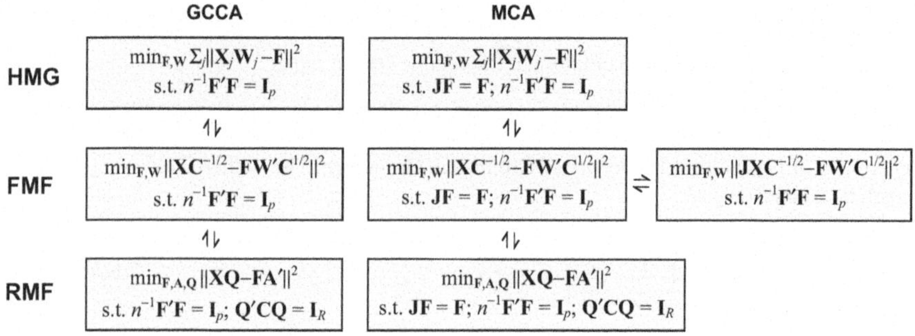

Fig. 1 Overview of the formulations of GCCA and MCA whose equivalence is to be shown

$$\min f_{\mathrm{R}}(\mathbf{F}, \mathbf{A}, \mathbf{Q}) = \left\|\mathbf{XQ} - \mathbf{FA}'\right\|^2 = \sum_{j=1}^{m}\left\|\mathbf{X}_j\mathbf{Q}_j - \mathbf{FA}_j'\right\|^2,$$

s.t.

$$\frac{1}{n}\mathbf{F}'\mathbf{F} = \mathbf{I}_p \text{ and } \mathbf{Q}'\mathbf{CQ} = \mathbf{I}_R \text{ or, equivalently, } \mathbf{Q}_j'\mathbf{C}_j\mathbf{Q}_j = \mathbf{I}_{R_j}. \quad (3)$$

Here, $R = \sum_j R_j$, $\mathbf{A} = \left[\mathbf{A}_1', \ldots, \mathbf{A}_m'\right]'$ is an $R \times p$ matrix whose jth block is an $R_j \times p$ matrix $\mathbf{A}_j$, and:

$$\mathbf{Q} = \begin{bmatrix} \mathbf{Q}_1 & & \\ & \ddots & \\ & & \mathbf{Q}_m \end{bmatrix}$$

is a $K \times R$ block diagonal matrix whose j-th block is a $K_j \times R_j$ matrix $\mathbf{Q}_j$. We suppose $R_j \leq \min\left(p, K_j\right)$ and $R \geq p$. We refer to (3) as a *reduced matrix* factorisation (RMF) problem in contrast to our calling (2) the FMF problem, since the number of columns in the matrix $\mathbf{XC}^{-1/2}$ factorised in (2) is the same K as in $\mathbf{X}$, while that number for $\mathbf{XQ}$ in (4) is reduced to $R \leq \sum_j \min\left(p, K_j\right) \leq K$.

The model part $\mathbf{FA}'$ fitted to data-based $\mathbf{XQ}$ in the RMF problem (3) takes the same form as the model part in the principal component analysis (PCA) formulated as a lower rank approximation of a data matrix (Eckart and Young 1936). In that formulation of PCA, a data matrix is approximated by the lower rank matrix $\mathbf{FA}'$, with $\mathbf{F}$ and $\mathbf{A}$ called PC score and loading matrices, respectively. Analogously, $\mathbf{XQ}$ is approximated by $\mathbf{FA}'$ in (3). However, $\mathbf{XQ}$ is the product of data and unknown parameter matrices.

MCA is performed for $\mathbf{X} = [\mathbf{X}_1, \ldots, \mathbf{X}_m]$, when $\mathbf{X}$ is a binary matrix with $\mathbf{X}_j\mathbf{1}_{K_j} = \mathbf{1}_n$ (Benzécri 1973; Greenacre 1984). Here, $\mathbf{1}_n$ is the $n \times 1$ vector of ones. In parallel with our discussions of GCCA, how MCA is formulated in terms of the HMG, FMF, and RMF problems will be discussed in this paper, as illustrated on the right of Fig. 1. There we can find two differences in the MCA versions from their GCCA counterparts. One difference is that MCA has two FMF formulations, and the loss function in one of them (furthest on the right of Fig. 1) differs from its GCCA counterpart in that $\mathbf{XC}^{-1/2}$ is pre-multiplied by the $n \times n$ centring matrix $\mathbf{J} = \mathbf{I}_n - n^{-1}\mathbf{1}_n\mathbf{1}_n'$. This formulation was derived by Adachi (2004) using Greenacre's (1988) formulation of the correspondence analysis for contingency tables. Another difference is that the additional constraint $\mathbf{F} = \mathbf{JF}$ is included in the problems for MCA except the above one of FMF formulations. The HMG formulation of MCA has been presented in Gifi (1990) and also described in ten Berge (1993), while the RMF formulation was proposed by Murakami et al. (1999). The formulation of Murakami's (2020) procedure can also be considered as a variant of the RMF one.

Nishisato's (1980) dual scaling is also performed for the above binary $\mathbf{X}$ and the resulting solution is equivalent to the MCA solution. However, the formulation of dual scaling is different from that of MCA to be described in this paper, and dual scaling can be applied to paired comparison and rank order data other than $\mathbf{X}$ treated in this paper (Nishisato 1978).

GCCA and MCA are treated in the following two sections, respectively. In each of them, we show how the HMG, FMF, and RMF problems are equivalent, but they provide different goodness-of-fit (GOF) indices. In the final section, we discuss the implications of the matrix factorisation problems (2) and (3) with those of the MCA counterparts.

2 Generalised Canonical Correlation Analysis

In this section, we first show that GCCA can be reformulated as the FMF problem (2), which is followed by presenting the explicit form of the GCCA solution. Next, we will show that GCCA can be reformulated as the RMF problem (3). Finally, we will discuss how goodness-of-fit (GOF) indices and their behaviours differ among formulations (1), (2), and (3). Here, we let $r(\mathbf{X}) = K$ and $r(\mathbf{X}_j) = K_j$ with $r(\mathbf{X})$ being the rank of $\mathbf{X}$. It may be considered that GCCA is typically performed for a column-centred $\mathbf{X}$ (Gifi 1990), but the facts to be shown in this section are independent of whether $\mathbf{X}$ is column-centred or not.

The next theorem shows that GCCA can be reformulated as (2):

Theorem 1 *The solution of* (1) *is equivalent to that of* (2).

Proof The loss function in (1) is expanded as:

$$f_{\mathrm{H}}(\mathbf{F}, \mathbf{W}) = \mathrm{tr}\left(\sum_j \mathbf{W}_j'\mathbf{X}_j'\mathbf{X}_j\mathbf{W}_j\right) - 2\mathrm{tr}\left(\sum_j \mathbf{W}_j'\mathbf{X}_j'\mathbf{F}\right) + nmp$$
$$= n\,\mathrm{tr}(\mathbf{W}'\mathbf{C}\mathbf{W}) - 2\mathrm{tr}(\mathbf{W}'\mathbf{X}'\mathbf{F}) + nmp, \tag{4}$$

while the function in (2) is expanded as:

$$f_{\mathrm{F}}(\mathbf{F}, \mathbf{W}) = nK + n\mathrm{tr}(\mathbf{W}'\mathbf{C}\mathbf{W}) - 2\mathrm{tr}(\mathbf{W}'\mathbf{X}'\mathbf{F}), \tag{5}$$

where $\|\mathbf{X}\mathbf{C}^{-1/2}\|^2 = \mathrm{tr}(\mathbf{C}^{-1}\mathbf{X}'\mathbf{X}) = n\mathrm{tr}(\mathbf{I}_K)$ and the constraint $n^{-1}\mathbf{F}'\mathbf{F} = \mathbf{I}_p$ has been used. Here, (4) and (5) are identical except for the constants independent of parameter matrices. Furthermore, (4) and (5) have the same constraint. This completes the proof.

The FMF problem (2) is the approximation of $\mathbf{X}\mathbf{C}^{-1/2}$ by the lower rank matrix $\mathbf{F}\mathbf{W}'\mathbf{C}^{1/2}$ with $r(\mathbf{X}\mathbf{C}^{1/2}) = \min(n, K) \geq p \geq r(\mathbf{F}\mathbf{W}'\mathbf{C}^{1/2})$. Thus, the optimal $\mathbf{F}\mathbf{W}'\mathbf{C}^{1/2}$ for (2) is given explicitly through the singular value decomposition (SVD) of $\mathbf{X}\mathbf{C}^{-1/2}$ defined as:

$$\mathbf{X}\mathbf{C}^{-1/2} = \mathbf{K}\boldsymbol{\Lambda}\mathbf{L}'. \tag{6}$$

See Adachi (2020, pp. 402–403), and Eckart and Young (1936). Here, $\mathbf{K}$ ($n \times K$) and $\mathbf{L}$ ($K \times K$) satisfy $\mathbf{K}'\mathbf{K} = \mathbf{L}'\mathbf{L} = \mathbf{I}_K$, and $\boldsymbol{\Lambda}$ is the $K \times K$ diagonal matrix whose diagonal elements are the singular values of $\mathbf{X}\mathbf{C}^{1/2}$ and arranged in descending order. The optimal $\mathbf{F}\mathbf{W}'\mathbf{C}^{1/2}$ is given by $\hat{\mathbf{F}}\hat{\mathbf{W}}'\mathbf{C}^{1/2} = \mathbf{K}_p\boldsymbol{\Lambda}_p\mathbf{L}_p'$, where $\mathbf{K}_p$ $n \times p$ and $\mathbf{L}_p$ $K \times p$ contain the first p columns of $\mathbf{K}$ and $\mathbf{L}$, respectively, and $\boldsymbol{\Lambda}_p$ is the first $p \times p$ diagonal block of $\boldsymbol{\Lambda}$. The above $\hat{\mathbf{F}}\hat{\mathbf{W}}'\mathbf{C}^{1/2} = \mathbf{K}_p\boldsymbol{\Lambda}_p\mathbf{L}_p'$ can be decomposed as:

$$\hat{\mathbf{F}} = n^{1/2}\mathbf{K}_p\mathbf{T}, \tag{7}$$

$$\hat{\mathbf{W}} = n^{-1/2}\mathbf{C}^{-1/2}\mathbf{L}_p\boldsymbol{\Lambda}_p\mathbf{T},$$

i.e.

$$\hat{\mathbf{W}}_j = n^{-1/2}\mathbf{C}_j^{-1/2}\mathbf{L}_p\boldsymbol{\Lambda}_p\mathbf{T}, \tag{8}$$

for $j = 1, \ldots, m$, so that $\hat{\mathbf{F}}$ satisfies $n^{-1}\hat{\mathbf{F}}'\hat{\mathbf{F}} = \mathbf{I}_p$, with $\mathbf{T}$ being an arbitrary $p \times p$ orthonormal matrix and $\hat{\mathbf{W}}_j$ corresponding to $\mathbf{W}_j$. Thus, the optimal $\mathbf{F}$ and $\mathbf{W}$ in GCCA are given by (7) and (8), respectively.

Recall that $R_j \le \min(p, K_j)$ was defined in Sect. 1. We replace $R_j \le \min(p, K_j)$ by more restrictive $R_j = \min(p, K_j)$ in this section. The next theorem shows that GCCA can be reformulated as the RMF problem (3).

Theorem 2 *The solution of* (1) *is equivalent to that of* (3), *and the optimal* $\mathbf{F}$, $\mathbf{Q}_j$, *and* $\mathbf{A}_j$ *in* (3) *are given by* (7):

$$\hat{\mathbf{Q}}_j = \mathbf{U}_j\mathbf{R}_j\boldsymbol{\Delta}_j^{-1}\mathbf{T}_j \text{ and } \hat{\mathbf{A}}_j = \mathbf{T}_j'\boldsymbol{\Delta}_j\mathbf{R}_j'\boldsymbol{\Theta}_j\mathbf{V}_j' \tag{9}$$

respectively, with $\mathbf{T}_j$ *an* $R_j \times R_j$ *orthonormal matrix. Here,* $\mathbf{U}_j$ $(K_j \times R_j)$, $\mathbf{R}_j$ $(R_j \times R_j)$, $\boldsymbol{\Delta}_j$ $(R_j \times R_j)$, $\boldsymbol{\Theta}_j$ $(R_j \times R_j)$, *and* $\mathbf{V}_j$ $(p \times R_j)$ *are obtained from the SVD of* $\hat{\mathbf{W}}_j$ *and the eigenvalue decomposition (EVD) of* $\mathbf{U}_j'\mathbf{C}_j\mathbf{U}_j$ *as follows:*

$$\hat{\mathbf{W}}_j = \mathbf{U}_j\boldsymbol{\Theta}_j\mathbf{V}_j' \text{ and } \mathbf{U}_j'\mathbf{C}_j\mathbf{U}_j = \mathbf{R}_j\boldsymbol{\Delta}_j^2\mathbf{R}_j', \tag{10}$$

with $\boldsymbol{\Theta}_j$ *and* $\boldsymbol{\Delta}_j$ *being diagonal and* $\mathbf{U}_j'\mathbf{U}_j = \mathbf{V}_j'\mathbf{V}_j = \mathbf{R}_j'\mathbf{R}_j = \mathbf{R}_j\mathbf{R}_j' = \mathbf{I}_{R_j}$.

Proof We start with showing that (3) is equivalent to:

$$\min f_{\mathrm{H}}^{*}(\mathbf{F}, \mathbf{A}, \mathbf{Q}) = \sum_{j=1}^{m}\|\mathbf{X}_j\mathbf{Q}_j\mathbf{A}_j - \mathbf{F}\|^2$$

s.t.

$$\frac{1}{n}\mathbf{F}'\mathbf{F} = \mathbf{I}_p \text{ and } \mathbf{Q}'\mathbf{C}\mathbf{Q} = \mathbf{I}_R \text{ or equivalently } \mathbf{Q}_j'\mathbf{C}_j\mathbf{Q}_j = \mathbf{I}_{R_j}. \tag{11}$$

This equivalence can be found as follows: Using the constraints $n^{-1}\mathbf{F}'\mathbf{F} = \mathbf{I}_p$ and $\mathbf{Q}'\mathbf{C}\mathbf{Q} = \mathbf{I}_R$ in (3) and (11), we can rewrite the loss function in (3) as:

$$f_{\mathrm{R}}(\mathbf{F}, \mathbf{A}, \mathbf{Q}) = nR + n\mathrm{tr}(\mathbf{A}\mathbf{A}') - 2\mathrm{tr}(\mathbf{Q}'\mathbf{X}'\mathbf{F}\mathbf{A}') \tag{12}$$

and the function in (11) as:

$$\begin{aligned} f_{\mathrm{H}}^*(\mathbf{F}, \mathbf{A}, \mathbf{Q}) &= n\sum_j \mathrm{tr}(\mathbf{A}_j'\mathbf{Q}_j'\mathbf{C}_j\mathbf{Q}_j\mathbf{A}_j) - 2\sum_j \mathrm{tr}(\mathbf{A}_j'\mathbf{Q}_j'\mathbf{X}_j'\mathbf{F}) + nmp \\ &= n\sum_j \mathrm{tr}(\mathbf{A}_j'\mathbf{A}_j) - 2\mathrm{tr}(\mathbf{A}'\mathbf{Q}'\mathbf{X}'\mathbf{F}) + nmp \\ &= n\,\mathrm{tr}(\mathbf{A}\mathbf{A}') - 2\mathrm{tr}(\mathbf{Q}'\mathbf{X}'\mathbf{F}\mathbf{A}') + nmp. \end{aligned}$$

This equation and (12) are identical except the constants independent of parameter matrices. Thus, our remaining task is to show the equivalence of (1) and (11).

We can rewrite $\mathbf{Q}_j\mathbf{A}_j$ in (11) as $\mathbf{W}_j$, so that the loss function in (11) is the same as that in (1). However, $\mathbf{Q}_j$ in (11) is constrained as $\mathbf{Q}_j'\mathbf{C}_j\mathbf{Q}_j = \mathbf{I}_{R_j}$ $(j = 1, \ldots, m)$, while $\mathbf{W}_j$ in (1) is unconstrained. That is, (11) is a more strongly constrained problem than (1), which implies $f_{\mathrm{H}}^*(\mathbf{F}, \mathbf{A}, \mathbf{Q}) \geq f_H(\hat{\mathbf{F}}, \hat{\mathbf{W}})$. The equality between these two functions holds for $\mathbf{F} = \hat{\mathbf{F}}$ and $\mathbf{Q}_j\mathbf{A}_j = \hat{\mathbf{W}}_j$, i.e. the equivalence of (1) and (11) can be found if $\hat{\mathbf{F}}$ can be substituted into $\mathbf{F}$ in $f_{\mathrm{H}}^*(\mathbf{F}, \mathbf{A}, \mathbf{Q})$, and the solution of $\hat{\mathbf{W}}_j$ in (8) can be decomposed as $\hat{\mathbf{W}}_j = \mathbf{Q}_j\mathbf{A}_j$ whose $\mathbf{Q}_j$ satisfies constraint $\mathbf{Q}_j'\mathbf{C}_j\mathbf{Q}_j = \mathbf{I}_{R_j}$. The substitution of $\hat{\mathbf{F}}$ into $\mathbf{F}$ follows from the equivalence of the constraint on $\mathbf{F}$ between (1) and (11). The latter decomposability of $\hat{\mathbf{W}}_j$ is shown next.

Since $R_j = \min(p, K_j)$ and $\hat{\mathbf{W}}_j$ is of size $K_j \times p$ then $r(\hat{\mathbf{W}}_j) \leq R_j$, and so we can decompose $\hat{\mathbf{W}}_j$ as $\mathbf{U}_j\boldsymbol{\Theta}_j\mathbf{V}_j'$; see (10). This can be rewritten as $\hat{\mathbf{W}}_j = \mathbf{U}_j\mathbf{S}_j\mathbf{S}_j^{-1}\boldsymbol{\Theta}_j\mathbf{V}_j'$ with $\mathbf{S}_j$ an $R_j \times R_j$ non-singular matrix. Here, $\mathbf{U}_j\mathbf{S}_j$ and $\mathbf{S}_j^{-1}\boldsymbol{\Theta}_j\mathbf{V}_j'$ can be set to be equal to $\mathbf{Q}_j$ and $\mathbf{A}_j$:

$$\mathbf{Q}_j = \mathbf{U}_j\mathbf{S}_j \text{ and } \mathbf{A}_j = \mathbf{S}_j^{-1}\boldsymbol{\Theta}_j\mathbf{V}_j', \tag{13}$$

which lead to $\hat{\mathbf{W}}_j = \mathbf{Q}_j\mathbf{A}_j$, or equivalently, $\hat{\mathbf{W}} = \mathbf{Q}\mathbf{A}$. Furthermore, $\mathbf{Q}_j = \mathbf{U}_j\mathbf{S}_j$ can satisfy $\mathbf{Q}_j'\mathbf{C}_j\mathbf{Q}_j = \mathbf{I}_{R_j}$, if $\mathbf{S}_j$ satisfies $\mathbf{S}_j'\mathbf{U}_j'\mathbf{C}_j\mathbf{U}_j\mathbf{S}_j = \mathbf{I}_{R_j}$. Such $\mathbf{S}_j$ is given by $\mathbf{S}_j = \mathbf{R}_j\boldsymbol{\Delta}_j^{-1}\mathbf{T}_j$ through the EVD of $\mathbf{U}_j'\mathbf{C}_j\mathbf{U}_j$ in (10). The use of $\mathbf{S}_j = \mathbf{R}_j\boldsymbol{\Delta}_j^{-1}\mathbf{T}_j$ in (13) gives (9). Here, the non-singularity of $\boldsymbol{\Delta}_j$ follows from the fact that $\mathbf{C}_j$ is non-singular, thus $r(\mathbf{U}_j'\mathbf{C}_j\mathbf{U}_j) = r(\mathbf{U}_j) = R_j$. This completes the proof.

Theorems 1 and 2 showed the equivalence of the solution among the HMG, FMF, and RMF problems. However, GOF indices and their behaviours are different among the problems, as discussed in the remainder of this section.

In order to derive the indices, we start by considering the minimum of the loss function in each problem. Using the GCCA solutions (7) and (8) in (4) and (5), we can find that the minimums in the HMG and FMF problems are expressed as $f_{\mathrm{H}}\left(\hat{\mathbf{F}}, \hat{\mathbf{W}}\right) = nmp - \mathrm{tr}\left(\mathbf{\Lambda}_p^2\right)$ and $f_{\mathrm{F}}\left(\hat{\mathbf{F}}, \hat{\mathbf{W}}\right) = nK - \mathrm{tr}\left(\mathbf{\Lambda}_p^2\right)$, respectively. The minimum function value in the RMF problem can be obtained from the following facts: $\mathbf{Q}'\mathbf{C}\mathbf{Q} = \mathbf{I}_R$ in (3) and $\hat{\mathbf{W}}_j = \mathbf{Q}_j\mathbf{A}_j$ (i.e. $\hat{\mathbf{W}} = \mathbf{Q}\mathbf{A}$) in the above proof allow (12) to be expressed as:

$$\begin{aligned} f_{\mathrm{R}}(\mathbf{F}, \mathbf{Q}, \mathbf{A}) &= nR - 2\mathrm{tr}\left(\mathbf{Q}'\mathbf{X}'\mathbf{F}\mathbf{A}'\right) + n\mathrm{tr}\left(\mathbf{A}\mathbf{A}'\mathbf{Q}'\mathbf{C}\mathbf{Q}\right) \\ &= nR - 2\mathrm{tr}(\widehat{\mathbf{W}}'\mathbf{X}'\hat{F}) + n\mathrm{tr}\left(\widehat{\mathbf{W}}'\mathbf{C}\widehat{\mathbf{W}}\right). \end{aligned}$$

Using (7) and (8) in the above equation, the minimum function value is found to be $f_{\mathrm{R}}\left(\hat{\mathbf{F}}, \hat{\mathbf{Q}}, \hat{\mathbf{A}}\right) = nR - \mathrm{tr}\left(\mathbf{\Lambda}_p^2\right)$. The equations for $f_{\mathrm{H}}\left(\hat{\mathbf{F}}, \hat{\mathbf{W}}\right)$, $f_{\mathrm{F}}\left(\hat{\mathbf{F}}, \hat{\mathbf{W}}\right)$, and $f_{\mathrm{R}}\left(\hat{\mathbf{F}}, \hat{\mathbf{Q}}, \hat{\mathbf{A}}\right)$ allow us to define their GOF indices as:

$$GOF_{\mathrm{H}}(p) = \frac{\mathrm{tr}\left(\mathbf{\Lambda}_p^2\right)}{nmp} \text{for the HMG problem (1)}, \tag{14}$$

$$GOF_{\mathrm{F}}(p) = \frac{\mathrm{tr}\left(\mathbf{\Lambda}_p^2\right)}{nK} \text{for the FMF problem (2)}, \tag{15}$$

$$GOF_{\mathrm{R}}(p) = \frac{\mathrm{tr}\left(\mathbf{\Lambda}_p^2\right)}{nR} \text{for the RMF problem (3)}, \tag{16}$$

each of which lies within the interval [0, 1]. It should be noted that the denominator $R = \sum_j R_j$ in (16) is a function of dimensionality p, i.e. R depends on p, since $R_j = \min\left(p, K_j\right)$.

The next theorem shows that (14)–(16) behave differently with changes in p:

Theorem 3 *The following inequalities and equalities hold:*

$$GOF_{\mathrm{F}}(p) \le GOF_{\mathrm{F}}(p+1), \tag{17}$$

$$GOF_{\mathrm{H}}(p) \ge GOF_{\mathrm{H}}(p+1), \tag{18}$$

$$GOF_{\mathrm{R}}(p) \ge \max(GOF_{\mathrm{H}}(p), GOF_{\mathrm{F}}(p)), \tag{19}$$

$$GOF_{\mathrm{R}}(p) = GOF_{\mathrm{H}}(p) \text{ for } p \le \min_{1\le j\le m} K_j, \tag{20}$$

$$GOF_{\mathrm{R}}(p) = GOF_{\mathrm{F}}(p) \text{ for } p \geq \max_{1\leq j\leq m} K_j. \tag{21}$$

Proof Since $\mathrm{tr}\left(\mathbf{\Lambda}_{p+1}^2\right) > \mathrm{tr}\left(\mathbf{\Lambda}_p^2\right)$, (17) follows from (15). On the other hand, (18) can be derived as follows:

$$\begin{aligned} GOF_{\mathrm{H}}(p) - GOF_{\mathrm{H}}(p+1) &= \frac{\mathrm{tr}\left(\mathbf{\Lambda}_p^2\right)}{nmp} - \frac{\mathrm{tr}\left(\mathbf{\Lambda}_{p+1}^2\right)}{nm(p+1)} \\ &= \frac{(p+1)\mathrm{tr}\left(\mathbf{\Lambda}_p^2\right) - p\mathrm{tr}\left(\mathbf{\Lambda}_{p+1}^2\right)}{nmp(p+1)}. \end{aligned}$$

The numerator on the right side of this is rewritten as:

$$p\sum_{s=1}^{p}\lambda_s^2 + \sum_{s=1}^{p}\lambda_s^2 - \left(p\sum_{s=1}^{p}\lambda_s^2 + p\lambda_{p+1}^2\right) = \sum_{s=1}^{p}\lambda_s^2 - p\lambda_{p+1}^2 = \sum_{s=1}^{p}\left(\lambda_s^2 - \lambda_{p+1}^2\right) \geq 0$$

Inequalities (19)–(21) are derived from $R_j = \min(p, K_j)$. That is, $R = \sum_j R_j = \sum_j \min(p, K_j) \leq \sum_j p = mp$ and $R = \sum_j \min(p, K_j) \leq \sum_j K_j = K$ imply:

$$GOF_{\mathrm{R}}(p) = \frac{\mathrm{tr}\left(\mathbf{\Lambda}_p^2\right)}{nR} \geq GOF_{\mathrm{H}}(p) = \frac{\mathrm{tr}\left(\mathbf{\Lambda}_p^2\right)}{nmp},$$

$$GOF_{\mathrm{R}}(p) \geq GOF_{\mathrm{F}}(p) = \frac{\mathrm{tr}\left(\mathbf{\Lambda}_p^2\right)}{nK}$$

so that we have (19). If $p \leq \min_{1\leq j\leq m} K_j$, then $R = \sum_j \min\left(p, K_j\right) = mp$, thus:

$$GOF_{\mathrm{H}}(p) = \frac{\mathrm{tr}\left(\mathbf{\Lambda}_p^2\right)}{nmp} = GOF_{\mathrm{R}}(p) = \frac{\mathrm{tr}\left(\mathbf{\Lambda}_p^2\right)}{nR},$$

i.e. (20). If $p \geq \max_{1\leq j\leq m} K_j$, then $R = \sum_j \min\left(p, K_j\right) = K$, thus:

$$GOF_{\mathrm{F}}(p) = \frac{\mathrm{tr}\left(\mathbf{\Lambda}_p^2\right)}{nK} = GOF_{\mathrm{R}}(p),$$

i.e. (21). This completes the proof.

Inequalities (17) and (18) imply that an increase in dimensionality p increases the value of $GOF_{\mathrm{F}}(p)$ in (15), but decreases the value of $GOF_{\mathrm{H}}(p)$ in (14). The latter decrease can be viewed as peculiar, if the value of a rational statistical GOF index should increase with the dimensionality of parameter matrices. By comparing (20) and (21) with (17) and (18), we can find that the value of $GOF_{\mathrm{R}}(p)$ in (16) changes with p in an *non-monotonous* manner when $\min_{1\leq j\leq m} K_j \geq 2$: an increase in p decreases the $GOF_{\mathrm{R}}(p)$ value for $p \le \min_{1\le j\le m} K_j$ but increases that value for $p \ge \max_{1\le j\le m} K_j$.

3 Multiple Correspondence Analysis

MCA is performed if $\mathbf{X} = [\mathbf{X}_1, \ldots, \mathbf{X}_m]$ consists of binary indicator matrices; each row of $\mathbf{X}_j$ $(n \times K_j)$ for $j = 1, \ldots, m$ is filled with zeros except only one element taking one. This implies that:

$$\mathbf{X}_j \mathbf{1}_{K_j} = \mathbf{1}_n \text{ and } \mathbf{X}\mathbf{1}_K = m\mathbf{1}_n \tag{22}$$

and $\mathbf{C}_j = n^{-1}\mathbf{X}_j'\mathbf{X}_j$ is diagonal, thus $\mathbf{C}$ is also diagonal. These properties further lead to:

$$n\mathbf{C}_j \mathbf{1}_{K_j} = \mathbf{X}_j'\mathbf{1}_n \text{ and } n\mathbf{C}\mathbf{1}_K = \mathbf{X}'\mathbf{1}_n. \tag{23}$$

Because of (22) and $\mathbf{J}\mathbf{1}_n = \left(\mathbf{I}_n - \frac{1}{n}\mathbf{1}_n\mathbf{1}_n'\right)\mathbf{1}_n = \mathbf{0}_n$ with $\mathbf{0}_n$ the $n \times 1$ zero vector, the K_j columns of $\mathbf{J}\mathbf{X}_j$ are linearly dependent. From this property and $n > K_j$, we can find that $r\left(\mathbf{J}\mathbf{X}_j\right) \le K_j - 1$ and $r(\mathbf{J}\mathbf{X}) \le \sum_j r\left(\mathbf{J}\mathbf{X}_j\right) \le K - m$. In this section, we assume that $\mathbf{X}$ does not include a zero column with $\mathbf{C}$ being positive definite and $r\left(\mathbf{J}\mathbf{X}\mathbf{C}^{-1/2}\right) = K - m \ge p$. On this assumption, we show the equivalence of the four problems for MCA in Fig. 1, then discuss how GOF indices and their behaviours differ among the problems, and finally discuss the role of the additional constraint $\mathbf{F} = \mathbf{J}\mathbf{F}$.

As in Fig. 1, the HMG problem for MCA is formulated as:

$$\min f_{\mathrm{H}}(\mathbf{F}, \mathbf{W}) = \sum_{j=1}^{m} \left\|\mathbf{X}_j\mathbf{W}_j - \mathbf{F}\right\|^2 \text{ s.t. } \mathbf{F} = \mathbf{J}\mathbf{F} \text{ and } \frac{1}{n}\mathbf{F}'\mathbf{F} = \mathbf{I}_p. \tag{24}$$

For the distinction of the two FMF problems for MCA in Fig. 1, we refer to:

$$\min f_{\mathrm{F}}(\mathbf{F}, \mathbf{W}) = \left\|\mathbf{X}\mathbf{C}^{-1/2} - \mathbf{F}\mathbf{W}'\mathbf{C}^{1/2}\right\|^2 \text{ s.t. } \mathbf{F} = \mathbf{J}\mathbf{F} \text{ and } \frac{1}{n}\mathbf{F}'\mathbf{F} = \mathbf{I}_p \tag{25}$$

simply as an FMF problem and call:

$$\min f_{\mathrm{FC}}(\mathbf{F}, \mathbf{W}) = \left\|\mathbf{J}\mathbf{X}\mathbf{C}^{-1/2} - \mathbf{F}\mathbf{W}'\mathbf{C}^{1/2}\right\|^2 \text{ s.t. } \frac{1}{n}\mathbf{F}'\mathbf{F} = \mathbf{I}_p \tag{26}$$

a full centred-matrix factorisation (FCMF) problem, as the factored matrix $\mathbf{J}\mathbf{X}\mathbf{C}^{-1/2}$ is column-centred. The equivalence of the three problems is now proven.

Theorem 4 *The solution to* (24)–(26) *are identical.*

Proof Since the FCMF problem (26) is the approximation of $\mathbf{JXC}^{-1/2}$ by the lower rank matrix $n^{-1}\mathbf{FW}'\mathbf{C}^{1/2}$ with $r\left(\mathbf{JXC}^{-1/2}\right) = K - m \geq p \geq r\left(n^{-1}\mathbf{FW}'\mathbf{C}^{1/2}\right)$, the solution for (26) is given through the SVD of $\mathbf{JXC}^{-1/2}$ defined as:

$$\mathbf{JXC}^{-1/2} = \tilde{\mathbf{K}}\tilde{\boldsymbol{\Lambda}}\tilde{\mathbf{L}}'. \tag{27}$$

Here, $\tilde{\mathbf{K}}$ $(n \times (K - m))$ and $\tilde{\mathbf{L}}$ $(K \times (K - m))$ satisfy $\tilde{\mathbf{K}}'\tilde{\mathbf{K}} = \tilde{\mathbf{L}}'\tilde{\mathbf{L}} = \mathbf{I}_{K-m}$ and $\tilde{\boldsymbol{\Lambda}}$ is the $(K - m) \times (K - m)$ diagonal matrix, whose diagonal elements are the singular values of $\mathbf{JXC}^{-1/2}$ and arranged in descending order. The optimal $\mathbf{FW}'\mathbf{C}^{1/2}$ is given by $\tilde{\mathbf{F}}\tilde{\mathbf{W}}'\mathbf{C}^{1/2} = \tilde{\mathbf{K}}_p\tilde{\boldsymbol{\Lambda}}_p\tilde{\mathbf{L}}_p'$, where $\tilde{\mathbf{K}}_p$ $(n \times p)$ and $\tilde{\mathbf{L}}_p$ $(K \times p)$ contain the first p columns of $\tilde{\mathbf{K}}$ and those of $\tilde{\mathbf{L}}$, respectively, and $\tilde{\boldsymbol{\Lambda}}_p$ is the first $p \times p$ diagonal block of $\tilde{\boldsymbol{\Lambda}}$. The above $\tilde{\mathbf{F}}\tilde{\mathbf{W}}'\mathbf{C}^{1/2} = \tilde{\mathbf{K}}_p\tilde{\boldsymbol{\Lambda}}_p\tilde{\mathbf{L}}_p'$ can be decomposed as:

$$\tilde{\mathbf{F}} = n^{1/2}\tilde{\mathbf{K}}_p\mathbf{T}, \tag{28}$$

$$\tilde{\mathbf{W}} = n^{-1/2}\mathbf{C}^{-1/2}\tilde{\mathbf{L}}_p\tilde{\boldsymbol{\Lambda}}_p\mathbf{T}, \text{ i.e. } \tilde{\mathbf{W}}_j = n^{-1/2}\mathbf{C}_j^{-1/2}\tilde{\mathbf{L}}_p\tilde{\boldsymbol{\Lambda}}_p\mathbf{T}, \tag{29}$$

for $j = 1, \ldots, m$, so that $n^{-1}\tilde{\mathbf{F}}'\tilde{\mathbf{F}} = n\mathbf{I}_p$ is satisfied. We should note that (28) also satisfies $\tilde{\mathbf{F}} = \mathbf{J}\tilde{\mathbf{F}}$, because of $\mathbf{JJ} = \mathbf{J}$ and the fact that (28) can be rewritten as $\tilde{\mathbf{F}} = n^{1/2}\tilde{\mathbf{K}}\tilde{\boldsymbol{\Lambda}}\tilde{\mathbf{L}}'\tilde{\mathbf{L}}_p\tilde{\boldsymbol{\Lambda}}_p^{-1} = n^{1/2}\mathbf{JXC}^{-1/2}\tilde{\mathbf{L}}_p\tilde{\boldsymbol{\Lambda}}_p^{-1}$ from (27). The equality $\tilde{\mathbf{F}} = \mathbf{J}\tilde{\mathbf{F}}$ implies that (26) is equivalent to:

$$\min f_{\mathrm{FC}}(\mathbf{F}, \mathbf{W}) = \left\|\mathbf{JXC}^{-1/2} - \mathbf{FW}'\mathbf{C}^{1/2}\right\|^2 \text{ s.t. } \mathbf{F} = \mathbf{JF} \text{ and } \frac{1}{n}\mathbf{F}'\mathbf{F} = \mathbf{I}_p \tag{30}$$

In the remainder, we show that (28) and (29) are also the solutions for (24) and (25).

Using $n^{-1}\mathbf{F}'\mathbf{F} = \mathbf{I}_p$, $\mathbf{F} = \mathbf{JF}$, and $\mathrm{tr}\left(\mathbf{X}'\mathbf{XC}^{-1}\right) = n\mathrm{tr}(\mathbf{I}_K)$, the loss functions in (24) and (25) are rewritten as:

$$f_{\mathrm{H}}(\mathbf{F}, \mathbf{W}) = n\mathrm{tr}\left(\mathbf{W}'\mathbf{CW}\right) - 2\mathrm{tr}\left(\mathbf{X}'\mathbf{JFW}'\right) + nmp, \tag{31}$$

$$f_{\mathrm{F}}(\mathbf{F}, \mathbf{W}) = nK + n\mathrm{tr}\left(\mathbf{W}'\mathbf{CW}\right) - 2\mathrm{tr}\left(\mathbf{X}'\mathbf{JFW}'\right), \tag{32}$$

respectively. Using $n^{-1}\mathbf{F}'\mathbf{F} = \mathbf{I}_p$ and $\mathbf{J} = \mathbf{I}_n - n^{-1}\mathbf{1}_n\mathbf{1}_n'$, the loss function in (26) or (30) is rewritten as:

$$\begin{aligned} f_{\mathrm{FC}}(\mathbf{F}, \mathbf{W}) &= \mathrm{tr}\left(\mathbf{X}'\mathbf{JXC}^{-1}\right) - 2\mathrm{tr}\left(\mathbf{X}'\mathbf{JFW}'\right) + n\mathrm{tr}\left(\mathbf{W}'\mathbf{CW}\right) \\ &= \mathrm{tr}\left(\mathbf{X}'\mathbf{XC}^{-1}\right) - \frac{1}{n}\mathrm{tr}\left(\mathbf{X}'\mathbf{1}_n\mathbf{1}_n'\mathbf{XC}^{-1}\right) - 2\mathrm{tr}\left(\mathbf{X}'\mathbf{JFW}'\right) + n\mathrm{tr}\left(\mathbf{W}'\mathbf{CW}\right). \end{aligned}$$

This can further be rewritten as:

$$f_{\text{FC}}(\mathbf{F}, \mathbf{W}) = n\text{tr}(\mathbf{I}_K) - n\text{tr}\big(\mathbf{C}\mathbf{1}_K\mathbf{1}'_K\mathbf{C}\mathbf{C}^{-1}\big) - 2\text{tr}\big(\mathbf{X}\mathbf{J}\mathbf{F}\mathbf{W}'\big) + n\text{tr}\big(\mathbf{W}'\mathbf{C}\mathbf{W}\big)$$
$$= nK - nm + n\text{tr}\big(\mathbf{W}'\mathbf{C}\mathbf{W}\big) - 2\text{tr}\big(\mathbf{X}'\mathbf{J}\mathbf{F}\mathbf{W}'\big), \tag{33}$$

using (23), $\text{tr}\Big(\mathbf{X}'\mathbf{X}\mathbf{C}^{-1}\Big) = n\text{tr}(\mathbf{I}_K)$, and $\mathbf{1}'_K\mathbf{C}\mathbf{1}_K = n^{-1}\mathbf{1}_K\mathbf{X}'\mathbf{1}_n = n^{-1}m\mathbf{1}'_n\mathbf{1}_n = m$ derived from (22) and (23). We can find that (31)–(33), i.e. the rewritten versions of the functions in (24)–(26), are identical except for the constants independent of parameter matrices. Furthermore, (24), (25), and the Eq. (30) which is equivalent to (26) have identical constraints. This completes the proof.

Here, we present two equations associated with (27)–(29). The first equation follows from (28) and (29) satisfying $\tilde{\mathbf{W}}_j = n^{-1}\mathbf{C}_j^{-1}\mathbf{X}'_j\tilde{\mathbf{F}}$ (Gifi 1990). This Eqs. (22), (23), $\mathbf{C}_j = n^{-1}\mathbf{X}'_j\mathbf{X}_j$ and $\mathbf{J}\mathbf{F} = \mathbf{F}$ lead to $\mathbf{1}'_{K_j}\mathbf{C}_j\tilde{\mathbf{W}}_j = n^{-1}\mathbf{1}'_{K_j}\mathbf{X}'_j\tilde{\mathbf{F}} = n^{-1}\mathbf{1}'_n\mathbf{J}\tilde{\mathbf{F}} = \mathbf{0}'_p$, which implies $r\big(\mathbf{W}_j\big) \leq \min(p, K_j - 1)$. The second equation is:

$$\left\|\tilde{\mathbf{\Lambda}}\right\|^2 = \left\|\mathbf{J}\mathbf{X}\mathbf{C}^{-1/2}\right\|^2 = n(K - m), \tag{34}$$

since

$$\left\|\mathbf{J}\mathbf{X}\mathbf{C}^{-1/2}\right\|^2 = \text{tr}\Big(\mathbf{C}^{-1/2}\mathbf{X}'\mathbf{X}\mathbf{C}^{-1/2}\Big) - \frac{1}{n}\text{tr}\big(\mathbf{C}^{-1/2}\mathbf{X}'\mathbf{1}_n\mathbf{1}'_n\mathbf{X}\mathbf{C}^{-1/2}\big)$$
$$= n\text{tr}(\mathbf{I}_K) - n\text{tr}\big(\mathbf{C}\mathbf{1}_K\mathbf{1}'_K\mathbf{C}\mathbf{C}^{-1}\big).$$

Here, we have used $\text{tr}\Big(\mathbf{X}'\mathbf{X}\mathbf{C}^{-1}\Big) = n\text{tr}(\mathbf{I}_K)$ and $\mathbf{1}'_K\mathbf{C}\mathbf{1}_K = m$ following from (22).

As in Fig. 1, the RMF problem for MCA is formulated as:

$$\min f_{\text{R}}(\mathbf{F}, \mathbf{A}, \mathbf{Q}) = \left\|\mathbf{X}\mathbf{Q} - \mathbf{F}\mathbf{A}'\right\|^2 = \sum_{j=1}^{m}\left\|\mathbf{X}_j\mathbf{Q}_j - \mathbf{F}\mathbf{A}'_j\right\|^2$$
$$\text{s.t.}\quad \mathbf{J}\mathbf{F} = \mathbf{F},\ \frac{1}{n}\mathbf{F}'\mathbf{F} = \mathbf{I}_p,\ \text{and}\ \mathbf{Q}'\mathbf{C}\mathbf{Q} = \mathbf{I}_R \text{ or } \mathbf{Q}'_j\mathbf{C}_j\mathbf{Q}_j = \mathbf{I}_{R_j} \tag{35}$$

See Murakami et al. (1999) for more on this formulation. The condition $R_j \leq \min\big(p, K_j\big)$ in Sect. 1 is replaced by more restrictive $R_j = \min\big(p, K_j - 1\big)$ in this section. This differs from $R_j = \min\big(p, K_j\big)$ in Sect. 2. The equivalence of the RMF and HMG problems for MCA is shown in the next theorem:

Theorem 5 *The solution of* (24) *is equivalent to that of* (35), *and the optimal* $\mathbf{F}$, $\mathbf{Q}_j$, *and* $\mathbf{A}_j$ *in* (35) *are given by* (28),

$$\tilde{\mathbf{Q}}_j = \tilde{\mathbf{U}}_j\tilde{\mathbf{R}}_j\tilde{\mathbf{\Delta}}_j^{-1}\tilde{\mathbf{T}}_j,\ \text{and}\ \tilde{\mathbf{A}}_j = \tilde{\mathbf{T}}'_j\tilde{\mathbf{\Delta}}_j\tilde{\mathbf{R}}'_j\tilde{\mathbf{\Theta}}_j\tilde{\mathbf{V}}'_j, \tag{36}$$

respectively, with $\tilde{\mathbf{T}}_j$ *an* $R_j \times R_j$ *orthonormal matrix. Here,* $\tilde{\mathbf{U}}_j$ $\big(K_j \times R_j\big)$, $\tilde{\mathbf{R}}_j$ $\big(R_j \times R_j\big)$, $\tilde{\mathbf{\Delta}}_j$ $\big(R_j \times R_j\big)$, $\tilde{\mathbf{\Theta}}_j$ $\big(R_j \times R_j\big)$, *and* $\tilde{\mathbf{V}}_j$ $\big(p \times R_j\big)$ *are obtained from the*

SVD of $\tilde{\mathbf{W}}_j$ *in* (29) *and EVD of* $\tilde{\mathbf{U}}_j'\mathbf{C}_j\tilde{\mathbf{U}}_j$ *as follows*:

$$\tilde{\mathbf{W}}_j = \tilde{\mathbf{U}}_j\tilde{\mathbf{\Theta}}_j\tilde{\mathbf{V}}_j' \text{ and } \tilde{\mathbf{U}}_j'\mathbf{C}_j\tilde{\mathbf{U}}_j = \tilde{\mathbf{R}}_j\tilde{\mathbf{\Delta}}_j^2\tilde{\mathbf{R}}_j', \tag{37}$$

with $\tilde{\mathbf{\Theta}}_j$ *and* $\tilde{\mathbf{\Delta}}_j$ *being diagonal and* $\tilde{\mathbf{U}}_j'\tilde{\mathbf{U}}_j = \tilde{\mathbf{V}}_j'\tilde{\mathbf{V}}_j = \tilde{\mathbf{R}}_j'\tilde{\mathbf{R}}_j = \tilde{\mathbf{R}}_j\tilde{\mathbf{R}}_j' = \mathbf{I}_{R_j}$.

Proof The proof is analogous to that for Theorem 2. At first, the equivalence of (35) to:

$$\min f_{\mathrm{H}}^*(\mathbf{F}, \mathbf{Q}, \mathbf{A}) = \sum_{j=1}^{m} \left\|\mathbf{X}_j\mathbf{Q}_j\mathbf{A}_j - \mathbf{F}\right\|^2 \text{ s.t. } \mathbf{JF} = \mathbf{F},\ \frac{1}{n}\mathbf{F}'\mathbf{F} = \mathbf{I}_p, \text{ and } \mathbf{Q}'\mathbf{CQ} = \mathbf{I}_R \tag{38}$$

is shown as follows: We can rewrite the loss function in (35) as:

$$f_{\mathrm{R}}(\mathbf{F}, \mathbf{Q}, \mathbf{A}) = nR + n\mathrm{tr}\left(\mathbf{A}'\mathbf{A}\right) - 2\mathrm{tr}\left(\mathbf{Q}'\mathbf{X}'\mathbf{JFA}'\right) \tag{39}$$

and $f_{\mathrm{H}}^*(\mathbf{F}, \mathbf{Q}, \mathbf{A})$ in (38) as $n\mathrm{tr}\left(\mathbf{A}'\mathbf{A}\right) - 2\mathrm{tr}\left(\mathbf{Q}'\mathbf{X}'\mathbf{JFA}'\right) + nmp$, using the constraints for **F** and **Q**. Thus, it remains to show the equivalence of (24) and (39). This task can be attained if $\tilde{\mathbf{W}}_j$ in (29) can be decomposed as $\tilde{\mathbf{W}}_j = \mathbf{Q}_j\mathbf{A}_j$ with $\mathbf{Q}_j$ satisfying $\mathbf{Q}_j'\mathbf{C}_j\mathbf{Q}_j = \mathbf{I}_{R_j}$, i.e. $\mathbf{Q}'\mathbf{CQ} = \mathbf{I}_R$ in (38). The decomposition of $\tilde{\mathbf{W}}_j$ can be found as follows: The inequality $r\left(\mathbf{W}_j\right) \leq \min(p, K_j - 1) = R_j$ shows that $\tilde{\mathbf{W}}_j$ can be decomposed as $\tilde{\mathbf{U}}_j\tilde{\mathbf{\Theta}}_j\tilde{\mathbf{V}}_j'$ in (37). This implies that $\tilde{\mathbf{Q}}_j$ and $\tilde{\mathbf{A}}_j$ in (36) can be substituted into $\mathbf{Q}_j$ and $\mathbf{A}_j$ in $\tilde{\mathbf{W}}_j = \mathbf{Q}_j\mathbf{A}_j$. Further, $\tilde{\mathbf{Q}}_j$ in (36) can be substituted into $\tilde{\mathbf{Q}}_j$ in $\mathbf{Q}_j'\mathbf{C}_j\mathbf{Q}_j = \mathbf{I}_{R_j}$, since we have $\tilde{\mathbf{T}}_j'\tilde{\mathbf{\Delta}}_j^{-1}\tilde{\mathbf{R}}_j'\tilde{\mathbf{U}}_j'\mathbf{C}_j\tilde{\mathbf{U}}_j\tilde{\mathbf{R}}_j\tilde{\mathbf{\Delta}}_j^{-1}\tilde{\mathbf{T}}_j = \tilde{\mathbf{T}}_j'\tilde{\mathbf{T}}_j = \mathbf{I}_{R_j}$ using (37). Here, the non-singularity of $\tilde{\mathbf{\Delta}}_j$ follows from the non-singularity of $\mathbf{C}_j$ implying $r\left(\tilde{\mathbf{U}}_j'\mathbf{C}_j\tilde{\mathbf{U}}_j\right) = r\left(\tilde{\mathbf{U}}_j\right) = R_j$. This completes the proof.

Murakami et al. (1999) stated Theorem 5 but did not provide a formal proof of it.

To derive the GOF indices for MCA, we consider the minimums of the loss functions in problems (24)–(26) and (35). First, using MCA solutions (28) and (29) in (31)–(33), we can find that the minima of the functions in (24), (25), and (26) are expressed as $f_{\mathrm{H}}\left(\tilde{\mathbf{F}}, \tilde{\mathbf{W}}\right) = nmp - \mathrm{tr}\left(\tilde{\mathbf{\Lambda}}_p^2\right)$, $f_{\mathrm{F}}\left(\tilde{\mathbf{F}}, \tilde{\mathbf{W}}\right) = nK - \mathrm{tr}\left(\tilde{\mathbf{\Lambda}}_p^2\right)$, and $f_{\mathrm{FC}}\left(\tilde{\mathbf{F}}, \tilde{\mathbf{W}}\right) = n(K - m) - \mathrm{tr}\left(\tilde{\mathbf{\Lambda}}_p^2\right)$, respectively. Next, the minimum of the function in (35) can be obtained from the following facts: $\tilde{\mathbf{W}}_j = \mathbf{Q}_j\mathbf{A}_j$, i.e.$\tilde{\mathbf{W}} = \tilde{\mathbf{Q}}\tilde{\mathbf{A}}$, and $\mathbf{Q}'\mathbf{CQ} = \mathbf{I}_R$ in the above proof allow (39) to be expressed as:

$$\begin{aligned} f_{\mathrm{R}}\left(\tilde{\mathbf{F}}, \tilde{\mathbf{Q}}, \tilde{\mathbf{A}}\right) &= nR - 2\mathrm{tr}\left(\mathbf{Q}'\mathbf{X}'\mathbf{JFA}'\right) + n\mathrm{tr}\left(\mathbf{AA}'\mathbf{Q}'\mathbf{CQ}\right) \\ &= nR - 2\mathrm{tr}(\tilde{\mathbf{W}}'\mathbf{X}'\tilde{\mathbf{F}}) + n\mathrm{tr}\left(\tilde{\mathbf{W}}'\mathbf{C}\tilde{\mathbf{W}}\right). \end{aligned}$$

Using (28) and (29) in the above equation, the minimum of the function is found to be $f_{\mathrm{R}}\left(\tilde{\mathbf{F}}, \tilde{\mathbf{Q}}, \tilde{\mathbf{\Lambda}}\right) = nR - \mathrm{tr}\left(\tilde{\mathbf{\Lambda}}_p^2\right)$. Those minimum function values allow the GOF indices to be defined as follows:

$$GOF_{\mathrm{H}}^*(p) = \frac{\mathrm{tr}\left(\tilde{\mathbf{\Lambda}}_p^2\right)}{nmp} \text{ for the HMG problem (24)}, \tag{40}$$

$$GOF_{\mathrm{F}}^*(p) = \frac{\mathrm{tr}\left(\tilde{\mathbf{\Lambda}}_p^2\right)}{nK} \text{ for the FMF problem (25)}, \tag{41}$$

$$GOF_{\mathrm{FC}}^*(p) = \frac{\mathrm{tr}\left(\tilde{\mathbf{\Lambda}}_p^2\right)}{n(K-m)} \text{ for the FCMF problem (26)}, \tag{42}$$

$$GOF_{\mathrm{R}}^*(p) = \frac{\mathrm{tr}\left(\tilde{\mathbf{\Lambda}}_p^2\right)}{nR} \text{ for the RMF problem (35)}, \tag{43}$$

Obviously $K > K - m$, thus $GOF_{\mathrm{F}}^*(p) < GOF_{\mathrm{FC}}^*(p)$. Other inequalities and equalities are proved next.

Theorem 6 *The following inequalities and equalities hold:*

$$GOF_{\mathrm{F}}^*(p) \le GOF_{\mathrm{F}}^*(p+1), \tag{44}$$

$$GOF_{\mathrm{FC}}^*(p) \le GOF_{\mathrm{FC}}^*(p+1), \tag{45}$$

$$GOF_{\mathrm{H}}^*(p) \ge GOF_{\mathrm{H}}^*(p+1), \tag{46}$$

$$GOF_{\mathrm{R}}^*(p) \ge \max\left(GOF_{\mathrm{H}}^*(p), GOF_{\mathrm{FC}}^*(p)\right), \tag{47}$$

$$GOF_{\mathrm{R}}^*(p) = GOF_{\mathrm{H}}^*(p) \text{ for } p \le \min_{1\le j\le m} K_j - 1, \tag{48}$$

$$GOF_{\mathrm{R}}^*(p) = GOF_{\mathrm{FC}}^*(p) \text{ for } p \ge \max_{1\le j\le m} K_j - 1 \tag{49}$$

Proof The proof of these results is analogous to that for Theorem 3. Inequalities (44)–(46) are proven in an analogous manner to how (17) and (18) are proven. We can derive (47)–(49) using $R_j = \min\left(p, K_j - 1\right)$ and $R = \sum_j R_j = \sum_j \min\left(p, K_j - 1\right)$ as follows: $R \le \sum_j p = mp$ and $R \le \sum_j \left(K_j - 1\right) = K - m$ implying:

$$GOF_{\mathrm{R}}^*(p) = \frac{\mathrm{tr}\left(\tilde{\mathbf{\Lambda}}_p^2\right)}{nR} \ge GOF_{\mathrm{H}}^*(p) = \frac{\mathrm{tr}\left(\tilde{\mathbf{\Lambda}}_p^2\right)}{nmp}$$

$$GOF_{\mathrm{R}}^{*}(p) \geq GOF_{\mathrm{FC}}^{*}(p) = \frac{\mathrm{tr}\left(\tilde{\boldsymbol{\Lambda}}_p^2\right)}{n(K-m)},$$

respectively, i.e. (47), while $R = mp$ for $p \leq \min_{1\leq j\leq m} K_j - 1$ and $R = K - m$ for $p \geq \max_{1\leq j\leq m} K_j - 1$ lead to (48) and (49), respectively. This completes the proof.

The MCA solution is given through the SVD of $\mathbf{JXC}^{-1/2}$ in (27) rather than the SVD of $\mathbf{XC}^{-1/2}$ that leads to the GCCA solution. The latter SVD for $\mathbf{X}$ in this section gives (7) and (8) whose first columns are trivially filled with ones. This fact is formally proved next.

Theorem 7 *If* $\mathbf{X}_j$ *satisfies* (22), *then* $\mathbf{K}$, $\boldsymbol{\Lambda}$, *and* $\mathbf{L}$ *defining the SVD of* $\mathbf{XC}^{-1/2}$ *in* (6) *are expressed as:*

$$\mathbf{K} = \left[\frac{1}{\sqrt{n}}\mathbf{1}_n, \tilde{\mathbf{K}}\right], \boldsymbol{\Lambda} = \begin{bmatrix} \sqrt{nm} & \\ & \tilde{\boldsymbol{\Lambda}} \end{bmatrix}, \mathbf{L} = \left[\frac{1}{\sqrt{m}}\mathbf{C}^{1/2}\mathbf{1}_K, \tilde{\mathbf{L}}\right], \tag{50}$$

with $\tilde{\mathbf{K}}$, $\tilde{\boldsymbol{\Lambda}}$, *and* $\tilde{\mathbf{L}}$ *defining the SVD of* $\mathbf{JXC}^{-1/2}$ *in* (27). *Through* (7) *and* (8), (50) *gives* $\mathbf{f}_1 = \mathbf{1}_n$ *and* $\mathbf{w}_1 = \mathbf{1}_K$, *with* $\mathbf{f}_1$ *and* $\mathbf{w}_1$ *being the first column's of* $\hat{\mathbf{F}}$ *and* $\hat{\mathbf{W}}$ *in* (7) *and* (8), *respectively.*

Proof Using (23) we have $\mathbf{JXC}^{-1/2} = \mathbf{XC}^{-1/2} - n^{-1}\mathbf{1}_n\mathbf{1}_n'\mathbf{XC}^{-1/2} = \mathbf{XC}^{-1/2} - \mathbf{1}_n\mathbf{1}_K'\mathbf{XC}^{1/2}$. This is decomposed using SVD as in (27): $\mathbf{XC}^{-1/2} - \mathbf{1}_n\mathbf{1}_K'\mathbf{C}^{1/2} = \tilde{\mathbf{K}}\tilde{\boldsymbol{\Lambda}}\tilde{\mathbf{L}}'$, which can be rewritten as $\mathbf{XC}^{-1/2} = \tilde{\mathbf{K}}\tilde{\boldsymbol{\Lambda}}\tilde{\mathbf{L}}' + \mathbf{1}_n\mathbf{1}_K{}'\mathbf{C}^{1/2} = \mathbf{K}\boldsymbol{\Lambda}\mathbf{L}'$, where $\mathbf{K}$, $\boldsymbol{\Lambda}$, and $\mathbf{L}$ are defined as (50). This being the SVD of $\mathbf{XC}^{-1/2}$ is proven from the following three facts:

1. Obviously $\|n^{-1/2}\mathbf{1}_n\| = 1$, and $\|m^{-1/2}\mathbf{C}^{1/2}\mathbf{1}_K\| = 1$ follows from (22) and (23) implying $m^{-1}\mathbf{1}_K'\mathbf{C}\mathbf{1}_K = (nm)^{-1}\mathbf{1}_n'\mathbf{X}'\mathbf{1}_K = n^{-1}\mathbf{1}_n'\mathbf{1}_n$.
2. $\mathbf{1}_n'\tilde{\mathbf{K}} = \left(\mathbf{C}^{1/2}\mathbf{1}_K\right)'\tilde{\mathbf{L}} = \mathbf{0}_{K-m}'$ is found as follows: We have $\mathbf{1}_n'\tilde{\mathbf{K}} = \mathbf{0}_{K-m}'$ since of $\mathbf{1}_n'\mathbf{J} = \mathbf{0}_n'$ and $\tilde{\mathbf{K}} = \mathbf{JXC}^{-1/2}\tilde{\mathbf{L}}\tilde{\boldsymbol{\Lambda}}^{-1}$ is derived from (27), while $\left(\mathbf{C}^{1/2}\mathbf{1}_K\right)'\tilde{\mathbf{L}} = \mathbf{0}_{K-m}'$ follows from the fact that (22) and (27) imply $\tilde{\mathbf{L}} = \mathbf{C}^{-1/2}\mathbf{X}'\mathbf{J}\tilde{\mathbf{K}}\tilde{\boldsymbol{\Lambda}}^{-1}$ and $\left(\mathbf{C}^{1/2}\mathbf{1}_K\right)'\mathbf{C}^{-1/2}\mathbf{X}'\mathbf{J}\,\tilde{\mathbf{K}}\tilde{\boldsymbol{\Lambda}}^{-1} = \mathbf{1}_K'\mathbf{X}'\mathbf{J}\,\tilde{\mathbf{K}}\,\tilde{\boldsymbol{\Lambda}}^{-1} = m\mathbf{1}_n'\mathbf{J}\,\tilde{\mathbf{K}}\,\tilde{\boldsymbol{\Lambda}}^{-1} = \mathbf{0}_{K-m}'$.
3. Let λ_1, $\mathbf{k}_1$, and $\mathbf{l}_1$ denote the first diagonal element of $\boldsymbol{\Lambda}$, the first column of $\mathbf{K}$, and that of $\mathbf{L}$ in (6), respectively. For $p = 1$, (7) and (8) show that the solution of (1) is given by $\hat{\mathbf{F}} = n^{1/2}\mathbf{k}_1$ and $\hat{\mathbf{W}} = n^{-1/2}\lambda_1\mathbf{C}^{-1/2}\mathbf{l}_1$. Substituting these into $\mathbf{F}$ and $\mathbf{W}$ in (1), we have $f_{\mathrm{H}}(\mathbf{F}, \mathbf{W}) = nm - \lambda_1^2 \geq 0$. This implies $\lambda_1 = (nm)^{1/2}$, $\mathbf{k}_1 = n^{-1/2}\mathbf{1}_n$, and $\mathbf{l}_1 = m^{-1/2}\mathbf{C}^{1/2}\mathbf{1}_K$.

This completes the proof.

Theorem 7 shows that the additional constraint $\mathbf{JF} = \mathbf{F}$ in (24), (25), and (35) serves to avoid a solution with trivial columns: $\mathbf{F}$ with $\mathbf{f}_1 = \mathbf{1}_n$ does not satisfy $\mathbf{JF} = \mathbf{F}$. On the other hand, $\mathbf{JF} = \mathbf{F}$ is not required in the FCMF problem (26) which is the lower rank approximation of $\mathbf{JXC}^{-1/2}$ rather than $\mathbf{XC}^{-1/2}$. Theorem 7 also allows us to find the following property of the GOF index (41): its maximum, though $n(K-m)/(nK) < 1$ from (34), can be one, if constraint $\mathbf{JF} = \mathbf{F}$ is excluded from (25). This fact is found by the comparison of (50) with (34), which shows $\|\mathbf{\Lambda}\|^2 = nK$; that is, the exclusion of $\mathbf{JF} = \mathbf{F}$ equalises (25) to (2) and the solution of (25) is given through (50), then the maximum of index (41) for (25) can be $\|\mathbf{\Lambda}\|^2/(nK) = 1$, though the solution includes trivial columns.

4 Discussion

In this paper, we have shown that GCCA and MCA, defined as homogeneity problems (Gifi 1990), can be reformulated as full matrix factorisation (FMF) and reduced matrix factorisation (RMF) problems. Each problem is regarded as approximating a transformed data matrix by a lower rank matrix, as illustrated in Fig. 2. Though the FCMF problem for MCA is not mentioned in the following discussions, they can cover that problem by pre-multiplying $\mathbf{X}_1\mathbf{C}_1^{1/2}, \ldots, \mathbf{X}_m\mathbf{C}_m^{1/2}$ by $\mathbf{J}$.

In the FMF problem, the transformed data matrix is $\mathbf{XC}^{-1/2} = \left[\mathbf{X}_1\mathbf{C}_1^{-1/2}, \ldots, \mathbf{X}_m\mathbf{C}_m^{-1/2}\right]$, an $n \times K$ matrix whose block $\mathbf{X}_j\mathbf{C}_j^{-1/2}$ corresponds to the jth set of variables. To be noted is that each block is column-orthonormal, i.e. $n^{-1}\left(\mathbf{X}_j\mathbf{C}_j^{-1/2}\right)'\mathbf{X}_j\mathbf{C}_j^{-1/2} = \mathbf{I}_{K_j}$, from the definition of $\mathbf{C}$. This property can be called within-blocks orthonormality, as shown at the centre in Fig. 2. In contrast, between-blocks are not orthogonal: the columns of $\mathbf{X}_j\mathbf{C}_j^{-1/2}$ are not orthogonal to those of $\mathbf{X}_k\mathbf{C}_k^{-1/2}(k \neq j)$ in general. As the orthogonality implies (inter-column linear) independence, the above contrast shows that between-blocks dependence exists, but within-blocks dependence does not exist in $\mathbf{XC}^{-1/2}$. This matrix is approximated by a lower rank matrix $\mathbf{FW}'\mathbf{C}^{1/2} = \left[\mathbf{FW}_1'\mathbf{C}_1^{1/2}, \ldots, \mathbf{FW}_m'\mathbf{C}_m^{1/2}\right]$. As a result, the between-blocks dependence, which can be restated as relationships among the m sets of variables, can be expected to be summarised by $\mathbf{FW}'\mathbf{C}^{1/2}$.

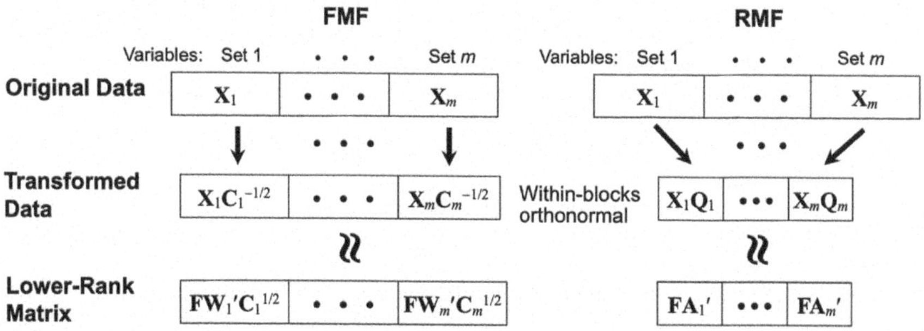

Fig. 2 Graphical illustration of FMF and RMF problems

Also for the RMF problem, the argument in the last paragraph holds with $\mathbf{C}_j^{-1/2}$ and $\mathbf{W}_j'\mathbf{C}_j^{1/2}$ replaced by $\mathbf{Q}_j$ and $\mathbf{A}_j'$, respectively; that is, each block of the transformed data matrix $\mathbf{XQ} = [\mathbf{X}_1\mathbf{Q}_1, \ldots, \mathbf{X}_m\mathbf{Q}_m]$ is within-blocks orthonormal as in (3) and (35), but the columns of $\mathbf{X}_j\mathbf{Q}_j$ are not orthogonal to $\mathbf{X}_k\mathbf{Q}_k (k \neq j)$ in general. Thus, the matrix $\mathbf{XQ}$, which has between-blocks dependence without within-blocks dependence, is approximated by lower rank $\mathbf{FA}' = \left[\mathbf{FA}_1', \ldots, \mathbf{FA}_m'\right]$, so that the relationships among the sets of variables can be expected to be summarised by $\mathbf{FA}'$.

However, the RMF problem differs from the FMF one in the following three points, as found in Fig. 2: (1) The columns of the transformed data matrix are reduced from those of the original data matrix; (2) $\mathbf{Q}_j$ in the transformed data matrix is the parameter to be estimated; and (3) the lower rank matrix is the product of two parameter matrices $\mathbf{F}$ and $\mathbf{A}'$ as in PCA, which differs from the FMF problem with the lower rank matrix being the product including data-based $\mathbf{C}_j^{1/2}$. These points show that $\mathbf{F}, \mathbf{A}$, and $\mathbf{Q}_1, \ldots, \mathbf{Q}_m$ are jointly estimated so that the reduced transformed matrix $\mathbf{XQ}$ is well approximated by the model part $\mathbf{FA}'$ in the RMF problem.

Finally, we must remark on a slight difference of MCA from GCCA in the transformation from the original $\mathbf{X} = [\mathbf{X}_1, \ldots, \mathbf{X}_m]$ into $\mathbf{XC}^{-1/2} = \left[\mathbf{X}_1\mathbf{C}_1^{-1/2}, \ldots, \mathbf{X}_m\mathbf{C}_m^{-1/2}\right]$ or $\mathbf{XQ} = [\mathbf{X}_1\mathbf{Q}_1, \ldots, \mathbf{X}_m\mathbf{Q}_m]$, namely, that $\mathbf{X}_j$ is not column-orthogonal in GCCA, but $\mathbf{X}_j$ is column-orthogonal in MCA with $\mathbf{X}_j'\mathbf{X}_j$ diagonal from $\mathbf{X}_j$ being binary and (22). That is, the transformation in GCCA is within-blocks orthonormalisation, while that in MCA is to normalise orthogonal $\mathbf{X}_j$ into $\mathbf{X}_j\mathbf{C}_j^{-1/2}$ and $\mathbf{X}_j\mathbf{Q}_j$, i.e. the within-blocks normalisation.

References

Adachi, K.: Oblique promax rotation applied to multiple correspondence analysis. Behaviormetrika **31**, 1–12 (2004)

Adachi, K.: Matrix-Based Introduction to Multivariate Data Analysis, 2nd edn. Springer, Singapore (2020)

Benzécri, J.P.: L'analyses des Données: Tome (Vol.) 1, la Taxinomie; Tome. 2, l'Analyses des Correspondances. Dunod, Paris (1973)

Carroll, J.D.: Generalization of canonical correlation analysis to three or more sets of variables. In: Proceedings of the 76th Annual Convention of the American Psychological Association, pp. 227–228. American Psychological Association, Washington, DC (1968)

Dahl, T., Næs, T.: A bridge between Tucker-1 and Carroll's generalized canonical analysis. Comput. Stat. Data Anal. **50**, 3086–3098 (2006)

Eckart, C., Young, G.: The approximation of one matrix by another of lower rank. Psychometrika **1**, 211–218 (1936)

Gifi, A.: Nonlinear Multivariate Analysis. Wiley, Chichester (1990)

Greenacre, M.J.: Theory and Applications of Correspondence Analysis. Academic Press, London (1984)

Greenacre, M.J.: Correspondence analysis of multivariate categorical data. Biometrika **75**, 457–467 (1988)

Kettenring, J.R.: Canonical analysis of several sets of variables. Biometrika **58**, 433–460 (1971)

Murakami, T.: Orthonormal principal component analysis for categorical data as a transformation of multiple correspondence analysis. In: Imaizumi, T., Nakayama, A., Yokoyama, S. (eds.) Advanced Studies in Behavior Metrics and Data Science: Essays in Honor of Akinori Okada, pp. 211–231. Springer, Singapore (2020)

Murakami, T., Kiers, H.A.L., ten Berge, J.M.F.: Non-metric principal components analysis for categorical variables with multiple quantifications. Unpublished manuscript (1999)

Nishisato, S.: Optimal scaling of paired comparison and rank order data: an alternative to Guttman's formulation. Psychometrika **43**, 263–271 (1978)

Nishisato, S.: Analysis of Categorical Data: Dual Scaling and its Applications. University of Toronto Press, Toronto (1980)

Takane, Y., Hwang, H., Abdi, H.: Regularized multi-set canonical correlation analysis. Psychometrika **73**, 753–775 (2008)

ten Berge, J.M.F.: Least Squares Optimization in Multivariate Analysis. DSWO Press, Leiden, The Netherlands (1993)

Tenenhaus, A., Tenenhaus, M.: Regularized generalized canonical correlation analysis. Psychometrika **76**, 257–284 (2011)

Van de Geer, J.P.: Linear relations among *k* sets of variables. Psychometrika **49**, 79–94 (1984)

Van de Velden, M., Bijmolt, T.H.A.: Generalized canonical correlation analysis of matrices with missing rows: a simulation study. Psychometrika **71**, 323–331 (2006)

Van de Velden, M., Takane, Y.: Generalized canonical correlation analysis with missing values. Comput. Stat. **27**, 551–571 (2012)

van der Burg, E., de Leeuw, J., Dijksterhuis, G.: OVERALS: nonlinear canonical correlation with *k* sets of variables. Comput. Stat. Data Anal. **18**, 141–163 (1994)

High-Dimensional Mixed-Data Regression Modelling Using the Gifi System with the Genetic Algorithm and Information Complexity

Suman Katragadda and Hamparsum Bozdogan

1 Introduction and Purpose

Statistical analysis is very much dependent on the quality and type of the data set. There are three types of data sets: continuous, categorical and a mix of the two. A continuous data set is one in which all the variables are in continuous form. A categorical data set is one in which all the variables are either ordinal (ordering of the categories exist) or nominal (no specific ordering of the categories exist). A particular form of categorical data set (e.g. yes/no, presence/absence) are coded as a 0 or a 1. A mixed data set is a data set in which some of the variables are in continuous form and the remainder of the variables are in categorical form. In other words, a mixed data set is a combination of continuous and categorical data variables. Statistical analysis would have been easy if data set is purely continuous or purely categorical. In reality, most of the data sets are neither purely continuous nor purely categorical but are in mixed form which makes the statistical analysis and modelling quite difficult. For instance, most of the data sets in the finance, insurance and medical sectors are of the mixed type.

Researchers in statistical data analysis usually face problems if the data are of the mixed data type. Most of the classical univariate and multivariate statistical concepts deals with continuous data or with categorical data but not mixed data. The usual statistical analysis done with the presence of many qualitative variable(s) in the data set containing other quantitative variables might not give accurate results. For instance, in the medical sector where the classification of the data is very important,

S. Katragadda
HEAPS.ai, Bengaluru, Karnataka, India
e-mail: suman@heaps.ai

H. Bozdogan (✉)
Department of Business Analytics and Statistics, The University of Tennessee, Knoxville, TN, USA
e-mail: bozdogan@utk.edu

E. J. Beh et al. (eds.), *Analysis of Categorical Data from Historical Perspectives*, Behaviormetrics: Quantitative Approaches to Human Behavior 17,
https://doi.org/10.1007/978-981-99-5329-5_25

the presence of many categorical and continuous predictors or variables results in poor modelling since the underlying probability distributional assumptions are violated. In the insurance and finance sectors, lots of categorical and continuous data are collected on customers for targeted marketing, detection of suspicious claims, actuarial modelling, risk analysis, modelling of financial derivatives, detection of profitable zones, credit scoring, etc.

In this paper, we address the problem of discovering interesting patterns from a mixed data set. Since a mixed data set is a combination of continuous and categorical variables, we transform the non-linear categorical variables to a linear scale by a mechanism called the "Gifi system" or "Gifi transformations" (Gifi 1990). Once the non-linear variables are transformed to a linear scale (in an Euclidean space), we carry out regression modelling using the genetic algorithm (GA) and information complexity (ICOMP) criterion of Bozdogan (1988, 1990a, b, 2004) (Bozdogan 2024) as our model selection criteria to choose the best subset of variables. The advantage of this transformation is that it has a one-to-one mapping property. In other words, the scaling is preserved and it is invariant, unlike the usual Reproducing Kernel Hilbert Space (RKHS)-based methods in machine learning. Hence, the transformed set of continuous value(s) can be remapped to a unique set of categorical value(s) in the original space. In this paper, we show the implementation of multiple regression to generate good models using the Gifi system.

2 What Is the Gifi System and Transformation?

As discussed in Katragada and Bozdogan (2008), the Gifi system is a clever and flexible technique that transforms categorical variables into continuous variables. If the data is of a mixed type, the Gifi system maps the categorical variables to a continuous scale so that all the data become continuous for the subsequent analysis. The Gifi system uses two main algorithms. They are:

(a) *The Optimal Scaling Method* (*OSM*) which optimally scales the categorical variables, thus making the data set purely continuous. Hence for the optimally scaled version, p-dimensional categorical variables are transformed to a p-dimensional continuous variables, and
(b) *Linear Combination Method* (*LCM*) takes the linear combination of the categories of the categorical variables and thus maps it into a 1-dimensional continuous space.

Hence, for the *LCM*, p-dimensional categorical variable is transformed to a 1-dimensional continuous space. The *OSM* might be useful when there are a few categorical variables in the data set whereas the *LCM* is useful when the number of the categorical variables is very large.

2.1 Coding of Categorical Data

A categorical variable (also known as a qualitative variable) is a type of data which may be divided into categories or groups. For instance, the variable gender has only two categories; namely "male" and "female". Categorical variables are discrete in nature. There are two types of categorical variables namely ordinal and nominal. An ordinal variable is a type of categorical variable where the categories of that variable can be ordered or ranked (e.g. "disagree", "neutral", "agree") (Tamhane and Dunlopet 2003). A nominal variable is a type of categorical variable whose categories simply represent distinct labels (e.g. "red", "green", "black").

2.2 Data Representation

Let us assume that there is a finite number of m categorical variables h_j, for $j = 1, 2, \ldots, m$. Also assume that each variable h_j has k_j distinct categories. Suppose that these m categorical variables are observed on a finite set of n objects (or individuals). We represent the data matrix $\mathbf{H}$ as an $n \times m$ matrix with elements h_{ij} giving the category of variable h_j for object i. That is:

$$\mathbf{H} = \begin{bmatrix} h_{11} & h_{12} & \cdots & h_{1m} \\ h_{21} & h_{22} & \cdots & h_{2m} \\ \vdots & \vdots & \ddots & \vdots \\ h_{1n} & h_{2n} & \cdots & h_{nm} \end{bmatrix}. \tag{1}$$

2.2.1 Indicator Matrix

An $n \times k_j$ binary matrix $\mathbf{G}_j$ for each variable h_j is defined as $G_j\ (i,\ t) = 1$, for $i = 1, \ldots, N$ and $t = 1, \ldots, k_j$, if object i belongs to category t, and $G_j\ (i,\ t) = 0$ if it belongs to some other category. $\mathbf{G}_j$ is called the *indicator matrix* of h_j. The matrix $\mathbf{G} = \left(\mathbf{G}_1, \ldots, \mathbf{G}_j, \ldots, \mathbf{G}_m\right)$ of dimension $n \times \sum_{j=1}^{m} k_j$ is a collection of such matrices and is also called a super-indicator matrix. Now, we illustrate an example of a super-indicator matrix. Consider a data matrix $\mathbf{H}$, with 10 observations (or objects) and 3 categorical variables, given in Table 1. Each of the 3 categorical variables has two categories. The frequency of the data matrix $\mathbf{H}$ is given in Table 2 and the summary of the frequency of $\mathbf{H}$ is given in Table 3. The indicator matrix $\mathbf{G}$ is given in Table 4.

Table 1 Data matrix **H**

p	x	a
p	x	b
p	x	a
q	x	a
q	x	b
p	y	b
q	y	a
p	x	b
q	x	a
p	x	a

Table 2 Frequency of **H**

p	x	a	3
p	x	b	2
p	y	a	0
p	y	b	1
q	x	a	2
q	x	b	1
q	y	a	1
q	y	b	0

Table 3 Summary of frequency of **H**

p	x	a	3
p	x	b	2
p	y	b	1
q	x	a	2
q	x	b	1
q	y	a	1

2.2.2 Complete Indicator Matrix

The indicator matrix $\mathbf{G}_j$ is said to be complete if each row of $\mathbf{G}_j$ has only one element equal to unity and zeros elsewhere, so that row sums of $\mathbf{G}_j$ are equal to unity Gifi (1990). In vector form, we can represent this as $\mathbf{G}_j u = u$ where u is a $n \times 1$ vector of length n. If all $\mathbf{G}_j$ are complete, their combined super-indicator matrix $\mathbf{G}$ is also said to be complete. In vector form, we can write $\mathbf{G}u = mu$ since the rows of $\mathbf{G}$ add up to m. For further properties of a complete indicator matrix we refer the readers to Gifi (1990).

Table 4 Indicator matrix **G** for the data matrix **H**

p	q	x	y	a	b
1	0	1	0	1	0
1	0	1	0	0	1
1	0	1	0	1	0
0	1	1	0	1	0
0	1	1	0	0	1
1	0	0	1	0	1
0	1	0	1	1	0
1	0	1	0	0	1
0	1	1	0	1	0
1	0	1	0	1	0

2.2.3 Quantification

The quantification of a categorical variable h_j is a process of converting its categorical value to a continuous scale so that the classical techniques of multivariate analysis can be applied. The quantification of categories of the variable h_j implies that these k_j categories are mapped as the k_j numerical values of a vector $\mathbf{y}_j$. Let the quantified variable, $\mathbf{q}_j = \mathbf{G}_j\mathbf{y}_j$ be a single vector which gives a numerical result for each object with respect to h_j.

Let us define $\mathbf{x} = m^{-1}\sum_j \mathbf{q}_j$, the mean vector of all $\mathbf{q}_j$'s. Then $\mathbf{x}$ contains the quantification of the objects or, in other words, the induced score of objects. We define the quantification of a category as the averaging of the scores of those objects that are mapped into that category. Mathematically, we write it as $\mathbf{y}_j = \mathbf{D}_j^{-1}\mathbf{G}_j'\mathbf{x}$. The vector $\mathbf{x}$ is of size $n \times 1$ and the vector $\mathbf{y}_j$ is of size $k_j \times 1$.

3 Homogeneity Analysis: HOMALS

Categorical PCA (HOMALS) is a particular form of non-linear PCA that is based on a categorical coding of variables using their indicator matrix form. As described in Sect. 2.2.1, $\mathbf{G}_j$ is an indicator matrix for variable h_j. The quantification of objects and of categories for a set of complete indicator matrices $\{\mathbf{G}_1, \ldots, \mathbf{G}_j, \ldots, \mathbf{G}_m\}$ should satisfy the following:

$$\mathbf{x} \propto \frac{1}{m}\sum_{j=1}^{m}\mathbf{G}_j\mathbf{y}_j \quad (2)$$

$$\mathbf{y}_j \propto \mathbf{D}_j^{-1}\mathbf{G}_j'\mathbf{x}, \quad (3)$$

where $\mathbf{D}_j = \mathbf{G}_j'\mathbf{G}_j$ is the $k_j \times k_j$ diagonal matrix containing the univariate marginals of variable h_j. In (2) and (3), $\mathbf{x}$ is the vector of object scores and $\mathbf{y}_j$ is the vector of the quantifications of the categories of variable h_j.

We note that (2) and (3) are very much the same as the formulae of Professor Nishisato—see, for example, Nishisato (1994)—in his seminal work who developed his dual scaling, and other similar methods like reciprocal averaging. This shows the parallels that exist between the Gifi system and dual scaling.

Let $\mathbf{X}$ be the $n \times p$ matrix (usually $p \le m$) containing the object scores and $\mathbf{Y}_j$ be the $k_j \times p$ matrix containing the category quantification of variable h_j. Since the quantification process incurs some loss of information, a typical loss function is given as:

$$\sigma\left(\mathbf{X};\ \mathbf{Y}_1, \ldots, \mathbf{Y}_m\right) = m^{-1}\sum_{j=1}^{m} \mathrm{SSQ}\left(\mathbf{X} - \mathbf{G}_j\mathbf{Y}_j\right)$$
$$= m^{-1}\mathrm{trace}\left[\left(\mathbf{X} - \mathbf{G}_j\mathbf{Y}_j\right)'\left(\mathbf{X} - \mathbf{G}_j\mathbf{Y}_j\right)\right], \tag{4}$$

where SSQ ($\circ$) denotes the sum of squares of the elements of the matrix $\mathbf{X}$. The loss function in (4) is at the heart of the Gifi (1990) system. We want to minimise the above loss function simultaneously over $\mathbf{X}$ and $\mathbf{Y}_j$'s. By imposing various restrictions on the category quantifications $\mathbf{Y}_j$ and, in some cases, the coding of the data, different types of analysis can be derived.

In the process of minimising the loss function of (4), we impose two further constraints in order to avoid the trivial solution corresponding to $\mathbf{X} = 0$, and $\mathbf{Y}_j = 0$ for every j. The two constraints are:

$$\mathbf{X}'\mathbf{X} = n\mathbf{I} \text{ and } \mathbf{X}\mathbf{u} = 0, \tag{5}$$

where $\mathbf{u}'$ is a vector of ones with dimension $p \times 1$. The constraint (5) standardises the squared length of the object scores to be equal to n. Further, in two or higher dimensions it requires the columns of $\mathbf{X}$ to be orthogonal. In addition, the second constraint basically makes the plot to be centred around the origin.

We minimise the above loss function simultaneously over $\mathbf{X}$ and $\mathbf{Y}_j$'s by employing an *Alternating Least Squares* (ALS) algorithm. We start the process with a uniformly random choice of $\mathbf{X}$ ($\mathbf{X} \neq 0$), with a mean zero, normalise it so that its sums of squares is n (rather than 1), so that the scores have variance 1. We compute a first set of category quantification Y_j by:

$$\widehat{\mathbf{Y}}_j = \mathbf{D}_j^{-1}\mathbf{G}_j'\mathbf{X}, \tag{6}$$

where $\mathbf{D}_j = \mathbf{G}_j'\mathbf{G}_j$ is the $k_j \times k_j$ diagonal matrix containing the univariate marginals of variable h_j.

In the second step of the algorithm, the loss function in (4) is minimised with respect to $\mathbf{X}$ for fixed $\mathbf{Y}_j$'s . It is given by:

$$\widehat{\mathbf{X}} = \frac{1}{m}\sum_{j=1}^{m}\mathbf{G}_j\mathbf{Y}_j\,. \tag{7}$$

In the third step of the algorithm the object scores $\mathbf{X}$ are column centred by setting $\mathbf{B} = \widehat{\mathbf{X}} - \mathbf{u}\left(\mathbf{u}'\mathbf{X}/n\right)$, and then orthonormalised by the modified Gram-Schmidt procedure $\mathbf{X} = \sqrt{n}\,\mathrm{GRAM}\,(\mathbf{B})$, so that both the normalisation constraints in (5) is satisfied. See Trefethen and Bau (1997) for more details.

The usual normalisation condition used in ALS is given by:

$$\mathbf{X} = \mathbf{X}\left(\mathbf{X}'\mathbf{X}\right)^{-1/2}. \tag{8}$$

The problem with the usual normalisation condition in (8) might arise when p is large. When p is large this method could become quite expensive from a computational point of view. It can be replaced with the cheaper Gram-Schmidt method. The Gram-Schmidt method starts with unit normalising the first column of $\mathbf{X}$, then projects the second column of $\mathbf{X}$ onto the space orthogonal to the first column, replaces the second column by the unit normalised anti-projection, next projects the third column of $\mathbf{X}$ onto the space orthogonal to the new second column, and so on. This process can be summarised by stating that $\mathbf{X}$ is decomposed as $\mathbf{X} = \mathbf{UT}$, with $\mathbf{U}'\mathbf{U} = \mathbf{I}$ and where $\mathbf{T}$ is an upper triangular matrix. The matrix $\mathbf{U}$ is scaled by the $\sqrt{n}$ and the resulting matrix is taken as the new $\mathbf{X}$.

The ALS algorithm cycles through these three steps until the convergence criterion is met. Equation (6) expresses the first centroid principle (a category quantification is in the centroid of the object scores they belong to it), while (7) shows that an object score is the average of the quantifications of the categories it belongs to. Hence, this solution accomplishes the goal of producing a graph with objects close to the categories they fall in and categories close to the objects belonging in them. One can review Gifi (1990) and Michailidis and de Leeuw (1996) for more details on the working of HOMALS algorithm.

4 Optimal Scaling and Linear Combination Algorithms

4.1 Optimal Scaling Algorithm

The Optimal Scaling Algorithm (OSA) fits a multiple regression model for a continuous response and a mixed set of predictors, $\mathbf{X}$. It also selects the optimal predictors that explain most of the variation in the response. We briefly explain the steps of the OSA using the genetic algorithm (GA) for a five variable regression model as follows:

(1) Run the Gifi transformation on the data $\mathbf{X}$ and optimally scale the categorical variables. A categorical variable h_j is optimally scaled by multiplying its indi-

cator matrix, $\mathbf{G}_j$, with its optimal weight (score) vector, $\mathbf{y}_j$. Suppose if the data contains variables $[\mathbf{x}_1,\ \mathbf{x}_2,\ \mathbf{x}_3,\ \mathbf{x}_4,\ \mathbf{x}_5]$. Let $\mathbf{x}_1$ and $\mathbf{x}_4$ be continuous and $\mathbf{x}_2$, $\mathbf{x}_3$, and $\mathbf{x}_5$ be categorical. Let $\mathbf{G}_2$ and $\mathbf{y}_2$ be the indicator matrix and the optimal weight vector for the categorical variable $\mathbf{x}_2$, respectively. Similarly, $\mathbf{G}_3$, $\mathbf{G}_5$ and $\mathbf{y}_3$, $\mathbf{y}_5$ are the indicator matrices and optimal weight vectors of the categorical variables $\mathbf{x}_3$ and $\mathbf{x}_5$, respectively. Therefore, the data matrix in the Gifi space will be of the form $\left[\mathbf{x}_1,\ \mathbf{G}_2\mathbf{y}_2,\ \mathbf{G}_3\mathbf{y}_3,\ \mathbf{x}_4,\ \mathbf{G}_5\mathbf{y}_5\right]$.

(2) Generate a random population of size N and dimension p, where p is the number of predictors in the model. Consider each row of the population to be a chromosome.

(3) For each chromosome in the population:
   - Build a new predictor data matrix, $\mathbf{X}_{\text{new}}$.
   - Perform multiple regression analysis with $\mathbf{y}$ as the response and $\mathbf{X}_{\text{new}}$ as predictors and compute the desired information criteria.

(4) Sort the chromosomes in the population according to the ascending order of their information score. The chromosome with the lowest information score is considered to be the best chromosome among the $N-1$ other chromosomes.

(5) Stop if the stopping criteria is met and return the best model from the current population or else:
   - Perform crossover and mutation with pCrossover, pMutation and the crossover type to generate a new population. Always include the best model in the new population.
   - Go to Step (3).

4.2 Linear Combination Algorithm

The Linear Combination Algorithm (LCA) fits a multiple linear regression model to a continuous response given a mixed set of predictors, $\mathbf{X}$. It also selects the optimal predictors that explain most of the variation in the response and works as follows:

(1) Generate a random population of size N and dimension p, where p is the number of predictors in the model. Consider each row of the population to be a chromosome.

(2) For each chromosome in the population:
   - Build a new predictor data matrix, $\mathbf{X}_{\text{new}}$. Since $\mathbf{X}_{\text{new}}$ might be a mixed data set, we split the $\mathbf{X}_{\text{new}}$ matrix into $\mathbf{X}_{\text{con}}$ and $\mathbf{X}_{\text{cat}}$ where $\mathbf{X}_{\text{con}}$ is the data on the continuous predictors and $\mathbf{X}_{\text{cat}}$ is a 1-dimensional continuous data of the categorical predictors. Hence $\mathbf{X}_{\text{new}}$ can be represented as $\mathbf{X}_{\text{new}} = [\mathbf{X}_{\text{con}}\ \mathbf{X}_{\text{cat}}]$. Suppose, if the current chromosome selects $\mathbf{x}_1$, $\mathbf{x}_2$, $\mathbf{x}_3$, $\mathbf{x}_4$ which are a subset of the original set of predictors $\mathbf{x}_1, \ldots, \mathbf{x}_p$ where $p \geq 5$. Suppose $\mathbf{x}_1$ and $\mathbf{x}_3$ are continuous and $\mathbf{x}_2$, $\mathbf{x}_4$ and $\mathbf{x}_5$ are categorical. We perform the Gifi

transformation on $\mathbf{x}_2$, $\mathbf{x}_4$ and $\mathbf{x}_5$ and transform it to a 1-dimensional continuous variable, $\mathbf{X}_{\text{cat}}$. Since $\mathbf{x}_1$ and $\mathbf{x}_3$ are continuous, $\mathbf{X}_{\text{con}} = \left[\mathbf{x}_1\ \mathbf{x}_3\right]$. Therefore, $\mathbf{X}_{\text{new}} = [\mathbf{X}_{\text{con}}\ \mathbf{X}_{\text{cat}}]$.
- Perform multiple regression analysis with y as the response and $\mathbf{X}_{\text{new}}$ as the set of predictors and compute the desired information criteria for each chromosome.

(3) Sort the chromosomes in the population according to the ascending order of their information score. The chromosome with the lowest information score is considered to be the best chromosome among the $N - 1$ other chromosomes.
(4) Stop if the stopping criteria is met and return the best model from the current population or else:

- Perform crossover and mutation with pCrossover, pMutation and the crossover type to generate a new population. Always include the best model in the new population.
- Go to Step (2).

5 Regression Modelling Using the Gifi System

In this section, we present the regression modelling of data using the OSA and LCA by making the data purely continuous.

Following Bozdogan (2004, 2024), we consider the multiple linear regression model given by:

$$\mathbf{y} = \beta_0 + \beta_1\mathbf{X}_1 + \cdots + \beta_k\mathbf{X}_k + \varepsilon\,, \tag{9}$$

or, in a more compact matrix form, we have:

$$\mathbf{y} = \mathbf{X}\boldsymbol{\beta} + \varepsilon \tag{10}$$

where

$$\mathbf{y} = \begin{bmatrix} y_1 \\ \vdots \\ y_n \end{bmatrix}, \quad \mathbf{X} = \begin{bmatrix} x_{11} & \dots & x_{1k} \\ \vdots & \ddots & \vdots \\ x_{n1} & \dots & x_{nk} \end{bmatrix}, \quad \boldsymbol{\beta} = \begin{bmatrix} \beta_1 \\ \vdots \\ \beta_k \end{bmatrix}, \quad \text{and} \quad \boldsymbol{\varepsilon} = \begin{bmatrix} \varepsilon_1 \\ \vdots \\ \varepsilon_n \end{bmatrix},$$

so that:

- $\mathbf{y}$ is a vector of $(n \times 1)$ observations on a dependent variable,
- $\mathbf{X}$ is a full rank $(n \times q)$ matrix of nonstochastic predetermined predictor variables,
- $\boldsymbol{\beta}$ is a $(q \times 1)$ coefficient vector, and $\boldsymbol{\varepsilon}$ is an $(n \times 1)$ vector of unknown random error (or noise) term.

Often the first column of $\mathbf{X}$ consists of ones. That is, $(x_{11}, \ldots, x_{n1})' = (1, \ldots, 1)'$ to signify the presence of an *intercept* in the regression model.

Given the model in (9), under the *assumption of normality*, we can analytically express the density function of regression model as:

$$f\left(\mathbf{y}_i \mid \mathbf{x}_i, \boldsymbol{\beta}, \sigma^2\right) = \left(2\pi\sigma^2\right)^{-\frac{1}{2}} \exp\left[-\frac{\left(\mathbf{y}-\mathbf{x}_i'\boldsymbol{\beta}\right)^2}{2\sigma^2}\right]. \tag{11}$$

The likelihood function for a random sample of n observations is:

$$L\left(\boldsymbol{\beta}, \sigma^2 \mid \mathbf{y}, \mathbf{X}\right) = \left(2\pi\sigma^2\right)^{-\frac{n}{2}} \exp\left[-\frac{(\mathbf{y}-\mathbf{X}\boldsymbol{\beta})'(\mathbf{y}-\mathbf{X}\boldsymbol{\beta})}{2\sigma^2}\right], \tag{12}$$

and the log likelihood function is:

$$l\left(\boldsymbol{\beta}, \sigma^2\right) = -\frac{n}{2}\log(2\pi) - \frac{n}{2}\log\sigma^2 - \frac{(\mathbf{y}-\mathbf{X}\boldsymbol{\beta})'(\mathbf{y}-\mathbf{X}\boldsymbol{\beta})}{2\sigma^2}. \tag{13}$$

Using matrix differential calculus, the maximum likelihood estimates (*MLE's*) $(\hat{\boldsymbol{\beta}}, \hat{\sigma}^2)$ of $\left(\boldsymbol{\beta}, \sigma^2\right)$ are given by:

$$\hat{\boldsymbol{\beta}} = \left(\mathbf{X}'\mathbf{X}\right)^{-1}\mathbf{X}'\mathbf{y}, \tag{14}$$

and

$$\hat{\sigma}^2 = \frac{\left(\mathbf{y}-\mathbf{X}\hat{\boldsymbol{\beta}}\right)'\left(\mathbf{y}-\mathbf{X}\hat{\boldsymbol{\beta}}\right)}{n} = \frac{\text{RSS}}{n}. \tag{15}$$

The *maximum likelihood (ML) covariance matrix of the estimated regression coefficients* is given by:

$$\widehat{\text{Cov}}\left(\hat{\boldsymbol{\beta}}\right)_{\text{MLE}} = \hat{\sigma}^2\left(\mathbf{X}'\mathbf{X}\right)^{-1} \tag{16}$$

without centreing and scaling.

To study the sampling performance of $\hat{\boldsymbol{\beta}}$ and $\hat{\sigma}^2$, we compute the estimated *Fisher information matrix* (*FIM*) given by:

$$\hat{\mathcal{F}} = \begin{bmatrix} \frac{\mathbf{X}'\mathbf{X}}{\hat{\sigma}^2} & \mathbf{0} \\ \mathbf{0} & \frac{n}{2\hat{\sigma}^4} \end{bmatrix}. \tag{17}$$

To derive the information complexity, *ICOMP*, of the estimated *inverse Fisher information matrix (IFIM)*, we have:

$$\widehat{\text{Cov}}\left(\hat{\boldsymbol{\beta}}, \hat{\sigma}^2\right) = \hat{\mathcal{F}}^{-1} = \begin{bmatrix} \hat{\sigma}^2\left(\mathbf{X}'\mathbf{X}\right)^{-1} & \mathbf{0} \\ \mathbf{0}' & \frac{2\hat{\sigma}^4}{n} \end{bmatrix} \tag{18}$$

which is a biased covariance matrix of the estimated parameters. The covariance of the unbiased estimators $\hat{\boldsymbol{\beta}}$ and $\hat{\sigma}^2$ is:

$$\widehat{\text{Cov}}\left(\hat{\boldsymbol{\beta}}, \hat{\sigma}^2\right) = \hat{\mathcal{F}}^{-1} = \begin{bmatrix} \hat{\sigma}^2 \left(\mathbf{X}'\mathbf{X}\right)^{-1} & \mathbf{0} \\ \mathbf{0}' & \frac{2\hat{\sigma}^4}{n-k-1} \end{bmatrix}. \tag{19}$$

When the model is misspecified, we define a consistent estimator of the covariance matrix Cov $\left(\boldsymbol{\theta}_k^*\right)$ given by:

$$\widehat{\text{Cov}}\left(\hat{\boldsymbol{\theta}}\right)_{\text{Misspec}} = \hat{\mathcal{F}}^{-1}\hat{\mathcal{R}}\hat{\mathcal{F}}^{-1}. \tag{20}$$

This is often called the *"sandwich covariance"* or *"robust covariance"* estimator, since it is a correct variance regardless whether of the assumed model is correct or not.

In (20) $\hat{\mathcal{R}}$ is the meat and the two $\hat{\mathcal{F}}^{-1}$'s are the two slices of the bread. When the model is correct we get $\hat{\mathcal{F}} = \hat{\mathcal{R}}$, and the formula reduces to the usual *inverse Fisher information matrix*, $\hat{\mathcal{F}}^{-1}$ (White 1982).

The estimated *outer-product form of the Fisher information matrix* is given by:

$$\hat{\mathcal{R}} = \begin{bmatrix} \frac{1}{\hat{\sigma}^4}\mathbf{X}'\mathbf{D}^2\mathbf{X} & \mathbf{X}'\mathbf{1}\frac{Sk}{2\hat{\sigma}^3} \\ \left(\mathbf{X}'\mathbf{1}\frac{Sk}{2\hat{\sigma}^3}\right)' & \frac{(n-q)(Kt-1)}{4\hat{\sigma}^4} \end{bmatrix}, \tag{21}$$

where $\mathbf{D}^2 = \text{diag}\left(\hat{\varepsilon}_1^2, \ldots, \hat{\varepsilon}_n^2\right)$ and $\mathbf{X}$ is a $(n \times q)$ matrix of regressors or model matrix, Sk is the estimated residual skewness, Kt the kurtosis, and $\mathbf{1}$ is a $(n \times 1)$ vector of ones. That is:

$$Sk = \frac{\left(\frac{1}{n}\sum_{i=1}^{n}\hat{\varepsilon}_i^3\right)}{\hat{\sigma}^3} \tag{22}$$

and

$$Kt = \frac{\left(\frac{1}{n}\sum_{i=1}^{n}\hat{\varepsilon}_i^4\right)}{\hat{\sigma}^4}. \tag{23}$$

Hence, the *"sandwich covariance"* or *"robust covariance"* estimator is given by:

$$\begin{aligned} \widehat{\text{Cov}}\left(\hat{\boldsymbol{\theta}}\right)_{\text{Misspec}} &= \hat{\mathcal{F}}^{-1}\hat{\mathcal{R}}\hat{\mathcal{F}}^{-1} \\ &= \begin{bmatrix} \hat{\sigma}^2 \left(\mathbf{X}'\mathbf{X}\right)^{-1} & \mathbf{0} \\ \mathbf{0}' & \frac{2\hat{\sigma}^4}{n} \end{bmatrix} \begin{bmatrix} \frac{1}{\hat{\sigma}^4}\mathbf{X}'\mathbf{D}^2\mathbf{X} & \mathbf{X}'\mathbf{1}\frac{Sk}{2\hat{\sigma}^3} \\ \left(\mathbf{X}'\mathbf{1}\frac{Sk}{2\hat{\sigma}^3}\right)' & \frac{(n-q)(Kt-1)}{4\hat{\sigma}^4} \end{bmatrix} \begin{bmatrix} \hat{\sigma}^2 \left(\mathbf{X}'\mathbf{X}\right)^{-1} & \mathbf{0} \\ \mathbf{0}' & \frac{2\hat{\sigma}^4}{n} \end{bmatrix}. \end{aligned} \tag{24}$$

Note that this covariance matrix should impose greater complexity than the *inverse Fisher information matrix (IFIM)*. It also takes into account presence of skewness and kurtosis.

Next, we show several derived forms of *ICOMP* and other information criteria for model selection in multiple regression models.

5.1 Derived Forms of Information Criteria for the Regression Model

Here, we briefly give the derived forms of various information criteria for the multiple regression model with k predictors.

5.1.1 AIC and AIC-Type Criteria

Akaike (1973) AIC for the regression model to be used as fitness function in the *GA* is given by:

$$\text{AIC(Regression)} = n\log(2\pi) + n\log\left(\hat{\sigma}^2\right) + n + 2(k+1). \tag{25}$$

Applying the finite sample correction to AIC, we have:

$$\text{AIC}_c\text{(Regression)} = n\log(2\pi) + n\log\left(\hat{\sigma}^2\right) + n + 2\frac{n(k+1)}{n-k-2}. \tag{26}$$

The consistent AIC (CAIC) of Bozdogan (1987) for the regression model is given by:

$$\text{CAIC(Regression)} = n\log(2\pi) + n\log\left(\hat{\sigma}^2\right) + n + k\left[\log(n) + 1\right]. \tag{27}$$

Similar to CAIC, the Bayesian criterion (SBC) proposed by Schwarz (1978) is defined by:

$$\text{SBC(Regression)} = n\log(2\pi) + n\log\left(\hat{\sigma}^2\right) + n + k\log(n). \tag{28}$$

5.1.2 ICOMP Based on IFIM

If we use the estimated *inverse Fisher information matrix* (*IFIM*), then we define *ICOMP(IFIM)* as:

$$\begin{aligned}\text{ICOMP(IFIM)}_{\text{Regression}} &= -2\log L\left(\hat{\theta}\right) + 2C_1\left(\hat{\mathcal{F}}^{-1}\right)\\ &= n\log(2\pi) + n\log\left(\hat{\sigma}^2\right) + n + 2C_1\left(\hat{\mathcal{F}}^{-1}\right),\end{aligned} \tag{29}$$

where $C_1\left(\hat{\mathcal{F}}^{-1}\right)$ is the maximal entropic complexity of IFIM given by:

$$C_1\left(\hat{\mathcal{F}}^{-1}\right) = \frac{s}{2}\log\left[\frac{\operatorname{trace}\left(\hat{\mathcal{F}}^{-1}\right)}{s}\right] - \frac{1}{2}\log\left|\hat{\mathcal{F}}^{-1}\right|. \tag{30}$$

For the regression model, $C_1\left(\hat{\mathcal{F}}^{-1}\right)$ is given by:

$$C_1\left(\hat{\mathcal{F}}^{-1}\right) = \frac{(q+1)}{2}\log\left[\frac{\operatorname{trace}\left(\hat{\sigma}\left(\mathbf{X}'\mathbf{X}\right)^{-1} + \frac{2\hat{\sigma}^4}{n}\right)}{q+1}\right] - \log\left|\hat{\sigma}^2\left(\mathbf{X}'\mathbf{X}\right)^{-1}\right| - \log\left(\frac{2\hat{\sigma}^4}{n}\right). \tag{31}$$

Similarly, using the *second order equivalent measure of complexity* to the original $C_1(\circ)$ measure, that is:

$$\text{ICOMP(IFIM)}_{\text{Regression}} = n\log(2\pi) + n\log\left(\hat{\sigma}^2\right) + n + 2C_{1F}\left(\hat{\mathcal{F}}^{-1}\right), \tag{32}$$

where $C_{1F}\left(\hat{\mathcal{F}}^{-1}\right)$ is given by:

$$C_{1F}\left(\hat{\mathcal{F}}^{-1}\right) = \frac{1}{4\overline{\lambda}_a^2}\sum_{j=1}^{s}\left(\lambda_j - \overline{\lambda}_a\right)^2, \tag{33}$$

and where λ_j, $j = 1, 2, \ldots, s$, are the eigenvalues of $\hat{\mathcal{F}}^{-1}$ and $\overline{\lambda}_a$ is the arithmetic mean of the eigenvalues.

We note that $C_{1F}(\circ)$ is a *second order equivalent measure of complexity* to the original $C_1(\circ)$ measure. Also, we note that $C_{1F}(\circ)$ is *scale-invariant* and $C_{1F}(\circ) \geq 0$ with $C_{1F}(\circ) = 0$ only when all $\lambda_j = \overline{\lambda}_a$. Also, $C_{1F}(\circ)$ measures the *relative variation* in the eigenvalues.

Further in the *inverse Fisher information matrix (IFIM)*, as the number of parameters increases (i.e. as the size of $\mathbf{X}$ increases), the error variance $\hat{\sigma}^2$ gets smaller even though the complexity gets larger. Also, as $\hat{\sigma}^2$ increases, $\left(\mathbf{X}'\mathbf{X}\right)^{-1}$ decreases. Therefore, $C_1\left(\hat{\mathcal{F}}^{-1}\right)$ achieves a trade-off between these two extremes and helps to avoid *multicollinearity*.

5.1.3 ICOMP Under Misspecification

When the model is misspecified, we define *ICOMP* under misspecification as:

Table 5 Summary of model selection criteria used in genetic algorithm (GA)

1.	$\text{AIC(Regression)} = n\log(2\pi) + n\log(\hat{\sigma}^2) + n + 2(k+1)$
2.	$\text{CAIC(Regression)} = n\log(2\pi) + n\log(\hat{\sigma}^2) + n + k[\log(n) + 1]$
3.	$\text{SBC(Regression)} = n\log(2\pi) + n\log(\hat{\sigma}^2) + n + k\log(n)$
4.	$\text{ICOMP(IFIM)} = n\log(2\pi) + n\log(\hat{\sigma}^2) + n + 2C_1\left(\hat{\mathcal{F}}^{-1}\right)$
5.	$\text{ICOMP(IFIM)} = n\log(2\pi) + n\log(\hat{\sigma}^2) + n + 2\left[\frac{1}{4\overline{\lambda}_a^2}\sum_{j=1}^{q}(\lambda_j - \overline{\lambda}_a)^2\right]$
6.	$\text{ICOMP(IFIM)}_{\text{Misspec}} = n\log(2\pi) + n\log(\hat{\sigma}^2) + n + 2C_1\left(\widehat{\text{Cov}}(\hat{\theta})_{\text{Misspec}}\right)$

$$\text{ICOMP(IFIM)}_{\text{Misspec}} = n\log(2\pi) + n\log(\hat{\sigma}^2) + n + 2C_1\left(\widehat{\text{Cov}}(\hat{\theta})_{\text{Misspec}}\right), \tag{34}$$

where

$$\widehat{\text{Cov}}(\hat{\theta})_{\text{Misspec}} = \hat{\mathcal{F}}^{-1}\hat{\mathcal{R}}\hat{\mathcal{F}}^{-1} \tag{35}$$

is a consistent estimator of the covariance matrix $\text{Cov}(\boldsymbol{\theta}_k^*)$. For further details on information criteria, we refer the readers Bozdogan (1988, 1990a, b, 2004, 2024).

In summary, we score any one of the following information criteria using the genetic algorithm (GA) to carry out subset selection of variables in multiple regression model (Table 5).

6 Genetic Algorithm

A Genetic Algorithm (GA) is a stochastic approach search algorithm which is based on concepts of biological evolution and natural selection that can be applied to solving problems where a vast number of possible solutions exist. Unlike conventional optimisation techniques, the GA requires no calculation of the gradient of the objective function and is not restricted to local optima (Goldberg 1989). GA treats information as a series of codes on a binary string, where each string represents a different solution to a given problem. These strings are analogous models to the genetic information coded by genes on chromosome. A string can be evaluated according to some fitness value, for its particular ability to solve the problem. In our case, we use the information criteria as our fitness function. On the basis of the fitness values, strings are either retained or removed from the analysis after each run so that, after many runs, the best solution has been identified. One important difficulty with any GA is in choosing an appropriate fitness function as the basis for evaluating each solution. For a detailed review on GA we refer the readers to Goldberg (1989), Forrest (1993), and Marczyk (2004).

In GA, mating of chromosomes is performed as a crossover process (Bozdogan 2004). A model chosen for crossover is controlled by the crossover probability or the crossover rate. During the crossover process, we randomly pick a position along each pair of parent models (strings) as the crossover point. For any pair of parents, the strings are broken into two pieces at the crossover point and the portions of the two strings to the right of this point are interchanged between the parents to form two off spring strings.

There are three different crossover methods including *single point crossover*, *two point crossover* and *uniform crossover*. In *single point crossover*, one crossover point is selected; a binary string from beginning of the chromosome to the crossover point is copied from one parent, the rest is copied from the second parent. In *two point crossover*, two crossover points are selected; binary string from the beginning of the chromosome to the first crossover point is copied from one parent, the part from the first to the second crossover point is copied from the second parent and the rest is copied from the first parent. Finally in the *uniform crossover*, the bits are randomly copied from the first or from the second parent. In our algorithm, the user has the option of choosing any one of the three crossovers.

Mutation

Mutation of models is used in GA as a way of creating new combinations of variables so that the searching process can jump to another area of the fitness function landscape instead of searching in a limited area. By mutation, a randomly selected locus can change from 0 to 1 or from 1 to 0. Thus, a randomly selected predictor variable is either added to or removed from the model.

6.1 Steps of GA: Pseudo Code

(1) Generate a random population of N random regression models $\mathbf{y} = \mathbf{X}\boldsymbol{\beta} + \boldsymbol{\varepsilon}$.
(2) Evaluate each model by using the fitness function (i.e. your favourite information criterion).
(3) Sort the models in increasing order of the fitness function; the model with the minimum fitness function is the first model considered.
(4) Since the population has N models, we choose the first $N/2$ models for the crossover operation.
(5) Crossover operation is performed on each of the models in the population from 2 to $N/2 - 1$ with the first model.
(6) The new population always contains the first model in the population; the model having the minimum information criteria used as the fitness function.
(7) The $N - 2$ offspring's produced in step 4 go in to the new population. At this point we have $N - 1$ models in the new population.
(8) To generate the new population of size N, we perform a crossover using the first model and the $N/2 + 1$ model from the old population. Since there will be two offspring's produced from this crossover operation, we randomly select

one offspring and place it in the new population. Hence the new population contains N models.

(9) Perform the mutation operation with a mutation probability on the models in the new population.

(10) If the stopping criteria is met, return the best solution (which is the first model) from the current population

(11) Go to Step (2).

7 A Real Numerical Example and the Computational Results

We use the Gifi transformation on a real mixed data set and transform the categorical predictor variables to create a continuous variable. Then we fit a multiple regression model using the genetic algorithm (GA) to select the best subsets of the predictor variables. We show the results LMC and OSM in Sects. 7.1.1 and 7.1.2, respectively.

7.1 Analysis of Beta-Carotene Data Set

Observational studies have suggested that low dietary intake or low plasma concentrations of retinol, beta-carotene, or other carotenoids might be associated with increased risk of developing certain types of cancer. However, relatively few studies have investigated the determinants of plasma concentrations of these micronutrients. A cross-sectional study was designed to investigate the relationship between personal characteristics and dietary factors, and plasma concentrations of retinol, beta-carotene and other carotenoids. Study subjects ($n = 315$) were patients who had an elective surgical procedure during a three-year period to biopsy or remove a lesion of the lung, colon, breast, skin, ovary or uterus that was found to be non-

Table 6 GA parameter input

Maximum iteration	maxIter
Probability of cross over	pCrossover
Probability of mutation	pMutation
Cross over type	Uniform, single point, two point
Population size	N
Predictor data	X
Continuous response data	y
Information score	AIC, ICOMP, ICOMPIFIM, CAIC, SBC

Table 7 Beta-carotene data set

Variable names	Description of variables
AGE	Age (years)
SEX	(1 = male, 2 = female)
SMOKSTAT	(1 = never, 2 = former, 3 = current smoker)
QUETELET	Quetelet $\left(\text{weight}/\left(\text{height}^2\right)\right)$
VITUSE	(1 = Yes, fairly often, 2 = Yes, not often, 3 = No)
CALORIES	Number of calories consumed per day
FAT	Grams of fat consumed per day
FIBRE	Grams of fibre consumed per day
ALCOHOL	Number of alcoholic drinks consumed per week
CHOLESTEROL	Cholesterol consumed (mg per day)
BETADIET	Dietary beta-carotene consumed (mcg per day)
RETDIET	Dietary retinol consumed (mcg per day)
BETAPLASMA	Plasma beta-carotene (ng/ml)
RETPLASMA	Plasma retinol (ng/ml)

cancerous. We display the data for only two of the analytes (BETAPLASMA and RETPLASMA) (Table 7). This data has not been published yet but a related reference is Nierenberg et al. (1989).

7.1.1 Beta-Carotene Data: Using Linear Combination Method

Since the variables SEX, SMOKSTAT and VITUSE are categorical, we use the Gifi transformation to generate an optimal weight (score) vector that is used for transforming each of the categorical variables to a continuous variable. Some of the pair-wise kernel density estimates of this data is shown in Fig. 1 using the Gaussian kernel with bandwidth, $h = 0.5$.

We fit a multiple regression model with $\mathbf{y} =$ RETPLASMA as the dependent variable and the variables AGE, SEX, SMOKSTAT, QUETELET, VITUSE, CALORIES, FAT, FIBRE, ALCOHOL, CHOLESTROL, BETADIET and RETDIET as independent variables. We also include the intercept term in this model. We assume that the residuals are normally distributed. We use GA for variable selection with maximum iterations of 100, population size of 20, probability of crossover of 0.75, probability of mutation of 0.10 and crossover type as uniform. ICOMP_{C1} is used as the fitness function.

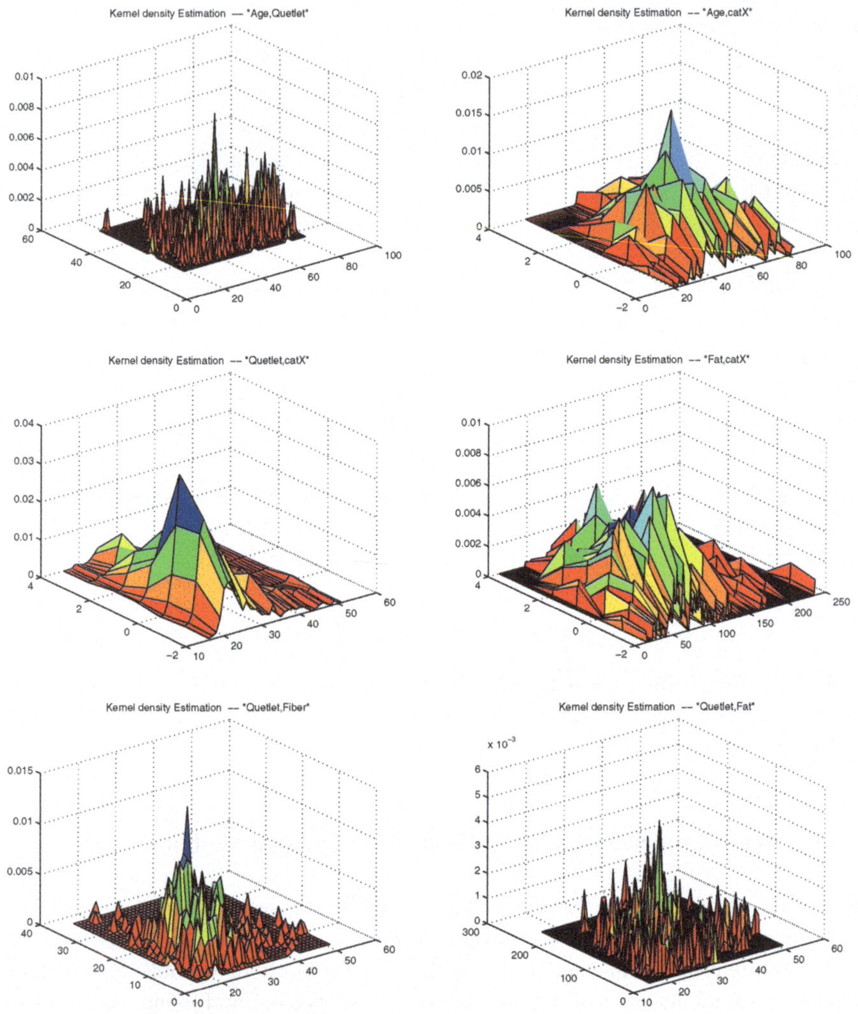

Fig. 1 Kernel density estimate of beta-carotene data

The best subset of variables selected by GA and its associated information score is given by:

Model:	Intercept	AGE	SEX	SMOKSTAT	QUETELET
		FAT	ALCOHOL	RETDIET	

where **ICOMP = 4233.376**. The estimated coefficients for the above set of variables is given by:

$$\hat{\beta} = \begin{bmatrix} 500.7981 \\ 2.5039 \\ 1.0760 \\ -0.5323 \\ -0.2685 \\ -0.0118 \\ 37.2665 \end{bmatrix}.$$

The RMSE for this model is 200.3348. The optimal weights (scores) associated with the categories of the variable SEX and SMOKSTAT are:

$$\text{wSEX} = \begin{bmatrix} 1.7298 \\ -0.2661 \end{bmatrix} \quad \text{wSMOKSTAT} = \begin{bmatrix} -0.6169 \\ 0.5241 \\ 0.8508 \end{bmatrix}.$$

Hence, the estimated regression model can be written as:

$$\begin{aligned} \hat{\mathbf{y}} &= \text{RETPLASMA} \\ &= 500.7981 + 2.5039 \times \text{AGE} + 1.0760 \times \text{QUETELET} - 0.5323 \times \text{FAT} \\ &\quad -0.2685 \times \text{ALCOHOL} - 0.0118 \times \text{RETDIET} + 37.2665 \times \text{catX}, \end{aligned}$$

where catX is the linear combination of the weights (scores) of the categories of the variables SEX and SMOKESTAT, respectively.

For instance, if the categorical variable SMOKSTAT takes value 1 and the categorical variable SEX takes value 0. The corresponding weight associated with a value of 1 for the categorical variable SMOKSTAT is 0.5241 and the corresponding weight associated with a value of 0 for the categorical variable SEX is 1.7298. Therefore, the value of catX would be:

$$0.5241 + 1.7298 = 2.2539\,.$$

The best value of ICOMP at the end of each iteration of the GA process is shown in the Fig. 2.

The best subset of variables selected by AIC with same set of GA parameters is given by:

Model:	Intercept	AGE	SEX	SMOKSTAT	FAT

The AIC score for this model is **4241.647**. The estimated regression coefficients for the model chosen by AIC is:

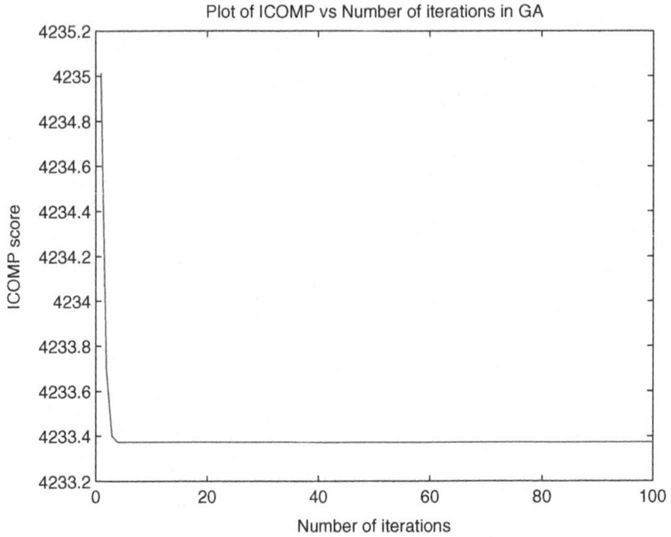

Fig. 2 Beta-carotene (RetPlasma): plot of ICOMP versus number of iterations in GA

$$\hat{\boldsymbol{\beta}} = \begin{bmatrix} 527.5606 \\ 2.4593 \\ -0.6243 \\ 36.5431 \end{bmatrix}.$$

We note that the AIC suggests the omission of two important predictor variables: ALCOHOL and RETDIET. We consider the model selected by ICOMP as our best fitting model for this data using the LCM with GA, since **ICOMP = 4233.376** is much smaller then the AIC value.

7.1.2 Beta-Carotene Data: Using Optimal Scaling Method

The categorical variables are transformed and optimally scaled. We fit a multiple regression model with RETPLASMA as the dependent variable and the variables AGE, SEX, SMOKSTAT, QUETELET, VITUSE, CALORIES, FAT, FIBRE, ALCOHOL, CHOLESTROL, BETADIET and RETDIET as independent variables. We also include the intercept term in this model. We assume that the random error is normally distributed. We use GA for variable selection with maximum iterations of 100, population size of 20, probability of crossover of 0.75, probability of mutation of 0.10 and crossover type as uniform. ICOMP$_{C1}$ is used as the fitness function.

The best subset of variables selected by GA and its associated information score is given by:

Model:	Intercept	AGE	SEX	SMOKSTAT	VITUSE

The ICOMP score for this model is **4244.857**. The estimated coefficients for the above set of variables is given by:

$$\hat{\boldsymbol{\beta}} = \begin{bmatrix} 469.2317 \\ 2.6634 \\ 41.0245 \\ 26.9100 \\ -22.4476 \end{bmatrix}.$$

Hence the estimated regression model is given by:

$$\begin{aligned} \hat{\mathbf{y}} &= \text{RETPLASMA} \\ &= 469.2317 + 2.6634 \times \text{AGE} + 41.0245 \times \text{SEX} \\ &\quad + 26.9100 \times \text{SMOKSTAT} - 22.4476 \times \text{VITUSE}\,. \end{aligned}$$

The best value of the above fitness function at the end of each iteration of the GA process is shown in the Fig. 3.

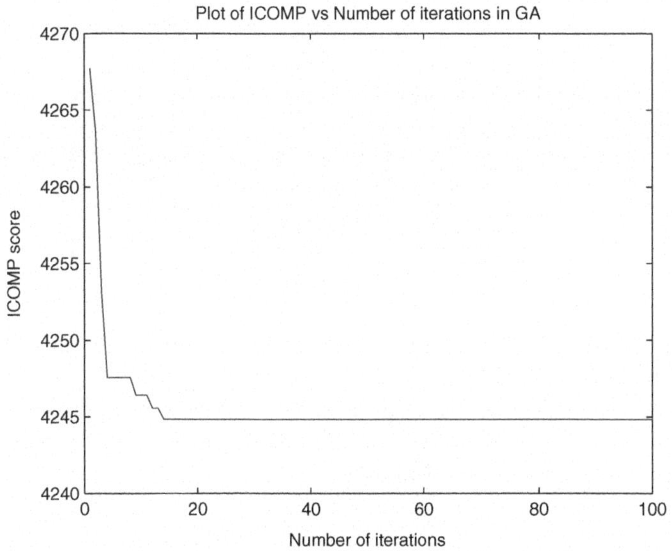

Fig. 3 Beta-carotene (RETPLASMA): plot of ICOMP versus number of iterations in GA

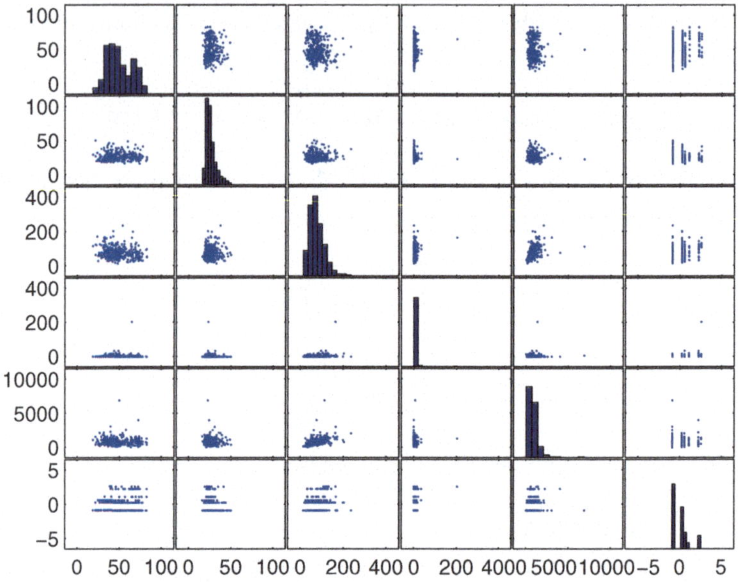

Fig. 4 Beta-carotene: plot matrix of the best predictors in the model RETPLASMA

The best subset of variables selected by using the AIC with the same set of GA parameters is given by:

Model:	Intercept	AGE	SEX	SMOKSTAT	FAT

The AIC for this model is **4243.2061**. The estimated coefficients for the above model is given by:

$$\hat{\beta} = \begin{bmatrix} 537.5733 \\ 2.2938 \\ 45.5978 \\ 27.1234 \\ -0.6466 \end{bmatrix}.$$

We note that the AIC allows to choose the same subset of variables using both the LCM and OSM with GA.

We choose the model selected by the ICOMP as our best fitting model since there is not much difference between AIC and ICOMP. Figure 4 shows the plot matrix of the best predictors in the model RetPlasma.

8 Conclusion

In this paper, we presented the idea of transforming the mixed data to a pure continuous data by a transformation known as Gifi transformation (Gifi, 1990). Since the transformed data set is purely continuous, one can implement the classical statistical analysis and modelling. We presented two algorithms by which one can perform multiple regression along with the genetic algorithm (GA) using the information criteria as the fitness function. The OSM optimally scales the categorical variables, thus making the data set purely continuous. The LCM generates a linear combination of the categories of the categorical variables thus making it a 1-dimensional continuous variable. The OSM is useful when there are a few categorical variables in the data set whereas LCM is useful when the number of the categorical variables is very large. We showed our results on a benchmark real data set. Our analysis shows that the results using the Gifi system is new when one has many categorical variables in their data sets.

Acknowledgements We express our thanks to Professor Nishisato for inviting Professor Bozdogan to make a contribution to the *Analysis of Categorical Data from Historical Perspectives—Essays in Honour of Professor Shizuhiko Nishisato*. We are grateful for the detailed comments by Professor Beh and his careful reading of the first draft of the paper which improved the current paper further.

References

Akaike, H.: Information theory and an extension of the maximum likelihood principle. In: Petrov, B.N., Csaki, F. (eds.) Second International Symposium on Information Theory, pp. 267–281. Academiai Kiado, Budapest (1973)

Bozdogan, H.: Model selection and Akaike's information criterion (AIC): the general theory and its analytical extensions. Psychometrika **52**(3), 345–370 (1987)

Bozdogan, H.: ICOMP: a new model selection criterion. In: Bock, H.H. (ed.) Classification and Related Methods of Data Analysis, pp. 599–608. North-Holland, Amsterdam (1988)

Bozdogan, H.: On the information based measure of covariance complexity and its application to the evaluation of multivariate linear models. Commun. Stat. Theo. Meth. **19**(1), 221–278 (1990a)

Bozdogan, H.: Multisample cluster analysis of common principal component model in K groups using an entropic statistical criterion. In: Invited Paper Presented at the International Symposium on Theory and Practice of Classification, December 16–19, Puschino, Soviet Union (1990b)

Bozdogan, H. (ed.): Intelligent Statistical Data Mining with Information Complexity and Genetic Algorithms. Chapman and Hall/CRC, Boca Raton, FL (2004)

Bozdogan, H.: Information Complexity and Multivariate Learning in High-Dimensions with Applications. Forthcoming Book (2024)

Forrest, S.: Genetic algorithms: principles of natural selection applied to computation. Science **261**(2.4), 872–878 (1993)

Gifi, A.: Nonlinear Multivariate Analysis. Wiley, Chichester (1990)

Goldberg, D.E.: Genetic Algorithms in Search, Optimization, and Machine Learning. Addison-Wesley, New York, NY (1989)

Katragadda, S., Bozdogan, H.: Multivariate mixed data mining with Gifi system using genetic algorithm and information complexity. Unpublished Manuscript (2008)

Marczyk, A.: Genetic algorithms and evolutionary computation. In: The TalkOrigins Archive, pp. 8077 (2004)
Michailidis, G., de Leeuw, J.: The Gifi System of Descriptive Multivariate Analysis. Technical Report. UCLA Statistics Program, Los Angeles, California, USA (1996)
Nierenberg, D.W., Stukel, T.A., Baron, J.A., Dain, B.J., Greenberg, E.R.: Determinants of plasma levels of beta-carotene and retinol. Am. J. Epidemiol. **130**, 511–521 (1989)
Nishisato, S.: Elements of Dual Scaling: An Introduction To Practical Data Analysis. Lawrence Erlbaum Associates, Hillsdale, NJ (1994)
Schwarz, G.: Estimating the dimension of a model. Ann. Stat. **6**, 461–464 (1978)
Tamhane, A.C., Dunlop, D.: Statistics and Data Analysis from Elementary to Intermediate. Prentice-Hall (2003)
Trefethen, L. N., Bau, D.: Numerical Linear Algebra, pp. 56–61 (1997)
White, H.: Maximum likelihood estimation of misspecified models. Econometrica, 1–25 (1982)

General Topics

Complex Difference System Models for Asymmetric Interaction

Naohito Chino

1 Introduction

Asymmetric relationships between objects are frequently observed in the phenomena observed in various branches of sciences. A typical example in psychology would be a one-sided affection among members of any informal group. The amount of migration from one region to another in geography is another example. These data can be arranged in matrix form called a *sociomatrix*. For such a matrix, its rows (say) indicate the raters, and its column denotes the ratees. Such a data matrix is generally asymmetric. We call such a relational data matrix an *asymmetric similarity matrix*. We shall hereafter abbreviate it as the ASM between objects. Here, objects are sometimes called nodes in graph theory. Suppose that we have a set of longitudinal asymmetric similarity matrices. This type of data belongs to *two-mode three-way data*, because the rows and columns belong to the same category, i.e., objects or nodes, and time belongs to the second category. Broadly speaking, this type of data can be said to be a *three-way data*.

Two-mode three-way data, including a set of longitudinal ASM's, can be analysed using several statistical and/or mathematical models; see, for example, Desarbo et al. (1992), Grorud et al. (1995), Okada and Imaizumi (1997, 2005), Zielman (1991), and Zielman and Heiser (1991). These models can be thought of as extensions of the individual differences MDS proposed by Carroll and Chang (1970) to ASM's. To be precise, these models reduce differences or changes in an ASM between objects or nodes to the individual differences. In other words, a major concern of these models can be said to obtain some *static structures* of asymmetric relationships among objects or nodes.

In contrast, there have been some models which are intended to obtain some *dynamic structures* of these asymmetric relationships among objects or nodes; see,

N. Chino (✉)
Aichi-Gakuin University, Nissin, Aichi, Japan
e-mail: chino@dpc.agu.ac.jp

E. J. Beh et al. (eds.), *Analysis of Categorical Data from Historical Perspectives*, Behaviormetrics: Quantitative Approaches to Human Behavior 17,
https://doi.org/10.1007/978-981-99-5329-5_26

for example, Chino (1990), Gregerson and Sailer (1993), Tobler (1976), and Yadohisa and Niki (1999). For example, Chino (1990) fitted a set of two-dimensional nonlinear differential equations to a set of longitudinal sociomatrices gathered by Newcomb (1961), and obtained several qualitative patterns of the trajectories of the vector fields in which members interact with each other. Here, the vector field at each point in time is estimated from the data. In other words, a major concern of this model is to obtain some dynamic structures of asymmetric relationships among members. Thus, this model can be said to be a dynamical system scaling and we call it DYNASCAL. Gregersen and Sailer (1990) examined a meta-model of two-person social systems described by a set of real two-dimensional nonlinear difference equations, and found curious chaotic behaviours. These equations include Mandelbrot's set. Tobler (1976) proposed a "wind" model for the interaction between geographical areas. In his model, the ASM is, for example, the amount of migration from place i to place j. The wind is interpreted as facilitating the interaction between geographical areas in particular directions. Tobler estimates a special vector filed on a map from the data, then decomposes it into divergence- and curl-free parts, and finally calculates the scalar and vector potentials. Yadohisa and Niki (1999) proposed a vector field representation of asymmetric proximity data, especially the scalar potential of the field.

Among these models, DYNASCAL has excellent features since it utilises qualitative theories of dynamical system, such as those of *singularities, structural stability*, and *bifurcations of vector field*. As a result, given the longitudinal AMS's between members, it draws a two-dimensional vector field on the estimated configuration of members at each time. Furthermore, it depicts singularities and several fundamental solution curves peculiar to each of the vector fields. This enables interpretation of global and local dynamical properties of the group structure at each time. However, DYNASCAL has several disadvantages, too, of which we describe four of them. Firstly, it presupposes asymmetric relationships between members but the estimated relationships are symmetric. Secondly, it might not be fully justified mathematically to administer the Procrustes rotations to the neighbouring pairs of configurations. The reason for this is that DYNASCAL assumes a deterministic, nonlinear solution curve of each member in a state space as an underlying dynamics which cannot be sometimes congruent with the Procrustes rotations. Thirdly, DYNASCAL will not capture the so-called chaotic behaviours since it is restricted to a two-dimensional differential system. Fourthly, it is not possible for DYNASCAL to examine the behaviours of the system theoretically, since it merely estimates the solution curves using spline functions (Chino 2005).

In this paper, we shall discuss complex difference system models for asymmetric interaction which were first proposed at "The International Conference on Measurement and Multivariate Analysis" held on May 2000 in Banfff, Canada in order to overcome those difficulties pointed out above and subsequently developed further by the author; see Chino (2000, 2001, 2002, 2003, 2005, 2006, 2014, 2015a, b).

2 Earlier Version of the Complex Difference System Models

The complex difference system models we proposed elsewhere (Chino, op. cit.) have several assumptions. Firstly, the *state space* in which we embed members (objects, nodes) is assumed to be a finite-dimensional Hilbert space or an indefinite-metric space. If we restrict our attention to a one-dimensional space, then an indefinite-metric space may be identified with a Hilbert space. This assumption can be justified by the *Hermitian form model* (abbreviated to *HFM*) which is underpinned by the Chino-Shiraiwa theorem (Chino 1993). In fact, for the HFM, any ASM, say, $\boldsymbol{S}$, is decomposed into two parts as follows:

$$\boldsymbol{S} = \frac{1}{2}(\boldsymbol{S} + \boldsymbol{S}^t) + \frac{1}{2}(\boldsymbol{S} - \boldsymbol{S}^t) = \boldsymbol{S}_s + \boldsymbol{S}_{sk}, \tag{1}$$

where $\boldsymbol{S}$ is a square asymmetric matrix of order n which is the number of objects, and $\boldsymbol{S}_s$ and $\boldsymbol{S}_{sk}$ are called the symmetric part and the asymmetric part (to be precise, the skew-symmetric part), respectively. This decomposition has been used extensively in the literature; see, for example, Beh et al. (2022), Bove (1992), Constantine and Gower (1978), Escoufier and Grorud (1980), Gower (1977), and Greenacre (2000).

The HFM is deduced by reinterpreting the eigenvalue problem of the Hermitian matrix $\boldsymbol{H}$, which is constructed uniquely from the observed real square asymmetric matrix $\boldsymbol{S}$, from the view point of asymmetric MDS, or, stated another way, from a geometric view point. Here, the Hermitian matrix $\boldsymbol{H}$ is simply computed as follows:

$$\boldsymbol{H} = \boldsymbol{S}_s + i\, \boldsymbol{S}_{sk}, \tag{2}$$

where i is the imaginary number, that is, a square root of -1. Equation (2) is nothing but the definition of the Hermitian matrix. If $\boldsymbol{H}$ is Hermitian, then the conjugate transpose of $\boldsymbol{H}$ is $\boldsymbol{H}$; see, for example, Wilkinson (1965). It should be noted that, in general, $\boldsymbol{S}_{sk}^t$ is equal to $-\boldsymbol{S}_{sk}$. $\boldsymbol{H}$ is thought of as a complexification of a real matrix $\boldsymbol{S}$, and there is a one-to-one correspondence between them. Escoufier and Grorud (1980) also utilises this equation in their asymmetric MDS. However, they do not solve the eigenvalue problem of $\boldsymbol{H}$ defined by this equation directly. Instead, they solve it by defining a real symmetric matrix of order $2n$ such that:

$$\boldsymbol{H} = \begin{pmatrix} \boldsymbol{S}_s & -\boldsymbol{S}_{sk} \\ \boldsymbol{S}_{sk} & \boldsymbol{S}_s \end{pmatrix}.$$

Let us rewrite the eigenvalue problem of $\boldsymbol{H}$, so that $\boldsymbol{H}\boldsymbol{u}_j = \lambda_j \boldsymbol{u}_j$, as follows:

$$\boldsymbol{H} = \boldsymbol{U}_1 \boldsymbol{\Lambda}_p \boldsymbol{U}_1^*. \tag{3}$$

Here, the $p \times n$ matrix $\boldsymbol{U}_1^*$ is the *conjugate transpose* of the $n \times p$ matrix $\boldsymbol{U}_1$. Of course, the $p \times p$ matrix $\boldsymbol{\Lambda}_p$ is a real diagonal matrix $\boldsymbol{\Lambda}_p = \mathrm{diag}\left(\lambda_1, \ldots, \lambda_p\right)$ consisting of the non-zero eigenvalues of $\boldsymbol{H}$ arranged in descending order. The matrix

U_1 consists of the p eigenvectors corresponding to these non-zero eigenvalues. If we define an n by n matrix $\boldsymbol{U}$ which is composed of the eigenvectors associated with all the eigenvalues including zeros of $\boldsymbol{H}$ as:

$$\boldsymbol{U} = \{\underbrace{\boldsymbol{u}_1, \ldots, \boldsymbol{u}_p}_{p}, \underbrace{\boldsymbol{u}_{p+1}, \cdots, \boldsymbol{u}_n}_{n-p}\} = (\boldsymbol{U}_1, \boldsymbol{U}_2), \tag{4}$$

then $\boldsymbol{U}_1$ is the first part of the *unitary matrix* $\boldsymbol{U}$ corresponding to the non-zero eigenvalues.

Let us now rewrite (3) as:

$$h_{jk} = \varphi\left(\boldsymbol{\tau}_j, \boldsymbol{\tau}_k\right) = \boldsymbol{\tau}_j \boldsymbol{\Lambda}_p \boldsymbol{\tau}_k^*, \tag{5}$$

then $\varphi\left(\boldsymbol{\tau}_j, \boldsymbol{\tau}_k\right)$ satisfies the properties of *Hermitian form* (Cristescu 1977; Lancaster and Tismenetsky 1985), where $\boldsymbol{\tau}_j$ is a p-dimensional row vector corresponding to the jth row of $\boldsymbol{U}_1$. Furthermore, (5) associates h_{jk} with a Hermitian form.

Chino (1993) proved that n objects are embedded in a finite-dimensional complex (f.d.c.) *Hilbert space* if $\boldsymbol{H}$ is positive semi-definite (p.s.d.) (or negative semi-definite (n.s.d.)), whereas they are embedded in an *indefinite-metric space* if $\boldsymbol{H}$ is indefinite.

Another assumption is composed of the following basic principles of interpersonal behaviours:

(1) The asymmetric sentiment relationships among members make their affinities change.
(2) If a member has a positive sentiment towards another member, then he or she approaches to the target member.
(3) If a member has a negative sentiment towards another member, then he or she departs from the target member.

There exist two minor principles in this family, as listed below:

(1) The magnitude of change in coordinate of members is proportional to the *sine* of the difference in angles (arguments) between two members in a complex plane.
(2) The magnitude of change in coordinate of members is proportional to the *norm* of the coordinate in a complex plane.

The complex difference system models were defined under the above assumptions as follows:

$$z_{j,n+1} = z_{j,n} + \sum_{m=1}^{q} \sum_{k \neq j}^{N} \boldsymbol{D}_{jk,n}^{(m)} \boldsymbol{f}^{(m)}\left(z_{j,n} - z_{k,n}\right) + z_0, \quad j = 1, 2, \ldots, N, \tag{6}$$

where

$$\boldsymbol{f}^{(m)}(z_{j,n} - z_{k,n}) = \left(\left(z_{j,n}^{(1)} - z_{k,n}^{(1)}\right)^m, \left(z_{j,n}^{(2)} - z_{k,n}^{(2)}\right)^m, \ldots, \left(z_{j,n}^{(p)} - z_{k,n}^{(p)}\right)^m\right)^t, \tag{7}$$

and

$$\boldsymbol{D}_{jk,n}^{(m)} = \text{diag}\left(w_{jk,n}^{(1,m)},\ w_{jk,n}^{(2,m)},\ \ldots,\ w_{jk,n}^{(p,m)}\right), \tag{8}$$

$$w_{jk,n}^{(l,m)} = a_n^{(l,m)} r_{j,n}^{(l,m)} r_{k,n}^{(l,m)} \sin\left(\theta_{k,n}^{(l,m)} - \theta_{j,n}^{(l,m)}\right), \tag{9}$$

for $l = 1, 2, \ldots, p$ and $m = 1, 2, \ldots, q$. Here, $z_{j,n}$ denotes the coordinate vector of member j at time n in a p-dimensional *Hilbert space* or a p-dimensional *indefinite-metric space*. Moreover, m denotes the degree of the vector function $\boldsymbol{f}^{(m)}\left(z_{k,n} - z_{j,n}\right)$ in (7), which is assumed to have the maximum value of q. z_0 is a complex constant. Furthermore, $a_n^{(l,m)}$ is a real constant coefficient of the term $\left(z_{j,n}^{(l)} - z_{k,n}^{(l)}\right)^m$, $r_{j,n}^{(l,m)}$ and $\theta_{j,n}^{(l,m)}$ are, respectively, the *norm* and the *argument* of $z_{j,n}$ at time n on dimension l. Usually, both of $r_{j,n}^{(l,m)}$ and $\theta_{j,n}^{(l,m)}$ are independent of m.

At this point, we shall briefly explain how these results in (6) through (9) relate to $\boldsymbol{S}$, $\boldsymbol{H}$, $\boldsymbol{U}$ (especially, $\boldsymbol{U}_1$), and $\boldsymbol{\tau}$'s introduced previously. The matrix $\boldsymbol{S}$ in (1) consists of *observed similarities*, s_{jk}, between objects, and thus, it is a *real matrix*. On the other hand, the matrix $\boldsymbol{H}$ in (2) consists of *hypothetical similarities*, h_{jk}, between objects, and it is a *complex matrix*. It is apparent that there is a one-to-one correspondence between $\boldsymbol{S}$ and $\boldsymbol{H}$.

In HFM, we decompose $\boldsymbol{H}$ into $\boldsymbol{\Lambda}_p$ and $\boldsymbol{U}_1$ which are composed of the non-zero eigenvalues and eigenvectors corresponding to these eigenvalues of $\boldsymbol{H}$, as shown in (3). According to the Chino-Shiraiwa theorem, objects are embedded in a p-dimensional Hilbert space if $\boldsymbol{H}$ is p.s.d., and are embedded in an indefinite-metric space if $\boldsymbol{H}$ is indefinite which means that $\boldsymbol{H}$ has both positive and negative eigenvalues. In any case, $\boldsymbol{\tau}_j$ and $\boldsymbol{\tau}_k$ in (5) are p-dimensional row vectors corresponding to the jth row and kth row, respectively, of $\boldsymbol{U}_1$ in (3). Therefore, $\boldsymbol{\tau}_j$ and $\boldsymbol{\tau}_k$ are coordinate vectors of objects j and k, respectively in a p-dimensional Hilbert space if $\boldsymbol{H}$ is p.s.d. From (1) through (5), it is apparent that these coordinate vectors (eigenvectors) and eigenvalues explain the hypothetical similarities, h_{jk}, and corresponding observed asymmetric similarities, s_{jk}, between objects.

In our complex difference system models, we model the changes in observed asymmetric similarities, s_{jk}, over time. Since there is a one-to-one correspondence between $\boldsymbol{S}$ and $\boldsymbol{H}$, and since the eigenvalue problem of $\boldsymbol{H}$ gives us the complex coordinate vectors of objects in a Hilbert space if $\boldsymbol{H}$ is p.s.d., we consider these vectors as *state vectors* of objects which change over time. Here, we assume that *hypothetical asymmetric interactions* between objects exist which cause the changes in state vectors over time. The $w_{jk,m}^{(l,m)}$ are parameters concerned with these hypothetical asymmetric interactions. The $z_{j,n}$ in (6) is nothing but these state vectors at time n in a p-dimensional Hilbert space. It should be noticed that the second right-hand side of (6) is a vector function since $\boldsymbol{D}_{jk,n}^{(m)}$ defined in (8) is a $p \times p$ diagonal matrix and $\boldsymbol{f}^{(m)}\left(z_{k,n} - z_{j,n}\right)$ is a p-dimensional column vector defined in (7). As a result, each element of the vector function represented by the second right-hand side of (6) is a complex polynomial function of $\left(z_{j,n} - z_{k,n}\right)$ whose degree is q. Finally, z_0 is a

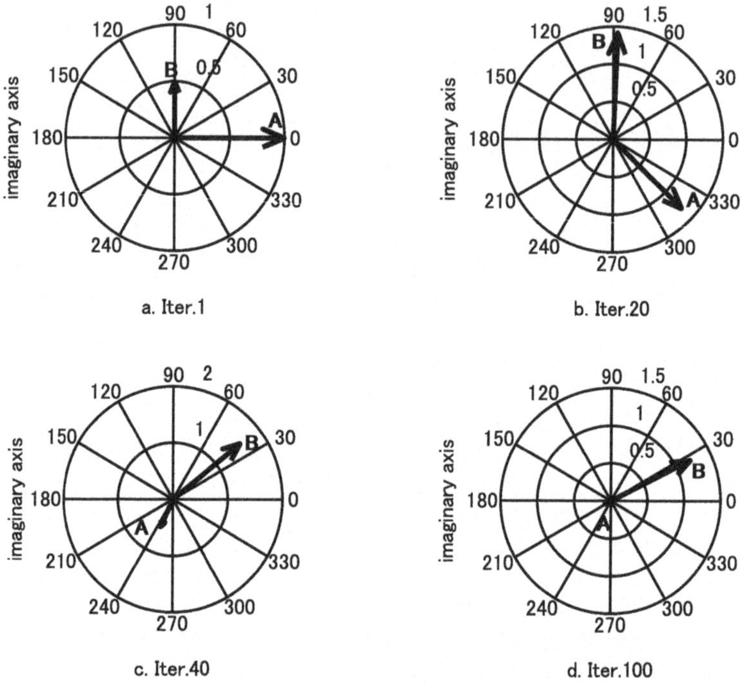

Fig. 1 Changes in configurations of two members in a one-dimensional Hilbert space at iterations, 1, 20, 60, and 100 in a simulation study

complex constant since the location of object j, $z_{j,n}$, in (6), which is embedded in a p-dimensional Hilbert space, is a complex vector.

Figure 1 shows an example of simulations using a special case of the above difference systems in which we show changes in configurations of two members in a one-dimensional Hilbert space. This special case is written as follows for n-iteration:

$$\begin{cases} w_{jk,n} = a\left(|z_{j,n}||z_{k,n}|\right)^a \sin\left(\theta_{k,n} - \theta_{j,n}\right), \\ z_{j,n+1} = z_{j,n} + w_{jk,n}\left(z_{j,n} - z_{k,n}\right)^2, \\ z_{k,n+1} = z_{k,n} + w_{jk,n}\left(z_{j,n} - z_{k,n}\right)^2, \end{cases}$$

where a is a scaling factor of the configuration which controls the domain (i.e., the coordinate at time n) and range (i.e., the coordinate at time $n+1$) of the configuration and is a special case of $a_n^{(l,m)}$ in (9). In this simulation, it was set equal to $1/50$. Moreover, the degree of the polynominal of $\mathbf{z}_{j,n} - \mathbf{z}_{k,n}$ in (6) was assumed to be 2, as can be seen from the above equations. Furthermore, in this case the p-dimensional vector $\mathbf{z}_{j,n}$ in (6) becomes a scalar $z_{j,n}$, because we assume here that $p = 1$. Finally, we set the initial configuration $(z_{j,1}, z_{k,1})$ of the above difference systems equal to $(1, i/2)$. The reason for setting this configuration is that the skewness of the similarities

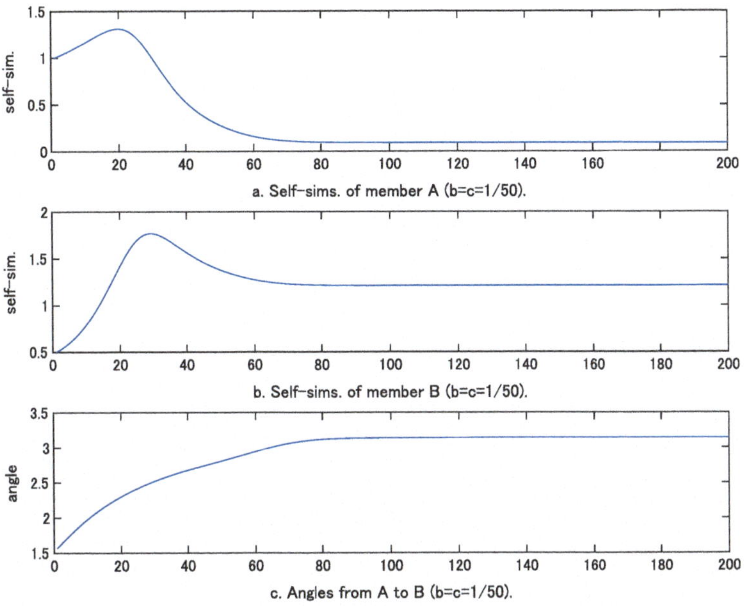

Fig. 2 Changes in self-similarities of two members as well as the angles between them in a one-dimensional Hilbert space over 200 iterations in a simulation study

between two members j and k is theoretically the largest of all, if the angle between two members in the complex space is $\pi/2$; see, for example, Chino (2020a). In Fig. 1, we denote $z_{1,n}$ and $z_{2,n}$ simply by A and B, respectively. Moreover, *time* is identified with *iteration*. Thus, for example, A and B in Fig. 1a indicate $z_{1,1}$ and $z_{2,1}$, respectively, in the initial configuration of members. Since the angle between two members at iteration 1 is $\pi/2$, this means that member B likes member A very much but member A does not like member B at all at iteration 1. As for the interpretation of the configuration of objects in HFM; see Chino (2020a). Finally, the complex constant z_0 in (6) was set equal to zero.

Figure 2 shows the changes in self-similarities of two members and those in angles over 200 iterations. In Fig. 2c, one can see that the angle between two members approaches π as the iterations increase. Figure 3 illustrates changes in locations of two members over 200 iterations in a one-dimensional Hilbert space. In this figure, A_1 and B_1 indicate initial points of members j $(= 1)$ and k $(= 2)$, respectively. To be precise, coordinates of A_1 and B_1 in this complex plane are $(1, 0)$ and $(0, i/2)$, respectively.

Similarly, if we set a non-zero value to z_0 in (6), we can obtain more curious patterns of changes in locations of members over iteration than those in Fig. 3. However, there is a serious drawback in the complex system described by (6), (7), (8), and especially in (9). That is, the function $w_{jk,n}^{(l,m)}$ in (9) is not *holomorphic* (Chino 2014), since both r and θ are the functions of z and the conjugate of z, i.e., $\bar{z}$. Here,

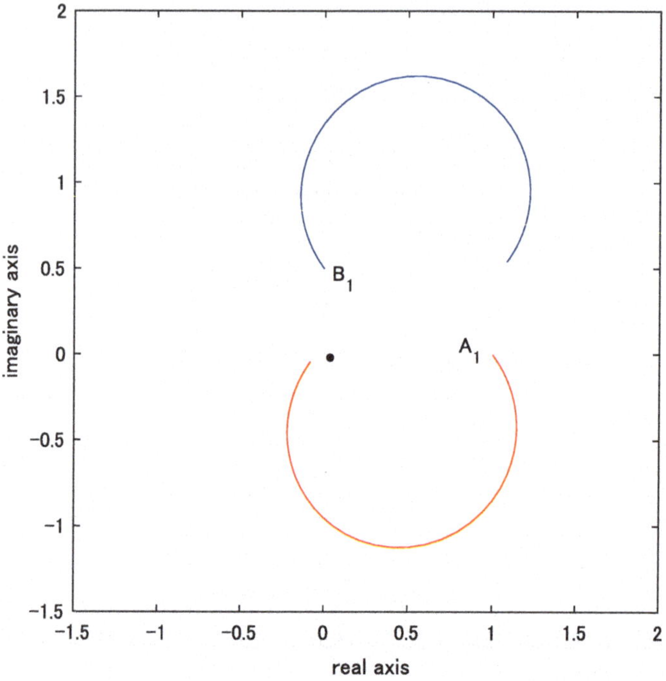

Fig. 3 Changes in locations of two members over iterations in a one-dimensional Hilbert space

holomorphic means *complex differentiable*; see, for example, Ebeling (2007). The complex differentiability of a complex-valued function is a natural extension of the differentiability of a real-valued function in a real space to that of a complex-valued function in a complex space. As a result, we cannot examine mathematical properties of the above different system models using complex differential calculus. The late K. Shiraiwa (personal communication, March 3, 2014), who had long been one of my colleagues, pointed out this drawback. Therefore, we have discarded (9) in our complex difference system models since then. The next section discusses a revised version of the complex difference system models which are composed of holomorphic functions.

3 Revised Version of the Complex Difference System Models

The revised version of the complex difference system models which are composed of holomorphic functions (Chino 2016a, b, 2017) is nothing but a simplified version of the earlier version without (9) in the previous section. As a result, all the $w_{jk,n}^{(l,m)}$ in (8) become *complex constants*, and the corresponding minor principles in the previous section are no longer necessary. Thus, we have the following complex difference system models in a strict sense:

$$z_{j,n+1} = z_{j,n} + \sum_{m=1}^{q} \sum_{k \neq j}^{N} \boldsymbol{D}_{jk,n}^{(m)} \boldsymbol{f}^{(m)} \left(z_{j,n} - z_{k,n} \right) + z_0, \quad j = 1, 2, \ldots, N. \quad (10)$$

Here

$$\boldsymbol{f}^{(m)}(z_{j,n} - z_{k,n}) = \left(\left(z_{j,n}^{(1)} - z_{k,n}^{(1)} \right)^m, \left(z_{j,n}^{(2)} - z_{k,n}^{(2)} \right)^m, \ldots, \left(z_{j,n}^{(p)} - z_{k,n}^{(p)} \right)^m \right)^t, \quad (11)$$

and

$$\boldsymbol{D}_{jk,n}^{(m)} = \mathrm{diag} \left(w_{jk,n}^{(1,m)}, w_{jk,n}^{(2,m)}, \ldots, w_{jk,n}^{(p,m)} \right). \quad (12)$$

Equations (10) through (12) are the same as (6) through (8), but no constraints are imposed on the elements of the diagonal matrix $\boldsymbol{D}_{jk,n}^{(m)}$ in (12). In other words, in the revised version, we have discarded the weight constraints (9) in the earlier version. Therefore, $w_{jk,n}^{(1,m)}, w_{jk,n}^{(2,m)}, \ldots, w_{jk,n}^{(p,m)}$ are considered as free parameters in the revised version. This means that we may assume any values in these parameters.

Chino (2017) added a control term $\boldsymbol{g} \left(\boldsymbol{u}_{j,n} \right)$ to the right-hand side of (10). In general, a control term in a control theory (e.g., Elaydi 1996) is a forcing term which controls a (difference or differential) system from its outside. In (10), the control can be applied to affect directly each of the state variables $z_{1,n}, z_{2,n}, \ldots, z_{N,n}$. In this case, (10) is revised as follows:

$$z_{j,n+1} = z_{j,n} + \sum_{m=1}^{q} \sum_{k \neq j}^{N} \boldsymbol{D}_{jk,n}^{(m)} \boldsymbol{f}^{(m)} \left(z_{j,n} - z_{k,n} \right) + \boldsymbol{g} \left(\boldsymbol{u}_{j,n} \right) + z_0, \quad j = 1, 2, \ldots, N, \quad (13)$$

where $\boldsymbol{g} \left(\boldsymbol{u}_{j,n} \right)$ is a *control*; see Elaydi (1996) and Ott, Grebogi and Yorke (1990).

Moreover, we assume in the revised version that members obey only the three basic principles of interpersonal behaviours discussed in Section 2. It should be noted that we discarded the two minor principles discussed there. Figure 4 shows another example of simulations using a special case of the above difference systems, in which we show changes in configurations of two members in a one-dimensional Hilbert space (Chino 2016a). This case is written as follows for n-iteration:

$$\begin{cases} w_{jk,n} = 0.01\,(1 + i), \; w_{kj,n} = -0.02\,(1 + i), \\ z_{j,n+1} = z_{j,n} + w_{jk,n} \left(z_{j,n} - z_{k,n} \right), \\ z_{k,n+1} = z_{k,n} + w_{kj,n} \left(z_{k,n} - z_{j,n} \right). \end{cases}$$

The initial coordinates of two members, $z_{j,1}$ and $z_{k,1}$, were set equal to 1 and $i/2$, which means that the initial configuration of members is the same as that in the example shown in Section 2. However, in this example, the system is linear in contrast

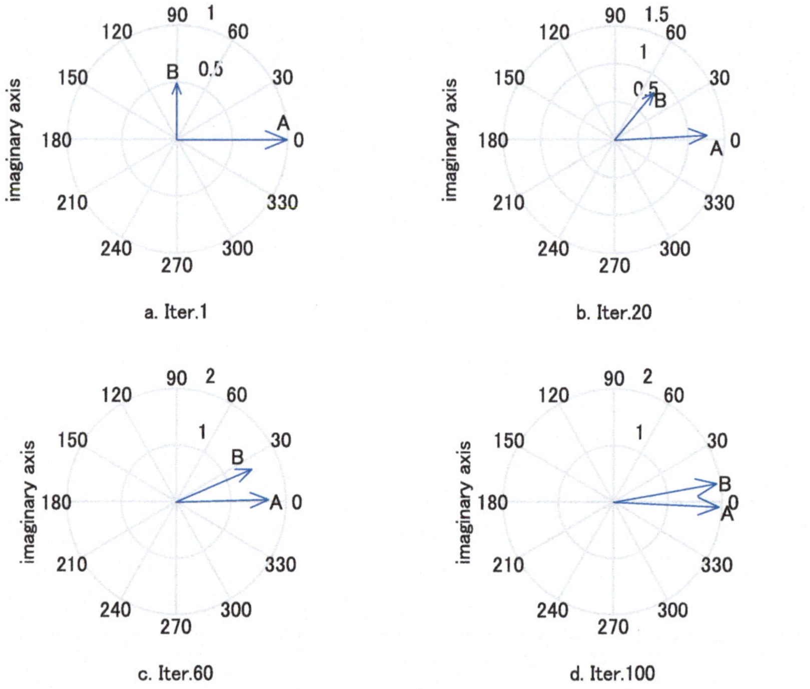

Fig. 4 Changes in configurations of two members in a one-dimensional Hilbert space at iterations, 1, 20, 60, and 100, in another simulation study

with the system shown in Section 2. In general, a wide class of linear difference equations can be solved explicitly and the qualitative behaviours of the solution curves in these equations are simple. However, most nonlinear difference equations cannot be solved explicitly; see, for example, Cull (2005) and Elaydi (1996). Moreover, the elements of the diagonal matrix $\boldsymbol{D}_{jk,n}^{(m)}$ in (12) are assumed to be (complex) constants in marked contrast to those of the diagonal matrix in the earlier version. Note that in the earlier version, the elements of the diagonal matrix vary with time n according to (9). Finally, the reason why we consider here a linear system as an example of the above revised version is that we can solve this kind of linear system analytically. In fact, we can prove that, for example, the above system has a *fixed point* using a familiar method called the *Putzer algorithm* in difference equations; see, for example, Cull (2005) and Elaydi (1996). If we apply this algorithm to the above system, we can compute its fixed point as $2 - 0.5i$, although we shall not show its proof here because it is beyond the scope of this paper. In the following, we shall check whether the fixed point of the above system approaches to this value.

Figure 5 shows the changes in self-similarities of two members as well as angles over 1000 iterations. In Fig. 5c, one can see that the angle between two members approaches 0 as iteration proceeds. Figure 6 illustrates changes in locations of two members over 1000 iterations in a one-dimensional Hilbert space. In this figure, A_1

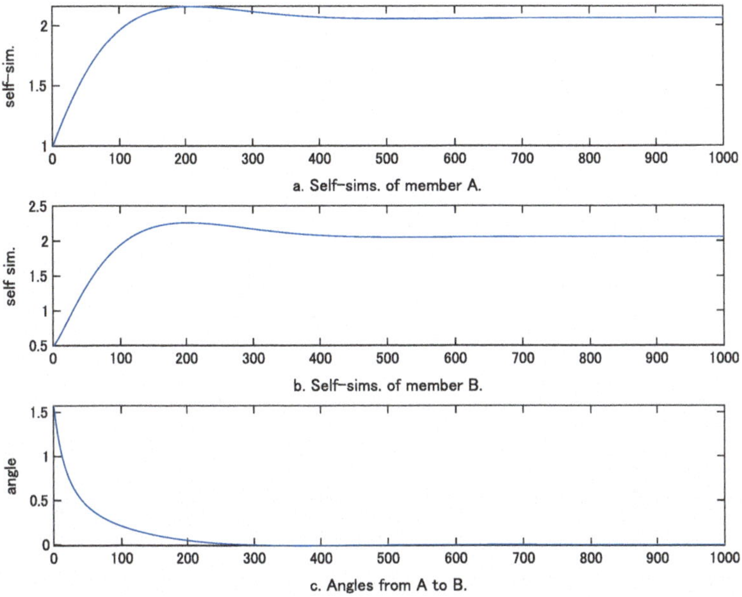

Fig. 5 Changes in configurations of two members in a one-dimensional Hilbert space over 1000 iterations in another simulation study

and B_1 indicate initial points of members j and k, respectively, as in Fig. 3 in the previous section. As can be seen in this figure, the speed of convergence became slower and slower as locations of two members approach the fixed point. Even after 500 iterations these locations did not reach the fixed point. However, after 1000 iterations, those of members A and B reached $2.0 - 0.5000i$ and $2.0 - 0.5001i$, respectively. This means that two members become deeply in love with each other as the iteration proceeds.

In this way, we can find various patterns of dynamics which are generated by the asymmetric interactions among members. Such a job might be said to be a *classification of dynamics* generated by the complex interactions among objects. This type of classification of dynamics may be contrasted with a *classification of the static structures* among members obtained by applying a traditional two-mode three-way asymmetric MDS to a longitudinal set of asymmetric matrices.

4 Discussion

The complex difference system models for asymmetric interaction discussed in this paper were first proposed by the author at "The International Conference on Measurement and Multivariate Analysis" held on May, 2000, in Banff, Canada, and have been revised since then, as introduced in Sects. 2 and 3. In these sections, we have

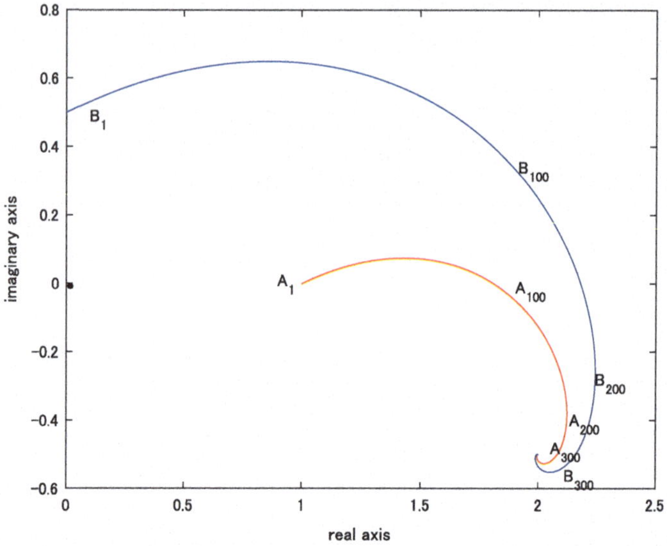

Fig. 6 Changes in locations of two members in a one-dimensional Hilbert space over 1000 iterations in the above difference system

been mainly concerned with social interactions. However, asymmetric interaction can be observed ubiquitously not only in our daily lives but also in vivo, in vitro, and studies in the field, in various disciplines of science. For example, pecking order among hens and cocks is a special asymmetric interaction in ethology [e.g. Masure and Allee (1934)]. Biosynthetic pathway of proteins in mammals has one-sided paths and cycles [e.g. Imai and Guarente (2014)], which can be considered as an asymmetric interaction among proteins. Weight matrix among hidden layers in neural networks represents asymmetric interactions in the brain [e.g. Goodfellow et al. (2016)].

Considering these phenomena as well as the relation between weight matrix and directed graph (abbreviated as *digraph*), we have recently renamed our complex difference system models with holomorphic functions *dynamic weighted digraph* (abbreviated as *DWD)* (Chino 2018a, b, 2019, 2020, 2021). Here, if a number is associated with an edge of a graph, these numbers are called *weights*, and a matrix with these numbers is called a *weight matrix*. In a digraph, the weight matrix is generally asymmetric. Therefore, in DWD asymmetric interactions are no longer restricted to social interactions. As discussed in Chino (2018a, b), the weighted digraph in DWD is a digraph with weights specified at time n, which are attached to each *directed arc* (or edge, link) between *nodes* (or vertices) as well as each *loop* of the digraph. Moreover, our elementary theory of DWD assumes that the weight matrix denotes the proximity strengths among nodes at any instance of time, and that it varies as time proceeds. As a result, we obtain a set of longitudinal ASM introduced in the introductory section.

As in the complex difference system models with holomorphic functions, the state space in which we embed members (objects, nodes) is assumed to be a finite-dimensional Hilbert space or an indefinite-metric space. It should be noted here that the state space is a *hypothetical or latent space* and cannot be observed directly. Furthermore, we assume that the configuration of nodes varies according to the *mutual interactions among nodes* as time proceeds. Parameters related to these mutual interactions are specified *a priori* as certain functions of $\alpha_{jk}^{(1,m)}$, $\alpha_{jk}^{(2,m)}$, ..., $\alpha_{jk}^{(p,m)}$, described by (12) in which $w_{jk}^{(p,m)}$ are replaced by $\alpha_{jk}^{(p,m)}$.

As also pointed out in Chino (2018a, b), the purpose of DWD is two-fold. One is *theoretical*, and the other *practical*. For the theoretical purpose, we compute the trajectories of the nodes using (10), by setting an arbitrary initial configuration of the nodes. Then, we recover the longitudinal digraphs associated with these similarity matrices. We can classify the patterns of changes in digraphs over time (or iteration) according to the patterns of trajectories of nodes over time (or iteration). For practical purposes, it will be possible to demonstrate the ideas above using empirical examples. For example, if we apply HFM to an observed asymmetric similarity matrix at a point in time, we can compute a p-dimensional configuration of members (objects, nodes). If the Hermitian matrix $\boldsymbol{H}$ computed from the observed similarity matrix is p.s.d., we can embed a p-dimensional configuration of members in a Hilbert space. Then, we can use the configuration thus obtained as initial values of DWD, if we assume the hypothetical complex weights in (12). Finally, if we apply (10) to the configuration of members with initial values and these hypothetical weights, we can examine various scenarios of solution curves like those in Fig. 6. These tasks remain to be done for future works. We shall go further with details of DWD in a book to be published in the near future.

Acknowledgements I would like to acknowledge Gregory Lewis Rohe for the proofreading of the earlier version of this paper. I would also like to acknowledge editors of this *Festschrift* for providing many valuable comments on the earlier version.

References

Beh, E.J., Lombardo, R.: Visualising departures from symmetry and Bowker's χ^2 statistic. Symmetry **14**, 1103, 25 (2022)

Bove, G.: Asymmetric multidimensional scaling and correspondence analysis for square tables. Statistica Applicata **4**, 587–598 (1992)

Carroll, J.D., Chang, J.-J.: Analysis of individual differences in multidimensional scaling via an N-way generalization of "Eckart-Young" decomposition. Psychometrika **35**, 283–319 (1970)

Chino, N.: Complex space models for the analysis of asymmetry. Paper presented at The International Conference on Measurement and Multivariate Analysis, Banff, Alberta, Canada, May 11 – 14 (2000)

Chino, N.: Complex difference system models for the analysis of asymmetry. In: Yanai, H., Okada, K., Shigemasu, K., Kano, Y., Meulman, J.J. (eds.) New Developments in Psychometrics, pp. 479–486. Springer-Verlag, Tokyo (2001)

Chino, N.: Complex space models for the analysis of asymmetry. In: Nishisato, S., Baba, Y., Bozdogan, H., Kanefuji, K. (eds.) Measurement and Multivariate Analysis, pp. 107–114. Springer-Verlag, Tokyo (2002)

Chino, N.: Fitting complex difference system models to longitudinal asymmetric proximity matrices. Paper presented at the 13th International Meeting and the 68th Annual Meeting of the Psychometric Society. University of Cagliari, Sardinia, Italy (2003)

Chino, N.: Abnormal behaviors of members predicted by a complex difference system model. Bull. Faculty Psychol. Phys. Sci. **1**, 69–73 (2005)

Chino, N.: Asymmetric multidimensional scaling and related topics. Manuscript for the invited talk at the Weierstrass Institute for Applied Analysis and Stochastics, Berlin, Germany (2006)

Chino, N.: A general non-Newtonian n-body problem and dynamical scenarios of solutions. In: Proceedings of the 42nd Annual Meeting of the Behaviormetric Society of Japan, pp. 48–51, Sendai, Japan (2014)

Chino, N.: A simulation study of a Hilbert state space model for changes in affinities among members in informal groups. J. Inst. Psychol. Phys. Sci. **7**, 31–47 (2015)

Chino, N.: Time series analyses of changes in asymmetric relationships among members over time. In: Proceedings of the 43rd Annual Meeting of the Behavior Metric Society of Japan, pp. 398–401, Tokyo, Japan (2015b)

Chino, N.: Time series analyses of changes in asymmetric relationships among members over time (2). Proceedings of the 44th Annual Meeting of the Behavior Metric Society of Japan, pp. 84–87. Japan (2016a)

Chino, N.: A general non-Newtonian n-body problem and dynamical scenarios of solutions. Paper presented at the 31st International Congress of Psychology. Yokohama, Japan (2016b)

Chino, N.: Dynamical scenarios of changes in asymmetric relationships on a Hilbert space. Handout presented at the 45th Annual Meeting of the Behavior Metric Society of Japan, pp. 1–29. Shizuoka, Japan (2017)

Chino, N.: An elementary theory of a dynamic weighted digraph. Proceedings of the 46th Annual Meeting of the Behavior Metric Society of Japan, pp. 26–29. Tokyo, Japan (2018a)

Chino, N.: An elementary theory of a dynamic weighted digraph (2). Bull. Faculty Psychol. Phys. Sci. **14**, 23–31 (2018b)

Chino, N.: An elementary theory of a dynamic weighted digraph (2). Proceedings of the 47th Annual Meeting of the Behavior Metric Society of Japan, pp. 56–59. Osaka, Japan (2019)

Chino, N.: How to use the Hermitian Form Model for asymmetric MDS. In: Imaizumi, T., Nakayama, A., Yokoyama, S. (eds.) Advanced Studies in Behavior Metrics and Data Science: Essays in Honor of Akinori Okada, pp. 19–41. Springer Nature, Singapore (2020a)

Chino, N.: An elementary theory of a dynamic weighted digraph (3). Proceedings of the 48th Annual Meeting of the Behavior Metric Society of Japan, pp. 54–57. Tokyo, Japan (2020b)

Chino, N.: An elementary theory of a dynamic weighted digraph (4). Proceedings of the 49th Annual Meeting of the Behavior Metric Society of Japan, pp. 86–89. Tokyo, Japan (2021)

Chino, N., Nakagawa, M.: A bifurcation model of change in group structure. Jpn. J. Exper. Soc. Psychol. **29**, 25–38 (1990)

Chino, N., Shiraiwa, K.: Geometrical structures of some non-distance models for asymmetric MDS. Behaviormetrika **20**, 37–42 (1993)

Constantine, A.G., Gower, J.C.: Graphical representation of asymmetric matrices. Appl. Stat. **27**, 297–304 (1978)

Cristescu, R.: Topological Vector Spaces. Noordhoff International Publishing, Leiden, The Netherlands (1977)

Cull, P., Flahive, M., Robson, R.: Difference Equations from Rabbits to Chaos. Springer, New York (2005)

Desarbo, W.S., Johnson, M.D., Manrai, A.K., Manrai, L.A., Edwards, E.A.: TSCALE: a new multidimensional scaling procedure based on Tversky's contrast model. Psychometrika **57**, 43–69 (1992)

Ebeling, W.: Functions of Several Complex Variables and Their Singularities. American Mathematical Society, Providence (2007)
Elaydi, S.N.: An Introduction to Difference Equations. Springer-Verlag, New York (1996)
Escoufier, Y., Grorud, A.: Analyse factorielle des matrices carrees non symetriques [Factor analysis of square asymmetric matrices]. In: Diday, E. (ed) Data Analysis and Informatics, pp. 263–276. North-Holland, Amsterdam (1980)
Goodfellow, I., Bengio, Y., Courville, A.: Deep Learning. The MIT Press, Cambridge, MA (2016)
Gower, J.C.: The analysis of asymmetry and orthogonality. In: Barra, J.R., Brodeau, F., Romer, G., van Cutsem, B. (eds.) Recent Developments in Statistics, pp. 109–123. North Holland, Amsterdam (1977)
Greenacre, M.: Correspondence analysis of square asymmetric matrices. Appl. Stat. **49**, 297–310 (2000)
Gregerson, H., Sailer, L.: Chaos theory and its implications for social science research. Human Rel. **46**, 777–802 (1993)
Grorud, A., Chino, N., Yoshino, R.: Complex analysis for three-way asymmetric relational data. Proceedings of the 23rd Annual Meeting of thee Behavior Metric Society of Japan, pp. 292–295. Osaka, Japan (1995)
Imai, S., Guarente, L.: NAD+ and sirtuins in aging and disease. Trends Cell Biol. **24**, 464–471 (2014)
Lancaster, P., Tismenetsky, M.: The Theory of Matrices, 2nd edn. Academic Press, New York (1985)
Masure, R.H., Allee, W.C.: The social order in flocks of the common chicken and the pigeon. Auk **51**, 306–327 (1934)
Newcomb, T.M.: The Acquaintance Process. Holt, Rinehart and Winston, New York (1961)
Okada, A., Imaizumi, T.: Asymmetric multidimensional scaling of two-mode three-way proximities. J. Classif. **14**, 195–224 (1997)
Okada, A., Imaizumi, T.: External analysis of two-mode three-way asymmetric multidimensional scaling. In: Weihs, C., Gaul, W. (eds.) Classification—The Ubiquitous Challenge, pp. 288–295. Springer-Verlag, Berlin (2005)
Ott, E., Grebogi, C., Yorke, J.A.: Controlling chaos. Phys. Rev. Lett. **64**, 1196–1199 (1990)
Tobler, W.: Spatial interaction patterns. J. Environ. Syst. **6**, 271–301 (1976/77)
Wilkinson, J.H.: The Algebraic Eigenvalue Problem. Clarendon Press, Oxford (1965)
Yadohisa, H., Niki, N.: Vector field representation of asymmetric proximity data. Commun. Stat. Theor. Method **28**, 35–48 (1999)
Zielman, B.: Three-way scaling of asymmetric proximities. Research Report RR91-01. Department of Data Theory, Leiden University, Leiden, The Netherlands (1991)
Zielman, B., Heiser, W.J.: Analysis of asymmetry by a slide-vector. Research Report RR91-05. Department of Data Theory, Leiden University, Leiden, The Netherlands (1991)

Introduction to the "s-concordance" and "s-discordance" of a Class with a Collection of Classes

Edwin Diday

In Remembrance

Edwin Diday

2 March 1940–28 April 2023

It is with great sadness that during the final stages of preparing Nishi's Festschrift we learned of the passing of Edwin Diday; he died on April 28, 2023. Edwin was aged 83 years. We extend our heartfelt condolences to his family, friends and colleagues. Edwin was an outstanding researcher who dedicated his life to the development of Classification and Data Analysis leaving an unforgettable impact on future generations.

E. Diday (Deceased) (✉)
CEREMADE Laboratory, University of Paris Dauphine, Paris, France
e-mail: ericb@uow.edu.au

E. J. Beh et al. (eds.), *Analysis of Categorical Data from Historical Perspectives*, Behaviormetrics: Quantitative Approaches to Human Behavior 17,
https://doi.org/10.1007/978-981-99-5329-5_27

1 Introduction

In the seventeenth century, Galileo Galilei set himself the task of measuring what is measurable and making measurable what is not. Our goal here is to make measurable the notions of "concordance" and "discordance" often used between behaviours, events, ideas, results, etc. Each entity or object is subject to internal or external variations in time and/or space. We thus consider that each object is a class of units of a population Ω. These units are described by classical numerical or qualitative variables, but the description of the classes of units requires taking into account the variability of the units that constitute them. This is why, to express this variability, we are led to use variables with interval values, histograms, probability distributions, character sequences and the like. These data are called "symbolic" because they cannot be manipulated as numbers. For example, anyone knows what the multiplication between two or more numbers is, but the multiplication between intervals or histograms is not evident. Therefore, the description of a class is a vector of symbolic values expressing the variability of each variable for this class.

On the other hand, what is the representation of a class? The representation x of a class c is any entity obtained from the descriptive variables of the units, for which a fit can be measured with the class c. Hence, a class can be represented, for example, by a category, a mode, a mean, a probability density, a cumulative distribution, a regression, a hierarchy or a pyramidal clustering, a factorial axis from a principal component analysis (PCA), a dual analysis (also known as correspondence analysis), a canonical analysis and the like. There are several methods able to provide the representations of the best fit with the classes of a given collection of classes of Ω such as the expectation–maximisation (EM) algorithm (Dempster et al. 1977). For example, the dynamic clustering method (DCM) (Diday 1971, 1973; Diday and Schroeder 1976; Diday and Simon 1980) provides a family of algorithms associated with each form of representation, tending to improve in successive steps the fit of each class, of a collection of classes of empty intersection of the population Ω, to its representation. In the particular case of the popular clustering algorithm called K-means, the representation is a mean. Many (DCM) methods were then developed based on this duality between classes and representation; see, for example, regression (Charles 1977; Spath 1979), probability density (Diday et al. 1979) and canonical analysis (Diday 1986).

A broad overview of the s-concordance consists in a class c that is all the more "concordant" (i.e. in concordance) with a collection of classes P for a given representation that the proportion of classes of P, having a fit and a representation close to that of c, is large. The "discordance" can be measured in different ways. For example, by considering that it is all the greater, the weaker is the concordance (this will be the generally our choice) or by considering the proportion of classes c' of P that the representation and (or) fit to c' deviate too much from the representation and (or) fit to c.

Suppose, for example, that we have a qualitative variable whose categories express different causes of mortality including that of COVID-19. Class is represented by this category "COVID-19". Let Ω be the population of the European countries and define P to be a partition of this population consisting of each European country. To what extent can it be said, over a given period, that a European country with a proportion of the COVID-19 category denoted f_c (COVID-19) is in "concordance" or in "discordance" with the other European countries (i.e. with P)? In this case, the concordance (resp. discordance) measure can be considered to be proportional to the proportion of European countries $c \in P$ having a proportion of deaths $f_{c'}$ (COVID-19) close (resp. distant) to f_c (COVID-19). This proportion is denoted g_c (COVID-19, P). An illustrative example using data from an international teaching and learning survey is summarised in Korenjak-Černe et al. (2022).

Since we want to know if a class is "concordant" with a collection of classes for a given representation, we are led to use the symbolic data that describe each of these classes and their representation. The "concordance" and "discordance" used by Kendall (1975) aim to compare the ranking of two ordinal variables. Therefore, both kinds of concordance and discordance have nothing to see together as they measure two things completely differently; the first compares a class to a collection of classes while the second compares two ordinal variables. That is why we will use the terms "s-concordance" and "s-discordance" ("s" for symbolic) for the kind of concordance measure discussed in this paper.

Similarities and dissimilarities are defined by Axioms that express our perception of reality. Analogously, we define "s-concordance" and "s-discordance" by Axioms. We do this in order to measure the concordance or discordance of a class c with a collection of classes, P, for a representation defined by the value x of a given variable.

We first introduce in Sect. 1, two basic functions f_c and g_x, where $f_c(x)$ expresses the fit of the representation x with c and $g_x(c, P)$ expresses the proportion of classes c' of P having a fit and a representation close to that of c. Sections 2 and 3 give the axiomatic definitions of a s-concordance and a s-discordance. Section 4 gives examples of s-concordance and s-discordance families. Section 5 shows the links between concordances and copulas. Section 6 gives a general formulation of the classical likelihood taking into account the concordance when a collection of P classes is given (like in the case with the European countries). In the case where P is unknown at the beginning, Sect. 7 gives a general formulation of mixture decomposition (by the DCM method) taking into account the concordance or discordance and allowing to construct P and the probability density representation of each of its classes. Section 8 gives a way to visualise (in 2D or 3D) clusters of classes of P.

Regarding useful publications for a better understanding of s-concordance and s-discordance, let us mention Afonso et al. (2018), the review article published by Diday (2016) and the first article where the notions of s-concordance and s-discordance are introduced, Diday (2020).

2 The Two General Basic Functions

The first basic function we define, denoted $f_c(x) \in [0, 1]$, expresses the fit between a representation x and a class c (like f_c (COVID-19) in the preceding example). In the case of a categorical variable, $f_c(x)$ expresses the proportion of the category x in the class c. In the case of a numerical variable, f_c is a probability density and $f_c(x)$ is the density of the pair (c, x).

The second basic function is denoted by $g_x(c, P)$, where P is a collection of given classes and is in the case of qualitative variables, the proportion of classes $c\prime$ of P whose representation $x\prime$ and fit $f_{c\prime}(x\prime)$ are close to those of c (x and $f_c(x)$), like g_x (COVID-19, P).

In the case of a numerical variable, g_x is a probability density, and $g_x(c, P)$ is the density of the pair (c, x). To help simplify our discussion, we impose the assumption that $f_c(x)$ and $g_x(c, P)$ vary between 0 and 1.

Knowing these two basic functions, the s-concordance and s-discordance can be built. For a class c, a collection of classes P, and a representation x, the s-concordance and the s-discordance are denoted by $S_{\text{conc}}(c, P, x)$ and $S_{\text{disc}}(c, P, x)$, respectively. In other words, S_{conc} and S_{disc} are functions defined from $P(\Omega)P(P(\Omega))M$ to the set of positive numbers, where $P(\Omega)$ is the power set of Ω (i.e. the set of all classes of Ω) and M is the set of all possible representation of any class.

3 Axiomatic Definition of Similarities and Dissimilarities

Similarities and dissimilarities are notions which exists since they have long been understood, for example, in the Organon of Aristotle around 4 centuries BC. Other famous names can be mentioned including Buffon (1707–1788), Adanson (1727–1806), Lamarck (1744–1829), Cuvier (1769–1832) and Darwin (1809–1882). More recently, Gower (1971), Sneath and Sokal (1973), Anderberg (1973) provide a thorough review of measures of association including "similarities" of different kinds. The formal Axioms which define the similarities can be found in Benzécri (1973, p. 72); the similarities and dissimilarities are formally defined in Diday et al. (1982) and in Jain and Dubes (1988). The Axioms that define a similarity (denoted "sim") or a dissimilarity (denoted "diss") associate to a similarity for any pair of units (i, j) a positive number which satisfy the following Axioms:

$$\text{sim}(i, j) \geq 0, \text{sim}(i, i) = \text{sim}(j, j) = 0, \text{sim}(i, j) \geq \text{sim}(j, i)(\text{symmetry}).$$

These Axioms are as follows for a dissimilarity:

$$\text{diss}(i, j) \geq 0, \text{diss}(i, i) \leq \text{diss}(j, j), \text{diss}(i, j) = \text{diss}(j, i)$$

Intuitively, the similarity Axioms express mainly the fact that someone is more similar to themself than to any other. The dissimilarity Axioms express the fact that someone is less dissimilar to themself than to anyone else.

The Axiomatic definition of s-concordance and s-discordance will imply properties similar to those of the Axioms which define similarities and the dissimilarities. Moreover, it is shown (at the end of Sect. 4) that a similarity (resp. dissimilarity) can become (under some special conditions), a case of s-concordance (resp. s-discordance).

Before diving into the Axiomatic definition of the s-concordance and s-discordance, we give the following practical example.

Example. The population is constituted by the European people described by socio-demographic variables and the classes are the subpopulation of each European country. Our aim is to use dual scaling analysis (Nishisato 1994, 2014), related to correspondence analysis (Lebart et al. 1995) to find the concordance of each European country with other European countries in order to obtain their ranking from the highest to the lowest concordance or discordance. We start with a contingency table of n countries described by the p categories of a categorical socio-demographic variable. A dual analysis is applied to the subpopulation of each European country (i.e. class). The representation of each class are the factors associated to the two first axes with highest inertia of the dual analysis. The factors, denoted V_{1c} and V_{2c}, are for $i = 1, 2$ vectors $V_{ic} = \left(v_{ic1}, \ldots, v_{icp}\right)$ which allows for the projection of any class into the plane defined by the axis with the highest inertia associated to these both factors. The similarity between two classes c and c' can be defined, for example, by:

$$\mathrm{sim}(c, c\prime) = \sum_{l=1}^{p} \sum_{i=1}^{2} |v_{icl} - v_{ic'l}|.$$

Knowing the collection of classes P (i.e. the European countries), the fit of each class to its representation (i.e. factor) and the similarity between representations, we have all that we need to get the s-concordance and the s-discordance, as a result of their Axiomatic definition (given in the next Section). For example, S_{conc} and S_{disc} can be defined by $S_{\text{conc}}(c, P, x) = f_c(x)g_x(c, P)$ which means that the concordance of a European country to the other European countries increases with $f_c(x)$ and $g_x(c, P)$. The discordance can be defined by $S_{\text{disc}}(c, P, x) = f_c(x)/(1 + g_x(c, P))$ which means that the concordance of a European country to the other European countries increases with $f_c(x)$ and decreases when $g_x(c, P)$ increases. This provides a concordance (or discordance) ranking of all European countries following their socio-demographic description.

4 Axiomatic Definition of s-concordance and s-discordance

4.1 Axiomatic Definition of s-concordance

As for standard similarities and dissimilarities (or the Euclidean geometry, the set theory, etc.), the s-concordance denoted S_{conc} and the s-discordance denoted S_{disc} also satisfy natural Axioms, derived from our intuition. They are expressed for any triplet $(c, P, x) \in P(\Omega)P(P(\Omega))M$ in the form of the following three Axioms, for s-concordance:

(a) $S_{\text{conc}}(c, P, x) \geq 0$.
(b) $S_{\text{conc}}(c, P, x)$ is a function of $g_x(c, P)$ and can also be a function of $f_c(x)$ and $g_x(c)$.
(c) S_{conc} is an increasing function of $g_x(c, P)$.

These Axioms are based on our intuitive perception of the word's "concordance" and "relevance" in this context. The higher our intuitive "concordance" of c to P for a representation x, the higher is the proportion $g_x(c, P)$ of P classes c' having close fit $f_{c'}(x')$ and representation x' to the c ones ($f_c(x)$ and x). A first consequence is given in Axiom (b) that says $S_{\text{conc}}(c, P, x)$ must be expressed as a function of $g_x(c, P)$. Moreover, the s-concordance can be a function of $f_c(x)$ and $g_x(c, P)$ in order to take care of our intuitive perception of the $f_c(x)$ "relevance" which increases with $f_c(x)$. The second consequence is expressed in Axiom (c) where our intuitive "concordance" of a class c to a collection of classes P for a representation x is an increasing function of $g_x(c, P)$. Notice that a measure of s-concordance S_{conc} is very different from a measure of similarity "sim" since S_{conc} aims to measure the relation between a class and a collection of classes, whereas "sim" aims to measure the relation between two units. Therefore, the Axioms which define a similarity are not satisfied by a s-concordance save in very particular cases (where c is reduced to a unit and P is reduced to a unique class) that we consider here under in this section.

Axiom (a) is compatible with the condition that $S_{\text{conc}}(c, P, x) \in [0, 1]$, $f_c(x) \in [0, 1]$ and $g_x(c, P) \in [0, 1]$. In the following, it is assumed for simplicity that the density functions f_c and g_x take their values in the interval [0, 1]. This is the assumption that we will generally accept to be satisfied.

Axiom (b) means that there is a function denoted "conc", $[0, 1]^2 \rightarrow [0, 1]$, such as:

$$S_{\text{conc}}(c, P, x) = \text{conc}(f_c(x), g_x(c, P))$$

or, more precisely,

$$S_{\text{conc}_i}(c, P, x) = \text{conc}_i(f_c(x), g_x(c, P)).$$

Subsequently, this application will be called "concordine" if and only if $\text{conc}(f_c(x), g_x(c, P))$ satisfies the three Axioms that define a concordance. When

$f_c(x)$ does not appear, (e.g. when $f_c(x)$ is constant), the concordine is denoted as $\text{conc}(g_x(c, P))$.

Axiom (c) means that:

$$\begin{aligned}&\text{conc}(f_c(x), g_x(c, P)) \le \text{conc}\left(f_c(x), g_x\left(c', P\right)\right)\\&\text{if} \quad g_x(c, P) \le g_x\left(c', P\right).\end{aligned}$$

To say that S_{conc} is an increasing function of $f_c(x)$ means that:

$$\text{conc}(f_c(x), g_x(c, P)) \le \text{conc}(f_{c'}(x), g_x(c, P)) \text{ if } \quad f_x(c, P) \le f_x\left(c', P\right).$$

As the concordance becomes more significant when $f_c(x)$ increases and approaches 1, we say that the "concordance is relevant" when S_{conc} is also an increasing function of $f_c(x)$.

From these three Axioms, we can deduce the following two properties showing that in the very special case where P is reduced to a unique class, an s-concordance satisfies Axioms that define a standard similarity:

The first property states that $S_{\text{conc}}(c, \{c\}, x) \ge S_{\text{conc}}\left(c, \left\{c'\right\}, x\right)$ for any $x \in M$, $c \in P(\Omega)$ and $c' \in P(\Omega)$. This property means that the concordance of a class $c \in P(\Omega)$ with itself (i.e. with a collection of classes P reduced to the class c itself) is greater than the concordance of c with any other class c'. It can be proved as follows:

We have, $g_x(c, \{c\}) = 1$ because P is reduced to a single class which is the class c itself. Then, we have $1 = g_x(c, \{c\}) \ge g_x\left(c, \left\{c'\right\}\right)$. Therefore, from Axiom (c), we get:

$$S_{\text{conc}}(c, \{c\}, x) \ge S_{\text{conc}}\left(c, \left\{c'\right\}, x\right).$$

Note that, unlike the case of similarities, the condition $S_{\text{conc}}(c, \{c\}, x) = S_{\text{conc}}\left(c', \left\{c'\right\}, x\right)$ is not necessarily satisfied for any $c' \in P(\Omega)$. This seems consistent with our intuitive perception of the words "concordance" and "similarity" where it seems normal that the concordance of an individual with themself can differ from one individual to another, while the value of the similarity of an individual with themself remains the same for all individuals.

The second property concerns "symmetry". We have the following symmetry property:

$S_{\text{conc}}\left(c, \left\{c'\right\}, x\right) \ge S_{\text{conc}}\left(c', \{c\}, x\right)$ under the conditions that $f_c(x)$ and x are, respectively, close to $f_{c'}\left(x'\right)$ and to x' or the condition that $f_c(x)$ or x are, respectively, not close to $f_{c'}\left(x'\right)$ and to x'.

This can be proved as follows: if $f_c(x)$ and x are respectively close to $f_{c'}\left(x'\right)$ and to x', then $1 = g_x\left(c, \left\{c'\right\}\right) \ge g_x\left(c', \{c\}\right)$ and therefore:

$$S_{\text{conc}}\left(c, \left\{c'\right\}, x\right) \ge S_{\text{conc}}\left(c', \{c\}, x\right)$$

due to Axiom (c).

If $f_c(x)$ and x are, respectively, *not* close to $f_{c'}(x')$ or to x', then $0 = g_x(c, \{c'\}) = g_x(c', \{c\})$ and therefore:

$$S_{\text{conc}}(c, \{c'\}, x) = S_{\text{conc}}(c', \{c\}, x).$$

Therefore, the symmetry condition is satisfied in the considered cases.

4.2 Axiomatic Definition of a s-discordance

Analogous to s-concordance, the s-discordance is defined by three Axioms. The first two Axioms (a') and (b') that define discordance are the same as Axioms (a) and (b) which define the concordance. The third Axiom becomes:

(c') S_{disc} is a decreasing function of $g_x(c, P)$.

Axiom (b') is equivalent to saying that there is an application $[0, 1]^2 \rightarrow [0, 1]$, denoted "disc" and called "discordine" such that:

$$S_{\text{disc}}(c, P, x) = \text{disc}(f_c(x), g_x(c, P)),$$

satisfies the Axioms that define a discordance. When $f_c(x)$ does not appear, the discordine is denoted $\text{disc}(g_x(c, P))$. In the case where S_{disc} is increasing when $f_c(x)$ is decreasing, we say that the "discordance is irrelevant" since, when $f_c(x)$ decreases, the relevance decreases too.

From Axiom (c'), we can deduce the following property: $S_{\text{disc}}(c, \{c\}, x) \leq S_{\text{disc}}(c, \{c'\}, x)$ which means that the discordance of a class with itself is smaller than its discordance with any other class. It can be proved as follows: we have: $1 = g_x(c, \{c\}) \geq g_x(c, \{c'\})$. Therefore, Axiom (c') leads to:

$$S_{\text{disc}}(c, \{c\}, x) \leq S_{\text{disc}}(c, \{c'\}, x).$$

The symmetry property $S_{\text{disc}}(c, \{c'\}, x) = S_{\text{disc}}(c', \{c\}, x)$ is satisfied in the same condition than in the case of a s-concordance.

5 Examples of Families of s-concordances and s-discordances

A family of s-concordance is defined by a measure of concordance $S_{\text{conc}_i}(c, P, x) = \text{conc}_i(f_c(x), g_x(c, P))$ and its different possible variants satisfying the three Axioms (a), (b), (c). Each s-concordance can be associated with at least the s-discordance

Table 1 Six examples of concordines and their associated discordines

Concordine	Discordine
$\mathrm{conc}_1(f_c(x), g_x(c, P)) = f_c(x) \cdot g_x(c, P)$	$\mathrm{disc}_1(f_c(x), g_x(c, P)) = 1 - f_c(x) \cdot g_x(c, P)$
$\mathrm{conc}_2(f_c(x), g_x(c, P)) = 1 - \frac{f_c(x)}{1+g_x(c,P)}$	$\mathrm{disc}_2(f_c(x), g_x(c, P)) = \frac{f_c(x)}{1+g_x(c,P)}$
$\mathrm{conc}_3(f_c(x), g_x(c, P)) = 1 - \frac{f_c(x)}{1+g_x(c,P)} + \frac{f_c(x)}{2}$	$\mathrm{disc}_3(f_c(x), g_x(c, P)) = \frac{f_c(x)}{1+g_x(c,P)} - \frac{f_c(x)}{2}$
$\mathrm{conc}_4(f_c(x), g_x(c, P))$ $= 1 - (f_c(x) + g_x(c, P) - f_c(x) \cdot g_x(c, P))$	$\mathrm{disc}_4(f_c(x), g_x(c, P))$ $= f_c(x) + g_x(c, P) - f_c(x) \cdot g_x(c, P)$
$\mathrm{conc}_5(f_c(x), g_x(c, P))$ $= \frac{f_c(x) \cdot g_x(c,P)}{f_c(x)+g_x(c,P)-f_c(x) \cdot g_x(c,P)}$	$\mathrm{disc}_5(f_c(x), g_x(c, P))$ $= 1 - \frac{f_c(x) \cdot g_x(c,P)}{f_c(x)+g_x(c,P)-f_c(x) \cdot g_x(c,P)}$
$\mathrm{conc}_6(g_x(c, P)) = g_x(c, P)$	$\mathrm{disc}_6(f_c(x), g_x(c, P)) = 1 - g_x(c, P)$

obtained by: $S_{\mathrm{disc}_i}(c, P, x) = 1 - S_{\mathrm{conc}_i}(c, P, x)$ (which is our choice hereunder). Other possibilities are for example by inversion $S_{\mathrm{disc}_i}(c, P, x) = \left(S_{\mathrm{conc}_i}(c, P, x)\right)^{-1}$ or by:

$$S_{\mathrm{disc}_i}(c, P, x) = e^{-S_{\mathrm{conc}_i}(c,P,x)}.$$

Table 1 gives six families of concordances and discordances that can be considered.

In the case of conc_1 and conc_6 where, moreover, all the fit between each class of P and its representation are equal the s-concordance $S_{\mathrm{conc}}(c, \{c'\}, x)$ (resp. $S_{\mathrm{disc}}(c, \{c'\}, x)$) become proportional to a similarity (resp. dissimilarity) between the representation x of c and the representation x' of c'. Therefore, $S_{\mathrm{conc}}(c, \{c'\}, x) = \mathrm{sim}(x, x')$ is a solution for these two concordines. This means that under these conditions the similarities constitute a case of s-concordance. Analogously, under special conditions dissimilarities become a case of s-discordance.

6 s-concordance and Copulas

6.1 The Random Variables f_{Z_f}, f_{Z_g} and their Link

Let be a random variable be denoted by Z_f and defined from Ω to $P(\Omega)M$ with density f_{Z_f} which is associated with $(c, x) \in P(\Omega)M$ so that $f_{Z_f}(c, x) = f_c(x)$. Let be Z_g the random variable from Ω to $P(\Omega)M$ with density f_{Z_g} which is associated with $(c, x) \in P(\Omega)M$ so that $f_g(c, x) = g_x(c, P)$. The density f_{Z_f} (resp. f_{Z_g}) can be itself considered as a random variable from $P(\Omega)M$ to [0, 1] associating to any couple (c, x) with the value $f_c(x)$ (resp. $g_x(c, P)$). Let $f_{Z_{f_g}}$ be the random variable from [0, 1] to [0, 1], which associates to each $f_c(x)$ value its $g_x(c, P)$ density. Hence,

$f_{Z_{fg}}(f_c(x)) = g_x(c, P)$. Therefore, the random variables $f_{Z_{fg}}$ link to the random variables f_{Z_f} and f_{Z_g} by $f_{Z_{fg}}\left(f_{Z_f}(c, x)\right) = f_{Z_g}(c, x)$ which gives $f_{Z_g} = f_{Z_{fg}}\left(f_{Z_f}\right)$.

6.2 *The "Copuline" and the "Copulas Concordance"*

The joint probability of the two random variables f_{Z_f} and f_{Z_g} is an application $H\colon [0, 1]^2 \to [0, 1]$ such that:

$$H(f_c(x), g_x(c, P)) = \Pr\left(f_{Z_f} \le f_c(x), f_{Z_g} \le g_x(c, P)\right).$$

Sklar (1959) demonstrated that under certain conditions, there is a unique function called a "copula" connecting the joint distribution to its marginal distributions; see also Nelsen (2006). In the case of the random variables Z_f and Z_g, this result is written as follows:

$$H(f_c(x), g_x(c, P)) = \Pr\left(f_{Z_f} \le f_c(x), f_{Z_g} \le g_x(c, P)\right) = CC(u, v),$$

where H is the joint distribution of $u = \Pr\left(f_{Z_f} \le f_c(x)\right)$ and $v = \Pr\left(f_{Z_g} \le g_x(c, P)\right)$, and CC is a special copula since it is obtained from the random variables f_{Z_f} and f_{Z_g}. CC stands for "concordant copula" and is an application $[0, 1]^2 \to [0, 1]$ called "copuline", which associates to any couple $(f_c(x), g_x(c, P))$ the value $H(f_c(x), g_x(c, P))$.

Proposal 1

A copuline is a concordine but a concordine is not necessarily a copuline.

Proof To demonstrate this proposition, it is sufficient to verify that a copuline induces a function defined on $P(\Omega)P(P(\Omega))M$ which is a s-concordance. Let $S(c, P, x) = CC(u, v)$. The Axioms (a) and (c) are satisfied by S since from the definition of a copula $CC(u, v) = \Pr\left(F_{Z_f} \le f_c(x), F_{Z_g} \le g_x(c, P)\right)$ and therefore S is indeed an increasing function of $g_x(c, P)$. Axiom (b) is also satisfied since by definition, u is a function of $f_c(x)$ and v is a function of $g_x(c, P)$. Therefore, a copuline is a concordine.

A concordine is not necessarily a copuline since, by definition, a copuline increases with $f_c(x)$ which is not necessarily the case for a concordine.

The s-concordance which has been induced by a copuline CC and is denoted by S (in the proof of the Proposal 1) can be denoted S_{copconc} and called a "copulas concordance". It is characterised by:

$$\begin{aligned} S_{\text{copconc}}(c, P, x) &= H(f_c(x), g_x(c, P)) \\ &= \Pr\left(f_{Z_f} \le f_c(x), f_{Z_g} \le g_x(c, P)\right) \\ &= \text{CC}(u, v). \end{aligned}$$

Hence, the "copulas concordance" is the joint probability of the random variables Z_f and Z_g.

The copulas satisfy a number of properties that apply, of course, to the case of the concordant copula. This is the case with the Fréchet-Hoeffding inequality:

$$\mathrm{Max}(u + v - 1, 0) \le CC(u, v) \le \mathrm{Min}(u, v).$$

This can be written as:

$$\begin{aligned}\mathrm{Max}\big(\Pr\big(f_{Z_f} \leqslant f_c(x)\big) + \Pr\big(f_{Z_g} \leqslant g_x(c, P)\big) - 1, 0\big) &\leqslant S_{\mathrm{copconc}}(c, P, x)\\ &\leqslant \mathrm{Min}\big(\Pr\big(f_{Z_f} \leqslant f_c(x)\big), \Pr\big(f_{Z_g} \leqslant g_x(c, P)\big)\big).\end{aligned}$$

This inequality constitutes the concordance related version of the Fréchet-Hoeffding inequality.

7 A Generalisation of the Standard Likelihood Theory to the Case Where Underlying Classes Exist

In this section, a collection of P classes is known and the objective is to take them into account when calculating a likelihood function and when estimating its parameters. This is the case where, for example, Ω is the European population and the classes are the populations of each European country. From a sample $\mathbf{s} = (w_1, \ldots, w_n)$ of the given population Ω where the class of w_i is $C(w_i) = c_i$, its representation is $X(w_i) = x_i$, the parameter of f_c (resp. g_x) is α_{c_i} (resp. β_{c_i}) and the likelihood function generalised by s-concordance is written as:

$$L_{\mathrm{conc}}\big(w_1, \ldots, w_n; \boldsymbol{\alpha}_c, \boldsymbol{\beta}_c\big) = \prod_{i=1}^{n} S_{\mathrm{conc}}\big(c_i, P, x_i; \alpha_{c_i}, \beta_{c_i}\big),$$

where $\boldsymbol{\alpha}_c = \big(\alpha_{c_1}, \ldots, \alpha_{c_n}\big)$ and $\boldsymbol{\beta}_c = \big(\boldsymbol{\alpha}_{c_1}, \ldots, \boldsymbol{\alpha}_{c_n}\big)$. Using the concordine:

$$\mathrm{conc}_1\big(f_c(x; \boldsymbol{\alpha}_c), g_x\big(c, P; \boldsymbol{\beta}_c\big)\big) = f_c(x; \boldsymbol{\alpha}_c) \cdot g_x\big(c, P; \boldsymbol{\beta}_c\big),$$

one obtains:

$$L_{\mathrm{conc}_1}\big(w_1, \ldots, w_n; \boldsymbol{\alpha}_c, \boldsymbol{\beta}_c\big) = \prod_{i=1}^{n} f_c\big(x_i; \alpha_{c_i}\big) \cdot g_x\big(c_i, P; \beta_{c_i}\big).$$

The following proposition proves that the likelihood function L_{conc_1} becomes the standard likelihood function when there are no underlying classes.

Proposal 2

The classic likelihood function is a special case of the likelihood with concordance: L_{conc_1}.

Proof Recall that Z_f is the random variable from Ω to $P(\Omega)M$ with density F_{Z_f}, which is associated to $(c, x) \in P(\Omega)M$ so that $F_{Z_f}(c, x) = f_c(x; \alpha_c)$. In the case where $c_i = \Omega$ for $\forall i$, this is transformed into the random variable Z_f of Ω in ΩM with density $F_{Z_f}(\Omega, x) = f_\Omega(x; \alpha)$, denoted $f(x; \boldsymbol{\alpha})$ which is the density with the vector of parameters $\boldsymbol{\alpha}$ of length n of x over the whole population Ω.

In the case where $c_i = \Omega$ for $\forall i$, we have $P = \Omega$ and the likelihood function L_{conc_1} is written as:

$$
\begin{aligned}
L_{\text{conc}_1}\left(w_1, \ldots, w_n; \boldsymbol{\alpha}_c, \boldsymbol{\beta}_c\right)
&= \prod_{i=1}^{n} f_\Omega\left(x_i; \alpha_{c_i}\right) \cdot g_{x_i}\left(\Omega, \{\Omega\}; \beta_{c_i}\right). \\
&= \prod_{i=1}^{n} f_\Omega(x_i; \boldsymbol{\alpha}) \\
&= \prod_{i=1}^{n} f(x_i; \boldsymbol{\alpha}),
\end{aligned}
$$

because $g_{x_i}\left(\Omega, \{\Omega\}; \beta_{c_i}\right) = 1$. Hence:

$$
L_{\text{conc}_1}(w_1, \ldots, w_n; \boldsymbol{\alpha}, \boldsymbol{\beta}) = \prod_{i=1}^{n} f(x_i; \boldsymbol{\alpha}),
$$

which is the standard likelihood function $L(w_1, \ldots, w_n; \boldsymbol{\alpha})$ where $\boldsymbol{\alpha}$ is a vector of unique parameters as there are no underlying classes (i.e. there are all identical to Ω).

So, the standard likelihood function, L, is the special case of the likelihood function resulting from the concordance of concordine conc_1 with $c_i = \Omega$ for $\forall i$, (i.e. P is identical to Ω).

The estimate of $\boldsymbol{\alpha} = \left(\alpha_{c_1}, \ldots, \alpha_{c_n}\right)$ and $\boldsymbol{\beta} = \left(\boldsymbol{\alpha}_{c_1}, \ldots, \boldsymbol{\alpha}_{c_n}\right)$ is obtained by maximising $L_{\text{conc}_1}(w_1, \ldots, w_n; \boldsymbol{\alpha}, \boldsymbol{\beta})$ with respect to $\boldsymbol{\alpha}$ and $\boldsymbol{\beta}$. When there are no underlying classes (i.e. there are all identical to Ω), the vector of unique parameters $\boldsymbol{\alpha}$ is obtained by maximising the standard likelihood function $L(w_1, \ldots, w_n; \boldsymbol{\alpha})$.

Having thus obtained $\boldsymbol{\alpha}$ and $\boldsymbol{\beta}$ when there are underlying classes, there are four applications available from this result:

1. A density law can be built from the $\text{conc}_1(f_c(x; \boldsymbol{\alpha}), g_x(c, P; \boldsymbol{\beta}))$ values when x varies in a representative sample of Ω and c in P. From this law, a test can be built to say whether or not a class c is concordant with a collection of classes P for a

value x, at a given threshold. More empirically, we can proceed as follows: start with a representative sample $\mathbf{s} = (w_1, \ldots, w_n)$ of Ω and calculate the percentage of elements of this sample having a smaller value than $f_c(x; \boldsymbol{\alpha}_c) \cdot g_x(c, P; \boldsymbol{\beta}_c)$. If this percentage is below a given threshold ε, then the concordance of c with P for the value x is rejected. The choice of ε can be more or less severe depending on whether it takes the value 1/10, 1/100 or 1/1000 and so on.

2. The second result is that this approach takes into account concordances by giving a weight to the values $f_{c_i}(x_i; \boldsymbol{\alpha}_c)$ proportional to the value taken by $g_{x_i}(c_i, P; \boldsymbol{\beta}_c)$. This expresses the fact that $f_{c_i}(x_i; \boldsymbol{\alpha}_c)$ is more relevant and significant if many classes c' of P take values $f_{c'_i}(x_i; \boldsymbol{\alpha}_{c'})$ close to $f_c(x; \boldsymbol{\alpha}_c)$. Intuitively, by giving more weight to $f_{c_i}(x_i; \boldsymbol{\alpha}_c)$ than $g_{x_i}(c_i, P; \boldsymbol{\beta}_{c_i})$, the estimate is supposed to be more relevant which means that the parameters values of laws of high density (i.e. not flat) will tend to be surrounded by parameters values of laws of other classes of P of close parameter values. Taking this information into account when there are underlying classes can lead to an estimate quite different from that which would be obtained if it were not taken into account by using a standard likelihood function. Much must be done in that direction of research in order to compare the likelihood function with concordance and the standard likelihood function.
3. The third result is to consider, in any population (without classes known a priori), a sample cut into a collection P of classes of the same size and drawn at random. We can then use a likelihood function with concordance consistent with this P, in the hope of obtaining more relevant (insightful) and faster results than by the standard approach.
4. The fourth result is to be able to compare $f_c(x; \boldsymbol{\alpha}_c)$, $g_x(c, P; \boldsymbol{\beta}_c)$ and $f(x; \boldsymbol{\alpha})$ using different informative criteria. For example, consider the following criterion:

$$I(w) = (f_c(x; \boldsymbol{\alpha}_c) - f(x; \boldsymbol{\alpha}))g_x(c, P; \boldsymbol{\beta}_c),$$

where $C(w) = c$ is the class of w and its value is $X(w) = x$. The largest positive values of $I(w)$ means that the density of x in class c is greater than in the population Ω and, moreover, that the concordance of c with P for the x value is large, in the sense of the concordine conc_6.

8 The Case of Large Masses of Data

In the case of large masses of data, Beranger et al. (2023) had the economic idea of cutting M into k blocks $B_1, \ldots, B_k$ of values in order to make an estimate based on k blocks rather than on a sample s of size n. The advantage is that one can get the estimate much faster with a k much smaller than n. For this purpose, it is necessary to set:

$$p_i(\boldsymbol{\alpha}) = \int\limits_{B_i} f_c(x; \boldsymbol{\alpha}_c)\mathrm{d}x.$$

The classic likelihood for the purpose of saving calculations is then so written:

$$\begin{aligned} L(w_1, \ldots, w_n; \boldsymbol{\alpha}) &= \prod_{i=1}^{n} p_i(\boldsymbol{\alpha}) \\ &= (p_1(\boldsymbol{\alpha}))^{n_1} \cdots (p_k(\boldsymbol{\alpha}))^{n_k}, \end{aligned}$$

where n_i is the number of times an element w_i of the sample belongs to B_i.

This approach can be generalised by using the likelihood from concordine conc_1 by posing this time:

$$p_i(\boldsymbol{\alpha}, \boldsymbol{\beta}) = \frac{\int\limits_{B_i} f_{c_i}\left(x_i; \boldsymbol{\alpha}_{c_i}\right) \cdot g_{x_i}\left(c_i, P; \boldsymbol{\beta}_{c_i}\right)\mathrm{d}c\mathrm{d}x}{\int\limits_{P(\Omega)M} f_{c_i}\left(x_i; \boldsymbol{\alpha}_{c_i}\right) \cdot g_{x_i}\left(c_i, P; \boldsymbol{\beta}_{c_i}\right)\mathrm{d}c\mathrm{d}x},$$

from which results:

$$\begin{aligned} L_{\mathrm{conc}_1}(w_1, \ldots, w_n; \boldsymbol{\alpha}, \boldsymbol{\beta}) &= \prod_{i=1}^{n} p_i\left(\boldsymbol{\alpha}_{c_i}, \boldsymbol{\beta}_{c_i}\right) \\ &= \left(p_1\left(\boldsymbol{\alpha}_{c_1}, \boldsymbol{\beta}_{c_1}\right)\right)^{n_1} \cdots \left(p_k\left(\boldsymbol{\alpha}_{c_k}, \boldsymbol{\beta}_{c_k}\right)\right)^{n_k}, \end{aligned}$$

where n_i is the number of times an element w_i of the sample belongs to B_i.

As Beranger et al. (2023) says, if we assume that the sample s follows a multinomial distribution of parameters $(p_1(\boldsymbol{\alpha}), \ldots, p_k(\boldsymbol{\alpha}))$, then the probability of obtaining $(n_1, \ldots, n_k)$ for the parameter $\boldsymbol{\alpha}$ is:

$$\begin{aligned} \Pr(n_1, \ldots, n_k; \boldsymbol{\alpha}) &= \frac{n!}{n_1! \ldots n_k!}(p_1(\boldsymbol{\alpha}))^{n_1} \cdots (p_k(\boldsymbol{\alpha}))^{n_k} \\ &= \frac{n!}{n_1! \ldots n_k!} L(w_1, \ldots, w_n; \boldsymbol{\alpha}). \end{aligned}$$

This is generalised using the likelihood from conc_1 by:

$$\begin{aligned} \Pr(n_1, \ldots, n_k; \boldsymbol{\alpha}) &= \frac{n!}{n_1!...n_k!}(p_1(\boldsymbol{\alpha}, \boldsymbol{\beta}))^{n_1} \cdots (p_k(\boldsymbol{\alpha}, \boldsymbol{\beta}))^{n_k} \\ &= \frac{n!}{n_1! \ldots n_k!} L_{\mathrm{conc}_1}(w_1, \ldots, w_n; \boldsymbol{\alpha}, \boldsymbol{\beta}), \end{aligned}$$

from which by optimisation of the parameters $n_1, \ldots, n_k$, $\boldsymbol{\alpha}$ and $\boldsymbol{\beta}$ can be obtained. Note that the third application given above can be used in the economic approach of Beranger et al. (2023), with the hope to obtain an estimate even more relevant and even faster.

9 Generalisation of Mixing Decomposition by Concordance or Discordance

In this section, we are interested in the case where P is unknown and is to be constructed by an iterative improvement of the fit between each class and its representation. More precisely, the objective of mixture decomposition, is to find a partition $P = (P_1, \ldots, P_k)$ whose classes P_ℓ are in good fit with the probability densities $f(\cdot, \mathbf{a}_\ell)$ defined by a given model (Gaussian, for example) of a vector of parameters $\mathbf{a}_\ell$ that represent them. Let there be a sample $\mathbf{s} = (w_1, \ldots, w_k)$ of the given population Ω where the class of w_i is $C(w_i) = c_i$ and its value is $X(w_i) = x_i$.

In this context, the dynamic clustering method (Diday 1971, 1973, 1979; Billard and Diday 2020) provides the partition P by an algorithm based on a phase of representation of each class P_ℓ (drawn at random at the beginning, for example) by a law $f(\cdot, \mathbf{a}_\ell)$ whose parameters $\mathbf{a}_\ell$ are estimated by maximising the likelihood. This is followed by an allocation phase where each w_i of the given sample $\mathbf{s}$ is allocated to the class ℓ of highest density $f(x_i, \mathbf{a}_\ell)$.

This algorithm can be generalised by introducing s-concordances as follows. The representation space of a partition $P = (P_1, \ldots, P_k)$ is the set $\mathbf{A}_k = (\mathbf{a}_1, \ldots, \mathbf{a}_k)$ of the possible values of the parameters on which the densities depend. Note A an element of $\mathbf{A}_k$. The criterion to be maximised is:

$$W(A, P) = \sum_{\ell=1}^{k} \log\big(L_{\text{conc}_1}(P_\ell, \mathbf{a}_\ell)\big),$$

where

$$L_{\text{conc}_1}(\mathrm{P}_\ell, \mathbf{a}_\ell) = \prod_{x_i \in P_\ell} f_{P_\ell}(x_i; \mathbf{a}_\ell) \cdot g_{x_i}(P_\ell, P; \mathbf{b}_\ell),$$

where $f_{P_\ell}(x_i; \mathbf{a}_\ell)$ measures the s-concordance of the class P_ℓ with the classes of the partition P in x_i and where g_{x_i} is the probability density $g_{x_i}(P_\ell, P; \mathbf{b}_\ell)$ of parameter $\mathbf{b}_\ell$.

Note that we come back to the standard method in the special case where in the formula $g(x_i; \mathbf{b}_\ell) = g_{x_i}(P_\ell, P; \mathbf{b}_\ell)$, P is replaced by $\{P_\ell\}$ since in this case we get $g_{x_i}(P_\ell, \{P_\ell\}; \mathbf{b}_\ell) = 1$.

The justification of the use of likelihood with concordance in mixture decomposition is based on the assertion that $f_c(x; \boldsymbol{\alpha}_c)$ is more relevant (useful, pertinent) if it is high than if it is low. Moreover, $f_{P_\ell}(x_i; \mathbf{a}_\ell) \cdot g_{x_i}(P_\ell, P; \mathbf{b}_\ell)$ is considered more relevant than $f_{P_\ell}(x_i; \mathbf{a}_\ell)$ since we give more weight to $\mathrm{f}_{P_\ell}(x_i; \mathbf{a}_\ell)$ if there are numerous classes $P' \in P$ such that $f_{P'}(x_i; \mathbf{a}_\ell)$ is close to $f_{P_\ell}(x_i; \mathbf{a}_\ell)$. Hence, if $f_{P_\ell}(x_i; \mathbf{a}_\ell)$ is isolated, it is considered to be less relevant. Intuitively, the iterative process of the DCM and EM methods will advantage the close laws of high density and tend to agglomerate them. Therefore, the parameters value of such laws will tend to be close and so more relevant. It can be interesting to apply alternatively EM and DCM to see if the agglomerative effect converges towards a reduced number of laws and classes. Much remains to be done in this direction in order to compare the concordant and the standard approach of mixture decomposition.

10 Clustering of Classes Characterised by their Concordance or Discordance in 2D or 3D Visualisation

First, we recall the assumption that $f_c(x)$ and $g_x(c, P)$ vary in the interval [0, 1] and that x is fixed. In this way, we can construct a $[0, 1] \times [0, 1]$ square whose abscissa (resp. the ordinate) represents the values $f_c(x)$ (resp. $g_x(c, P)$). When c varies, we can represent in this square at all the points $(f_c(x), g_x(c, P))$. When two points are close in the square, it implies that their concordances are close. Suppose we have the concordine conc_1: $\mathrm{conc}_1(f_c(x), g_x(c, P)) = f_c(x) \cdot g_x(c, P)$ and $\mathrm{conc}_1(f_{c\prime}(x), g_x(c\prime, P)) = f_{c\prime}(x) \cdot g_x(c\prime, P)$ are close if $(f_c(x), g_x(c, P))$ and $\left(f_{c'}(x), g_x\left(c', P\right)\right)$ are close points in the square. We can thus see (if there exist) clusters of points associated with a high or low concordance. For example, a cluster at the top right of the square gathers classes c that for the value x have a large concordance with P. This concordance can be represented in 3D by adding a concordance axis associating to each pair $(f_c(x), g_x(c, P))$ its concordance $f_c(x) \cdot g_x(c, P)$ value. We can also calculate the s-concordance of a cluster by defining it as the surface of the smallest rectangle (parallel to the axis) containing this cluster. If this rectangle is defined by its lowest point on the left denoted (a_1, b_1) and the highest point on the right denoted (a_2, b_2) where $a_i = f_{c_i}(x)$ and $b_i = g_{x_i}(c_i, P)$, then it is easy to see using the 2D visualisation that the concordance of this cluster is given by:

$$a_2 b_2 - a_1 b_2 - a_2 b_1 + a_1 b_1,$$

where $a_i b_i$ is the concordance associated with the concordine conc_1 of class c_i with the collection of classes P for the value x.

11 Conclusion and Outlook

We gave an Axiomatic definition of s-concordance and s-discordance and then showed their links with copula theory, estimation by maximisation of likelihood and decomposition of mixtures. Other links exist, for example, with the Tf-Idf which constitutes a special case of discordance, with the Latent Dirichlet Allocation (LDA) which makes it possible to “create” a class having a good concordance with a collection of given classes, with the Kullback–Leibler divergence in order to construct hierarchical or pyramidal classifications of concordant or discordant classes. As a result of any method leading to local classes and models, we now have a tool, taking into account s-concordances or s-discordances, that allows for one to obtain an estimate of the parameters more in accordance with the data than by conventional approaches, in the case where the classes are known or by generalised mixing decomposition when they are not. The classes can then be ordered according to their concordance or discordance with the parameters thus estimated. All this offers new avenues of research in data science with great potential for applications, theoretical and practical development.

Acknowledgements With the sad passing of Edwin shortly before the publication of Nishi’s *Festschrift*, he was unable to provide final input on changes and revisions during the final preparation of this paper. We, the editors, would like to extend our sincere thanks to Gilbert Saporta for kindly providing some help in finalising Edwin’s paper. We would also like to extend our sincere condolences to Edwin’s family and thank them for allowing us to include the picture that appears at the start of his paper.

References

Afonso, F., Diday, E., Toque, C.: Data Science par Analyse des Données Symboliques: Une Nouvelle Façon d’Analyser les Données Classiques. Complexes et Massives à Partir des Classes. Éditions Technip, Paris (2018)

Anderberg, M.R.: Cluster Analysis for Applications. Academic Press, New York (1973)

Beranger, B., Lin, H., Sisson, S.: New models for symbolic data analysis. Adv. Data Anal. Classif. **17**, 659–699 (2023)

Benzécri, J.P.: L’Analyse des Données, Tome1: La Taxonomie. Dunod, Paris (1973)

Billard, L., Diday, E.: Clustering Methodology for Symbolic Data. Wiley, Hoboken, NJ (2020)

Charles, C.: Régression typologique et reconnaissance des formes. Thèse de doctorat 3ème cycle, Université Paris IX-Dauphine, Juin, (1977)

Dempster, A., Laird, N., Rubin, D.: Maximum likelihood from incomplete data with the EM algorithm. J. R. Stat. Soc. Ser. B (Methodol.) **39**, 1–38 (1977)

Diday, E.: Une nouvelle méthode en classification automatique et reconnaissance des formes la méthode des nuées dynamiques. Revue De Statistique Appliquée **19**(2), 19–33 (1971)

Diday, E.: The dynamic clusters method in nonhierarchical clustering. Int. J. Comput. Inform. Sci. **2**, 61–88 (1973)

Diday, E.: Canonical analysis from the automatic classification point of view. Control Cybern. **15**(2), 115–137 (1986) [English reprint of “Analyse canonique du point de vue de la classification automatique”. Rapport Laboria n°293. INRIA, Rocquencourt, France (1978)]

Diday, E.: Thinking by classes in data science: the symbolic data analysis paradigm. Wires Comput. Stat. Symbolic Data Anal. **8**, 172–205 (2016)
Diday, E.: Explanatory tools for machine learning in the symbolic data analysis framework. In: Diday, E., Guan, R., Saporta, G., Wang, H. (eds.) Advances in Data Science: Symbolic, Complex and Network Data, pp. 1–30. ISTE-Wiley (2020)
Diday, E., Schroeder, A.: A new approach in mixed distributions detection. RAIRO Operations Research - Recherche Opérationnelle **10**(6), 75–106 (1976)
Diday, E., Simon, J.C.: Clustering analysis. In: Fu, K.S. (ed.) Communication and Cybernetics Digital Pattern Recognition, pp. 47–94. Springer Verlag, Berlin (1980)
Diday, E., Lemaire, J., Pouget, J., Testu, F.: Elements d'Analyse des Données. Dunod, Paris (1982)
Diday, E., et al.: Optimisation en Classification Automatique. INRIA, Rocquencourt (1979)
Gower, J.C.: A general coefficient of similarity and some of its properties. Biometrics **27**(4), 857–871 (1971)
Jain, A.K., Dubes, R.C.: Algorithm for Clustering Data. Prentice Hall, NJ (1988)
Kendall, M.: Rank Correlation Method. Griffin, London (1975)
Korenjak-Černe, S., Japelj-Pavešić, B., Diday, E.: Symbolic concordance and discordance illustrated on data from an international teaching and learning survey. Presented at the 17th Conference of the International Federation of Classification Societies (IFCS2022), Porto, Portugal, 19–23 July (2022)
Lebart, L., Morineau, A., Piron, M.: Statistique Exploratoire Multidimensionnelle. Dunod, Paris (1995)
Nelsen, R.B.: An Introduction to Copulas, 2nd edn. Springer, New York (2006)
Nishisato, S.: Analysis of Categorical Data: Dual Scaling and its Applications. University of Toronto Press, Toronto (1980)
Nishisato, S.: Elements of Dual Scaling: An Introduction to Practical Data Analysis. Lawrence Erlbaum Associates, Hillsdale, NJ (1994)
Nishisato, S.: Multidimensional Nonlinear Descriptive Analysis. Chapman & Hall/CRC, Boca Raton, FL (2014)
Sklar, A.: Distribution function at n dimensions and their margins. Publ. Inst. Stat. Univ. Paris **8**, 229–231 (1959)
Sneath, P.H.A., Sokal, R.R.: Numerical Taxonomy. Freeman and Company, San Francisco (1973)
Spath, H.: Algorithm 39: Clusterwise linear regression. Computing **22**(4), 367–373 (1979)

Discrete Functional Data Analysis Based on Discrete Difference

Masahiro Mizuta

1 Introduction

Functional data analysis (FDA) began with a paper by Ramsay with the challenging title "When the data are functions" (Ramsay 1982). Since then, various studies have been published, and books on FDA have been published; see, for example, Ramsay (2002) and Ramsay and Silverman (2005). Many theoretical studies, methodological developments, and applications have been promoted up to the present day. Functional data are rarely directly available. Therefore, in the ordinary approach of FDA, analysis is performed after functionalization, in which discretely obtained numerical values are transformed into functions. In the functionalization, objects are represented as functions using various basis functions. For these functions, registration, computation of the basic statistics as functions, and the application of extended methods of ordinary multivariate analysis are performed. In particular, differentiating the functions allows the characteristics possessed by the objects to be examined from a different perspective. Furthermore, it may be possible to describe the structure of an object with differential equations. So far, numerous methods have been developed and actually used as functional versions of conventional multivariate analysis methods, such as functional regression analysis, functional discriminant analysis, functional principal component analysis, and functional multidimensional scaling.

In this paper, we focus on the fact that the input data are discrete and examine how to treat it as a discrete function without converting it into a (continuous) function.

M. Mizuta (✉)
The Institute of Statistical Mathematics, Tachikawa, Tokyo, Japan
e-mail: mizuta@ism.ac.jp

E. J. Beh et al. (eds.), *Analysis of Categorical Data from Historical Perspectives*,
Behaviormetrics: Quantitative Approaches to Human Behavior 17,
https://doi.org/10.1007/978-981-99-5329-5_28

2 Overview of Functional Data Analysis

Suppose n objects $\{x_{ij},\ i = 1, 2, \ldots, n\}$ with multiple variables are obtained. If the number of each variable is constant (say p), then this is ordinary p-dimensional data of size n. However, if each variable depends on time or other variables, it is inappropriate to treat it as conventional multidimensional data. In the following, for the sake of simplicity, we assume that the data take multiple values that depend on time t. However, this need not be limited to time, but can be spatial coordinates, the order of experiments or observations, or conditions expressed numerically. From now on, the data set is denoted $\{x_i(t_j)\}$.

In FDA, $\{x_i(t_j)\}$ is transformed into a function $\{x_i(t)\}$ for each object i. This is called functionalisation. In Ramsay and Silverman (2005), it is shown that trigonometric functions, Legendre polynomials, wavelets, etc., can be used as basis functions. Originally, the degrees of freedom (dimension) of the general functions are infinite, but once the basis function system is determined and functionalised, the function can be represented by a finite number of real numbers. Furthermore, if the basis functions are orthonormal, those finite real numbers can be treated in the same way as ordinary Euclidean space points.

For example, in functional multiple regression analysis, where the scalar is the objective variable, the usual calculation methods of multiple regression analysis can be applied by considering the coefficient vector as the explanatory variables (Shimokawa et al. 2000; Yamanishi and Tanaka 2001). In functional principal components analysis, each object is represented as a point in the low-dimensional space by finding the variance–covariance matrix for the coefficient vector of each object and solving its eigenvalue problem. Functional principal component analysis was applied to temperatures in Canada to derive factors that can be interpreted as a mode of variation, a measure of uniformity, etc. (Ramsay and Silverman 2005). Other than that, most methods of multidimensional data analysis, such as discriminant analysis, cluster analysis (Mizuta 2002, 2003a, b; Tarpey and Kinateder 2003) and multidimensional scaling methods (Mizuta 2000, 2005) can be extended to functional data analysis methods. However, methods of use, interpretation of results, and statistical tests must be devised for each.

The use of the derivative of a function is an effective approach in FDA. Functions based on linear combinations of basis functions are higher-order differentiable, and the differentiated functions are also new functional data. Deriving the relationship between these functions yields the differential equation. In many cases, the structure of the data can be clarified.

3 Discrete Difference and Uncorrelated Discrete Difference

The derivative of a function is a useful tool in FDA. In this chapter, we introduce the method of treating a function as a discrete function instead of a continuous function.

The concept that corresponds to the derivative of a continuous function for a discrete function is discrete difference. We explain the discrete difference and then propose an uncorrelated discrete difference. The uncorrelated discrete difference in this paper is an improved version of Mizuta (2006).

We consider discrete difference operators to the discrete functional data. Discrete differences include forward, backward, and central finite differences, but for the sake of simplicity, we will use forward differences here. However, in the latter part of this section, we will follow the central approach. In the following, again for simplicity, we will examine one function from the functional data.

Let us consider $x(t);\ t \in \{\ldots, t_{-1}, t_0, t_1, t_2, \ldots\}$ as a discrete function. We assume that $t_i - t_{i-1}$ are constant (equally spaced) for convenience. Usually, the first-order discrete differences $d^{(1)}(t)$ are defined as $d^{(1)}(t_i) = x(t_i) - x(t_{i-1})$. The second-order discrete differences are defined as the discrete differences of the first-order discrete differences and so on:

$$\begin{aligned}
d^{(0)}(t_i) &= x(t_i) \\
d^{(1)}(t_i) &= x(t_i) - x(t_{i-1}) \\
d^{(2)}(t_i) &= x(t_i) - 2x(t_{i-1}) + x(t_{i-2}) \\
d^{(3)}(t_i) &= x(t_i) - 3x(t_{i-1}) + 3x(t_{i-2}) - x(t_{i-3}) \\
d^{(4)}(t_i) &= x(t_i) - 4x(t_{i-1}) + 6x(t_{i-2}) - 4x(t_{i-3}) + x(t_{i-4})\ .
\end{aligned}$$

In the case that we treat them statistically, however, a problem arises in these discrete differences.

Consider the most random situation. Suppose $x(t_i)$ follows a standard normal distribution with i.i.d. $x(t)$ or $d^{(0)}(t)$, $\{d^{(k)}\}$ are expected to be uncorrelated. However, $\mathrm{Cov}\left(d^{(0)}(t_i), d^{(2)}(t_i)\right) \neq 0$ and $\mathrm{Cov}\left(d^{(1)}(t_i), d^{(3)}(t_i)\right) \neq 0$ etc. The pairwise scatter plot of $x = \left\{d^{(0)}, d^{(1)}, \ldots, d^{(4)}\right\}$ is shown in Fig. 1. In other words, using conventional discrete difference of $x(t)$, which is completely random, misleads us as if some structure exists.

We modify the conventional discrete differences to be mutually independent under the previous condition. The proposed discrete difference (we call them k-th uncorrelated discrete differences) $d^{(k)*}(t_i)$ is a linear combination of k-th conventional discrete differences:

$$d^{(k)*}(t_i) = \sum_{l=i-m}^{i+m} \omega_l d^{(k)}(t_l),$$

where m is a positive integer, the variance of $d^{(k)*}$ is one, and $d^{(k)*}(t_i),\ k = 0, 1, \ldots$ are uncorrelated if $x(t_i)$ follows is an i.i.d. standard normal random variable.

The following is a brief outline of how to determine ω_l. Create a simultaneous equation for ω_l, given that it is uncorrelated with $d^{(k_1)*}$ for which $k_1 < k$. Solve the simultaneous equations sequentially under m as small as possible.

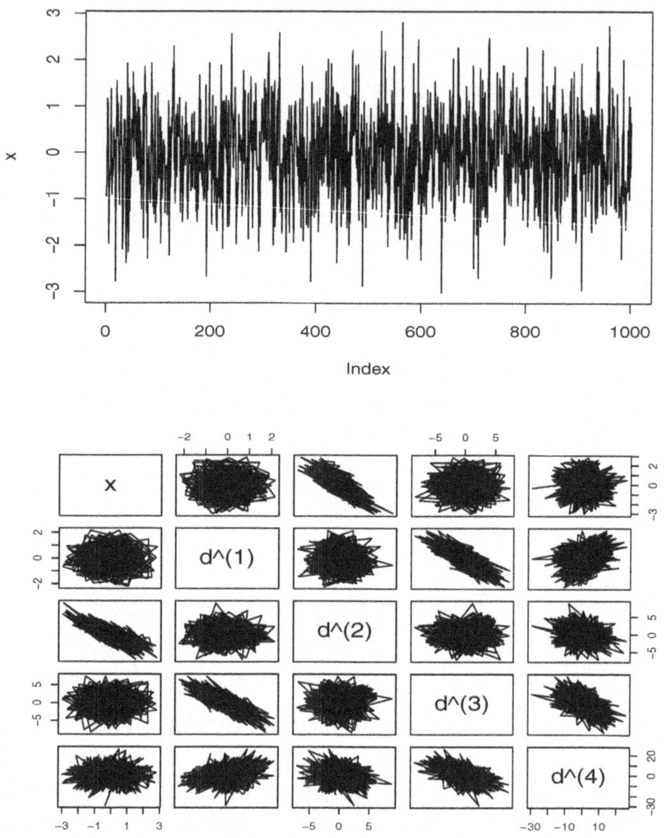

Fig. 1 Discrete function and its pairwise scatter plot of conventional discrete differences

Here are concrete *uncorrelated discrete differences*;

$$
\begin{aligned}
d^{(0)*}(t_i) &= x(t_i) \\
d^{(1)*}(t_i) &= \frac{x(t_{i+1}) - x(t_{i-1})}{\sqrt{1^2 + 1^2}} \\
d^{(2)*}(t_i) &= \frac{x(t_{i+2}) - x(t_{i+1}) - x(t_{i-1}) + x(t_{i-2})}{\sqrt{1^2 + 1^2 + 1^2 + 1^2}} \\
d^{(3)*}(t_i) &= \frac{2x(t_{i+3}) - 3x(t_{i+2}) + 3x(t_{i-2}) - 2x(t_{i-3})}{\sqrt{2^2 + 3^2 + 3^2 + 2^2}} \\
d^{(4)*}(t_i) &= \frac{7x(t_{i+5}) - 16x(t_{i+4}) + 9x(t_{i+3}) + 9x(t_{i-3}) - 16x(t_{i-4}) + 7x(t_{i-5})}{\sqrt{7^2 + 16^2 + 9^2 + 9^2 + 16^2 + 7^2}}.
\end{aligned}
$$

The pairwise scatter plot of $x, d^{(1)*}, \ldots, d^{(4)*}$ of the discrete functions in Fig. 1 is shown in Fig. 2.

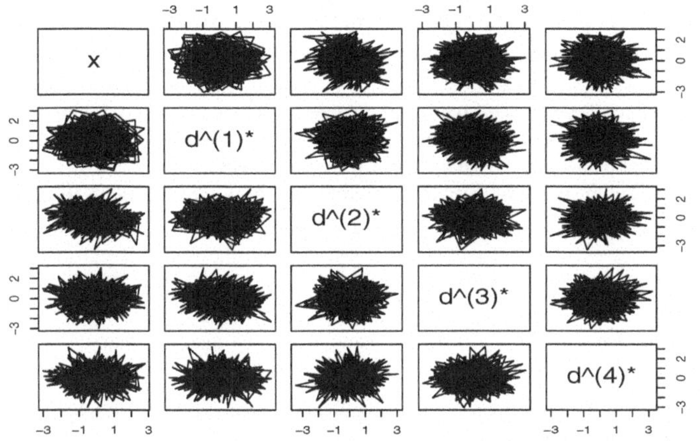

Fig. 2 Pairwise scatter plot of uncorrelated discrete differences

4 Detection of Relations Among Discrete Differences

Uncorrelated discrete differences of discrete functions are defined in the previous section. If we can find out relations among discrete function x and discrete differences $d^{(1)*}, d^{(2)*}, d^{(3)*}, d^{(4)*}, \ldots$, the characteristic of the functional data can be detected. Ramsay and Silverman (2005) described *Derivative and functional linear models* (in chapter "17"), *Differential equation and operators* (in chapter "18"), and *Principal differential analysis* (in chapter "19"). They use differential operators and high-order differential operators effectively.

These ideas can be applied to discrete difference operators. We presented a method for detection of relations among discrete uncorrelated differences with principal component analysis (Mizuta 2006). By defining discrete uncorrelated differences appropriately as mentioned before, x, $d^{(1)*}, \ldots, d^{(k)*}$ become independent under the condition that $x(t)$ are completely random. We regard $x(t_i), d^{(1)*}(t_i), \ldots, d^{(k)*}(t_i)$, for $i = 1, \ldots, n$, as $(k+1)$-dimensional data and apply principal component analysis to them. The structure can be found out with this method, in the case that there are some relations among x, $d^{(1)*}, \ldots, d^{(k)*}$. Principal components with small eigenvalues represent the structure of the linear combination of x, $d^{(1)*}, \ldots, d^{(k)*}$.

5 Concluding Remarks

In this paper, we have improved upon the paper of Mizuta (2006) titled "Discrete Functional Data Analysis". In this paper, we showed that discrete difference is a powerful tool. However, ordinary discrete difference has a correlation even if the discrete function is completely random. Therefore, we proposed uncorrelated discrete difference. Other types of discrete difference are the subject of future work.

Acknowledgements Dr. Nishisato, who is an alumnus of Hokkaido University like myself, embodies the frontier spirit, which is a motto of the university. I have been greatly inspired by him in the field of categorical data analysis. This current research is also influenced by his teachings. This work was supported by Japan Society for the Promotion of Science KAKENHI [grant numbers 23K11023].

References

Mizuta, M.: Functional multidimensional scaling. In: Proceedings of the Tenth Japan and Korea Joint Conference of Statistics, 77–82 (2000)

Mizuta, M.: Cluster analysis for functional data. In: Proceedings of the 4th Conference of the Asian Regional Section of the International Association for Statistical Computing, 219–221 (2002)

Mizuta, M.: Hierarchical clustering for functional dissimilarity data. In: Proceedings of the 7th World Multiconference on Systemics, Cybernetics and Informatics, vol. V, 223–227 (2003)

Mizuta, M.: K-means method for functional data. Bull. Int. Stat. Inst. 54th Session Book **2**, 69–71 (2003)

Mizuta, M.: Multidimensional scaling for dissimilarity functions with several arguments. Bull. Int. Stat. Inst. 55th Session, 244 (2005)

Mizuta, M.: Discrete functional data analysis. In: Proceedings in Computational Statistics 2006, Edited by Alfredo Rizzi and Maurizio Vichi, Physica-Verlag, A Springer Company, 361–369 (2006)

Ramsay, J.O.: When the data are functions. Psychometrika **47**, 379–396 (1982)

Ramsay, J.O., Silverman, B.W.: Applied Functional Data Analysis—Methods and Case Studies. Springer-Verlag, New York (2002)

Ramsay, J.O., Silverman, B.W.: Functional Data Analysis, 2nd edn. Springer-Verlag, New York (1997)

Shimokawa, M., Mizuta, M., Sato, Y.: An expansion of functional regression analysis. Jpn. J. Appl. Stat. **29**(1), 27–39 (2000). (in Japanese)

Tarpey, T., Kinateder, K.K.J.: Clustering functional data. J. Classif. **20**, 93–114 (2003)

Yamanishi, Y., Tanaka, Y.: Geographically weighted functional multiple regression analysis: a numerical investigation. In: Proceedings of the International Conference on New Trends in Computational Statistics with Biomedical Applications, 287–294 (2001)

Probability, Surprisal, and Information

James Ramsay

1 Introduction

For those of us like my great friend Nishi and I embedded in social science departments, choice data are of primary importance by virtue of the number of tests and scales that appear each year, and by the millions of people, young and old, who complete them. Choice is an easy crop to sow, but the yield is rather limited. This is in part due to using only counts of right answers for tests, and the subjectivity of *a priori* weights assigned each choice limits what the data can tell us and raises serious bias concerns. We both asked ourselves whether we could do better, and we both spent much of our careers trying to discover how.

Nishi and I met often at conferences, and I really admired (and envied) his gentle and kindly way of communicating his ideas. My favourite recollection of our meeting involves a train to Toronto that was delayed three hours by an accident and left me wandering along Bloor Street at 11PM wondering how I could secure a hotel given a big football game the next day. There he was, no doubt after committing his evening to preparing for the meeting the next day! What a lovely time this hapless guest enjoyed in his beautiful home!

1.1 Information Theory

Information theory is central to signal processing where an error-prone system transmits a message and the receiver must contend with the static. A recent and readable text is Cover and Thomas (2006). In multiple choice testing, the message is the test taker's choice and the noise arises the lack of certainty about which choice is right.

J. Ramsay (✉)
Department of Psychology, McGill University, Montreal, QC, Canada
e-mail: james.ramsay@mcgill.ca

E. J. Beh et al. (eds.), *Analysis of Categorical Data from Historical Perspectives*, Behaviormetrics: Quantitative Approaches to Human Behavior 17,
https://doi.org/10.1007/978-981-99-5329-5_29

In my earlier career as a telegrapher on the railway, there were 36 choices for a telegraph key stroke, and it was by no means certain that what was sent or received was correct for this neophyte, given that hotshots on the line like my Dad were capable of speeds of 60 words per minute, or about eight key strokes per second.

Let $\mathbf{P} = (p_1, \ldots, p_M)$ be a probability vector with no zeros. The central concept in information theory is *entropy*:

$$I(\mathbf{P}) = -\sum_{m=1}^{M} p_m \log p_m . \quad (1)$$

Entropy is a measure of information, in the sense that $I(\mathbf{P})$ measures the amount of information that is required to perfectly predict or identify a transmitted message. Entropy is maximised when $p_m = 1/M$ for all m and is minimised to zero when all but one of the probability values are 0.

1.2 Surprisal

Note that if we define:

$$s_m(p_m) = \log_a p_m \quad \text{so that} \quad p_m(s_m) = a^{-s_m}, \quad m = 1, \ldots, M, \quad (2)$$

then entropy is simply the expected value of s in $\mathbf{P}$, and therefore of necessity itself an information value.

We define $\mathbf{S}(\mathbf{P}) = -\log \mathbf{P}$ and, in principle, the base of the logarithm can be any number larger than one. It will be convenient to use the length of $\mathbf{P}$ as the base and use the notation:

$$\mathbf{S}_M(\mathbf{P}) = -\log_M \mathbf{P} \quad \text{and} \quad \mathbf{P} = M^{-\mathbf{S}_M} . \quad (3)$$

For example, for $\mathbf{P} = (0.05, 0.95)$ we have that $\mathbf{S}_2(\mathbf{P}) = (4.322, 0.074)$ and for $\mathbf{P} = (0.01, 0.99)$ that $\mathbf{S}_2(\mathbf{P}) = (6.644, 0.0145)$.

In information theory s_m is called *self-information* (Cover and Thomas 2006), but also *surprisal* since, as the probability goes to zero, surprisal increases without limit, and for probability one is at its lower limit zero. Figure 1 displays how surprisal increases as the probability goes to zero for a variety of values of the length M of the multinomial vector $\mathbf{P}$.

Imagine a roulette wheel with M pockets. Then, $S_M(1/M) = 1$ indicates the probability that the ball will land in a specific pocket for a single spin of the wheel, and $S_M = 2$ the probability of same pocket landing in two consecutive spins. Surprisal is, therefore, a *count* of a sequence of surprising events. More generally, for any positive real P value, $S_M(P)$ can be defined as the expected count:

$$S_M(P) = \lambda S_{M-1}(P) + (1 - \lambda) S_M(P), \; 0 \leq \lambda \leq 1, \quad (4)$$

where $M - 1$ is the largest integer such that $1/(M-1) \geq P$.

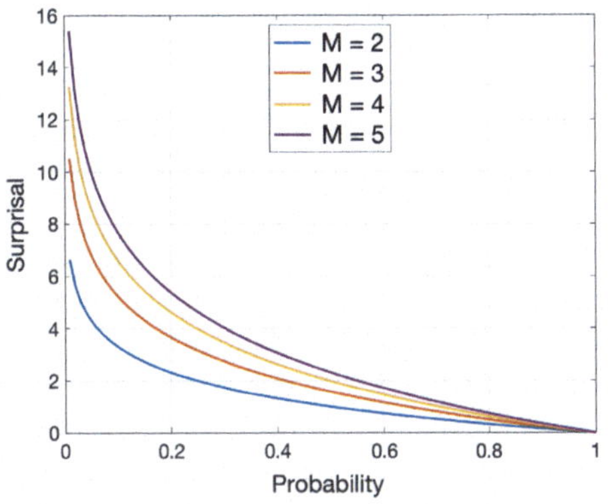

Fig. 1 Relation between surprisal or information and probability for various values of $\log_M$

1.3 The Surprisal Metric

As a continuous counting variable, S_M has all the metric qualities of a ratio scale or any other magnitude. Its unit is $S_M(1/M) = 1$, and can be called the *M-bit*, extending the 2-bit used in computer science. Surprisal values can be multiplied by any positive real number, added, subtracted if the difference is non-negative, and most importantly of all a fixed difference means the same thing at all positions of the surprisal scale. That is, surprisals think like we do and are therefore much easier for us to interpret, unlike probabilities with their fading differences at each boundary.

The surprisal transform is already extensively used by statisticians in the form of the log-odds transform, negative log-likelihood and deviance. There is a substantial literature using surprisal in other fields: early work within statistics by Kullback (1959), in thermodynamics where the term was introduced by Tribus (1961), and in the theory of choice in mathematical psychology in Luce (1959) where surprisal is referred to as the "strength" of choice.

This paper attempts to demonstrate that switching from probability to surprisal/information brings many benefits to the study of social science data defined by choice data. The most important of these is the ratio scale quality of information.[1] Useful judgements of the size of magnitudes requires that, no matter where one directs one's focus, a specified interval has the same meaning. Clearly, a count of independent events satisfies this property.

In order to avoid subscript clutter, we will drop the M subscript on S and $\log_M$.

[1] The mathematical status of true measures or ratio scales in the social sciences has been as century-long debate stimulated by the writings Stevens (1946) and was studied in depth in the three volumes of Krantz et al. (1971). It appears that such a measure was within easy reach all along.

2 Probability and Surprisal Manifolds $\mathcal{M}_P$ and $\mathcal{M}_S$

A manifold is a low dimensional structure embedded within a high-dimensional space. Because $\mathbf{1}'\mathbf{P} = 1$, vectors $\mathbf{P}$ and $\mathbf{S}$ correspond to points within their respective manifolds of dimension $M - 1$. For example, for $M = 3$ the manifold is a flat equilateral triangle with vertices $(1, 0, 0)$, $(0, 1, 0)$, and $(0, 0, 1)$.

The parameterisation of a manifold requires a one-to-one mapping, called a *chart*, that indexes positions within the manifold. Manifolds $\mathcal{M}_P$ and $\mathcal{M}_S$ are conveniently charted as follows. Let the M by $M - 1$ matrix $\mathbf{Z}$ satisfy the conditions $\mathbf{Z}'\mathbf{1} = \mathbf{0}$ and $\mathbf{Z}'\mathbf{Z} = \mathbf{I}$, where $\mathbf{1}$ is a column vector of M ones, and $\mathbf{0}$ and $\mathbf{I}$ are a column of $M - 1$ zeros and the identity matrix of order $M - 1$, respectively. There are many ways to construct matrices $\mathbf{Z}$ with this orthonormal structure, including using the full QR-decomposition of $\mathbf{1}$ and the Fourier and polynomial orthonormal series.

Let $\mathbf{X}$ be an arbitrary real vector of length $M - 1$. Then, $\mathbf{P}$ can be defined in terms of $\mathbf{X}$ as:

$$\mathbf{P} = \frac{M^{\mathbf{ZX}}}{\mathbf{1}'M^{\mathbf{ZX}}}. \tag{5}$$

Division by $\mathbf{1}'M^{\mathbf{ZX}}$ is a *retraction* operation which pulls arbitrary positive M-vectors into the $M - 1$ dimensional manifold of probability vectors of length M.

The surprisal analogue of (5) for the surprisal manifold $\mathcal{M}_S$ is:

$$\mathbf{S} = -\mathbf{ZX} + \mathbf{1}\left(\log\left(\mathbf{1}'M^{\mathbf{ZX}}\right)\right), \tag{6}$$

and the retraction operation that maps a positive M-vector into surprisal space is now the addition of the second term.

Figure 2 displays a two-dimensional surprisal manifold within a three-dimensional ambient space. This was constructed by applying (6) to $\mathbf{X} = (x_1, x_2)'$ where one of the elements was fixed and other interval varied over $[-3, 3]$. The black coordinates lines are from one element being zero. The curved surprisal manifold hugs the planar boundaries of the positive three-dimensional orthant, and the point within the manifold closest to zero is $[0, 0, 0]$. Inside the manifold, the cartesian coordinate system is reproduced exactly within two dimensions instead of three.

3 Probability and Surprisal Functions $\mathbf{P}(\boldsymbol{\theta})$ and $\mathbf{S}(\boldsymbol{\theta})$

Item response theory (IRT) and many other methods require multinomial vectors that evolve over a score indexing variable of one or more dimensions, usually called a latent variable, and/or over one or more dimensions of observed values. These curves can be defined by converting the indexing vectors $\mathbf{X}$ to vector function $\mathbf{X}(\theta)$.

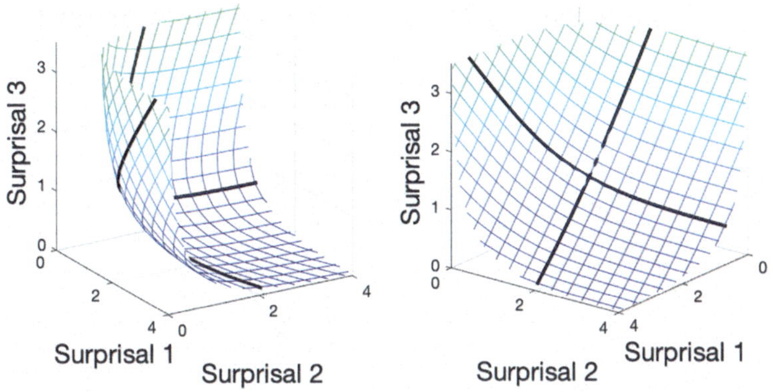

Fig. 2 Two views of the two-dimensional surprisal surface defined by the retraction (6) of a three-dimensional lattice constructed from 21 equally-spaced numbers spanning $[-3, 3]$. The black lines contain the transformed points on the coordinate lines in the original surface

3.1 Constructing Surprisal Functions with Splines

It is assumed here that θ is defined over the already familiar closed interval [0, 100]. A natural and convenient approach to parameterising these functions is to use a B-spline basis system $B_k(\theta), k = 1, \ldots, K$ and to define $\mathbf{X}(\theta)$ as weighted linear combinations $\sum_k c_k B_k(\theta)$ of these basis functions. The splines must have at least three basis functions along with an order of at least three in order to assure that resulting probability and surprisal functions are differentiable. My colleagues and I find that seven B-spline basis functions and order five provide about as much flexibility as required for large data sets, but smaller sample sizes, such as 200 or so, will benefit from using the fewer basis functions and an order closer to the lower limit of three. Figure 3 displays (3-basis/order-3) and (7-basis/order-5) B-spline basis functions.

3.2 Probability and Surprisal Functions for a Test Item

Figure 4 displays the probability and surprisal curves estimated from the choices of about 55,000 examinees on each of 80 items in a Swedish SAT test of quantitative aptitude. The correct answer curve is blue, and the others are red. Also shown are 53 point estimates constructed by binning the data. The vertical dashed lines specify locations of the 5, 25, 50, 75, and 95% percentiles for the examinee score index values. The near-zero probability curve and the flat surprisal curve with value about 4 are for missed or illegitimate responses.

This is a difficult question, only the top 25% of examinees are likely to get it right. The curves tell a complex storey. The bottom 10% or so of examinees tend to guess,

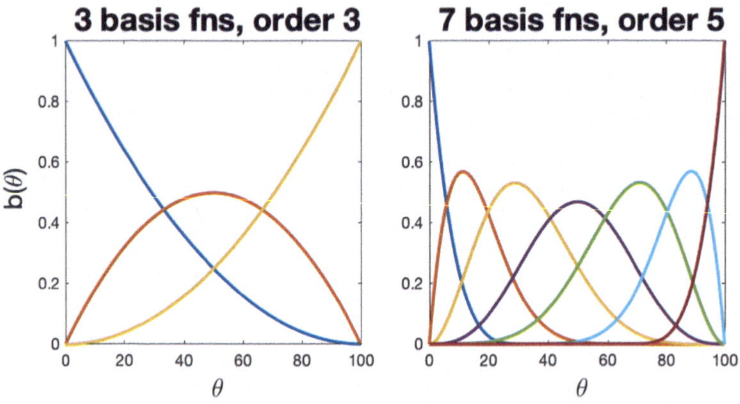

Fig. 3 Two spline bases. The left is appropriate for small subject sample sizes and the right for large samples

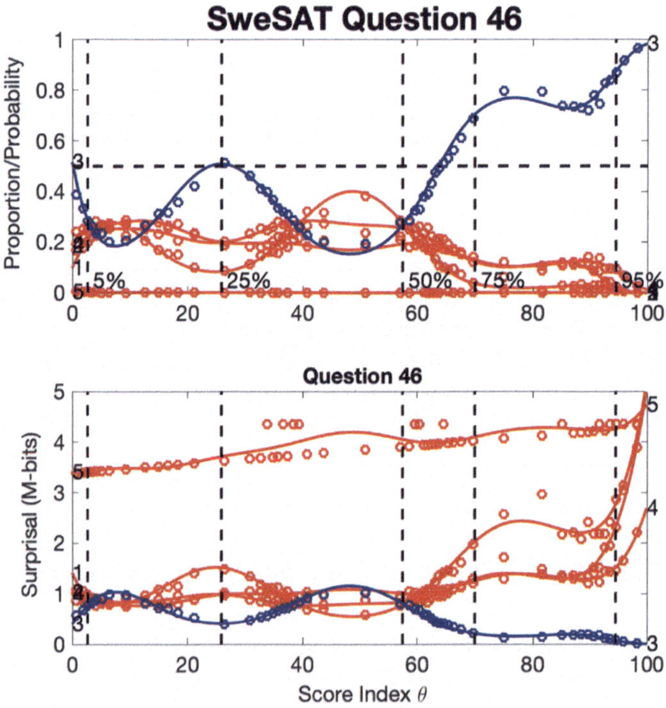

Fig. 4 Probability and surprisal curves for a test item along with point estimates. Blue curves are the correct answer and the others for wrong answers. Dashed lines indicate 5, 25, 50, 75, and 95 percentiles in the index distribution

so that all four options are chosen with equal values. Examinees tend to believe that choosing 3 or C is their best chance if they have no idea, and this produces an isolated peak in the right answer C around the 25% marker. On the other hand, just below the median score index a wrong answer seems especially attractive, producing a dip in the right answer choice. Examinees in the 75–95% range are not entirely convinced that choice C is right. The correct curve is not monotonic, and our analyses that strictly monotonic correct curves are more rare than not. All this detail is possible because all choices are modelled, rather than the usual procedure in IRT of modelling only the correct answer.

3.3 The Score Index Estimate $\hat{\theta}$ and Expected Test Scores $\mu(\hat{\theta})$

Maximum likelihood estimation was used for estimating test taker j's score index $\hat{\theta}_j$ conditioned on the estimated surprisal functions, assuming that, conditional on θ, option choices are independent. The negative log-likelihood H in terms of surprisal and its derivative with respect to θ are:

$$H(\theta) = \sum_{i=1}^{n} \mathbf{U}'_{ji}\mathbf{S}_i(\theta) \quad \text{and} \quad \frac{\mathrm{d}H}{\mathrm{d}\theta} = \sum_{i=1}^{n} \mathbf{U}'_{ji}\frac{\mathrm{d}\mathbf{S}_i}{\mathrm{d}\theta} = 0, \tag{7}$$

where $\mathbf{U}_{ji}$ is a 0/1 choice indicator vector of length M_i for examinee j and item i.

The simplicity of these equations in comparison with their counterparts for probability is an important by-product of switching to surprisal. Now, we see that the fitting criterion H is simply an inner product of binary indicator values U_{jim} and their surprisal counterparts. Thus, the estimation problem has the characteristics of a one-predictor linear model, and the gradient equation is a linear combination with weights $\mathrm{d}s_{im}/\mathrm{d}\theta$. The best value of θ is simply that in which, for the chosen items, the sums of the negative and positive weights cancel. By contrast, the probability equations involve the ratio $(\mathrm{d}p_{im}/\mathrm{d}\theta)/p_{im}(\theta)$, which can cause convergence problems if $p_{im}(\theta)$ approaches 0.

3.4 The Arc Length of Surprisal Functions

For any item, the function values $\mathbf{P}(\theta)$ and $\mathbf{S}(\theta)$ will fall along one-dimensional curves within the ambient space of dimension M but are actually vectors of dimension $M-1$ for all θ. For any smooth strictly monotone function with values $h(\theta)$, the use of the transformed item response functions $p^*_m(\theta) = p_m\left[h^{-1}(\theta)\right]$ and $s^*_m(\theta) = s_m\left[h^{-1}(\theta)\right]$ for all m, and thus leaves this space curve invariant. For this reason, the

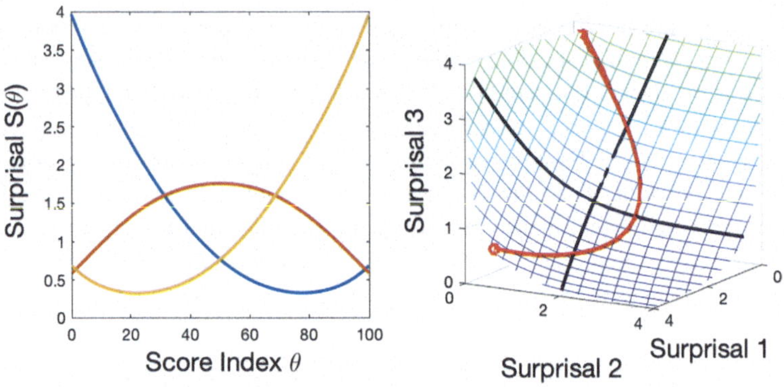

Fig. 5 Left panel contains three simple surprisal curves, with the blue curve behaving like a correct answer. The right panel shows how the three curve vary jointly within the surprisal surface. The initial point on the curve is indicated by a circle

phrase "indexing system" is to be preferred to "latent variable" since θ is neither a fixed variable nor latent.

Progress along the manifold provides a one-dimensional space curve for item i that can be called its *item information*. Progress is measured in M-bits, and *arc length* $d_{Si}(\theta)$ along the curve is defined by the indefinite integral:

$$d_{Si}(\theta) = \int_0^\theta \sqrt{\sum_{m=1}^{M_i} \left[\frac{\mathrm{d}s_{mi}}{\mathrm{d}u}\right]^2}\, \mathrm{d}u, \tag{8}$$

and is also invariant with respect to strictly monotonic transformations of θ.[2] Since $d_{Si}(\theta)$ is a measure of distance, it has the metric property that a fixed increase in distance means the same thing everywhere along the curve, so that arc length distances can be added, subtracted, and subjected to scale changes as required.

Figure 5 illustrates the information curve concept with a toy example. The left panel displays three surprisal curves, of which the blue one behaves like a correct answer since it declines to zero. The right panel displays how the red information curve varies within the surprisal surface defined by the three simple surprisal curves.

The maximum arc length $d_S(100)$ for the simultaneous evolution of all item curves is a measure of the information covered by the whole test and can be viewed as the perfect information score, and arc length $d_{Si}(100)$ restricted to item i is a natural measure of the amount of information required to get that item correct. The higher $d_{Si}(100)$ is, the more effective it is as a contribution to the total surprisal test score. We can also integrate over sub-intervals of θ, such as, for example, the top 10% of

[2] Information can also be normalised to a fixed value such as 100, and the resulting proportion manifold retains its status as a measure with unit *proportional bit* $max(d_S)/100$.

the test takers; the items with the longest arc lengths are those that provide the most information about relative rankings within this upper range.

It follows, too, that surprisal values can be compared legitimately across two different tests, two tests of different lengths, two different samples of examinees, and even across tests of different knowledge classes. A fixed surprisal interval can serve as a unit of effort required to advance information by that amount.

3.5 Estimating the Manifold $\mathcal{M}_S$ and Score Indices θ

While the number of items n is in most tests does not exceed 100, the number of examinees N can be in the thousands or even millions.[3] An alternating optimisation strategy common in high-dimensional problems both in psychometrics and elsewhere was used. Given initial estimates $\hat{\theta}$, an optimisation cycle involved estimating surprisal curves followed by re-estimating score index values conditional on current surprisal values. It was observed that about ten cycles were sufficient to reach near optimal results and to reveal the important structure in the data. Point-wise confidence limits for the surprisal curves for a test item using the delta method conditional on previously estimated values of $\hat{\theta}_j$ were found to reasonably match confidence regions computing using data simulation.

Further details on the estimation procedure can be find in Li et al. (2019), Ramsay and Wiberg (2017), Ramsay et al. (2020a), Ramsay et al. (2020b), and Ramsay et al. (2022).

4 The Swedish SAT Test Information Manifold

The Swedish SAT test had a total of 412 surprisal curves, and as they all evolved over θ, the test information curve was traced out, and the total arc length of the curve was about 75.

Figure 6 displays a histogram and smooth representations of the distribution of the arc length test information scores. These are strongly partitioned into five peaks, and the scores for the top 5% are much more spread out than those for bottom 5%. Only two examinees out of over 55,000 achieved perfect sum scores and no examinees achieved sum scores below 10, but by contrast 66 examinees were assigned information scores of 0 and 119 the top information score value of 75.

Figure 7 shows how arc length measure was related to the expected test scores on the left and to the score index on the right. The expected test score, which is the sum of an examinee's probabilities of choosing only the right answer, is severely biassed against the top test takers, and also biased in favour of the weakest cohort

[3] Few test users will be interested in scores more accurate than the first decimal place and, consequently, fast computation as opposed to accuracy is the prime consideration.

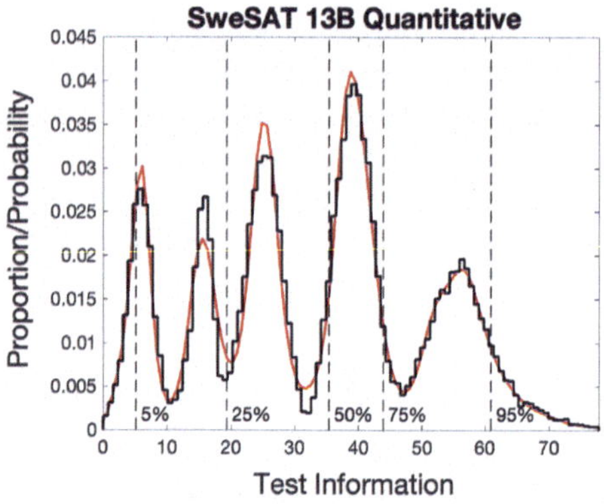

Fig. 6 Density function for the test information scores. The vertical lines indicate the locations of the five percentage markers. The open circles at the beginning and end of the scores indicate the proportions of students achieving the two boundary scores

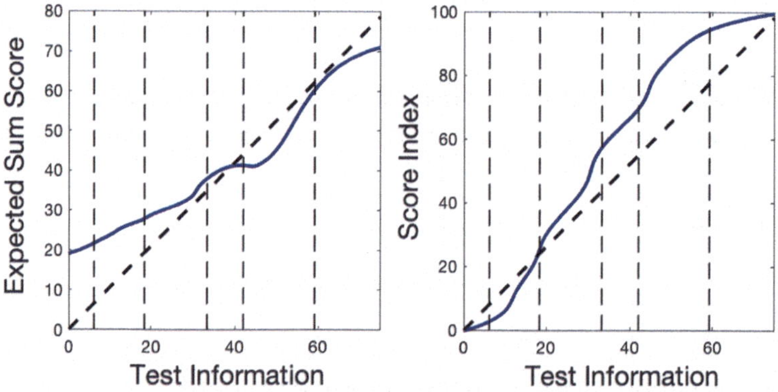

Fig. 7 Relations of expected test score and of score index to test information

by rewarding guessing. The examinees that the universities of Sweden are most interested in have scores that are both compressed relative to the test information score and underestimate their proficiency because the high end performers were negatively affected by deficient items.

The score index θ begins at zero as one would expect, and progresses roughly in proportion to test information up to the 90% percentile. But we see that it also compresses the variability of the top 10%, whereas the test information spreads these out. The dashed marker per cent lines indicate that the top 25% of examinees absorbed much more of the test information than those in the first three quartiles.

Although the test information manifold is embedded in a space of 412 surprisal vectors, 98.9% of its shape variation can be viewed by using the first two principal components of the space curve.[4] Figure 8 displays two test information manifolds

---

[4] The mean of the function values was not subtracted.

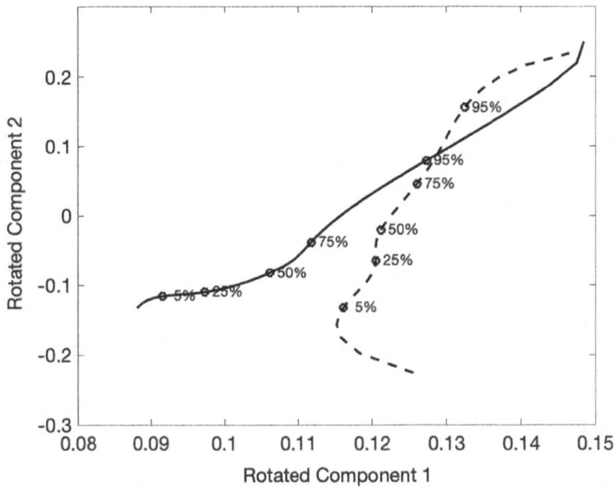

Fig. 8 Solid line is the test information space curve $\mathcal{M}_S$ viewed in the space spanned by its first two rotated singular value functions and using all of the data. The dashed line is the analogous result using only 200 randomly selected test takers. The dots indicate the percentages of the examinees at or below their position on the curve

$\mathcal{M}_S$ for the SweSAT-Q test in these two dominant dimensions, after some rotation to enhance graphical clarity. The solid curve was estimated from all the data and dashed from only 200 randomly chosen examinees, and we see that the basic features of full data curve are well reflected in its small sample version.

The test manifolds have four distinct sections. The first 5% of the examinees have command of only a tiny amount of the information covered by the test. The next 20% are moving from pure guessing to a point where the choices are correct for only the easiest items requiring rudimentary mathematical skills. There are changes in direction at the 50%, and this is the point at which examinees can begin to use mathematics to solve problems. The third quartile makes as much progress as the first two combined, and finally, the top 25% and especially the top 5% possess more information than the first three quartiles combined. What we see is what most of us who teach mathematics tend to believe; learning may be slow at first as basic literacy is acquired but is afterwards a positively accelerating process in terms of the information metric.

5 Discussion and Conclusions

The most important benefit of the information-based analysis is the test information curve, displayed for two sample sizes in Fig. 8. Its ratio scale metric makes information the ideal substrate for representing acquired knowledge. With this as the abscissa, we can see much more detail in plots such as Fig. 6, included a view of the accelerating learning speed among top performing examinees. As these plots are scanned horizontally, fixed differences can be compared across any set of locations. This ratio scale benefit is also observed in the surprisal plot in Fig. 4 for vertical scanning over surprisal, which is also has the information metric. These features are possible because all choices are used, including if necessary a choice not to respond.

The surprisal transformation is easily understood as a count of events, such as is already routine in describing climate risks in terms of one hundred-year events. Note, too, that by expressing (7) in terms of surprisal, we see a simple relation of performance variation to the shape of the first derivative of the surprisal functions. For a particular location, a decreasing curve acts to push a candidate θ upwards, but downwards for an increasing curve, and not at all for no slope. The optimal score index is at a location where these pushes and pulls cancel each other out to provide a zero total slope. Moreover, the effect of guessing is nullified because a guessed response near a location yields a flat surprisal interval.

5.1 Software Resources

We have developed a free open-source stand-alone application called TestGardener and a web-based version available at http://testgardener.azurewebsites.net/ that is suitable for teaching and analyses of small to moderate sample sizes. The application was introduced in Li et al. (2019). The stand-alone version (currently for Microsoft Windows systems) and the web version implement convenient interfaces and display results primarily by graphs. A book-length introduction to better test scoring for secondary school test takers and teachers that is also a manual on how to run the application is available from the Website. An R package called `TestGardener` is on the CRAN for advanced users who need more control over parameter settings and displays; see https://cran.r-project.org/web/packages/TestGardener/index.html. A MATLAB toolbox is also available on GitHub. All three versions processed the 55,000 SweSAT choices in about 20 min on a laptop computer.

References

Cover, T.M., Thomas, J.A.: Elements of Information Theory. Wiley-Interscience, New York, NY (2006)

Kullback, S.: Information Theory and Statistics. Wiley, New York (1959)

Krantz, D.H., Luce, R.D., Suppes, P., Tversky, A.: Foundations of Measurement, vol. I. Academic Press, New York, NY (1971)

Li, J., Ramsay, J.O., Wiberg, M.: TestGardener: A Program for Optimal Scoring and Graphical Analysis. In: Wiberg, M., Culpepper, S., Janssen, R., González, J., Molenaar, D. (eds.) Quantitative Psychology, pp. 87–94. Springer, Cham (2019)

Luce, R.D.: Individual Choice Behaviour: A Theoretical Analysis. Wiley, New York, NY (1959)

Ramsay, J.O., Wiberg, M.: A strategy to replace sum scoring. J. Educ. Behavioral Stat. **42**, 282–307 (2017)

Ramsay, J.O., Li, J., Wiberg, M.: Full information optimal scoring. J. Educ. Behavioral Stat. **45**, 297–315 (2020)
Ramsay, J.O., Li, J., Wiberg, M.: Better rating scale scores with information-based psychometrics. Psych **2**, 347–369 (2020)
Ramsay, J.O., Li, J., Wiberg, M.: An information manifold perspective on psychometrics. (in Review) (2022)
Shannon, C.: A mathematical theory of communication. Bell Syst. Techn. J. **27**, 379–423 (1948)
Stevens, S.S.: On the theory of scales of measurement. Science **107**, 677–680 (1946)
Tribus, M.: Thermodynamics and Thermostatistics: An Introduction to Energy, Information and States of Matter, with Engineering Applications. Van Nostrand, New York, NY (1961)
Wiberg, M., Ramsay, J.O., Li, J.: Optimal scores—an alternative to parametric item response theory and sum scores. Psychometrika **84**, 310–322 (2019)

The manufacturer's authorised representative in the EU is Springer Nature Customer Service Centre GmbH, Europaplatz 3, 69115 Heidelberg, Germany. If you have any concerns regarding our products, please contact ProductSafety@springernature.com

Printed and bound by CPI Group (UK) Ltd, Croydon, CR0 4YY
01/10/2026
02225896-0017